Menus for the TI-82

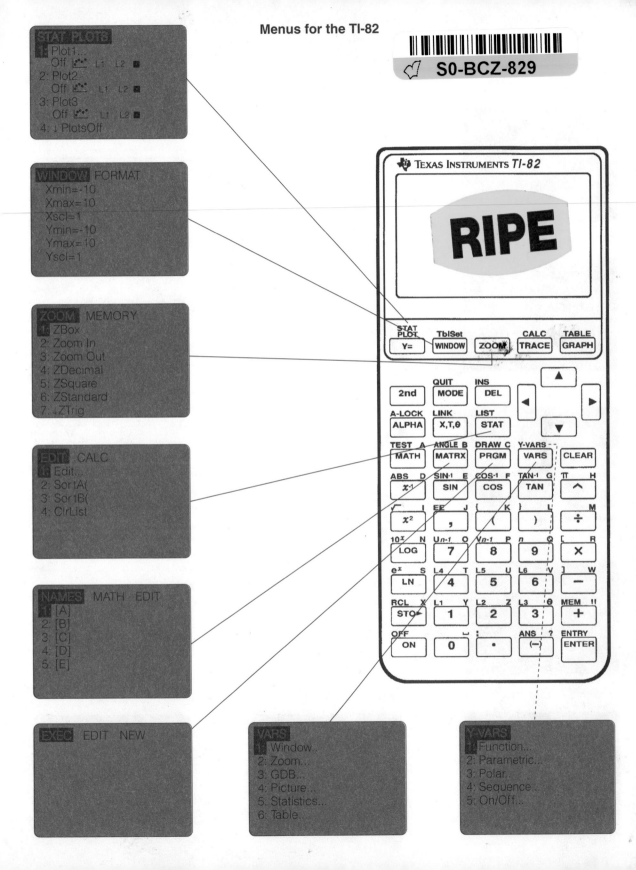

STAT PLOTS
1: Plot1...
   Off 🔛 L1 L2 ■
2: Plot2
   Off 🔛 L1 L2 ■
3: Plot3
   Off 🔛 L1 L2 ■
4: ↓ PlotsOff

WINDOW FORMAT
   Xmin=-10
   Xmax=10
   Xscl=1
   Ymin=-10
   Ymax=10
   Yscl=1

ZOOM MEMORY
1: ZBox
2: Zoom In
3: Zoom Out
4: ZDecimal
5: ZSquare
6: ZStandard
7: ↓ZTrig

EDIT CALC
1: Edit...
2: SortA(
3: SortB(
4: ClrList

NAMES MATH EDIT
1: [A]
2: [B]
3: [C]
4: [D]
5: [E]

EXEC EDIT NEW

VARS
1: Window...
2: Zoom...
3: GDB...
4: Picture...
5: Statistics...
6: Table...

Y-VARS
1: Function...
2: Parametric...
3: Polar...
4: Sequence...
5: On/Off...

Menus for the TI-81

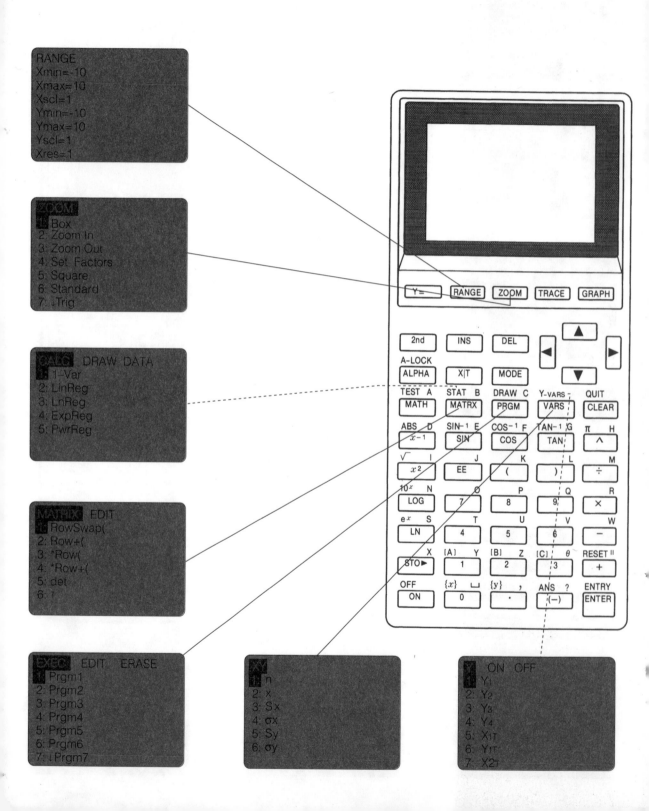

```
RANGE
Xmin=-10
Xmax=10
Xscl=1
Ymin=-10
Ymax=10
Yscl=1
Xres=1
```

```
ZOOM
1: Box
2: Zoom In
3: Zoom Out
4: Set Factors
5: Square
6: Standard
7: ↓Trig
```

```
CALC   DRAW   DATA
1: 1-Var
2: LinReg
3: LnReg
4: ExpReg
5: PwrReg
```

```
MATRIX  EDIT
1: RowSwap(
2: Row+(
3: *Row(
4: *Row+(
5: det
6: T
```

```
EXEC   EDIT   ERASE
1: Prgm1
2: Prgm2
3: Prgm3
4: Prgm4
5: Prgm5
6: Prgm6
7: ↓Prgm7
```

```
XY
1: n
2: x
3: Sx
4: σx
5: Sy
6: σy
```

```
Y   ON  OFF
1: Y₁
2: Y₂
3: Y₃
4: Y₄
5: X₁T
6: Y₁T
7: X₂T
```

Precalculus
with Trigonometry
Using the Graphing Calculator

Denny Burzynski
Wade Ellis, Jr.
Ed Lodi

West Valley College

PWS Publishing Company

Boston

PWS Publishing Company
20 Park Plaza, Boston, Massachusetts 02116-4324

PWS Publishing Company is a division of Wadsworth, Inc.

I(T)P
International Thomson Publishing
The trademark ITP is used under license

Library of Congress Cataloging-in-Publication Data

Burzynski, Denny.
    Precalculus with trigonometry using the graphing cal-
culator / Denny Burzynski, Wade Ellis, Jr., Ed Lodi.
        p.   cm.
    Includes Index.
    ISBN 0-534-18864-8
    1. Functions--Data processing. 2. Trigonometry--Data
processing. 3. Graphic calculators. I. Ellis, Wade, Jr.
II. Lodi, Ed. III. Title.
QA331.B88   1994   94-8250
515$'$.16–dc20   CIP

Printed in the United States of America

95 96 97 98--10 9 8 7 6 5 4 3 2

 This book is printed on recycled, acid-free paper

Editorial Assistant: Clark Benbow
Assistant Editor: Kelle Flannery
Developmental Editor: Bess Rogerson
Acquisitions Editor: David Dietz
Production Editor: Helen Walden
Manufacturing Coordinator: Ellen Glisker
Marketing Manager: Marianne Rutter

Contents

Appendix   A-1

Index   I-1

Preface

Goals and Philosophy

In 1990, the Mathematical Sciences Education Board publication, *A Challenge of Numbers*, stated that more of the new jobs emerging in the 90s will require more education in postsecondary mathematics. In 1989, the Council's report *Everybody Counts* stated that more than at any time in the past, Americans need to be able to think mathematically. It goes on to state that the mathematics curriculum is in need of an extensive overhaul and that overhaul should focus on essential mathematical ideas.

The authors have been part of mathematics education reform movements for many years and continue to be excited about the reform of the precalculus curriculum. We feel that this textbook represents a new direction in the evolution of that curriculum. The essential manipulative skills and basic function concepts that are required in any calculus course are presented in this text. The text, however, is especially appropriate in preparing students for reform calculus courses.

The authors are active participants in state and national organizations such as the MAA, AMATYC, and California Mathematics Council of Community Colleges, and most of our goals and philosophy are represented by the goals and philosophy of these organizations.

Some of our major goals in creating this text are

1. To make precalculus mathematics more accessible to students and to provide them with greater opportunities for success.
2. To provide a rich and broad mathematical experience that encourages independent thinking and mathematical exploration, and builds mathematical confidence.
3. To connect mathematics to other disciplines as well as to present it as a developing human discipline.

Success in mathematics and in fields that apply mathematics depends on a solid understanding of the function concept. This text is about functions and their use in mathematics and in investigating applied and theoretical problem situations. Students use graphing calculators in learning nearly every new topic or concept. The regular and consistent use of graphing calculators allows students to focus on and understand the power of symbolic representations and manipulations. Since students are able to find approximate solutions quickly, they are motivated to spend more time interpreting their results and in seeking a deeper understanding of the theory which, in turn, builds stronger problem-solving skills.

The text relies on the use of the graphing calculator in its presentation of mathematical concepts. We have tried to achieve a balance between the use of technology and symbol manipulation in concept presentation and problem-solving development.

Use of Technology

Our incorporation of graphing calculator technology into the text reflects several years of thinking about how to use technology to enhance the quality of student understanding. Much of the material that appears in the text is an outgrowth of our experiences using technology in our classrooms and in workshops we have given on how technology affects the curriculum. We use technology to provide a greater variety of examples for motivating concepts, to provide a computational platform for a more complete understanding of concepts, and to speed the problem-solving process. This use of technology allows students to investigate a wider class of applied and theoretical problems than is possible with pencil-and-paper methods.

Precalculus with Trigonometry Using the Graphing Calculator relies upon the power of graphing calculators in presenting mathematical concepts and ideas. Each student should have a graphing calculator to use on homework assignments, in-class activities, and tests. Though students will use graphing calculators often, the focus of this book is on precalculus *concepts*.

Major Themes

The text has four major themes that are designed to emphasize a broad understanding of when and how mathematical techniques can be used to solve applied and theoretical problems.

Describing Functions   We develop the student's ability to describe the behavior of functions. When is a function increasing, decreasing, constant, undefined, and where does it attain its maximums and minimums? Although the full explanation of these ideas requires calculus, they are presented here using approximations with considerations of error. The process of approximating maximum and minimum values of a function provides the student with the necessary background to appreciate the exactness that comes with the study of calculus. Also, it will provide them with some understanding that not all problems can be solved exactly and that error estimates are an important part of mathematical investigations.

Functions Model Applied Phenomena   Functions are used to model a wide variety of problem situations. Here, we take advantage of the close relationship between a function defined by an expression and the representation of a function as a graph. Although some functions may be defined for all real numbers, the domains of most functions arising in problem situations are restricted in some natural way. These restrictions are emphasized and students are encouraged to think about the domain of applicability of the functions they create and use. The generation of symbolic function representations from measurement data (tables) is presented early in the text to motivate the study of the function concept.

Multiple Representations of Functions   Multiple representations of a function are used throughout this text. A function can be represented by a sentence, a graph, an expression, or a table. Each of these representations has important and specific uses, and the interrelationships between them can expose additional information about the behavior of a function. As students learn to work with each of these representations and to exploit their connections, they become more aware of the nature of functions and more able to use them as instruments in investigating both applied and theoretical problems.

Mathematics Resource Bank   We encourage students to develop a large "resource bank" of mathematical ideas, concepts, and functions. Students learn to use these resources systematically in analyzing possible strategies in their attempt to solve problems. By the end of the book, a student should be able to understand a problem situation and decide upon the type of function that is likely to be useful in investigating that situation.

Pedogogical Features

Writing Requirements   We have been careful to write many of the exercises so that they require the students to convey their answers in sentence or paragraph form. To that end, we have included many examples in which problem situations are interpreted verbally. For example, when the student is asked to describe the behavior of a function, she might respond by writing "As the input, x, increases in value from $-\infty$ to 5, the output values, $f(x)$, increase and pass through a zero at $x = 2$. The function reaches a maximum at $x = 5$. Then as x goes from 5 to ∞, the function values decrease and pass through a zero at approximately $x = 30$. They continue to decrease forever."

Group Work   We have classroom tested the text using group activities that required students to work in groups on the more difficult problems from the exercise sets. Some group activities were homework assignments that were collected and graded. Other group activities were in class where the groups discussed a challenging problem from the exercises with the expectation that one group would present its solution to the entire class. We found these activities to be quite successful and most students were eventually pleased with the approach. This group approach also caused us to assign homework differently. Traditionally, we might assign every odd exercise; but using the group approach, we would carefully choose 4–6 exercises that required more thought and time than other, more routine exercises. Each group was encouraged to assure that each student in the group understood how to do each of these problems. **Chapter Projects** (described below) are more conceptual, thought-provoking exercises that can be assigned as group activities.

Objectives   Each section begins with a list of the topics that are examined in the section. Each topic is set off as a subsection to make reading, studying, and reviewing easier.

Example Sets   Example problems that are to be solved are set off from the main text as *Example Sets*. Example sets include one or more examples of similar type. They have been carefully chosen and developed to demonstrate concepts, calculator skills, and problem-solving techniques in the most instructive way.

Illuminator Sets   Example problems that are meant to illuminate concepts, rather than demonstrate a problem-solving technique are set off from the main text as *Illuminator Sets*.

Calculator Windows   We have included calculator windows to illustrate *TI* calculator screens. We have tried to make them reflect the appearance of the actual screen. The main difference is that the curves that appear in the windows are smooth and do not display the jagged, pixel appearance of curves that actually appear in the windows. We purposely constructed smooth curves so as to best illustrate specific concepts.

Section Exercises   The section exercises are graded in terms of difficulty, although they are not explicitly grouped into categories. In many cases, the exercises are odd-even paired. In most sections, we have included exercises that require students to convey their solutions in sentence or paragraph form. Answers to the odd-numbered exercises appear at the back of the text. These solutions have, when possible, been checked using both Mathematica and Maple.

Chapter Projects Each chapter has chapter projects. These projects are problems that require more time and thought than do the section exercises. They can be solved individually by students or in groups. Most projects require some calculator mechanics and a good understanding of previous theoretical concepts. We have tried to include problems that extend the students' mathematical knowledge and skills so as to develop their mathematical power.

Chapter Summaries Each chapter includes a chapter summary that we have written in an informal style, much the way we would hope a student would write the summary. Each summary touches on the main points of the chapter and serves as a good review of the chapter material.

Supplementary Materials

Instructor's Manual The instructor's manual contains an overview of important concepts in each chapter, a section-by-section instructor's guide, a test bank, answers to even exercises, and guides to the chapter projects. The test bank problems require the use of technology and frequently ask students to present answers in paragraph form.

Casio/HP Graphing Calculator Guidebook The guidebook shows keystrokes for Casio and Hewlett Packard graphing calculators. Specific keystrokes for TI-81 and TI-82 graphing calculators are covered in the text.

Student's Solution Manual The student's solution manual contains worked-out solutions to every other odd-numbered exercise.

Calculator Videotape Tutorial The calculator videotape tutorial presents an introduction to graphing calculators manufactured by Texas Instruments, Casio, and Hewlett Packard. The authors demonstrate the calculator features that are used in the text.

Acknowledgments

Books are rarely constructed alone and this one is no exception. Many talented people worked on the construction of this text and we most gratefully acknowledge their contributions.

We thank the staff at PWS Publishing for believing that we could produce a textbook using a production process that is significantly different from the standard one. Our sincere gratitude goes to our editors, Ann Scanlon-Rohrer, Tim Anderson, and David Dietz, our developmental editors Bess Rogerson and Kelle Flannery, and our production editor, Helen Walden. We thank them for their commitment and vision of the future. We believe PWS has taken a giant step forward in textbook production and are most pleased that they took that step with us.

Typesetting   This textbook was typeset by the authors using TeX. The design for the text was created by PWS and Wadsworth publishing companies. The macro package of TeX commands that implemented this design was created by Rachel Goldeen. The textbook, with the exception of the answer section, was created by the authors using the macro package and we are responsible for the finished product. The answer section was created by Laurel Technical Services of Redwood City, California, using Microsoft Word.

Proofreading   We express our thanks and appreciation to Helen Walden of PWS Publishing, and Amy Miller of Green Bean Graphics for proofreading the entire text. We also thank Kurt Norlin of Laurel Tutoring, Sandi Wiedemann of University of California at Santa Cruz, and Roz Pitchford for reading parts of the text. The errors that remain, however, are our own.

Graphics   The mathematical figures and calculator windows were produced by the authors using Adobe Illustrator and PSMathGraphs II, a graphics package created by John Jacob of the College of Marin and MaryAnn Software, San Rafael, California. Assistance with artistic elements was provided by Margie Moran, J. Ross Moran, and Laura Strong.

Supplementary Materials   The instructor's manual, student's solution manual, and guidebook were constructed by Terri Bittner of Laurel Technical Services. They worked hard to produce the solutions to the exercises we developed and politely gave us suggestions when some of our exercises were less than clear.

Reviewers   We are very grateful to the following mathematics instructors for their comments on the manuscript and recommendations on how it could be improved. They provided numerous insights and suggestions that markedly improved the quality of the book, and their help was invaluable and much appreciated.

Gayle M. Bush, *Dekalb College*
James J. Eckerman, *American River College*
Mercedes McGowen, *William Rainey Harper College*
Ted Nirgiotis, *Diablo Valley College*
Louise Raphael, *Howard University*
Stuart Thomas, *University of Oregon*

Dedication

To Sandi, Kristen, and Erin

Window DB

To Jane

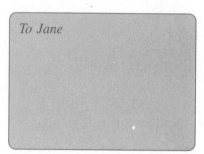

Window WEJr

To Rose Marie,
Kim, Dave, and Samantha,
Mark, Janice, Lisa,
John and Susie,
Rich and Kristin

Window EL

CHAPTER 1

Functions

1.1   An Intuitive Description of Functions

Introduction

Functions as Instruments for Describing Information

Function Notation

Introduction

The notion of function is of fundamental interest in the calculus, where you will study how it can be used to describe change and accumulation in physical and theoretical systems. This section is intended to provide you with an intuitive understanding of the function concept, while Section 2.2 explains the function concept on a more rigorous level. In addition, Chapters 1 and 2 provide an introduction to the features and power of the graphing calculator while reviewing some of the important ideas and mechanical techniques you studied in elementary and intermediate algebra.

It is often convenient, and many times necessary, to describe relationships between quantities of physical or theoretical phenomena. These relationships can be modeled by mathematical entities called functions. Functions can then be used to predict the future behavior of the system from which the phenomena develop.

We illustrate how information can be described using mathematical symbols, show how function notation is quite descriptive, and demonstrate how to use the notation in computation.

Functions as Instruments for Describing Information

Function

A **function** is a way to describe relationships and present information about those relationships. We say *My taxes are a function of my income, since the amount of my taxes depends on the amount of money I make.* This common usage of the word *function* conveys the idea that one thing depends on another. It is useful for you to think up some relationships yourself. For example, your water bill is a function of the size of your lawn; your interest in a lecture is a function of the quality of the lecture; your grade in a course is a function of your exam scores; your grade in a course is a function of your effort in the course; your telephone bill is a function of the number of long distance phone calls you make; your gasoline bill is a function of the number of miles you drive. Try a few yourself.

Not all these relationships or functions are well-defined or quantifiable, but they do indicate how one thing depends on another thing. Sometimes one thing can depend on more than one other thing. For example, the profit a company makes each year depends on the amount of money it spends on advertising *and* the number of salespeople it employs. When a definite, well-defined dependence of one thing on another can be developed, we may be able to express the relationship between those things mathematically (using mathematical symbols), and then extract information from that relationship.

Function Notation

It is cumbersome to describe a relationship between quantities using words and sentences. We will now develop the commonly used function notation.

How Functions Describe Relationships

If one thing depends on another, we have

> one thing depends on another thing

To distinguish the two things, we number them 1 and 2. Let's agree to say that the second thing is determined from the first, that is, that the second thing depends on the first. Then we have,

> *thing$_2$* depends on *thing$_1$*
>
> or
>
> *thing$_2$* is determined by *thing$_1$*

The idea is that once *thing$_1$* is known, it is placed into the rule that describes the dependence, and *thing$_2$* is produced. This is the *input-output* idea. Input *thing$_1$* and output *thing$_2$*.

Let's put names to *thing$_2$* and *thing$_1$*, and say that when something travels at a constant speed,

> *distance* depends on *time*
>
> or
>
> *distance* is determined by *time*

We can eliminate the words by using symbols. If we let d represent *distance* and t represent *time*, then we have

> d depends on t, or d is determined by t

We can compact the notation even more by using parentheses "( )" to represent the words *depends on* or *is determined by*. Thus,

> $d(t)$ means that *distance* depends on *time*
>
> or
>
> $d(t)$ means that *distance* is determined by *time*

We read the notation $d(t)$ as *dee of tee*, or as *dee at tee*, and think of it as shown in Figure 1.1. The notation is actually visually descriptive. The quantity inside the parentheses is the input quantity. (After all, it has been put into the parentheses.) The entire quantity $d(t)$ is the output quantity. *Rather than using only the quantity outside the parentheses to signify the output quantity, we include the parentheses to reinforce the fact that output depends on the noted input quantity.* In the case of our example, $d(t)$, t is the input quantity and $d(t)$ is the output quantity. When *time* is placed into the rule that describes the dependence, the *distance* traveled is output.

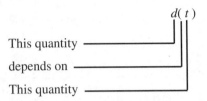

Figure 1.1

Table 1.1

t	$d(t)$
1	60
2	120
3	180
4	240
5	300
6	360

Suppose that for six hours we observe a car traveling at the constant speed of 60 miles per hour. Let's organize our observations into the table shown to the left. Careful observation of the table seems to indicate that the value of $d(t)$ is always 60 *times* the value of t. This observation helps us to establish the rule that describes the relationship between t and $d(t)$. It appears that t determines $d(t)$ by the rule $60 \cdot t$, that is $60t$. We express the relationship by writing an "=" sign between the output notation and the rule expressing what to do with the input. In this case,

$$\underbrace{d(t)}_{output} = \underbrace{60t}_{rule}$$

Thus, when we read this notation the left side tells us that two particular quantities are related, and the right side tells us precisely how one (the output) is determined from the other (the input). In this example, the left side tells us that distance and time are related, and the right side tells us that distance is determined by multiplying travel time by 60.

We can now use the function to produce the output values for any acceptable input values. For example, if $t = 8$, then $d(8) = 60 \cdot 8 = 480$. That is, if the time spent traveling is 8 hours, then the distance traveled is 480 miles. A value of t such as -3 is not an acceptable value since time and distance are positive quantities and $d(-3) = 60(-3) = -180$ has no physical meaning.

Before we proceed to several examples, we remind you of some terminology associated with functions. The set of values that are used as inputs to a function is called the **domain** of the function. The corresponding set of output values is called the **range** of the function. In Section 2.2, we define the domain and range of a function more rigorously. We now illustrate the use of function notation with several examples.

Domain
Range

ILLUMINATOR SET A

If we represent the radius of a circle by r and the circumference of the circle by C, then the relationship between the circumference and radius of a circle can be expressed as

$$C(r) = 2\pi r$$

The left side tells us that the circumference of a circle depends on the radius of the circle, and the right side tells us that the circumference is determined by multiplying the radius by 2π.

We can build a table that displays the relationship visually by inputting several values for the radius and computing the corresponding values of the circumference. Table 1.2 shows several exact values of the circumference and several approximations (using $\pi = 3.14$). Complete the table by filling in the missing parts. You may wish to use your calculator to carry out the multiplications. The table exhibits only seven of the infinitely many possible choices for input values and their corresponding output values. The table therefore represents only a partial description rather than a complete description of this function.

Table 1.2		
r	**Exact** $C(r)$	**Approximate** $C(r)$
1	2π	6.28
2	4π	12.56
3	6π	18.84
4	8π	25.12
5		
6		
10		

Judging by what we can compute, we might imagine that both the domain and the range of this function are the set of all real numbers. It is true that any real number—positive, negative, or zero—can be substituted for r and a value for $C(r)$ computed. However, this function describes some physical or theoretical situation and therefore its domain must be chosen to reflect the reality of that situation. To exist, a circle must have a positive radius and circumference. Therefore, both the domain and the range must be at most the set of all positive real numbers. In practical situations, circles also have some limit as to how big they can be. For example, an architect using this function to construct circles that represent trees and shrubs on the land surrounding a housing development might set the domain as $0 < r \le 60$, where r is in millimeters. The point is that *the domain of a function is chosen by the user of the function to describe a particular situation.*

When a particular situation is not known to us, we select the biggest domain that reflects the reality of the situation the function describes. In this case we say that the domain is some subset of the set of positive real numbers. We discuss how the range of a function is found, or estimated, later in the text.

ILLUMINATOR SET B

If we represent the length of the side of a square by s and the area of the square by A, then the relationship between the length of a side and the area of a square can be expressed as

$$A(s) = s^2$$

The left side tells us that the area of a square depends on the length of a side, and the right side tells us the area is determined by squaring the length of the side.

We can build a table that displays the relationship visually by inputting several values for the length of the side and computing the corresponding values of the area. Complete the table shown to the left by filling in the blank part of the output column. You may wish to use your calculator to carry out the multiplications.

For much the same reasoning as in the previous example, the domain and range of this function are some subsets of the set of positive real numbers. The table exhibits only seven input values and four output values, and therefore, presents only a partial description of this function.

Table 1.3

s	**Exact** $C(r)$
1	1
2	4
3	9
4	16
6	
10	
15	

ILLUMINATOR SET C

If we let m represent mass, v represent volume, and the Greek letter ρ (rho, read *row*) represent density, then the density of an object is related to the mass and the volume of the object by the function

$$\rho(m, v) = \frac{m}{v}$$

This function illustrates how one quantity depends on two other quantities.

The left side tells us that the density of an object depends on both the mass and the volume of the object, and the right side tells us that the density is determined by dividing the mass of the object by the volume of the object.

We can build a table that displays the relationship visually by inputting several values for the mass and volume and computing the corresponding value of the density. Complete the table by filling in the missing parts. You probably will not need your calculator to carry out these divisions.

	Table 1.4	
m	v	$\rho(m, v)$
1	6	1/6
3	4	3/4
5	5	1
8	2	4
10	25	
1/3	9	
12	1	

This table describes a function of the two variables m and v. Its domain consists of ordered pairs, one value for each of the input variables. To exist, an object must have both positive mass and volume, and, therefore, positive density. The domain of this function is then some subset of the set of ordered pairs in which both numbers are positive real numbers. The table presents only a partial description of this function.

ILLUMINATOR SET D

Let A represent the amount of money accumulated, P represent the amount of money invested, i represent the interest rate per year, and t represent the time (in years) of an investment. Then the amount of money accumulated in an investment is related to the amount of money invested, the interest rate, and the length of time for which the money is invested by the function

$$A(P, i, t) = P(1 + i)^t$$

This function illustrates how one quantity depends on three other quantities.

Write down the meaning of both the left and right sides of the equal sign. Then build a table that displays 3 or 4 values of A for some values of P, i, and t that you choose. One set of input values and the corresponding output value have been entered to get you started. (Use your graphing calculator to check that the three input values actually do generate the specified output value.)

		Table 1.5		
P	i	t	A
15,000	0.08	4	20,407.33

As in the previous examples, the table presents only a partial description of the function. The domain and range of this function are somewhat difficult to specify. Any real number, positive, negative, or zero, could be substituted for P, i, or t, and a value of A computed. But input values cannot be negative or zero, for in that case the function gives no meaningful information. They can only be certain positive numbers, and there must be some (practical) upper limit as to what they can be. Nor can they be irrational (you can't invest $\sqrt{6}$ dollars), so they must be rational. What upper bounds can be set on P, i, and t? Again, *the domain of a function is chosen by the user of the function to describe a particular situation.* For example, an investor who can invest at most $10,000 with a conventional bank and wants to collect his or her earnings sometime within the next 20 years would set the following domain: all ordered triples (P, i, t), where $0 < P \le 10,000$, $0 < i \le 100$, and $0 < t \le 20$, and where the numbers in each of these sets are rational numbers.

Table 1.6	
x	$f(x)$
-2	-4
0	0
5	10
20	40
a	$2a$
$2h$	$4h$
dot	$2 \cdot dot$
$7k$	
-3	
$10x$	
$15h$	
$b + x$	

ILLUMINATOR SET E

The function $f(x) = 2x$ can be thought of as a *doubling* function since it acts to double the value of the quantity that is input into it. Complete Table 1.6.

We might denote this function as *double*(x) rather than $f(x)$ since *double*(x) tells us what to do with the input quantity to get the output quantity—we double it! For example, *double*($a + b$) $= 2 \cdot (a + b)$. Since this function is not associated with any particular physical or theoretical situation, we describe the domain and range in the most general manner: the domain and range of this function is the set of all real numbers. As in the previous examples, the table presents only a partial description of the function.

EXERCISES

For exercises 1–7, use symbols to write the defining function expression for each situation. As best you can, specify the domain of your function.

1. The volume of a cube is a function of the length of an edge. Use V to represent volume and s to represent the length of an edge.

2. The area of a circle is a function of its radius.

3. The total cost of a certain number of the same item, each of which costs $0.20, is a function of the number of items that are produced and sold.

4. The perimeter of a square is a function of the length of a side of the square.

5. The total cost of n papers (if each paper costs $0.25) is a function of the number of papers purchased.

6. The area of a triangle with base 3 inches is a function of its altitude.

7. The area of a rectangle with base 3 inches is a function of its height.

8. Suppose $Cost(item) = 3(item)^2 - 2(item)$. Complete the following table. Does this table

Input	Output
2	
3	
4	
5	
a	
b	
x	
$x + 3$	
$x + h$	
$3x^2 + 2$	

present a complete or partial description of the function?

9. Suppose $Revenue(item) = (item)^2 - 4(item) - 2$. Complete the following table.

Input	Output
2	
3	
4	
5	
a	
b	
x	
$x + 3$	
$x + h$	
$x^2 - 4x - 2$	

10. Given the following table, determine a function definition that summarizes the information in the table.

Input	Output
1.2	1.44
1.3	1.69
1.4	1.96
1.5	2.25

11. Given the following table, determine a function definition that summarizes the information in the table.

Input	Output
2	5
3	10
4	17
5	26

In exercises 12–16, write the defining function expression for each situation using symbols. Specify the domain of your function and, using complete sentences, describe the reasons you choose that domain.

12. The area of a rectangle is a function of its base and height. (Remember the parentheses.)

13. The area of a triangle is a function of its base and height.

14. The volume of a cylinder with base 5 square inches is a function of its height. Use V to represent volume and h to represent height.

15. The volume of a sphere is a function of its radius.

16. The total cost of n items is a function of the price p and the number n of items purchased.

17. Think of two quantities that are related, choose letters to represent each of the quantities, and use function notation to express that relationship. Specify, as best you can, the domain of your function.

18. The number of cars produced at an automobile factory is a function of time and the rate r at which the cars are produced. Determine appropriate units of time and rate.

19. The number of buttons produced at a button factory is a function of time and the rate r at which the buttons are produced. Determine appropriate units of time and rate. These units should not be the same as the units in Exercise 18.

1.2   The Graphing Calculator

Introduction

The Calculator's Method of Constructing a Graph

The Graphing Keys

Creating a Graph

The ZOOM Key

The TRACE Key

Creating a Box

The RANGE Key

Introduction

The graphing calculator graphs only functions of one variable. To use it efficiently and effectively, you must be familiar with most of its keys and their various functions. In this section we explore how the calculator constructs graphs of functions and how you can use various calculator keys to obtain more detailed information about the graph of a function.

The Calculator's Method of Constructing a Graph

In Section 1.1, we considered an expression for distance as a function of time:

$$d(t) = 60t$$

You have described functions verbally and notationally in Section 1.1. In your notational descriptions of functions the left side of the defining equation gives the name of the function before the parentheses and the input variable (or variables) in the parentheses. The right side of the defining equation is an expression in terms of the input variable. The input variable is often called the **independent variable**.

Independent Variable

A table can be generated from a function definition. Using $d(t) = 60t$, we create the table of function values shown in Table 1.7. This table of values for the function is generated by using the values from the left column as input values in the right side expression $60t$. Your TI graphing calculator uses this idea to graph functions. First, you enter the expression that serves as the rule for determining the output value from the input value, in this case, $60t$. Then the calculator graphs this function from a table of values it generates using this expression. In Table 1.7, we arbitrarily chose four input values. The calculator *always* uses 96, 95, or 127 input values for the TI-81, 82, and 85, respectively. The values are evenly spaced starting at the leftmost x-value and ending with the rightmost x-value. For instance, the TI-81 uses 96 values because the viewing screen is made up of 96 points, called *pixels*, evenly spaced over 95 intervals.

2nd

You can use the 2nd along with the up and down arrow keys to brighten or darken the screen so you can see more easily the letters and graphs that appear on it.

Table 1.7	
time	**distance**
2	120
3	180
4	240
5	300

The Graphing Keys

The keys that appear just below the screen on your calculator are the graphing keys. Use these keys for the following purposes: (1) to enter the defining expression that prescribes how an output value is obtained from an input value, (2) to graph the function, (3) to change the input values the calculator uses to graph the function. Here is a brief description of what these keys do. They will be described more fully later. The window illustrations that follow are for the TI-81; the windows for the TI-82 and 85 are quite similar.

Y=

The Y= brings up a menu of four or more places where you can enter expressions that define functions: Y1=, Y2=, Y3=, and Y4=. Press this key now. Window 1.1 shows the Y= window.

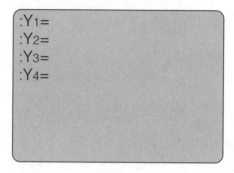

:Y1=
:Y2=
:Y3=
:Y4=

Window 1.1

It is important to note that the TI graphing calculator does not name functions using standard function notation. It names an output value with the letter Y. The subscripts specify any one of four functions. For example, Y1= specifies *function number 1*, or the *first function*. Thus, the calculator expresses our *time-distance* function, $d(t) = 60t$, as Y1=60X.

RANGE or WINDOW

The RANGE or WINDOW key brings up a menu that allows you to set the minimum and maximum input and output values for the graphing window. The graphing window is the portion of the rectangular coordinate system visible on the calculator display window.

ZOOM

The ZOOM key brings up a menu that allows you to quickly select a particular set of input and ouput values by zooming in or out by a preassigned factor or by visually setting a box around a portion of the graph.

TRACE

The TRACE key allows you to follow along certain points of a graph or graphs while coordinates of these points are displayed at the bottom of the screen.

GRAPH

The GRAPH key graphs the functions that are activated (highlighted) in the Y= menu. The black key marked X|T allows you to enter the input variable. Your only choice is X (or T with a different calculator setting) even though the variable in your original equation may be r, B, or t.

Creating a Graph

As a starting point, press ZOOM and then press the numeric key 6. This action selects the standard viewing window -10 to $+10$ for x and -10 to $+10$ for y. (See the calculator screen on the left of Window 1.2.) We will not always want this viewing window, but it is often a good place to start. Now press the RANGE key. You should see Xmin = -10, Xmax = 10, Xscl = 1, Ymin = -10, Ymax = 10, Yscl = 1, and Xres = 1. The calculator screen on the right of Window 1.2 shows this range-viewing screen.

```
ZOOM
1:Box
2:Zoom In
3:Zoom Out
4:Set Factors
5:Square
6:Standard
7:↓Trig
```

```
RANGE
Xmin=-10
Xmax=10
Xscl=1
Ymin=-10
Ymax=10
Yscl=1
Xres=1
```

Window 1.2

Let's graph the function $f(x) = \dfrac{1}{2}x - 1$. To enter this function press Y=. The flashing cursor should be just to the right of the "=" sign. If it is not, use the arrow keys to move it to this position.

Enter the defining expression $(1 \div 2)X - 1$. When you have completed typing these symbols, you will have entered the function $f(x) = \dfrac{1}{2}x - 1$ (although it will appear as Y1=(1/2)X-1. Window 1.3 displays this window.

```
:Y1=(1/2)X-1
:Y2=
:Y3=
:Y4=
```

Window 1.3

Press GRAPH. A line will appear in the viewing window (Window 1.4). Notice that the graph crosses the y-axis at the first tick mark below the origin. Where does the graph cross the x-axis? What is the value of the function when $x = 2$? Notice also that there are 10 tick marks starting at the origin and going to the horizontal and vertical edges of the screen.

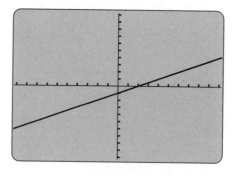

Window 1.4

The ZOOM Key

Press any of the arrow keys to locate the cross-hair cursor. Once you have located it, use the arrow keys to move it to the origin (the intersection of the two axes), if necessary. You will not always want the cursor located at the origin, but it provides a good starting point for a discussion of the ZOOM key. Now press ZOOM. Read the menu displayed in the viewing window to see your choices. Press 2 to select the ZOOM IN feature.

ENTER   Now press the ENTER key. It's easy to forget to press the ENTER key in this sequence, but we need to press this key because we have the option of changing the point at which we zoom. At this stage we are zooming about the origin (because that is where the cursor is located).

The graph zooms in toward the origin. Because the view is magnified, there are fewer tick marks and the space between them is larger. Window 1.5 displays this viewing screen.

The TRACE Key

Press the TRACE key. Notice that the cursor changes to a flashing square positioned on the line in the horizontal center of the viewing window. Look at the bottom of the screen. The numbers at the bottom of the screen are the x and y (first and second) coordinates of the point where the cursor is located. The x-value represents the input value and the y-value represents the output value. Since the point is on the line, its coordinates should satisfy the equation. You might try to verify this by substituting the *x*-value for *x* and the *y*-value for *y* in the equation $y = \frac{1}{2}x - 1$ and seeing if you get a true statement. (If the statement is not precisely true, it may be due to roundoff error. Roundoff occurs because of the pixel structure of the viewing screen.) See Window 1.6.

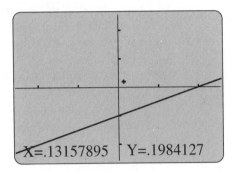

Window 1.5

Press the right-arrow key and hold it down for a second or two. Notice that the flashing cursor moves along the line to the right, and that the numbers at the bottom of the viewing window change. Turn off the trace feature now by pressing the GRAPH key. Use the arrow keys to move the cursor back to the origin.

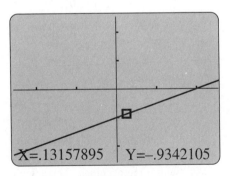

Window 1.6

Press ZOOM, then 2, then ENTER. The line is no longer visible because none of the points in the calculator-generated table are in the viewing screen.

We will now create another graph to demonstrate the TRACE and RANGE keys. Press the key combination ZOOM and then 6. This sequence always returns the screen to the standard graphing window with 10 tick marks in each direction from the origin.

Now press Y= so that you can enter the right-hand expression of a function definition. Press the CLEAR key to clear the previous function from the screen.

Type in $x\char`^2 - 3x + 14$ using the X|T key and the caret (ˆ) key. (You may be familiar with this notation for exponentiations if you have studied the BASIC, FORTRAN, or Pascal programming languages.) When you have finished typing these symbols, you will have entered the function $f(x) = x^2 - 3x + 14$.

Press GRAPH. Notice that no curve appears in the viewing screen. This is because there are no points of the curve in the standard viewing window. You will need to zoom out to get a bigger view of the coordinate system and to find the curve.

Press ZOOM and 3 to choose menu item 3. Press ENTER to activate the ZOOM OUT feature. The graph of a portion of the function should now be visible. See Window 1.7.

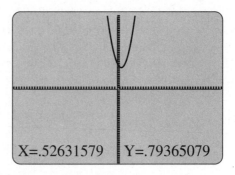

Window 1.7

Creating a Box

It is often convenient to zoom in on a particular region of a curve. For example, zooming in can help you to determine the coordinates of a point on a curve with more accuracy. The calculator's zoom-in feature provides you with a way of doing this. The zoom-in feature, however, magnifies a particular region at preset horizontal and vertical factors. (These ZOOM Set Factors settings are accessed by pressing ZOOM and, on the TI-82, using the right-arrow to access the MEMORY menu. We will not discuss them at this time.) Once the factors are set, the new viewing window is automatically determined. The ZOOM BOX feature of the TI graphing calculator, however, provides you with a method of defining the size of the new viewing window.

   To create a box around a particular part of a graph, press ZOOM and then 1. This will access the ZOOM BOX feature. Use the arrow keys to move the crosshairs to the left and then up to a position just below the lowest part of the graph. See the graph on the left of Window 1.8. Press ENTER to set one corner of the box. (Notice that the cursor changes to a blinking square.) Now press the right-arrow key 7 times and the up-arrow key 4 times. Notice that a box is created. If you set the first corner of the box at the position shown at the bottom of the left graph of Window 1.8, your box should look something like the graph on the right of Window 1.8.

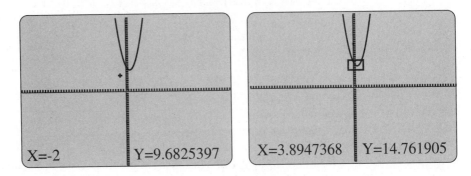

Window 1.8

   Press ENTER to complete the creation of the box. (The new viewing window is now the boxed region.) See the graph on the left of Window 1.9. Press TRACE. The flashing cursor should be visible on the curve. If it is not, it can be made visible by pressing the left- and right-arrow keys. Move the cursor to the lowest part of the graph using the left- and right-arrow keys. Because of the pixel structure of the calculator, the graph looks jagged and the lowest part looks flat, although in reality, it is not. Remembering that the coordinates of the cursor location appear at the bottom of the screen, try to get the cursor in the middle of the flat part and then specify the coordinates of the lowest part of the graph. Your coordinates should be close to X=1.5 and Y=11.75. The graph on the right of Window 1.9 shows this magnified view.

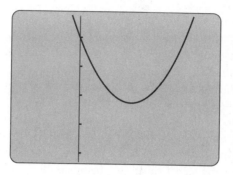
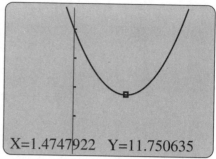

X=1.4747922 Y=11.750635

Window 1.9

You can now investigate any function of one variable that you can enter using the numerical keys, the X|T key, and the Y= key. For example, to investigate the graph of $f(x) = \dfrac{2x + 3}{5x - 4}$ enter the defining expression as

$$(2X + 3) \div (5X - 4)$$

The RANGE Key

Begin by resetting the screen to the standard window by pressing ZOOM and then 6.

Press the Y= key and clear the previous function by pressing CLEAR. Enter the function definition "$-.4x^2 + 950$". To enter the number $-.4$, you will have to use the *negative sign* key "(−)" rather than the *minus* sign key "−". After you have entered the function, press the GRAPH key to graph the function. Notice that no curve appears on the screen. The problem is that the points plotted by the calculator are outside the −10 to 10 for x and −10 to 10 for y window. We can readjust this window using the RANGE or WINDOW key. Press the RANGE key. We are interested in the Xmin=, Xmax=, Ymin=, and Ymax= lines. These lines allow us to adjust the boundaries of the viewing screen. With the appropriate boundaries, the graph will appear in the viewing screen. You may have to experiment with various values for x and y to find the ones you feel best display your curve.

As your study of precalculus progresses, your guesses for these values will improve. As a first attempt (and using the (−) key), change the Xmin= value to −500, the Xmax= value to 500, the Ymin= value to −500 and the Ymax= to 500. Press the GRAPH key to display the graph. Some type of curve now appears in the window. Try changing the range values again. Use Xmin= −250, Xmax= 250, Ymin= −1000, and Ymax= 1000, then press GRAPH again to display the graph. A parabola will appear in the viewing screen. Window 1.10 displays this graph.

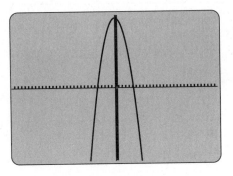

Window 1.10

The output values are usually the more difficult values to guess. One method is to graph the function using your best guesses for the values of Ymin and Ymax. If the graph does not appear in the window, press the TRACE key. Although the graph does not appear, the cursor's location on the graph is recorded at the bottom of the screen. The output value provides you with a good estimate of what y-values to expect and you can adjust the Ymin and Ymax values accordingly.

Now, still using the function $f(x) = -.4x^2 + 950$, suppose we wish to estimate which positive x-value produces the y-value 320. Press the TRACE key and the flashing cursor will appear on the curve. Press the right-arrow key to move the cursor along the curve in the positive x direction. Notice that the coordinates of the cursor appear at the bottom of the viewing screen. If you move the cursor below $y = 320$, use the left-arrow key to move it back up the curve. Notice that you cannot isolate $y = 320$. The best you can do is to locate the cursor above 320 at $y = 326.7313$ and below 320 at $y = 149.44598$. You can obtain more accuracy by creating a box around this part of the graph and magnifying it using the ZOOM key. Locate the cursor at one of these positions, say $y = 326.7313$. Press the ZOOM key and select the BOX option. Move the cursor slightly off the curve by pressing the left-arrow key once. Press ENTER to begin creating the box. Press the down-arrow key once and the right-arrow key two or three times (so that it crosses the graph) and then finish the box-creation process by pressing ENTER. A magnified portion of the curve should be visible on the viewing screen. If no curve appears, reset the range values and try again. Press TRACE. Use the arrow keys to move the cursor along the curve to $y = 320$. You may have to use the box feature again. A good estimate for the x-value that produces the y-value of 320 is 39.69.

EXERCISES

For exercises 1–4, write down what you would type into your calculator in the Y= menu to enter the given function. Then check your decision by trying it on your calculator.

1. $d(t) = 50t$.

2. $f(x) = 2x - 3$.

3. $g(x) = x^2 - 3$.

4. $A(r) = 3.14r^2$.

5. You are given the function $V(l, w, h) = lwh$, where l represents length, w width, and h height, and are told that the length is 2 and the width is 3 inches. How would you write this volume function as a function of height (h) alone? How would you type this function into your calculator?

6. The cost of 200 diskettes is given as a function of price, p, by the defining formula or equation $Cost(p) = 200p$. What would you type to enter this function in your calculator?

7. Given $f(x) = x^2 - 3x$, complete the table of values for this function.

x	$f(x)$
2	
3	
4	
5	

8. Given $g(x) = \dfrac{1}{x - 3}$, complete the table of values for this function.

x	$f(x)$
6	
5	
4	
3.5	
2.5	
2	
1	
0	

For exercises 9–16, write what you would type to enter each function.

9. $f(x) = \dfrac{2}{x + 1}$.

10. $f(x) = \dfrac{2}{x} + 1$.

11. $g(s) = \dfrac{2 + s}{3}$.

12. $g(s) = 2 + \dfrac{s}{3}$.

13. $h(a) = 1 + \dfrac{2}{a}$.

14. $h(a) = \dfrac{a + 2}{a}$.

15. $f(k) = \dfrac{1}{k + 2} + 3$.

16. $g(s) = \dfrac{2s^2 - 5s - 8}{s^2 + 2s + 1}$.

Graph the functions given in exercises 17–24. Then use complete sentences to describe how you arrived at your answers to the questions that follow.

17. $f(x) = 2x - 3$. What is the value of x where the graph crosses the x-axis? Where does it cross the y-axis?

18. $Quantity(x) = -3x + 5$. Where does this graph cross the x-axis? Where does it cross the y-axis?

19. $Demand(t) = -3t^2 + 4t - 3$. Does this graph cross the horizontal axis? What is the value of the function when the graph crosses the horizontal axis? Where does it cross the vertical axis?

20. $f(x) = x^2 - 2$. Where does this function cross the x-axis? Do you recognize these values? If so, what are they?

21. $g(u) = 2u - 18$. Where does the graph of this function cross the horizontal axis? Where does it cross the vertical axis? (Hint: You may need to use the ZOOM key.)

22. $f(x) = x^2 - 3x - 50$. Estimate the value where the function crosses the horizontal axis. Determine the value where the function crosses the vertical axis. Determine the value where the function has the value 25. (Hint: You may need to use the ZOOM key.)

23. $f(m) = 200m - 50$. What are the values of Xmin, Xmax, Ymin, and Ymax when the graph finally comes into view?

24. $p(t) = \dfrac{2t - 5}{2t + 4}$. Does this graph cross the horizontal axis? If so, where does it cross? Does this graph cross the line $y = 1$? If so, at what t value does it cross? Explain. Where does the graph cross the y-axis?

25. Consider $g(x) = 5000x - 11000$. Graph the function and write down in paragraph form the steps you would take to estimate the value of x where y is (a) 1105, and (b) 1250. For part (b) you may have to use the ZOOM BOX feature several times in order to get close to $y = 1250$.

Exercises 26–31 refer to Joe's business.

26. Joe has invented a new squirt gun and has decided to manufacture and market it himself. It costs Joe $500 to buy the machinery used to manufacture the squirt guns. He estimates that it costs him $0.75 to make one squirt gun. Write a function that gives the cost of producing x number of items.

27. Joe sells the squirt guns for $2 each. What is his income function? Write a function that represents the profit on x items. (Profit is the difference between income and cost.)

28. Graph the profit function and describe its behavior. Can you tell from the graph how many items Joe must sell in order to break even? How many items must he sell to make a profit of $1,000? (Graph Y1= $P(x)$ and Y2= 1000. Use TRACE to find the point of intersection.)

29. Joe wants to expand his business by advertising and has decided to spend 10% of the cost of producing the x items for advertising. Write a function representing his cost of advertising. What is the total expense function now?

30. What is the new profit function if Joe continues selling the squirt guns at the original price? Graph the function to find out how many he must sell to make a profit of $1,000. If he must sell more to make the same amount of profit, how might he justify not raising the price?

31. The ad campaign is very successful and Joe doubles the number of items he sold previously. If he previously made a profit of $1,000 per week, how much is he making now? (Solve by graphing and use the results of Exercise 28.)

32. The velocity of a boy riding a skateboard on a flat parking lot is given by $v(t) = 6.2 - 0.5t^2$ m/sec, where t is the time in seconds from the start. Graph the function and find out

when the velocity becomes zero. What was his initial velocity?

33. An auto manufacturer is claiming that their car can accelerate from 0 to 60 mph in six seconds. Suppose the velocity function is given by $v(t) = 14t$ ft/sec. Graph the function on your calculator to determine whether the manufacturer's claim is correct. (Hint: Convert 60 mph to ft/sec.)

Exercises 34–37 relate to Mr. Sims of Simple Sims Software Company.

34. Mr. Sims of Simple Sims Software Company calculates that his new product *Easy Algebra* will require two person-years of research and development, 1/2 person-year for documentation, 1/4 person-year for quality control and testing and one person-year for marketing and advertising. If each person-year is worth x dollars, what is the cost $C(x)$ of developing the product? (A person-year is the work done by one person in one year).

    If the average person-year is worth $50,000 for this company (that is, the average salary of all workers) what is the cost in dollars for the development and marketing of the product?

35. Suppose that n is the number of *Easy Algebra* packages produced. The cost of development is distributed among these n items. Write an expression for the cost per unit (using the results of the previous problem) if each unit also requires $2.50 for postage.

36. Mr. Sims can sell *Easy Algebra* for $105. Write an expression for the revenue if n items are sold. Graph the revenue function and the cost function. How many items must the company sell in order to break even?

37. If on the average the company sells 333 units per month, how long will it take to earn a profit of $1,000,000?

38. Pam has a hardware store and wants to earn 40% on her investment in the store. If she buys a tool for x dollars, write a function representing her selling price.

    In order to attract customers, Pam attaches a higher price tag to a tool and writes "30% off ticketed price" on the tag. What price does she put on the item to get the profit she wants?

39. Toni's T-Shirts is having a sale, with every item discounted 20% off the original price. If x denotes the original price on a shirt, what is the shirt's sale price? The sales tax in the city where Toni's T-Shirts is located is 8 1/2%. How much does the customer pay altogether?

40. (Refer to exercise 39.) A few days into the sale Toni announces that every item is now an additional 20% off the sale price. Find a formula for the new sale price. With the 8 1/2% tax, how much will the customer pay? If a customer purchases an item that originally retailed for $30, how much (with tax) does the item now cost?

1.3   **The Calculator and Approximation**

Introduction

Input Values from Output Values

Exact Values vs Approximate Values

Determining the Maximum Error in the Input Value

Determining the Maximum Error in the Output Value

Introduction

It is often necessary to know which input value of a function produces a particular output value. For example, suppose a function named $N(t)$ was created from data taken over a period of years to relate the number N of birds in a particular region of the country to the time t in years from now. You may wish to know in how many years from now there will be 400 birds in the region. As you saw in Section 1.2, the graphing calculator can be used to obtain the approximate coordinates of a point on a curve. (Because of the pixel nature of the calculator, it is difficult for it to produce exact values.) In this section you learn how to use the calculator to approximate, with almost any specified maximum error, the coordinates of a point on a curve. Figure 1.2 illustrates the input/output relationship pictorially while the graph on the left of Figure 1.3 illustrates the input/output relationship graphically.

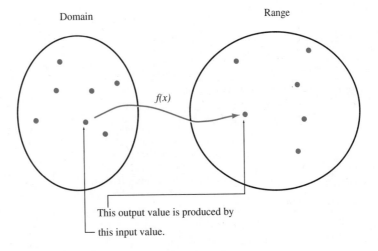

Figure 1.2

Input Values from Output Values

To determine the input value that produces the given output value, we begin by drawing a horizontal line from the output value on the output axis to the function. We then draw a vertical line from that point of intersection to the input axis. The point at which the vertical line intercepts the input axis is the sought-after input value. This situation is illustrated in the graph on the right of Figure 1.3.

EXAMPLE SET A

Suppose the function $N(t) = \dfrac{4t + 45{,}000}{t^2 + 45}$ was created by biologists from data observed over several years to approximate the number $N(t)$ of birds in a particular region. The purpose of the function is to give them a count of the number of birds, t years from now, over a period of 25 years. This situation is illustrated graphically in the left graph of Figure 1.3.

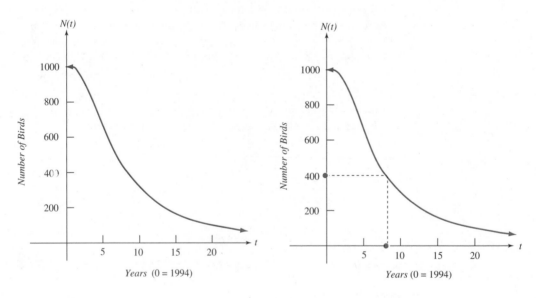

Figure 1.3

The graph indicates that there are currently 1000 of these types of birds in this particular region. How many years from now will there be only 400 birds in the region? That is, what input value produces the output value 400?

Solution:

The graph is constructed to display t-values from now ($t = 0$) to 25 years from now ($t = 25$). To determine the input value (the number of years from now) that produces the output value 400 (the number of birds in the region), construct a horizontal line from 400 on the output axis to the curve. Then, from this point of intersection, construct a vertical line down to the input axis. It is this input value that produces the output value of 400. (See the right graph of Figure 1.3.)

The input value appears to be between 5 years and 10 years (and close to 8 years).

Exact Values vs Approximate Values

The same approach can be used on your graphing calculator to estimate the input value that produces a particular output value. Before doing so, however, it is important to understand the difference between exact and approximate values.

First, recall that an equation is a statement that two expressions are equal, and that a solution to an equation is a value that, when substituted for the variable in the equation, results in a true statement. Some solutions to some equations can be found exactly and others only approximately. For example, the solutions to the equation $x^2 - x - 12 = 0$ can be found by factoring the left side of the equation, setting each factor equal to zero, and solving for x:

$$x^2 - x - 12 = 0$$

$$(x + 3)(x - 4) = 0$$

Then $x + 3 = 0$ and $x - 4 = 0$ produce the exact solutions -3 and 4. These solutions are exact because when each is substituted into the original equation, a true statement results. For example, using -3, $(-3)^2 - (-3) - 12 = 0$, exactly. (You should verify that 4 produces exactly 0.)

The equation $x^2 - 3x - 3 = 0$ can be solved exactly using the quadratic formula

$$x = \frac{-b \pm \sqrt{b^2 - 4ac}}{2a}$$

with $a = 1$, $b = -3$, and $c = -3$:

$$x = \frac{-(-3) \pm \sqrt{(-3)^2 - 4(1)(-3)}}{2(1)}$$

$$= \frac{3 \pm \sqrt{9 + 12}}{2}$$

$$= \frac{3 \pm \sqrt{21}}{2}$$

The values $\dfrac{3 + \sqrt{21}}{2}$ and $\dfrac{3 - \sqrt{21}}{2}$ are exact solutions to the equation

$$x^2 - 3x - 3 = 0$$

These solutions are exact because when each is substituted into the original equation, a true statement results. For example, using $\dfrac{3 + \sqrt{21}}{2}$,

$$\left(\frac{3 + \sqrt{21}}{2}\right)^2 - 3\left(\frac{3 + \sqrt{21}}{2}\right) - 3 = 0$$

exactly. Carry out the arithmetic to convince yourself. (You might also try to verify that $\dfrac{3 - \sqrt{21}}{2}$ produces exactly 0.)

However, suppose that this equation models some physical phenomenon and that we actually need to measure $\dfrac{3 + \sqrt{21}}{2}$ inches. The exact value can no longer be used. Since $\sqrt{21}$ is irrational, and no measuring device can measure exactly $\sqrt{21}$ inches, we must settle for an approximation. In the computation window on your calculator, you can approximate $\sqrt{21}$ using the $\sqrt{}$ key (this requires using the 2nd key). Some of the approximations we can choose are:

$$\sqrt{21} \approx 4.6$$

$$\sqrt{21} \approx 4.58$$

$$\sqrt{21} \approx 4.583$$

Using these values to approximate the solutions, we get

$$\frac{3 + \sqrt{21}}{2} \approx 3.8, \quad \frac{3 + \sqrt{21}}{2} \approx 3.79, \quad \text{and} \quad \frac{3 + \sqrt{21}}{2} \approx 3.7915$$

The values 3.8, 3.79, and 3.792 are approximate solutions since, when each is individually substituted into the original equation, $x^2 - 3x - 3 = 0$, an almost true statement results. For example, using 3.8 as an approximation to the exact input value $\dfrac{3 + \sqrt{21}}{2}$, we obtain $(3.8)^2 - 3(3.8) - 3 = 0.04$ rather than 0. The approximated input value produces an error of 0.04 in the output. (You should examine the results of substituting 3.79 and 3.792 into the equation.)

The fact is, however, that procedures for finding exact solutions exist for only a small class of equations. The Norwegian mathematician Niels Abel (1802-1829) proved that it is impossible to construct a procedure for finding exact solutions to general fifth-degree polynomial equations. Solutions to most equations must be approximated.

Determining the Maximum Error in the Input Value

Consider the approximation made in determining the input value that produces the output value of 400 in the bird example at the beginning of the section. (See Figure 1.4.) We can see from this graph that the exact input value that produces the output value 400 is between 5 and 10. If we use 5 as an approximation for this input value, we would make an error of at most $10 - 5 = 5$ units. Similarly, if we used 10 as an approximation for this input value, we would make an error of at most 5 units. In fact, if we use *any* value in the interval $[5, 10]$ as an approximation for this input value, we would make an error of at most 5 units (See Figure 1.4.) So, we could say that, with an error of at most 5 years, 8 years from now the number of birds will be 400. (This is a big error and we can certainly do better.)

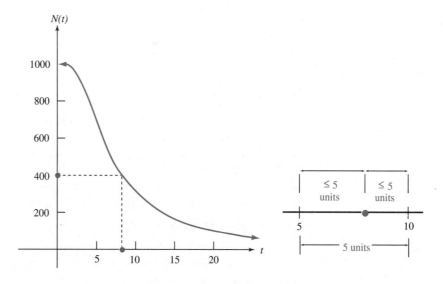

Figure 1.4

We now present a definition of an *error of at most r units*.

Error of at Most *r* Units

Error of at Most *r* Units

Suppose that the graph of the function $f(x)$ intersects a horizontal line constructed from a given output value d. Suppose further that the vertical line drawn from the intersection point to the input axis passes between two consecutive scale marks a and b that are r units apart (see Figure 1.5). Then, if c is any value between a and b, c iş said to be an *input value with an error of at most r units* that produces the output value d.

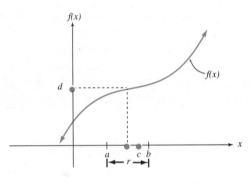

Figure 1.5

Next we demonstrate how the calculator can be used to obtain solutions to within this value of error using the bird example mentioned at the beginning of the section. We estimate, with an error of at most 0.01 units, the input value (number of years from now) that produces the ouput value 400 (number of birds in the region).

The following description of the calculator mechanics is somewhat detailed. You may wish to refer back to it as you progress through the text.

Construct the graph of $N(t) = \dfrac{4t + 45{,}000}{t^2 + 45}$, entering the defining expression for $N(t)$ using Y1. (Remember that your defining expression will involve x rather than t.) Be careful to use parentheses around both the numerator and denominator. Since both the number of years from now and the number of birds in the region must be nonnegative, set Xmin and Ymin equal to 0. Set

Xmax equal to 25, since the observation period is 25 years. When $t = 0$ (now), $N(t) = 1000$ as Figure 1.4 indicates, so set Ymax equal to 1000. Press GRAPH to construct the graph. Window 1.11 shows the graph of $N(t)$ in this viewing window.

Now, the given output value is 400 and we can have the calculator produce the horizontal line passing through 400 on the output axis by entering 400 for Y2 and pressing GRAPH. Window 1.12 displays this viewing window.

We will use the ZOOM Box feature to view the graph in a window in which it intercepts the horizontal line between two consecutive scale marks that are 0.01 units apart. (See "Creating a Box" in Section 1.2.) We may need to create several boxes before we have zoomed in enough to meet the

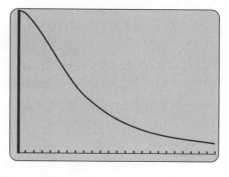

Window 1.11

condition that the error is at most 0.01, that is, that two consecutive scale marks are 0.01 units apart.

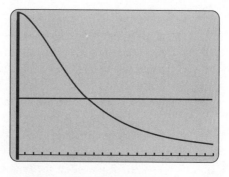

Window 1.12

Finding an Interval of Satisfactory Values   Two consecutive scale marks will be 0.01 units apart when the value at the right scale mark minus the value at the left scale mark equals 0.01. That is when, in our viewing window, the rightmost x-value (Xmax), minus the leftmost x-value, (Xmin), equals 0.01. We can set the calculator to alert us to this condition in the following way.

1. Press the Y= key and move the cursor down to Y3. Press the VARS key and use the right-arrow key to move the cursor along the top menu to RNG or select Window. Press 2 to select Xmax. This sequence of keystrokes *pastes* Xmax into the cursor's location at Y3. The calculator interprets this as the value of Xmax.

2. Press the minus key. Again press the VARS key, move the cursor to RNG and press 1 to select Xmin. This sequence of keystrokes *pastes* Xmin into the cursor's location at Y3. The calculator interprets this as the value of Xmin. Then place the cursor on the = sign in Y3 and press ENTER so that this expression is not graphed. Your viewing window should look like that illustrated in Window 1.13.

:Y1=(4X+45000)/(X^
2+45)
:Y2=400
:Y3=Xmax-Xmin
:Y4=

Window 1.13

3. Now create a box around the point of intersection of the curve and the horizontal line. Press GRAPH to access the graph of the function, and then press TRACE and move the cursor so that the y-value is near 400. Press ZOOM and select BOX by pressing 1. Move the cursor off the curve by pressing the right-arrow key once. Then press the up- or down-arrow key until the y-value is below 400 for the first time. Press ENTER to set the lower right corner of the box. Press the up-arrow key until the cursor goes above the line $y = 400$ for the first time. (You should see that the Y-value displayed at the bottom of the viewing screen is larger than 400.) Press the left-arrow key until the cursor appears on the other side of the

curve. You should now have a rectangle on the screen with the point of intersection inside. Press ENTER to complete the construction of the box. Window 1.14 displays a magnified view of the function.

4. When the graph is finished, press 2nd, Y-VARS, 3 to select Y3, and press ENTER to display the value of Y3 which equals Xmax − Xmin. The number Xmax − Xmin is not yet less than .01.

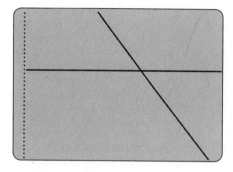

Window 1.14

5. Repeat steps 1–4 until the value of Xmax − Xmin is less than 0.01. (To quicken the process, make your boxes small. But if you make them too small, you can lose view of the graph.)

6. Press the RANGE or WINDOW key to view the value of the leftmost scale mark (Xmin, the minimum input value) and the value of the rightmost scale mark (Xmax, the maximum input value). According to the definition of error, we can choose any input value in this interval and achieve an error of at most 0.01 units. For the value to be within this interval, it must be larger than the Xmin value but smaller than the Xmax value. A satisfactory value is 8.2206.

Thus, we conclude that, with an error of at most 0.01 years, 8.2206 years from now the number of birds in the region will be 400. Of course, this is only a demonstration of a procedure that allows you to obtain information to within a certain error. It is very useful to understand this idea since it is used frequently in solving problems. Realistically, however, it would not make sense to calculate bird population information to within an error of 0.01 since the accuracy of the function created from data to approximate the bird population would almost certainly be less accurate.

Determining the Maximum Error in the Output Value

Exact input values produce correct output values. Approximations of input values result in errors in the corresponding output values. The better the approximation of the input value, the less error created in the corresponding output value.

Although approximations of input values result in errors in the output value, we can determine the maximum size of that error by knowing the maximum size of the error in approximating the input value (which we set). Figure 1.6 illustrates how this can be done.

The output value d is produced by some (unknown to us) input value that lies between a and b. If a is used as the input value, the corresponding output value is $f(a)$ and the error that results is the distance between d and $f(a)$, which is $|f(a) - d|$. Similarly, if b is used as the input value, the corresponding output value is $f(b)$ and the error that results is the distance between d and $f(b)$, which is $|f(b) - d|$. The larger of the two errors (or any values larger than these) can be used as a bound on the error in the output. In the case of the function in Figure 1.6, the maximum error in the output is $|f(b) - d|$ (as this distance is slightly larger than the distance $|f(a) - d|$).

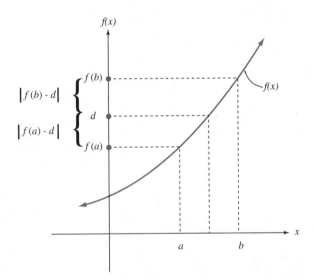

Figure 1.6

Now we present a method for determining the *maximum error of the output value* and then demonstrate how the calculator can be used to obtain solutions to within this value of error.

Maximum Error of
the Output Value

Maximum Error of the Output Value

If, between the input values a and b, the graph of $f(x)$ is always rising or always falling, the larger of the values $|f(a) - d|$ and $|f(b) - d|$ can be used as an estimate for the maximum error of the output value d when an input value is selected from the interval $[a, b]$. (See again Figure 1.6.)

ILLUMINATOR SET A

In the time vs birds example, suppose we use 8.2206 as an approximation to the exact input value that produces the value 400. The maximum error of the output value is the larger of $|f(b) - d|$ and $|f(a) - d|$. In our case, $|f(b) - d|$ is $|N(Xmax) - 400|$ and $|f(a) - d|$ is $|N(Xmin) - 400|$. Assuming your values for Xmax and Xmin are, respectively, 8.221052631 and 8.220586091, then

$$|N(8.221052631) - 400| = |399.9875798 - 400|$$

$$= |-0.0124202|$$

$$= 0.0124202$$

and

$$|N(8.220586091) - 400| = |400.014817 - 400|$$

$$= |0.014817|$$

$$= 0.014817$$

Since both these values are less than 0.015, 0.015 is a good approximation to the maximum error on the output value, $N(t)$, using the input value $t = 8.2206$. Since our example deals with counting birds and we can't have 0.015 birds, and 0.015 is less than 1, we could use 1 as a maximum error on the output.

So we could say that, with an error of at most 0.01 years, 8.2206 years from now, the number of birds in the region will be 400, plus or minus 1 bird. Figure 1.7 illustrates the meaning of this statement.

Although this approximation seems extraordinarily good, its reliability depends on the accuracy and reliability of the original function. How good a job does the original function do of describing the physical situation? We investigate this idea in Chapter 3.

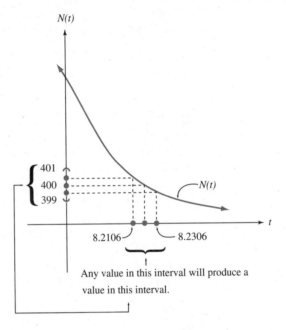

Figure 1.7

Writing Activity   Use your graphing calculator to determine, with an error of at most 0.01, the input value that produces the output value 11 for the function $f(x) = x^3 - 5x^2 + 2x - 4$. Be sure to write down in paragraph form the steps you take to arrive at your answer. Be prepared to discuss your solution in class.

EXERCISES

For exercises 1–4, determine, with an error of at most 0.01, the input value that produces the given output value for the given function. Then, using that value, determine the maximum error of the output.

1. $f(x) = 4.08x + 17.2$, output 5.88.

2. $f(x) = 3.7x - 1.9$, output 2.6.

3. $f(x) = x^2 - 4x + 2$, output 3.

4. $f(x) = x^2 - 5x + 4$, output -1.

For exercises 5–14, determine, with an error of at most 0.01, one input value that produces the given output value for the given function. When there is more than one input value, use the smallest. Then, using that value, determine the maximum error of the output. Be sure to write down the steps to your solution using complete sentences.

5. $f(x) = 6.14x^2 + 8.2x - 14.1$, output 24.3.

6. $f(x) = 0.97x^2 - 0.26x - 8.1$, output 12.

7. $f(x) = 0.86x^3 - 3.4x - 6.2$, output -25.

8. $f(x) = 5.1x^5 - 0.08x^4 + 2.2x^2 - 15.4$, output 37.9.

9. $f(x) = x^{2/3} + 2x - 5$, output -1. (Type parentheses around the $\frac{2}{3}$ when you enter it as the exponent on x.)

10. $f(x) = x^{2/5} + \sqrt{x-1} - 3$, output 10.2. (Type parentheses around the $\frac{2}{5}$ when you enter it as the exponent on x, and around the $x-1$ for the square root of $x-1$.)

11. $f(x) = \dfrac{x^{2/3}(x-3)}{x+5}$, output 6.

12. $f(x) = \dfrac{(2x-1)^2(x-6)^3}{(x+2)(x-5)}$, output 45.

13. $f(x) = \dfrac{x^2}{x-6}$, output 32.

14. $f(x) = \dfrac{x^3}{x+1.8}$, output 32.

15. An efficiency study of the morning shift of the electrical appliance assembly division of a large manufacturing company indicates that t hours after beginning work, the average worker will have assembled $N(t) = -t^3 + 7t^2 + 18t$ units. With an error of at most 0.01 hours, how many hours after beginning work will the average worker first assemble 30 units? What is the maximum error this approximation produces in the number of units assembled?

16. Real estate developers estimate that t years from now, the population of a particular town

will be $P(t) = 6.2t^3 + 5.6t^{3/2} + 8700$. With an error of at most 0.1 years, how many years from now will the population of the town be 12,500? What is the maximum error this approximation produces in the population?

17. A construction company is bidding on a project that involves laying cable from a building on a straight shoreline to a building on an island 900 feet off the shoreline and 3000 feet away. The company determines that the cost $C(x)$ of laying the cable is related to the amount of cable it lays along the ground x and how much it lays under water by the function $C(x) = 4x + 6\sqrt{900^2 + (3000 - x)^2} + 500$. With an error of at most 0.01 feet, how much cable must be layed along the ground to produce a cost of \$18,000? What is the maximum error this approximation produces in the cost?

18. The oxygen level in terms of purity units, $P(t)$, of a body of water t weeks after a particular type of toxic chemical is released into it is given by the function $P(t) = 765\left[1 - \dfrac{4.2}{x+4.2} + \dfrac{14.44}{(x+14.44)^2}\right]$. With an error of at most 0.01 weeks, how many weeks after a toxic release of this particular chemical into the water will the oxygen level be at 700 units? What is the maximum error this approximation produces in the oxygen level?

19. A particular state determines that its expenditures (in millions of dollars) on welfare $E(x)$ are related to the umemployment rate x by the function $E(x) = 0.0047x^4 + 0.103x^2 + 23.5$. With an error of at most 0.01, what unemployment rate will generate a state welfare expenditure of \$75,000,000? What is the maximum error this approximation produces in the expenditures?

1.4 Zeros of Functions and Roots of Equations

Introduction

Zeros vs Roots

Relating Zeros and Roots

Approximating Roots and Zeros with the Calculator

Introduction

An important activity in both precalculus and calculus is finding roots of equations and zeros of functions. In this section we investigate not only the difference between roots and zeros but also a relationship between them. We also see how the graphing calculator can be used to approximate both roots and zeros.

Zeros vs Roots

Roots of an Equation

Finding the solutions of an equation is a major activity in intermediate algebra. An equation is a statement that two expressions are equal. The numbers that, when substituted for the variable, make the statement true are called the solutions or **roots** of the equation. For example, the equation $x^2 - 7x = 8$ states that the expression $x^2 - 7x$ is equal to the expression 8, and the numbers 8 and -1, and only these numbers, make the equation true. They are the roots (or solutions) of the equation $x^2 - 7x = 8$. In precalculus and calculus we are concerned not only with finding the roots of an equation, but also with finding the zeros of a function. You can use your graphing calculator to approximate both. To begin, however, you must be able to distinguish a root of an equation from a zero of a function.

Zeros of a Function

A function, as we know from Sections 1.1 and 1.2, is a rule that prescribes how an output value is produced from an input value. The numbers that when used as inputs in a function produce zero as the output are called the **zeros** of the function. For example, consider the function $f(x) = x^2 - 2x - 24$. The numbers 6 and -4, and only these numbers, when input into the function $f(x)$ produce the output value zero. For example, inputting 6 gives

$$f(6) = 6^2 - 2 \cdot 6 - 24$$

$$= 36 - 12 - 24$$

$$= 0$$

(You should verify that -4 is zero of the function $f(x) = x^2 - 2x - 24$.) The example that follows illustrates an important relationship between zeros and roots.

Relating Zeros and Roots

Although roots of an equation and zeros of a function are two different entities, they are related. Consider the equation $x^2 - 7x = 8$. The roots of this equation are -1 and 8. By the techniques you investigated in elementary and intermediate algebra, you know that you can subtract 8 from each side of the equation without altering its set of solutions. This operation results in the equivalent equation $x^2 - 7x - 8 = 0$. The roots to this new equation are -1 and 8, the same as for the original equation. Suppose we use the expression on the left side of the "=" sign to define a function $f(x)$. We now have the equation $f(x) = 0$ (the expression "$f(x)$" is equal to the expression "0"). Since 8 and -1 are the roots of $x^2 - 7x - 8 = 0$, they are also the roots to the equation $f(x) = 0$. But -1 and 8 are also the zeros of the function $f(x) = x^2 - 7x - 8$. Thus, we have the following important relationship between the zeros of the function and the roots of an equation.

Relating Zeros and Roots
The zeros of the function $f(x)$ are the roots of the equation $f(x) = 0$.

Relating Zeros and Roots

ILLUMINATOR SET A

The roots of the equation $x^2 - 8x + 12 = 0$ are 6 and 2 (you should verify this). Thus the zeros of the function $f(x) = x^2 - 8x + 12$ are 6 and 2. Conversely, the zeros of the function $f(x) = x^2 - 8x + 12$ are 6 and 2, so that the roots of the equation $x^2 - 8x + 12 = 0$ are 6 and 2.

Consider again the equation $x^2 - 7x = 8$ and suppose that the expression on the left side of the "=" sign, $x^2 - 7x$, is assigned as the defining expression for the function $f(x)$; that is, $f(x) = x^2 - 7x$. Also imagine that the expression on the right side of the "=" sign, 8, is assigned as the defining expression for the function $g(x)$; that is, $g(x) = 8$. Now we have the equation $f(x) = g(x)$, and the statement that two functions are equal. Also, the roots of the equation $x^2 - 7x = 8$ are precisely the roots of the equation $f(x) = g(x)$. But $f(x)$ and $g(x)$ denote output values. Thus, the equation $x^2 - 7x = 8$ can be solved by finding the input values of both $f(x)$ and $g(x)$ that produce identical output values (-1 and 8 in this case). This is done by graphing both $f(x)$ and $g(x)$ on the same viewing window and locating their points of intersection. Figure 1.8 displays this situation *not* for these two functions, but rather for two general functions $f(x)$ and $g(x)$. Each of the input values a and b produce identical output values, $f(a)$ and $g(a)$, and $f(b)$ and $g(b)$, respectively.

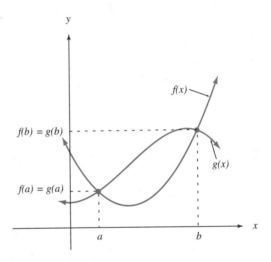

Figure 1.8

Approximating Roots and Zeros with the Calculator

The following examples illustrate how the roots of an equation can be approximated using three different approaches.

1. Approximate the roots of the equation $f(x) = 0$ by graphing the function $f(x)$ and finding the points at which it intercepts the horizontal (input) axis.

2. Approximate the input values for which $f(x) = g(x)$ by graphing each of the functions $f(x)$ and $g(x)$, finding the points at which they intersect, and approximating the input values of these points.

3. Approximate the roots of an equation using techniques from both methods 1 and 2.

The first approach is the same procedure discussed in Section 1.3 using 0 as the output value. When 0 is used as the output value, it can be assigned to Y2, or Y2 can be left unassigned. If 0 is entered for Y2, a horizontal line at 0 is constructed when Y2 is graphed. But this horizontal line is precisely the input axis. If no expression is entered for Y2, no graph is constructed for Y2, and there is essentially a default to the input axis. The next example illustrates just such a case.

Rather than just reading these examples, follow along with your calculator.

EXAMPLE SET A

1. With an error of at most 0.01, approximate the roots of the equation $x^2 - 7x = 8$ using method 1: approximate the roots of the equation $f(x) = 0$.

   Solution:
   Begin by subtracting 8 from each side of the equation to obtain the equation $x^2 - 7x - 8 = 0$. Now think of the expression $x^2 - 7x - 8$ as the defining expression for the function $f(x)$. This produces the function $f(x) = x^2 - 7x - 8$. Now, as you find the zeros of the function $f(x) = x^2 - 7x - 8$, you are simultaneously finding the roots of the equation $x^2 - 7x = 8$. Once you have entered the function using the Y= key, you can graph it. The graph of the function $f(x) = x^2 - 7x - 8$ is illustrated in Window 1.15.

   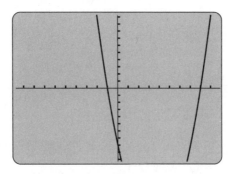

   Window 1.15

   Since the horizontal line constructed from 0 on the output axis corresponds to the horizontal axis, the zero-producing input values occur at points where the curve intercepts the horizontal axis, in this case at -1 and 8. Using the procedure outlined in Section 1.3 with output value 0, we approximate the zeros of the function $f(x) = x^2 - 7x - 8$ as -1.0006 and 8.0011, and therefore, the roots of the equation $x^2 - 7x = 8$ as -1.0006 and 8.0011.

2. This example illustrates how to approximate the roots of an equation using the idea of equality of functions. With an error of at most 0.01, approximate the roots of the equation $x^2 - x = 8$ using method 2: equality of functions.

Solution:

The procedure used here is also the same procedure described in Section 1.3 with the expression $x^2 - 7x$ assigned to Y1 and 8 to Y2. The graph appears in Window 1.16.

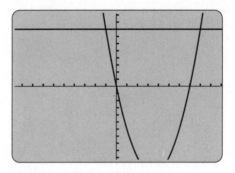

Window 1.16

Using the ZOOM feature and creating boxes as in the previous example we approximate, with an error of at most 0.01, the roots of the equation $x^2 - 7x = 8$ as -1.0006 and 8.0001.

3. A third approach to finding the roots of this equation uses ideas from both the two previous examples. You need to deactivate the graphing capability of a function by placing the cursor over the "=" sign at which the defining expression is located and pressing ENTER. The function is still defined, but it cannot be graphed. Approximate the roots of the equation $x^2 - 7x = 8$.

Solution:

As in the previous example, assign the expression $x^2 - 7x$ to Y1 and the expression 8 to Y2. Now turn off the graphing capability of both functions by placing the cursor on the "=" sign of each function and pressing ENTER. The ENTER key acts like a toggle switch, meaning you can turn the graphing capability back on by repeating this process. Now, move the cursor down to Y3. Move the cursor off the "=" sign with the right-arrow key. You are going to enter Y1 − Y2 for Y3. The calculator will interpret Y3 = Y1 − Y2 as $Y3 = \underbrace{x^2 - 7x}_{Y1} - \underbrace{8}_{Y2}$. For this to happen, you must paste the icons for Y1 and Y2 into the location of Y3's defining expression. Begin pasting the icons by selecting Y1 from the Y-VARS or Function menu. The icon Y1 will appear immediately to the right of Y1=. Press the subtraction key (not the (−) key), and then select Y2 from the Y-VARS or Function menu. Your viewing window should look like that pictured in Window 1.17.

```
:Y1=X^2-7X
:Y2=8
:Y3=Y2-Y1
:Y4=
```

Window 1.17

Now press ZOOM and 6 and the graph of Y3 = Y1 − Y2 (which is actually Y3 = $x^2 - 7x - 8$) will appear in the viewing window. The graph that appears is identical to Window 1.15. The zeros of the function Y3 are the roots of the equation $x^2 - 7x = 8$. As in the previous two examples, they are, with an error of at most 0.01, −1.0006 and 8.0001.

In parts 1, 2, and 3 of Example Set A, we examined points where the function intercepted the input axis and points where two functions intersected each other. You may be thinking that since we used only the standard viewing window, maybe there were points of intersection outside the viewing window that we missed, thereby missing additional roots of the equation. In this case, the nature of the equation we used guarantees exactly two roots. This equation is an example of a polynomial equation. In chapter 3, you will examine the properties and behavior of polynomial equations in detail.

Writing Activity   Write down in paragraph form the procedure you would use to find a zero of a function to the nearest hundredth. Then use your procedure to find the zeros of $f(x) = x^3 - 2x^2 - 5x - 7$ to the nearest hundredth.

EXERCISES

These exercises are designed in part to exercise your ability to use your graphing calculator. If you have a graphing calculator other than the TI, be sure that you have learned from the User's Manual how to perform the tasks required to do these exercises.

For exercises 1–9, approximate the root or roots to the nearest hundredth.

1. The roots of the equation $x^2 - 14x + 24 = 0$. (Hint: You may wish to zoom out to make sure that you have found both zeros of this polynomial.)

2. The roots of the equation $-2x^2 - 3x + 4 = 0$. Remember ZOOM 6.

3. The smaller root of the equation $-2x^2 + 4x - 1 = 0$.

4. The root closest to zero of the equation $x^3 - 5x^2 + 6x - 2 = 0$.

5. The root of the equation $200u = 5000$.

6. The greater zero of the function $f(x) = x^2 - 30x - 50$. (Hint: You may wish to zoom out first.)

7. (a) The zero of the function $f(x) = x^3 - 19x^2 - 377x + 6555$ that is closest to 0. (b) Check that 23 is a zero of this function. (c) Show that 0 is not a zero of this function. (Hint: You may not need to use your calculator.)

8. The roots of $\dfrac{x^2 + 4x - 165}{x^2 + 6} = 0$.

9. The roots of $\dfrac{2.06x^2 + 4.58x - 2.2}{2.2x^2 + 6} = 0$.

10. Approximate the zeros of the function $g(x) = x^2 + 2$. (Hint: What does it mean when the function graph does not cross the x or horizontal axis?)

11. Approximate a value x to within an error of 0.01 for which the function $g(x) = x^2 - 11x - 50$ has the value 25. Remember ZOOM 6.

12. Approximate a value x to within an error of 0.01 for which the function $g(x) = 5x^2 - 11x - 50$ has the value 30.

13. Write down a function that gives the score of a football game based on the number of touchdowns, extra points, and field goals scored by both teams. (A touchdown is worth 6 points, an extra point is worth 1 point, and a field goal is worth 3 points.)

14. Write down a function that gives the money collected at a concert if there are reserved seats costing $15 and unreserved seats costing $10. Be sure to give the name of the function and the names of the variables on the right side of the defining equation.

15. Use the TRACE feature to determine if the value of the function in exercise #7 is ever larger than 8000 when x is between -3 and 3.

16. Use the TRACE feature to determine if the value of the function in exercise #7 is ever less than -250 when x is between 15 and 25.

17. Use the RANGE key to set up a graphing window with x between -1000 and 1000 and y between -10000 and 10000. What does the graph of $f(x) = 2x + 50$ look like in this window?

18. What happens when you continue to move the flashing trace box to the right using the right-arrow keys for the function $f(x) = .3x - 2$?

Describe the function given by the table of values in exercises 19–20. Write down in paragraph form any and all approaches (including incorrect ones) that you considered. Taking several approaches to the solution of a problem before arriving at the correct one is very normal. In addition, it is very helpful to go over the thought processes used in solving a problem.

19.

Input	Output
8	3
4	2
2	1
1	0
1/2	−1
1/4	−2

20.

Input	Output
−2	1/4
−1	1/2
0	1
1	2
2	4
3	8

21. Graph the function of exercise 20 and determine the value of x to within an error of 0.01 for which this function has the value 128. (Hint: You may wish to adjust the graphing window using the RANGE key.)

1.5 Approximating Points of Intersection of Graphs of Functions

Introduction

Approximating the Output Value of the Point of Intersection

Introduction

In calculus, it is sometimes necessary to find the point where two curves intersect. For example, we are interested in computing the area of a region that is bounded by the two functions $f(x) = \frac{1}{2}x + 4$ and $g(x) = -x^2 + 8x - 5$. Figure 1.9 displays $f(x)$ and $g(x)$ and their points of intersection.

In this section we use the ZOOM, BOX, and TRACE features of our graphing calculator to approximate the coordinates of the points of intersection of two curves. The procedure is much the same as the procedure detailed in Section 1.3, the section in which we discussed approximation and controlling the error in the input variable when we were given a particular value for the output variable. Since we will not be given the value of the output variable, we will have to approximate it. As we need to control the error in the value of the input variable, we will also need to control the error in the value of the output variable.

Approximating the Output Value of the Point of Intersection

Recall that in approximating and controlling the error in the input variable that produced a particular value for the output variable (Section 1.3), we entered the function using Y1 under the Y= menu. We entered the given output value using Y2. We controlled the error of the *input* value by entering Xmax − Xmin for Y3. We used the value 0.01 as the maximum error in our approximation of the input value. We can follow a similar process for controlling the error in an *output* value. We control the error by entering, as Y4, Ymax − Ymin and seeing if it is less than or equal to some specified value. Ymax and Ymin are similarly entered using the RNG or Window menu of the VARS key.

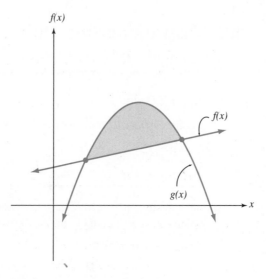

Figure 1.9

EXAMPLE SET A

With an error of at most 0.01, approximate the coordinates of the points of intersection of the curves $f(x) = x^2 - 6x + 11$ and $g(x) = -x^2 + 8x - 5$.

Solution:
Enter the expressions $x^2 - 6x + 11$ and $-x^2 + 8x - 5$ for Y1 and Y2, respectively. For Y3, enter Xmax − Xmin. (Recall that this can be done by setting the cursor at Y3 and selecting RNG under the VARS menu to access Xmax and Xmin).

For Y4, enter Ymax − Ymin in the same way.
Window 1.18 displays this calculator window.

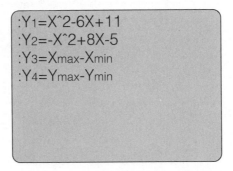

:Y1=X^2-6X+11
:Y2=-X^2+8X-5
:Y3=Xmax-Xmin
:Y4=Ymax-Ymin

Window 1.18

Press GRAPH to graph both functions. (We have selected the standard ZOOM 6 graphing window.) The curves are illustrated in Window 1.19 and appear to intersect at two points.

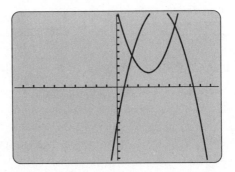

Window 1.19

First we approximate the point nearest the origin. Press the TRACE key and move the cursor near the point of intersection. Create a box around this point. Test to see if the errors in the approximations are less than 0.01 by pressing 2nd, Y-VARS, 3, and ENTER to test for X, and 2nd, Y-VARS, 4, and ENTER to test for Y. Remember, the conditions are true if both Y3 and Y4 are less than or equal to 0.01. Thus, the approximation process must continue until each test produces a number less than or equal to 0.01.

For the point of intersection nearest 0, we conclude that, with an error of at most 0.01, $x = 1.4384$ and $y = 4.4385$. (Read these values from the RANGE menu.)

Using the same procedure for the point of intersection farthest from the origin, we obtain, with an error of at most 0.01, $x = 5.5615$ and $y = 8.5615$.

Thus, with errors of at most 0.01, the functions $f(x) = x^2 - 6x + 11$ and $g(x) = -x^2 + 8x - 5$ intersect at the points $(1.4384, 4.4385)$ and $(5.5615, 8.5615)$.

It is interesting to note that the condition Xmax $-$ Xmin ≤ 0.01 was satisfied first. To satisfy the condition Ymax $-$ Ymin ≤ 0.01, the process had to continue. As it continued, the value of Xmax $-$ Xmin decreased even more. By the time the process ended, Xmax $-$ Xmin was not only less than 0.01, but less than 0.001.

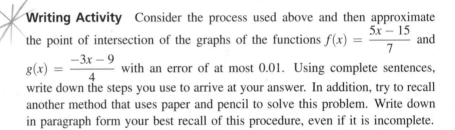

Writing Activity   Consider the process used above and then approximate the point of intersection of the graphs of the functions $f(x) = \dfrac{5x - 15}{7}$ and $g(x) = \dfrac{-3x - 9}{4}$ with an error of at most 0.01. Using complete sentences, write down the steps you use to arrive at your answer. In addition, try to recall another method that uses paper and pencil to solve this problem. Write down in paragraph form your best recall of this procedure, even if it is incomplete.

EXERCISES

For exercises 1–4, approximate the point(s) of intersection with an error of at most 0.01.

1. The graphs of the linear functions $f(x) = x + 3$ and $g(x) = -2x - 4$.

2. The graphs of the quadratic functions $f(x) = x^2 + 2$ and $g(x) = -x^2 + 4$.

3. The graph of $f(x) = x^2 - 2x + 3$ with the constant function $g(x) = 4$. When is the value of $f(x)$ larger than 4?

4. The graphs of the functions $f(x) = x^2 - 2x$ and $h(x) = x^2 - 3x + 4$.

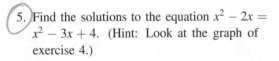

5. Find the solutions to the equation $x^2 - 2x = x^2 - 3x + 4$. (Hint: Look at the graph of exercise 4.)

For exercises 6–15, solve the problem. Be sure to write down your solution in paragraph form.

6. Find the solutions of the following system of equations with an error of at most 0.01.
$$\begin{cases} y = x^2 - 4x + 2 \\ y = -2x + 7 \end{cases}$$

7. Write the definition of the quantity of telephone handsets that are made as a function of time if the handsets are made at the rate of 200 per hour. Be sure to indicate what your function and variable names stand for.

8. Write the definition of the number of pages in a book as a function of the thickness of the book if books have 251 pages per inch.

9. Graph the function $f(x) = x^5 - 3x^2 + 5$ and, with an error of at most 0.01, approximate its zeros.

10. In the function $f(x) = x^5 - 3x^2 + 5$ from exercise 9, approximate the value of $f(3.12)$ with an error of at most 0.01.

11. What can be said about a function if its graph does not cross the x-axis?

12. Find a definition for a function that has the following table of values.

Input	Output
2	6
3	11
4	18
5	27

13. Find the value of the function given in exercise 12 when the value of the independent variable is 7. Is this called interpolation or extrapolation? (Hint: If you're not sure, look up the meaning of these words in the dictionary.)

14. Find the value of the function given in exercise 12 when the independent variable is 2.3. Is this called interpolation or extrapolation?

15. If the distance a boat travels downstream is a function of time given by $D(t) = \pi^2 t$, use the TRACE key to approximate when the boat will have gone 20 miles downstream.

16. Use the RANGE key to graph the function $Tough(q) = -100q + 300$. (Hint: You may wish to estimate the Xmin, Xmax, Ymin, and Ymax values from the function expression mentally, or using paper and pencil.)

17. The heart rate of a rabbit is determined by the function $H(d, s) = \dfrac{1}{2d}s^2$, where d is the distance that a predator is from the rabbit and s is the size of the predator. Write down the heart rate of the rabbit as a function of the size of the predator only, when $d = 5$.

18. Write down how you would enter the function $f(n) = \dfrac{n^2 - 3n}{n + 2} - 3n + \dfrac{1}{2 - n}$.

19. Complete the following table for the function $w(d) = \dfrac{1}{2d^2}$.

Input	Output
2	
3	
4	
5	

Are the values in the right column increasing or decreasing?

20. Suppose that the path of a comet is going to intersect the orbit of the earth. In order to find the points of intersection of the two orbits, astronomers have to solve the following two equations:

$$f(x) = \sqrt{1 - x^2}$$

$$g(x) = \sqrt{1.36(0.33 + x)}$$

which represent approximate distance given in Astronomical Units (AU). Graph the functions in the same window and find the value of x at the point of intersection.

1.6   Describing the Behavior of Functions

Introduction

Interval Notation and Types of Intervals

Increasing/Decreasing Functions

Relative Extreme Points of Functions

Points of Discontinuity

Introduction

When examined from left-to-right, that is, as the input value increases, the graph of a function can exhibit properties that allow us to describe the function's behavior. For example, the graph of the function can be

1. rising, falling, or staying constant,

2. opening upward or downward,

3. attaining a highest value or a lowest value relative to nearby values,

4. attaining a highest or lowest value on an entire interval, or

5. intersecting one or both or neither of the coordinate axes.

A function's behavior is usually described in terms of how its output values react to increases in its input values. Input values increase as they move from left-to-right along the horizontal axis. From the graph of a function, it may be possible to determine if an input value can be found for a particular output value. Since functions describe physical or theoretical phenomena, knowledge of these properties can be useful in analyzing the past, current, and future behavior of these phenomena.

Interval Notation and Types of Intervals

Interval

The behavior of a function is commonly described in terms of how output values react over particular sets, called **intervals**, of input values. Intervals are connected pieces of the input or output axes and are determined by two boundary points. Several types of intervals are possible. Suppose a and b are two real numbers.

Closed Interval Endpoints

1. The **closed interval** $[a, b]$, is the set of all numbers x between and including a and b. The numbers a and b are called the **endpoints** of the interval. Figure 1.10 shows a closed interval. A shaded circle at a point indicates that the point is included in the interval. In a closed interval, the endpoints

are included in the interval. In inequality notation, $[a, b]$ is equivalent to $a \leq x \leq b$.

Figure 1.10

Open Interval   2. The **open interval** (a, b), is the set of all numbers x between but not including a and b. Figure 1.11 shows an open interval. An unshaded circle at a point indicates that the point is not included in the interval. In an open interval, the endpoints are not included in the interval. In inequality notation, (a, b) is equivalent to $a < x < b$.

Figure 1.11

Half-Open Interval   3. The **half-open on the right interval** $[a, b)$, is the set of all numbers x between a and b in which the left endpoint a is included, but the right endpoint b is not. Figure 1.12 shows this half-open interval. In inequality notation, $[a, b)$ is equivalent to $a \leq x < b$.

Figure 1.12

The **half-open on the left interval** $(a, b]$, is the set of all numbers x between a and b in which the left endpoint a is not included, but the right endpoint b is.

Bounded/Unbounded Intervals   The four intervals listed above are **bounded intervals** and are finite in length.

4. Some intervals are infinite in length and are called **unbounded intervals.** Unbounded intervals involve $-\infty$ or ∞, which are symbols that indicate boundlessness. (The symbols $-\infty$ and ∞ do not represent real numbers.) The unbounded intervals are illustrated below.

(a, ∞) is equivalent to $x > a$. (See Figure 1.13).

a

Figure 1.13

$[a, \infty)$ is equivalent to $x \geq a$. (See Figure 1.14).

a

Figure 1.14

$(-\infty, a)$ is equivalent to $x < a$ and $(-\infty, a]$ is equivalent to $x \leq a$. Both of these intervals are similar to the two previous intervals but move in the opposite direction.

$(-\infty, \infty)$ represents all real numbers. (See Figure 1.15).

Figure 1.15

Increasing/Decreasing Functions

Informally, a function $f(x)$ is increasing on an interval (a, b) if the graph of $f(x)$ rises as x increases through the interval (a, b). Similarly, a function $f(x)$ is decreasing on an interval (a, b) if the graph of $f(x)$ falls as x increases through the interval (a, b). For example, the graph in Figure 1.16 shows a function $f(x)$ that is decreasing on the interval (a, b) and increasing on the interval (b, c). The output values decrease, then increase.

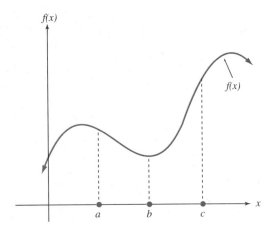

Figure 1.16

Calculator Limitations   When describing the behavior of a function from a graph produced by a graphing calculator, it is important to keep in mind that the resolution and/or size of the calculator viewing screen may keep us from getting an accurate picture of the function, and therefore, from producing an accurate summary of the function's behavior. For example, the graph of the function $f(x) = x^3 - 2.56x^2 + 2x + 1$, as it appears in the standard viewing window, is illustrated in Window 1.20.

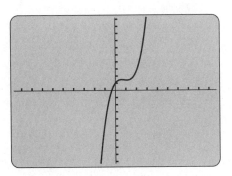

Window 1.20

However, when we create a box and zoom in on the function around $x = 1$, we see a small deviation in the function's behavior. Window 1.21 shows that the function actually decreases on some interval near 1. The resolution of the standard viewing window misled us.

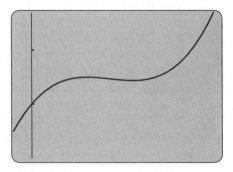

Window 1.21

The calculator screen on the left of Window 1.22 shows the graph of the function $f(x) = 0.13x^3 + 0.8x^2 + 0.4x + 15$ in the standard viewing window. It appears to be always increasing. However, as the calculator screen on the right of Window 1.22 shows, when we zoom out (ZOOM 3) just once, we get a different view of this function. It appears to increase, reach a high point, then decrease, reach a low point, then increase again. As before, we were misled by the standard viewing window.

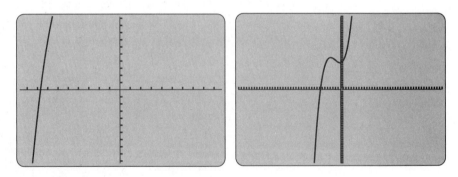

Window 1.22

An important point here is that although geometric methods are powerful, they may provide us with an incomplete or inaccurate picture of activity. The methods of algebra and calculus, although perhaps more cumbersome, are more powerful and provide the complete and accurate picture of activity. Courses in precalculus and calculus are designed to provide you with these powerful tools.

Relative Extreme Points of Functions

We can use the graphing calculator to describe, approximately, the behavior of a function in terms of where it is increasing or decreasing, and where it has high points or low points. A high point of a function is a point where a function changes from increasing to decreasing. A low point of a function is a point where a function changes from decreasing to increasing. Here we define high points and low points more precisely.

Relative Maximum and Relative Minimum

Relative Maximum and Relative Minimum

A function $f(x)$ has a *relative maximum* at $x = c$ if, for all values of x near and on either side of c, $f(c)$ exists and $f(c) > f(x)$. The point $(c, f(c))$ is called a relative maximum of $f(x)$.
A function $f(x)$ has a *relative minimum* at $x = c$ if, for all values of x near and on either side of c, $f(c)$ exists and $f(c) < f(x)$. The point $(c, f(c))$ is called a relative minimum of $f(x)$.

Figure 1.17 shows a function with a relative maximum at $(2.5, f(2.5))$, and a relative minimum at $(11.5, f(11.5))$. Notice that for all values near and on either side of $x = 2.5$, $f(2.5) > f(x)$. Also, for all values near and on either side of $x = 11.5$, $f(11.5) < f(x)$.

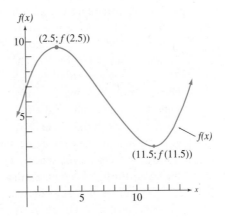

Figure 1.17

Use the Y= key to enter the function $f(x) = -x^2 - 4x + 3.5$. Press ZOOM 6 to graph the function in the standard viewing window. The graph appears in Window 1.23. Use the TRACE feature to move the cursor to the leftmost point on the graph. Press the right-arrow key again and again, observing the Y values at the bottom of the screen and the motion of the TRACE cursor. You can see from this activity that as the input values increase (as you move left-to-right), this function increases, then decreases. We'll summarize this

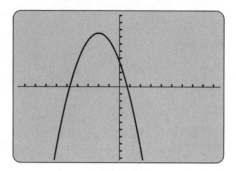

Window 1.23

function's behavior in terms of increasing values of x, and so that all errors in approximations are less than 0.01. Use the TRACE and left- and right-arrow keys to approximate the x-intercepts and the coordinates of the relative maximum. The x-coordinates of the x-intercepts are also zeros of the function. They are, with an error of at most 0.01, -4.74 and 0.74 (see Sections 1.3 and 1.4 to review the approximation process).

To approximate the coordinates of a point on a curve with a specified error, say 0.01, as we will do for this curve's relative maximum, use the same procedure that was outlined in Section 1.5. That is, enter Xmax - Xmin for Y3 and Ymax - Ymin for Y4, and magnify the region near the point using the ZOOM BOX feature until both Y3 and Y4 are less than 0.01. The boxes you create should be long and narrow to preserve the curvature of the function. Otherwise, the graph of the function may appear so flat you may be unable to distinguish the relative maximum (or minimum) from the other portions of the curve visible in the viewing rectangle. The graph on the left of Window 1.24 illustrates such a long, narrow curvature-preserving rectangle. The graph on the right illustrates the curve that results when the box is completed.

Continuing this procedure, we approximate, with an error of at most 0.01, the coordinates of the relative maximum as $(-2.0, 7.5)$.

The point at which the curve intercepts the y-axis is called the y-intercept (its first coordinate is 0). You can find its second coordinate easily by setting $x = 0$ and reading the value of y. The y-intercept is $(0, 3.5)$.

As the values of x increase from $-\infty$ up to -2, the function increases and passes through one of its zeros at $x \approx -4.74$. At $x \approx -2$, the function takes on its maximum value of about 7.5. Then, as the values of x increase beyond -2 toward $+\infty$, the function decreases. As it decreases, it intercepts

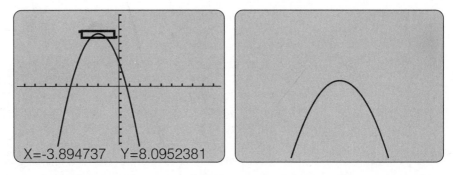

X=-3.894737   Y=8.0952381

Window 1.24

the output axis at 3.5 and then passes through its other zero at about $x = 0.74$. We summarize this function's behavior in Table 1.8. At this point in your

Table 1.8	
Interval/value	**Behavior of** $f(x)$
$-\infty < x < -2.0$	$f(x)$ is increasing
$x \approx -2.0$	Rel max at $\approx (-2.0, 7.5)$
$-2.0 < x < +\infty$	$f(x)$ is decreasing
x-intercept	Approx $(-4.74, 0)$
x-intercept	Approx $(0.74, 0)$
y-intercept	Approx $(0, 3.5)$

study of precalculus, you may not be sure that this summary completely and accurately describes the behavior of the function. After all, it may be that for larger or smaller values of x, the function turns around and increases or decreases again, or grows to values even larger than 7.5. Test whether this summary does a reasonably good job of describing this function's behavior: zoom out several times to see that the function exhibits the same basic shape, meaning that its behavior does not change.

Points of Discontinuity

Let's describe the behavior of another function, $f(x) = \dfrac{2}{x-2}$. Before graphing this function, we note that it is not defined for $x = 2$. (2, when substituted for x, produces division by 0.) Window 1.25 displays the graph (on the TI-81) in the standard viewing window.

This function appears to occur in two parts, that is, there appears to be a break at $x = 2$. The second evident feature of the graph is that it is, at least in the standard viewing window, always decreasing.

Press the TRACE key. The tracing cursor will appear at the point where the curve intercepts, approximately, the y-axis. Use the right-arrow key to slowly move the tracing cursor to the right along the curve. As the cursor moves to the right, you should notice that it is also moving downward, making it clear that the function is decreasing. As you move the cursor slowly toward $x = 2$, observe the cursor's coordinates at the bottom of the viewing screen. The x-values increase. At one instant, the cursor is at the bottom part of the left branch of the curve at about 1.789. With the next press of the right-arrow key, the tracing cursor moves off the viewing screen. You may suspect that the cursor is on the part of the curve that is below the viewing window.

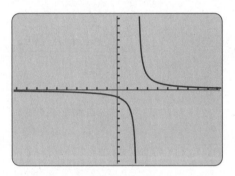

Window 1.25

However, notice (at the bottom of the viewing window) that the x-coordinate is 2 and there is no value displayed for the y-coordinate. The tracing cursor is not on any part of the curve. This is exactly what should happen, since this function is not defined at $x = 2$. There is a true break in the curve at $x = 2$ and we call this point a **point of discontinuity**.

The next press of the right-arrow key produces the coordinates $x = 2.2105263$ and $y = 9.5$, and the cursor appears in the viewing screen on the right branch of the curve. Press the right-arrow key, increasing the values of x, to slowly move the tracing cursor along the curve. As you do, notice

Point of Discontinuity

that the cursor moves downward. Again, this makes it clear that the function is decreasing. In fact, this function appears to be always decreasing.

There appears to be only one intercept, the y-intercept. The curve appears to approach the x-axis, but never to intercept it. Setting $x = 0$, we conclude that the y-intercept is $(0, -1)$. We summarize the behavior of $f(x) = \dfrac{2}{x - 2}$ in Table 1.9 and in paragraph form.

As the values of x increase from $-\infty$ to 2, the function decreases. Furthermore, when x is 0, the function intercepts the output axis 1 unit below the origin. At $x = 2$, the function is undefined and, therefore, is discontinuous there. Once the values of x are beyond 2, the function continues to decrease.

Another reasonable description would be: The function is decreasing as x increases from $-\infty$ toward 2. As the function decreases over this interval, it intercepts the output axis 1 unit below the origin. When $x = 2$, the function is undefined and, therefore, is discontinuous there. The function then continues to decrease as x increases beyond 2 to $+\infty$.

Table 1.9	
Interval/value	**Behavior of $f(x)$**
$-\infty < x < 2$	$f(x)$ is decreasing
$x = 2$	Discontinuous
$2 < x < +\infty$	$f(x)$ is decreasing
y-intercept	Approx $(0, -1.0)$

Set Factors

This summary accurately describes the behavior of this function. Try zooming out (after returning to the standard viewing window) several times to see that the curve does not change its shape. If on the first zoom-out, the curve appears too small and distorted to make out clearly, it is likely that the calculator has zoomed out too far. Press the ZOOM key (and choose MEMORY, if necessary), and select Set Factors. The standard, preset zoom factors are 4 for both X and Y. This means that each individual zoom-in and zoom-out magnifies or reduces the graph by a factor of 4. For example, a zoom-out from the standard -10 to 10 for x and -10 to 10 for y with a factor of 4 will change the window by a factor of 4 to about -40 to 40 for x and -40 to 40 for y.

To reset the zoom factors to any positive values you like, press the ZOOM key, then select the Set Factors feature, and enter your desired values. For our example, reset each zoom factor to 2. Press ZOOM 6 to return to the standard viewing window, then press ZOOM 3 ENTER to zoom out. Just one zoom-out (see Window 1.26) shows a new aspect of the curve, a line that connects the two branches of the graph.

   This line is *not* a part of the function. It represents a problem common to calculators. Calculators often do not know how to treat behavior at points of discontinuity ($x = 2$ in this case) and erroneously connect the branches of the curve. To solve the connecting problem at the expense of the quality of the graph, change from the Connected mode to the Dot mode. Press the MODE key. Use the arrow keys to scroll down and to the right to change the selection from Connected to Dot, and press ENTER. Window 1.27 shows

**Connected vs Dot
and the MODE Key**

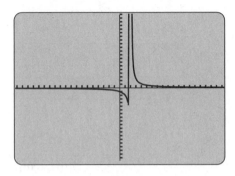

Window 1.26

this mode window. Pressing GRAPH will construct the graph of the function

Norm Sci Eng
Float 0123456789
Rad Deg
Function Param
Connected Dot
Sequence Simul
Grid Off Grid On
Rect Polar

Window 1.27

without connecting the calculated points (although many of the points are so close together, they may seem connected.) Window 1.28 shows this graph. The unwanted line is gone, but the quality of the graph has suffered. We can note here that although graphic methods are powerful methods for describing the behavior of a function, they are limited. Algebraic methods may be more cumbersome, but are usually more powerful. Return to the connected mode now (remember to press ENTER to activate your selection.) Let's examine

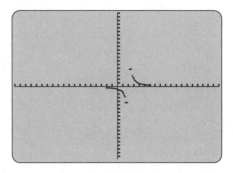

Window 1.28

another graph to describe the behavior of the function $f(x) = -0.3x^3 + 3x^2 + 2x + 2$. Enter this function and graph it using the standard viewing window. The graph appears on the left of Window 1.29.

The function appears to decrease, reach a minimum, then increase. But is this all it does? No! Zoom out 3 times (using Set Factors 2) to get a more global view of the function.

The graph on the right of Window 1.29 shows the expanded graph of this function. You can now see that the behavior is substantially different from what it appeared to be in the standard viewing window. Using the TRACE and

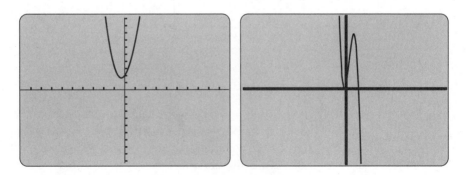

Window 1.29

the ZOOM/Box keys, we are able to summarize, with errors of at most 0.01, the behavior of this function in the following table.

Table 1.10	
Interval/value	**Behavior of $f(x)$**
$-\infty < x < -0.32$	$f(x)$ is decreasing
$x \approx -0.32$	Rel min at $\approx (-0.32, 1.68)$
$-0.32 < x < 6.98$	$f(x)$ is increasing
$x \approx 6.98$	Rel max at $\approx (6.98, 60.100)$
$6.98 < x < +\infty$	$f(x)$ is decreasing
y-intercept	Approx $(0, 2.00)$
x-intercept	Approx $(10.68, 0)$

Writing Activity   Use complete sentences to describe the behavior of the function $f(x) = -0.3x^3 + 3x^2 + 2x + 2$, whose summary table is given above.

We'll end this section by considering an application. Suppose a box with no top is to be constructed from a square piece of cardboard, 25 inches on each side, by cutting out squares of the same size from each corner. Find the dimensions of the squares that will produce the box with the maximum volume. Figure 1.18 shows the cardboard piece and the resulting box.

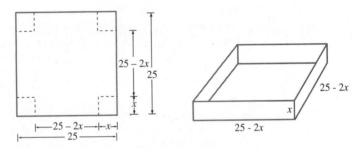

Figure 1.18

If we let x represent the length of the cut, then the volume $V(x)$ of the box is given by the volume function $V(x) = x(25 - 2x)(25 - 2x) = 625x - 100x^2 + 4x^3$. Enter this function. Before you graph it, notice that if $x = 1$, the value of the function (the volume) is about $625 - 100 = 525$. The standard $[-10, 10]$ for x and $[-10, 10]$ for y viewing window is just not going to do. Use the RANGE or Window key to reset the Xmin, Xmax, Ymin, and Ymax values. Since the volume cannot be negative, set the Ymin value at 0. Try setting the Ymax value at 600. What are reasonable choices for Xmin and Xmax? Since x is the length of the cut, the smallest it could be is 0 (meaning no cut at all). The biggest it could be is 12.5 (meaning that with cuts from each side you have cut all the way through the piece of cardboard). Set Xmin at 0 and Xmax at 12.5. Now graph the volume function. Part of the graph appears to be missing. Reset the Ymax value. Try 1000. Still too small. Try 1500. The volume function is graphed in Window 1.30. This works fairly well. The maximum volume of the box is the maximum value of the volume function. This is the y-coordinate of the high point on the graph. Use the TRACE and the arrow keys to move the tracing cursor to the high point on the graph and read the coordinates of this point at the bottom of the viewing screen. We conclude that cutting out squares with sides of approximately 4.17 inches will produce the box of maximum volume, that volume being about 1157.41 cubic inches.

Now, press the TRACE key and move the tracing cursor to the leftmost part of the curve, near the origin. The coordinates at the bottom of the viewing screen should read $x = 0$ and $y = 0$. Now, slowly move the tracing cursor to the right along the curve. Can you describe, in terms of the box itself, what the behavior of the graph illustrates? As the cursor moves to the right, the values of x are increasing. This means that the cut in the cardboard is getting longer. But as x moves to the right, the tracing cursor is moving upward. This means that the values of y are increasing, and thus that the volume of the box is increasing. So far, we can see that as the cut gets bigger, the volume of the resulting box gets bigger. Continue moving the tracing cursor to the top

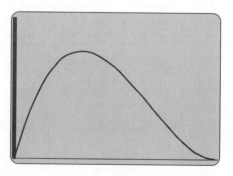

Window 1.30

of the curve. When x gets to about 4.179, the volume of the box becomes maximum at about 1157.41 cubic inches. Continue moving the tracing cursor to the right along the curve. As the cursor moves to the right, x continues to increase in value, meaning that the cut continues to get longer. However, as the cut gets longer, the volume decreases (since the function values decrease). Finally, at $x = 12.5$, the cardboard is cut in half, and there is no volume. We summarize this behavior in Table 1.11.

Table 1.11	
Length of cut	**Volume of box**
$0 < x < 4.17$	Volume is increasing
$x \approx 4.17$	Vol is max at ≈ 1157.41
$4.17 < x < 12.5$	Volume is decreasing

Writing Activity   Using complete sentences, describe the behavior of the volume of the box in the summary table above.

EXERCISES

For each of exercises 1–20, construct a summary table as well as a description for each function (in paragraph form) using 0.01 as the maximum error of the input.

1. $f(x) = x^2 - 2x - 48$.

2. $f(x) = 2x - 5$.

3. $f(x) = x^3 - 2x^2 + 5x - 1$.

4. $f(x) = 0.4x^3 - 1.5x^2 - 3.3x + 7$.

5. $f(x) = 1.8x^3 + 2.6x^2 + 0.8x - 12.5$.

6. $f(x) = x^4 - 5x^2 - 5$.

7. $f(x) = 1.07x^4 + 2.03x^3 + 1.09x^2 - 8.2x + 5.3$.

8. $f(x) = 0.7x^4 + 0.3x^3 + 19x^2 - .66x - 12$.

9. $g(x) = x^3 - 6x + 5$.

10. $f(x) = \dfrac{3}{x - 6}$.

11. $f(x) = \dfrac{-5}{x + 2}$.

12. $f(x) = \dfrac{x - 2}{x + 4}$.

13. $f(x) = \dfrac{0.07x - 2.3}{1.9x + 1}$.

14. $f(x) = \ln x^2 - 5.8$. Be careful as you use the LN and e^x keys at the left of your calculator to place parentheses around the input expression. (ln and e^x are the names of functions.)

15. $f(x) = \ln x^4 - 16$.

16. $f(x) = \sin 2x - 7$.

17. $f(p) = 160,000 \cdot 2^{-p/6250}$.

18. $f(x) = x^2 e^{-.3x}$.

19. $f(x) = x^2 e^{-.8x^2}$.

20. $f(x) = e^{-x^2}$.

For exercises 21–26, sketch the function and create a summary table. Use the table to describe the behavior of the function in terms of the physical situation.

21. A ball is thrown straight up into the air. The function relating the height $h(t)$ (in feet) of the ball and the time t (in seconds) after the ball is thrown is $h(t) = 3 + 48t - 16t^2$.

22. The cost of gasoline for a truck is approximated by the cost function $C(x) = \dfrac{7726 + 12.8x^2}{x}$, where $C(x)$ represents the cost of the gasoline and x represents the speed of the truck in miles per hour.

23. The percentage $P(t)$ of young children attending private schools in a particular county is approximated by the function $P(t) = \sqrt{35 + 8.6t - 1.2t^2}$, where t represents the number of years since 1985.

24. Efficiency experts have determined that at a particular business, t hours after beginning

work, a worker's efficiency is approximated by the function $E(t) = 0.85 + 0.07t - 0.026t^2$ percent.

25. A drug manufacturer claims that t hours after a particular drug is administered to a patient weighing between 100 and 150 pounds, the concentration $C(t)$ (in milligrams per liter) of the drug in the patient is approximated by the concentration function $C(t) = \dfrac{7.2t}{5.1t^2 + 3.98}$.

26. A political candidate's campaign officials believe that t weeks after the beginning of a campaign, the percentage $P(t)$ of the electorate favorably recognizing the candidate's name is given by the function $P(t) = \dfrac{46t}{t^2 + 15}$.

27. A 24-foot length of trimming is available to put around a rectangular window. If x denotes the length of the window, write down a function representing the area of the window. Graph the function and describe its behavior. What is the maximum possible area of the window?

28. A 24-square-foot piece of window protection material is available to cover a window. If the whole sheet is to be used up and if the width of the window cannot be less than 2 feet, find a function representing the perimeters of such a rectangular window. Graph the function and find the dimensions of the window with minimum perimeter.

Summary and Projects

Summary

Functions Describe Information   Functions are used to describe relationships between quantities and to present information about those relationships. Function notation is introduced as a means of describing relationships in a compact useful way. It is important to understand that the notation

$$f(x) = \text{some rule using } x$$

describes how an input value produces an output value. It is also important to understand that the expression on the left side of the $=$ symbol tells you that the output value f depends on the input value x, and that the expression on the right side is the expression that defines how to obtain the output value from the input value. The domain and range of a function are also introduced. They are the set of allowable input values for the function and the associated set of output values, respectively.

The Graphing Calculator   Graphing calculators are changing the way you learn mathematics. This section introduces the basics of constructing a graph on a calculator. It also introduces the ZOOM, Box, and TRACE graphing features that allow you to obtain more detail on any section of a graph.

Approximating Solutions   Learning approximating techniques is important to scientists, engineers, and others. It is a topic that has basically been skipped in courses like this. However, the graphing calculator allows you to learn some of the fundamentals of approximating techniques. For instance, you can use it to determine which input value of a function produces a particular output value. In most cases, the calculator cannot produce the exact input value, but rather only an approximation. The maximum error in an input value that produces a particular output value will be less than the width of the graphics window in which the graph appears. So if you continue to zoom in on a particular portion of the curve of a function, you will eventually arrive at an interval of input values that will produce the given output and will be within the desired error. The corresponding maximum error in the output value can be obtained by finding the maximum absolute value of the difference between the output value and the minimum and maximum output values corresponding to the interval of input values.

Zeros and Roots   There is a distinction between the zeros of a function and the roots of an equation. They are closely related, however, since the zeros of the function $f(x)$ are the roots of the equation $f(x) = 0$.

The Behavior of Functions   The graph of a function gives you a great deal of information about a function. You need to be able to describe how the function behaves as the domain values increase from left to right. Currently, the major aspects of the function's behavior involves describing where the function is increasing and decreasing, where the relative extreme points of a function are located, the location of the intercepts, and some idea regarding the continuity of the graph of a function.

Projects

1. It is sometimes desirable to have nice values for x when we use the TRACE feature. For instance, it is possible to change the Xmin value under the RANGE or Window menu so that the x values change by 0.2 each time. There are 96, 95, and 127 points, respectively, horizontally across the graphics windows of the TI-81, 82, and 85. This means there are 95 intervals on the TI-82 for instance. Use this information to write down in paragraph form how you would arrive at a choice of the Xmin value so that the x values would change by 0.2 each time. Be sure to try the TRACE feature on your solution with the graph of a function of your own choice.

2. The information in the following table gives the frequency of test scores on an economics quiz at a large state university. Refer to your calculator User's Guide or Guidebook to draw a histogram of this data, then determine

how to make the width of each group 2 points. Write an explanation of the procedure you used for a classmate giving an example of the result.

Data Number	Frequency	Score
1	3	0
2	15	1
3	25	2
4	40	3
5	35	4
6	55	5
7	78	6
8	92	7
9	60	8
10	42	9
11	30	10

3. Using your calculator User's Guide, construct two programs that will assign or store the RANGE or WINDOW values as follows:
Program1
Xmin=-10
Xmax=10
Xscl=2
Ymin=-100
Ymax=100
Yscl=20

Program2
Xmin=-100
Xmax=100
Xscl=20
Ymin=-1000000
Ymax=1000000
Yscl=100000
Use these two programs to investigate the following functions.
(a) $f(x) = x^3 - 3x^2 + 5x - 20$.
(b) $g(x) = x^3 - 30x^2 - 70x - 50$.

CHAPTER 2

Algebraic Relations and Functions

2.1   The Cartesian Coordinate System

Introduction

The Cartesian Coordinate System

The Distance Formula

The Midpoint Formula

Introduction

The function was introduced in an intuitive way in Section 1.1 so that we could review some of the concepts and techniques from intermediate algebra and discuss the mechanics of the calculator. In this section we put the concept of the function on a more solid foundation. In Section 2.2, we present the precise definition of a function and see, again, that a function can be described visually by means of a graph. One of the most commonly used structures for constructing graphs of functions in which one input value determines one output value is the rectangular coordinate system, or as it is often called, the **Cartesian coordinate system***. See Figure 2.1.

Cartesian Coordinate System

This system is important in that it allows us to draw connections between symbolic properties and geometric properties of functions. For example, in Chapter 1 we discussed the fact that a function is an instrument for describing information. When a function is described using a graph, it may reveal information that is not evident from a list of ordered pairs, or a table, or the symbolic form of the function.

The Cartesian Coordinate System

In the Cartesian coordinate system illustrated in the figure on the left of Figure 2.1, the perpendicular lines are called *axes* and they form a frame of reference for points in the plane. The point of intersection of the axes is called the *origin*. The horizontal axis is commonly called the *x*-axis and the vertical axis the *y*-axis, but any letters can be assigned to them. The horizontal axis corresponds to the input variable and the vertical axis corresponds to the output variable.

Cartesian Coordinates

As each point on the real number line is associated with a real number, each point in the plane is associated with an *ordered pair* of real numbers. This ordered pair of real numbers is called the **Cartesian coordinates** of the point.

---

*The Cartesian coordinate system and Cartesian coordinates are named in honor of the French philosopher and mathematician René Descartes (1596-1650). Descartes proposed the concept of the coordinate system as basic to the study of analytic geometry.

The term *ordered pair* makes it clear that a definite order exists in the notation. The notation for an ordered pair is (x, y), where x is the first component and represents an input value, and y is the second component and represents the corresponding output value. The x-coordinate of a point is measured from the y-axis to the point in a direction parallel to the x-axis. The y coordinate is measured from the x-axis to the point in a direction parallel to the y-axis. Since there is such a close relationship between points and ordered pairs, it is common to use the two terms interchangeably.

Graph of a Point

The geometric representation of an ordered pair is called the graph of the ordered pair, or the **graph of the point**. The figure on the right of Figure 2.1 shows the graphs of several points.

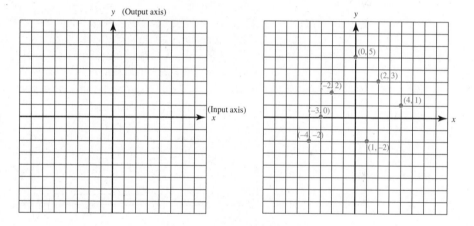

Figure 2.1

The Distance Formula

It is sometimes necessary in mathematical problem solving to know the distance between two points. Using the Cartesian coordinate system, we can derive a formula that produces the distance between any two points if their coordinates are specified. (A formula is simply a function of one or more variables that is used frequently in one or more fields.)

Suppose that P and Q are two points in the plane with respective coordinates (x_1, y_1) and (x_2, y_2). (Subscripts on letters indicate which one we wish to call the first, the second, the third, and so on.) We wish to determine the distance between P and Q. See Figure 2.2. The distance between P and Q is the length of the line segment joining P and Q. This measurement is described by $\overline{PQ}$. We begin by constructing the right triangle PQR with a right angle at

$R(x_2, y_1)$. We can measure both the horizontal and vertical distance between P and Q.

Horizontal distance $= |\overline{PR}| = |x_2 - x_1|$.

Vertical distance $= |\overline{RQ}| = |y_2 - y_1|$.

We now apply the Pythagorean Theorem:

$$|\overline{PQ}|^2 = |x_2 - x_1|^2 + |y_2 - y_1|^2$$

Taking the square roots of both sides produces

$$|\overline{PQ}| = \sqrt{|x_2 - x_1|^2 + |y_2 - y_1|^2}$$

But, since $|a|^2 = a^2, |x_2 - x_1|^2 = (x_2 - x_1)^2$ and $|y_2 - y_1|^2 = (y_2 - y_1)^2$. This gives us

$$|\overline{PQ}| = \sqrt{(x_2 - x_1)^2 + (y_2 - y_1)^2}$$

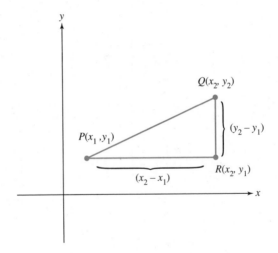

Figure 2.2

Thus, we have the *Distance Formula*.

The Distance Formula

The Distance Formula

If $P(x_1, y_1)$ and $Q(x_2, y_2)$ are two points in the plane, then the distance between them is given by

$$|\overline{PQ}| = \sqrt{(x_2 - x_1)^2 + (y_2 - y_1)^2}$$

EXAMPLE SET A

Find the distance between the given two points.

1. $(2, 3)$ and $(-3, 5)$.
2. $(-2, 5)$ and $(-2, -4)$.
3. $(64.25, 211.61)$ and $(85.63, 204.11)$.

Solution:

1. $(x_1, y_1):$ $(2, 3)$
   $(x_2, y_2):$ $(-3, 5)$

$$\text{Distance} = \sqrt{(x_2 - x_1)^2 + (y_2 - y_1)^2}$$

$$= \sqrt{(-3 - 2)^2 + (5 - 3)^2}$$

$$= \sqrt{25 + 4}$$

$$= \sqrt{29}$$

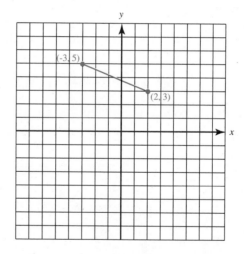

Figure 2.3

2. $(x_1, y_1):$ $(-2, 5)$
   $(x_2, y_2):$ $(-2, -4)$

$$\text{Distance} = \sqrt{(x_2 - x_1)^2 + (y_2 - y_1)^2}$$

$$= \sqrt{(-2 - (-2))^2 + (-4 - 5)^2}$$

$$= \sqrt{0^2 + (-9)^2}$$

$$= \sqrt{81}$$

$$= 9$$

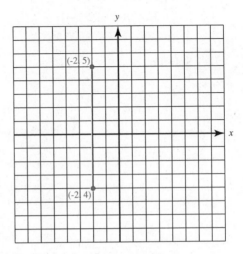

Figure 2.4

3. This is a good candidate for the calculator and an approximation.

$(x_1, y_1):$ $(64.25, 211.61)$
$(x_2, y_2):$ $(85.63, 204.11)$

$$\text{Distance} = \sqrt{(x_2 - x_1)^2 + (y_2 - y_1)^2}$$

$$= \sqrt{(85.63 - 64.25)^2 + (204.11 - 211.61)^2}$$

$$= \sqrt{457.1044 + 56.25}$$

$$= \sqrt{513.3544}$$

$$= 22.66 \quad \text{to two decimal places}$$

The Midpoint Formula

As the distance formula is a useful tool in mathematical problem solving, so is the midpoint formula. The midpoint formula provides us with a way of determining the coordinates of the point that is midway between two given points if their coordinates are specified. We can use the Cartesian coordinate system to help derive a midpoint formula.

To develop a formula for determining the coordinates of the midpoint (x, y) of a line segment joining the points (x_1, y_1) and (x_2, y_2), we construct the diagram pictured in Figure 2.5. Because (x, y) is the midpoint of the line

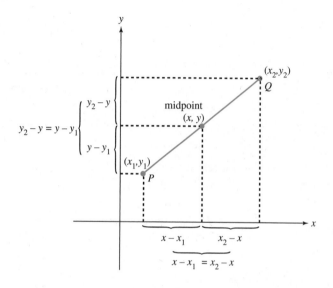

Figure 2.5

segment along the x-axis, the distance between x and x_1 is the same as the distance between x and x_2. That is, $x_2 - x = x - x_1$. We solve for x to obtain its value in terms of x_1 and x_2.

$$x_2 - x = x - x_1$$

$$x_2 + x_1 = 2x$$

$$x = \frac{x_1 + x_2}{2}$$

The value of y is determined similarly, and it is $\dfrac{y_1 + y_2}{2}$.

Midpoint Formula

Midpoint Formula
If $P(x_1, y_1)$ and $Q(x_2, y_2)$ are two points in the plane, then the coordinates of the midpoint of the line segment joining P and Q are $$\left(\frac{x_1 + x_2}{2}, \frac{y_1 + y_2}{2}\right)$$

EXAMPLE SET B

1. Find the midpoint of the line segment joining the points $(2, 3)$, and $(-4, -1)$.

2. Find the midpoint (to two decimal places) of the line segment joining the points $(-0.0048, 2.8815)$, and $(1.0089, 2.8882)$.

Solution:

1. $(x_1, y_1) : (2, 3), (x_2, y_2) : (-4, -1)$.

$$\text{Midpoint} = \left(\frac{x_1 + x_2}{2}, \frac{y_1 + y_2}{2}\right)$$

$$= \left(\frac{2 + (-4)}{2}, \frac{3 + (-1)}{2}\right)$$

$$= \left(\frac{-2}{2}, \frac{2}{2}\right)$$

$$= (-1, 1)$$

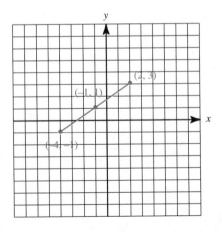

Figure 2.6

2. $(x_1, y_1) : (-0.0048, 2.8815), (x_2, y_2) : (1.0089, 2.8882)$. Use your calculator for this one.

$$\text{Midpoint} = \left(\frac{x_1 + x_2}{2}, \frac{y_1 + y_2}{2}\right)$$

$$= \left(\frac{-0.0048 + 1.0089}{2}, \frac{2.8815 + 2.8882}{2}\right)$$

$$= \left(\frac{1.0044}{2}, \frac{5.7697}{2}\right)$$

$$= \left(0.5022, \ 2.88485\right)$$

$$= \left(0.50, \ 2.88\right) \quad \text{to two decimal places}$$

Writing Activity   First recall and write down the distance formula and the midpoint formula for two arbitrary points $P_1(x_1, y_1)$ and $P_2(x_2, y_2)$ in the plane. Then try to write down in paragraph form an outline for deriving the distance formula without referring to the book.

EXAMPLE SET C

Graph the set of all points (x, y) in the plane that meet the condition $x^2 - 6x + 8 = 0$.

Solution:
We'll begin by finding the values of x that make this statement true. We can factor this equation:

$$x^2 - 6x + 8 = 0$$

$$(x - 2)(x - 4) = 0$$

Now we can see that this equation will only be true when $x = 2$ or 4. Furthermore, we want points (x, y) that satisfy the condition. Evidently, the only restrictions on the points is that their x-coordinates must be 2 or 4. The y-coordinates can be any numbers whatsoever. We can describe these points as $(2, y)$ and $(4, y)$. When we plot them, we'll fix x at 2, and let y run through every real number. Then, we'll fix x at 4 and let y run through every real number. The result is the two parallel vertical lines pictured in Figure 2.7.

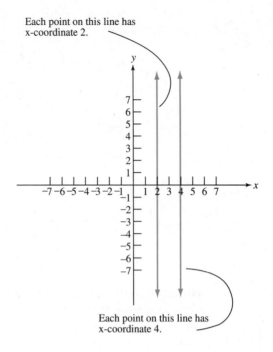

Each point on this line has
x-coordinate 2.

Each point on this line has
x-coordinate 4.

Figure 2.7

EXERCISES

For exercises 1–13, determine the distance between the given points. A calculator may be helpful for some of the exercises.

1. $(0,0)$ and $(2,3)$
2. $(2,4)$ and $(-5,1)$
3. $(1,7)$ and $(7,-2)$
4. $(-3,1)$ and $(6,-2)$
5. $(-4,-5)$ and $(3,7)$
6. $(-1,-2)$ and $(4,-2)$
7. $(-3,-6)$ and $(-2,-6)$
8. $(4,8)$ and $(4,-8)$
9. $(-1,9)$ and $(-1,12)$
10. $(6.97,56.29)$ and $(5.77,67.12)$
11. $(0.0135,-0.0065)$ and $(0.0089,0.0054)$
12. $(a,4b)$ and $(-5a,8b)$
13. $(3s,3y)$ and $(-9s,6y+4)$

For exercises 14–21, determine the midpoint of the line segment whose endpoints have the given coordinates.

14. $(2,3)$ and $(6,7)$
15. $(3,4)$ and $(5,-2)$
16. $(-3,-6)$ and $(-7,-4)$
17. $(-2,-5)$ and $(-3,-4)$
18. $(4,35)$ and $(4,21)$
19. $(-755,-216)$ and $(412,102)$
20. $(6a,-2b)$ and $(5a,5b)$
21. $(-8a,-12b)$ and $(-28a,0)$

In exercises 22–27, determine if the point P is equidistant from the points R and S.

22. $P(3,4)$, $R(6,3)$, $S(2,1)$

23. $P(2,6)$, $R(6,3)$, $S(2,1)$

24. $P(2,-4)$, $R(1,3)$, $S(-3,-9)$

25. $P(3,-2)$, $R(1,3)$, $S(-3,-9)$

26. $P(a+3,b-3)$, $R(a-5,b+6)$, $S(a+11,b-12)$

27. $P(2x^2 + 3x + 2, 3y^2 + 2y + 2)$, $R(x^2 + 4x + 1, 5y^2 + 5y + 4)$, $S(3x^2 + 2x + 3, 7y^2 + y)$

28. Construct a coordinate system and graph the points $(-2,6)$ and $(4,1)$. Draw a right triangle as in Figure 2.2 to show that the distance between these two points is $\sqrt{(-6)^2 + (5)^2}$

29. Construct a coordinate system and graph the points $(5,-2)$ and $(2,-5)$. Draw a right triangle as in Figure 2.2 to show that the distance between these two points is $\sqrt{(-3)^2 + (-3)^2}$.

In exercises 30–35, write an expression that describes the given set of points.

30. The points on the x-axis 6 units from $(4,3)$.

31. The points on the x-axis 15 units from $(2,-8)$.

32. The points on the y-axis 3 units from $(-10,5)$.

33. The points on the y-axis 7 units from $(25,62)$.

34. The points on the x-axis π units from $(6,-4)$.

35. The points on the y-axis $\sqrt{5}$ units from $(1,6)$.

In exercises 36–43, find a formula that describes the given set of points.

36. The points (x,y) that are 3 units away from the point $(4,8)$.

37. The points (x,y) that are 5 units away from the point $(7,1)$.

38. The points (x,y) that are 6 units away from the point $(-3,9)$.

39. The points (x,y) that are 12 units away from the point $(-3,-5)$.

40. The points (x,y) that are π units away from the point $(1,-1)$.

41. The points (x,y) that are π units away from the point $(-2,4)$.

42. The points (x,y) that are e units away from the point $(2a,8b)$.

43. The points (x,y) that are $\sqrt{5}$ units away from the point $(0,-2)$.

For exercises 44–55, construct a Cartesian coordinate system and graph the set of points satisfying each condition. Use your calculator when you need to approximate.

44. $x = 3$

45. $x = -2$

46. $y = 2.5$

47. $x = \sqrt{10}$

48. $x > 2$

49. $y \leq -4$

50. $y \geq \sqrt{3}$

51. $xy = 0$

52. $(x+2)(x-5) = 0$

53. $x^2 - x - 12 = 0$

54. $x^2 + x = 0$

55. $x^2 < 1$

Exercises 56-58 are related.

56. Martin's house is exactly 14 blocks to the east and 8 blocks to the north of his friend Michael's house. They decide to meet exactly midway. Assuming all the city blocks are of the same size, where do they meet?

57. Martin and Michael decide to meet midway because their friend Ron's house is exactly equidistant from both of their houses. Since their meeting place is not the only position that

satisfies the given condition, more information is needed to find the exact position of Ron's house. Set up coordinate axes with the origin at Michael's house and then find an equation describing the possible locations of Ron's house and graph it. (Use one block as the unit of distance.)

58. They later find out that Ron's house is exactly due north of Michael's house. Which block is Ron's house on?

2.2　Relations and Functions

Introduction

Relations

Functions

The Parts of a Function

Describing Relations and Functions

The Vertical Line Test

Determining the Domain and Range of a Function

Introduction

In this section we take a more rigorous look at functions. We will

1. present two alternate mathematical definitions of a function and its more general concept, the relation,
2. introduce the various parts of a function,
3. present four different ways of describing a function,
4. examine the vertical line test (a test that indicates if a graph does or does not represent a function) and
5. discuss how to determine the domain of a function.

Relations

Suppose that an assembly machine uses pressure to produce a particular shape of a piece of equipment. Consider the following relationship between the set of pressures (different pressure settings for the machine) and the set of defective pieces of equipment (number of defective pieces produced at each pressure setting). With each element in the set of pressures, the table associates

Table 2.1	
Pressure	**Number of Defectives Pieces**
30	0
35	1
40	2
45	2
50	3

an element from a set of defective pieces. This constitutes what is called a relation.

Relation, Domain, and Range

Relation, Domain, and Range

A **relation** is a process or method, such as a rule or procedure, of associating with each element of a first set, called the **domain**, one or more elements of a second set, called the **range**.

In the preceding example, if we call the relation *Pressures to Defectives*, R, then

Domain of $R = \{30, 35, 40, 45, 50\}$ and   range of $R = \{0, 1, 2, 3\}$

Notice that each domain element in the relation R is directed to precisely one range element.

$$30 \longrightarrow 0 \quad \text{(and only 0)}$$

$$35 \longrightarrow 1 \quad \text{(and only 1)}$$

$$40 \longrightarrow 2 \quad \text{(and only 2)}$$

$$45 \longrightarrow 2 \quad \text{(and only 2)}$$

$$50 \longrightarrow 3 \quad \text{(and only 3)}$$

Although both 40 and 45 are directed to 2, they each go only to 2 and nowhere else.

On the other hand, if we call the relation *Defectives to Pressures*, *S*, then

$$\text{Domain of } S = \{0, 1, 2, 3\} \quad \text{and} \quad \text{range of } S = \{30, 35, 40, 45, 50\}$$

The relation *S* does not have the property that each domain element is directed to precisely one range element. There is a domain element that is associated with more than one range element. (2 is associated with both 40 and 45.)

$$0 \longrightarrow 30 \quad \text{(and only 30)}$$

$$1 \longrightarrow 35 \quad \text{(and only 35)}$$

$$2 \longrightarrow 40 \quad \text{(but also)}$$

$$2 \longrightarrow 45 \quad \text{(a second value)}$$

$$3 \longrightarrow 50 \quad \text{(and only 50)}$$

Functions

In Chapter 1 we defined a function intuitively as a way to describe relationships and present information about those relationships. Now we define function more precisely.

Function

Function
A **function** is a relation in which each domain element is associated with precisely one range element.

A more practical or working definition of a function is as follows.

Working Definition of a Function

Working Definition of a Function
A **function** is two sets, called the domain and range, and a rule that prescribes how each domain element is assigned to exactly one range element.

The Parts of a Function

Notice that a function has three parts: a domain, a range, and a rule. To completely describe the function, all three must be precisely specified. If one part is not specified, the function is not described completely. The rule of a function need not be algebraic, and some nonalgebraic rules will be illustrated in the examples that follow. It is quite common when working with physical phenomena to describe a corresponding function only partially. We illustrate how functions are described later in the section.

In Chapter 1, we thought in terms of *input* and *output*. The domain of a function is composed of the input elements, and the range, the output elements. So, functions consist of three parts, the domain (the set of input values), the range (the set of output values), and the rule (or assignment maker) that assigns each input value to precisely one output value.

It is important to understand that the rule of a function assigns one input value to precisely one output value, as shown next.

ILLUMINATOR SET A

1. The relation R that assigns pressures to defectives is a function. Each pressure is assigned to one and only one number of defectives. The domain elements 40 and 45 are both assigned to the range element 2, but this does not violate the definition of function since each of the numbers, 40 and 45, are *individually* assigned to precisely one number. R assigns 40 to 2 and to no other number, and 45 to 2 and to no other number. Thus, $(40, 2)$ and $(45, 2)$ are both part of this function. For each input, R assigns one and only one output.

2. The relation S that assigns a number of defective to a pressure is not a function. There is one domain element that is assigned to more than one range element. S assigns the number 2 to both the numbers 40 and 45. The relation S is such that for each input, there can be more than one output. Thus, ordered pairs $(2, 40)$ and $(2, 45)$ are part of a relation, but not part of a function.

Describing Relations and Functions

Some of the more common ways of describing a function are:

1. tables,
2. ordered pairs,

3. symbolic or verbal rules, and

4. graphs.

It should be clear, after observing these methods, that it can be a difficult task to describe a function completely. Of the three parts of a function, the domain is usually the easiest to specify. The range, which is generally more difficult to specify, is commonly omitted. Also, when working with physical phenomena, it may be difficult or even impossible to develop a rule that assigns input values to output values. In such cases, the description of the function is only partial because the rule is omitted. The following is a standard assumption regarding domains.

Assumption Regarding Domains

Assumption Regarding Domains
If the rule of a function is specified but the domain is not, the domain of the function will be assumed to be the largest subset of the real numbers that assigns real numbers to the range.

The following example illustrates the four methods of description.

ILLUMINATOR SET B

Physicists have observed that as the temperature of helium approaches absolute zero ($-273°$ Celsius), it exhibits properties that are contrary to many of the conventional laws of physics. One such property is that solid helium allows a proportion of another solid to pass through it. (Physicists call the phenomenon of one solid passing through another *quantum tunneling*.) Suppose we call the relation that assigns temperature to proportion, Q. At each temperature of the helium, a particular proportion of another solid is allowed passage. The relation Q is therefore a function and can be described in any of the ways we noted above.

1. **Description by Tables**

The function Q can be described, at least partially, by a table. The domain values are listed in the left column of Table 2.2 and the range values in the right column. The rule is the nonalgebraic rule: *assign the element in the left column of a particular row to the element in the right column of that same row.* For example, $-270°$ C is assigned to 76.9%.

Table 2.2	
Temperature (°C)	Proportion (%)
−265	50.3
−270	76.9
−272	87.6
−272.9	92.4

Notice that the description is only partial since not *every* domain and range value has been listed.

2. **Description by Ordered Pairs**

The function Q can be described, at least partially, using ordered pairs (x, y). The domain values are listed as the first element in the ordered pair and the range values as the second element. The rule is the nonalgebraic rule, *assign the first element in an ordered pair to the second element of that same ordered pair.*

Letting the first element of an ordered pair represent temperature and the second element represent proportion, we can represent the function Q partially as the set

$$\{(-265, 50.3), \ (-270, 76.9), \ (-272, 87.6), \ (-272.9, 92.4)\}$$

For example, -272 is assigned to 87.6. Notice that the description is only partial since not *every* domain and range value has been listed.

Description of a relation or function by table or ordered pairs has the disadvantage that if the domain consists of a great number of elements, constructing the table or listing the ordered pairs can become tedious if not impossible. What's worse, the relationship may be very difficult to visualize if the list is long.

3. **Description by Rule**

A rule, if one can be found, that prescribes how an output value is obtained from an input value has the advantage that it may be concise and, when applied, can generate the output value associated with any chosen input value. The rule is commonly given in the symbolic form presented in Chapter 1. For example, for quantum tunneling with x representing temperature and $f(x)$ representing proportion, the rule of the function can be expressed in the form $f(x) = -0.0533x - 13.622$. To describe the function completely, both the domain and the range must be specified.

Helium exhibits quantum tunneling effects only when its temperature is near absolute zero ($-273°$ Celsius). A scientist working with the quantum tunneling effects of helium that has his equipment adjusted to produce temperatures between $-273°$ Celsius and $-255°$ Celsius has set the function's domain as $-273 \leq x \leq -255$. The function's range is $0 \leq f(x) \leq 100$, since anywhere from 0% to 100% of a solid can pass through the helium. Thus, for this scientist's purposes, the function is completely described as

$$f(x) = -0.0533x - 13.622, \quad \underbrace{-273 \leq x \leq -255}_{input\ values}, \quad \underbrace{0 \leq f(x) \leq 100}_{output\ values}$$

(Remember, $f(x)$ represents an output value.)

Notice two things. The function rule is symbolic (algebraic), and the function would be different if another scientist had adjusted the equipment so that it produced temperatures between $-270°$ Celsius and $-265°$ Celsius. In this case, the function would be completely described as

$$f(x) = -0.0533x - 13.622, \quad -270 \leq x \leq -265, \quad 0 \leq f(x) \leq 100$$

4. **Description by Graph**

Graph of a Function

The geometric representation (picture) of the input/output pairs of a function is called the **graph of the function**. The graphical description has an advantage over the other types of descriptions in that the graph may reveal information (such as trends) that may not be evident from the table, the list of ordered pairs, or the rule alone. The local behavior of a function is a consideration of the graph over a small portion of its domain while the global behavior considers the entire domain of the function. We begin by defining graphical completeness.

Complete Graph

Complete Graph

The graph of a function is **complete** if it suggests all of the possible input/output combinations and all the local and global behavior of the function.

Although we give more coverage of complete graphs later in the book, we begin the preliminary coverage.

EXAMPLE SET A

1. Graphically describe the relation $R = \{(x,y) | y = \pm\sqrt{x+3}\}$.

Solution:
(This notation is an example of set-builder notation and it is read as *R is the set of all ordered pairs (x,y) such that $y = \pm\sqrt{x+3}$.*)

The domain is not specified, so we must determine it from the rule. The domain of this relation is the set of all *x*-values that produce real numbers. This requires that $x + 3 \geq 0$, or $x \geq -3$. Thus, the domain of *R* is the set $\{x | x \geq -3\}$. We generate some ordered pairs by substituting several domain values into the rule and computing (see Figure 2.8). We plot these five ordered pairs and connect them with a smooth curve (see Figure 2.8). Notice that this relation is not a function since, in all cases but one $(x = -3)$, each *x*-value produces two *y*-values.

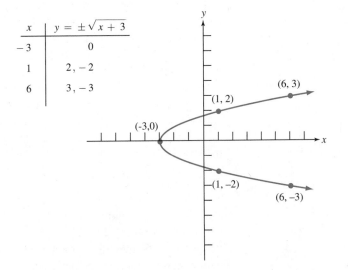

x	$y = \pm\sqrt{x+3}$
-3	0
1	$2, -2$
6	$3, -3$

Figure 2.8

Graphing Non-Functions Graphing calculators will only graph functions. $R(x) = \pm\sqrt{x+3}$ is not a function. Your calculator can graph this relation if you break it into two distinct pieces, each of which is a function. If we let $f(x)$ represent the positive piece, $+\sqrt{x+3}$, and $g(x)$ the negative piece, $-\sqrt{x+3}$, then $f(x)$ and $g(x)$ are both functions and can be graphed by the calculator. Figure 2.9 shows both $f(x)$ and $g(x)$ on separate coordinate systems. When these two pieces are placed together

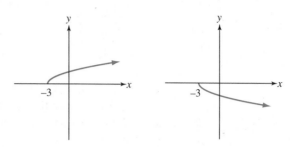

Figure 2.9

on the same coordinate system, they form the relation $R(x)$. (See again Figure 2.8.)

To graph the relation $R(x)$, press the Y= key and enter $\sqrt{x+3}$ for Y1 and $-\sqrt{x+3}$ for Y2. Press GRAPH and ZOOM 6. Each of the functions $f(x)$ and $g(x)$ will be drawn in the standard viewing window, $f(x)$ being drawn first and $g(x)$ second. The completed picture will be a complete description of $R(x)$ (since all criteria of a complete graph are met), and therefore a complete decription of the relation $R(x)$. On the calculator, you can turn a function on or off by highlighting or unhighlighting, respectively, the "="sign of the rule. Simply use the arrow keys to move the blinking cursor to an "=" sign and press the ENTER key. You should try this with the functions $f(x)$ and $g(x)$, turning one on and one off and then graphing the one that is on.

2. Describe the function $f = \{(x,f(x))|f(x) = x^3 + 1\}$ graphically.

Solution:
We first note that although the domain of f is not specified, it is the set of all real numbers. The table in Figure 2.10 shows some ordered pairs that satisfy this function. Plotting these five points and connecting them with a smooth curve gives the graph in Figure 2.10. Note that f is truly a function since each input value is assigned to precisely one output value. Notice also a trend that may not be quite as evident from observing the rule alone: As the input values increase, the output values increase. This is an example of an *increasing* function.

Try This Use your graphing calculator to graph (on the same coordinate system) some variations of the rule of this function. For example,

a. Graph $f(x) = 0.3x^3 + 1$ and $f(x) = 0.08x^3 + 1$. You should see that the branches of $f(x) = 0.08x^3 + 1$ open wider than the branches of $f(x) = 0.3x^3 + 1$. (What happens to the branches of the curve if you change the coefficient of x^3 to a number greater than 1?).

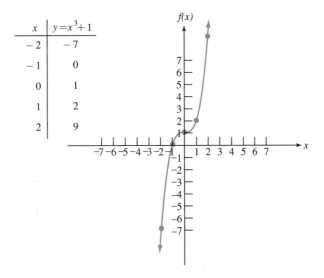

x	$y = x^3 + 1$
-2	-7
-1	0
0	1
1	2
2	9

Figure 2.10

b. Graph $f(x) = x^3 + 3$ and $f(x) = x^3 + 4$. You should see the same graph as $f(x) = x^3 + 1$, but it will be translated vertically upward. (What happens to the graph of $f(x) = x^3 + 1$ if you change the operation term to -2 or -4?).

c. Graph $f(x) = (x - 2)^3 + 1$ and $f(x) = (x - 5)^3 + 1$. You should see the same graph as $f(x) = x^3 + 1$, but it will be translated horizontally to the right. (What happens to the graph of $f(x) = x^3 + 1$ if you change the argument inside the parentheses to $x + 2$ or $x + 5$?)

We study these and other types of translations in Section 2.4.

For a function that describes a physical situation, the rule, taken alone, may have a larger domain than is warranted by the situation. In such cases, the graph of the function may not accurately describe the situation.

Situation Graph

Situation Graph

A **situation graph** is a subset of a complete graph and displays only the input/output combinations and behavior of the function that are pertinent to the situation.

ILLUMINATOR SET C

The function $f(x) = -0.0533x - 13.622$ is the rule used in quantum tunneling and its graph appears in Figure 2.11. The points are plotted using the

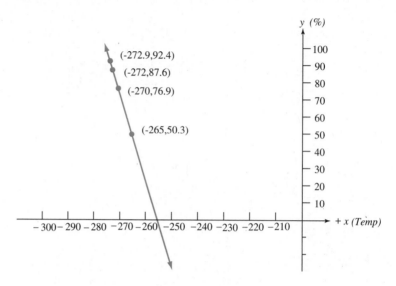

Figure 2.11

information provided in the table presented earlier in this section. The graph consists of only four points. Since both temperature and proportion are continuous, it is reasonable to assume that the points can be connected to produce a continuous, straight line.

    The complete graph is not the appropriate graph to describe quantum tunneling. It displays information that is not relevant to the situation. That is, it displays input values that are not in the domain and output values that are not in the range. For example, it indicates that percentage can be negative, and that temperature can be less than absolute zero. The appropriate graph is a situation graph, shown in Figure 2.12. If the function is specified in such a way that the domain and range correspond to the physical situation, the complete graph and the situation graph will be the same.

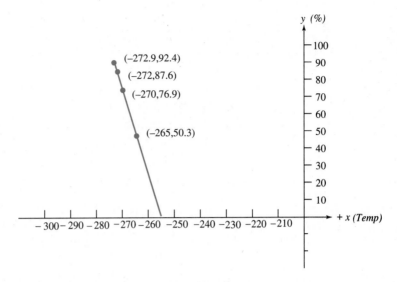

Figure 2.12

The Vertical Line Test

You can use the *vertical line test* to determine if a relation is or is **not a** function when it is described graphically. The test is carried out as **follows:** imagine a vertical line moving from left-to-right across the graph. If **all of** these vertical lines intersect the graph in at most one point, then the **graph** represents a function. The following examples illustrate the vertical line **test.**

ILLUMINATOR SET D

Use the vertical line test to determine whether or not the graph describes **a** function.

1. See the graph on the left of Figure 2.13. As the vertical line passes **from** left-to-right through the coordinate plane, it intersects the graph at at **most** one point at any one time. Thus it makes it visually clear that each **input** value in the domain produces precisely one output value.

2. See the graph on the right of Figure 2.13. As the vertical line passes **from** left-to-right through the coordinate plane, there are times when it **intersects** the graph in more than one point. This relation is not a function.

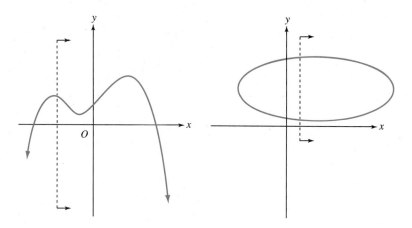

Figure 2.13

3. See the graph on the left of Figure 2.14. This relation is not a function since the vertical line intersects the graph in more than one point at $x = 3$.

4. See the graph on the right of Figure 2.14. This relation is a function since each input value is assigned to precisely one output value.

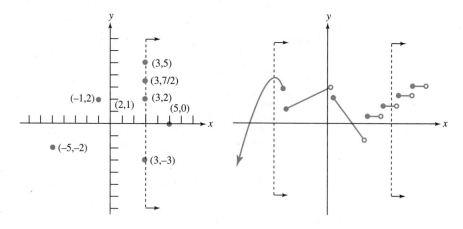

Figure 2.14

Determining the Domain and Range of a Function

The problems of the next example set illustrate how the domain and range of a function can be determined from a function rule that is given symbolically.

The domain of a function can usually be found by algebraic methods. The range can be rather difficult to determine algebraically but can be estimated from the graph of the function.

EXAMPLE SET B

Find the domain and (estimate if you must) the range of the function using the function rule.

1. $f(x) = 2x - 11$.

   Solution:

   Since real values of $f(x)$ result when any real number is substituted for x, the domain is the set of all real numbers.

   Use your graphing calculator to construct the graph of this function. From the graph of $f(x)$ you can estimate that the range of $f(x)$ is the set of all real numbers.

2. $f(x) = \dfrac{5}{x - 7}$.

   Solution:

   If $x = 7$, the expression $\dfrac{5}{x - 7}$ is not defined and $f(7)$ is therefore not a real number.

   The domain is $\{x | x \neq 7\}$. Use your graphing calculator to construct this graph and examine the behavior at $x = 7$. Although there should be no point on the graph corresponding to $x = 7$, there may be what appears to be a straight vertical line. It is not unusual for graphing calculators (and even computers) to have trouble graphing through undefined points. From the graph you can estimate that the range of $f(x)$ is the set of all real numbers except 0, but the graph does not tell us that $f(x)$ cannot be equal to 7.

3. $f(x) = \sqrt{2x + 9}$.

   Solution:

   $f(x)$ will be a real number only when $2x + 9 \geq 0$. Solve for x.

   $$2x + 9 \geq 0$$

   $$2x \geq -9$$

   $$x \geq \frac{-9}{2}$$

   The domain is $\{x | x \geq \dfrac{-9}{2}\}$.

Use your graphing calculator to construct this graph. Both the domain and the range should be clearly visible. From the graph you can estimate that the range is $\{f(x)|f(x) \geq 0\}$.

Writing Activity   Write down in paragraph form the approaches you consider to find the domain and range of the function $f(x) = \dfrac{\sqrt{8x + 5}}{2x - 1}$. Then write down the steps you actually take to find them.

EXERCISES

For exercises 1–15, determine if the relation is or is not a function.

1. $R = \{(9, 1), (9, 2)\}$
2. $T = \{(6, 0), (4, 1), (8, 3)\}$
3. $M = \{5, 2), (6, 2), (7, 2), (8, 2)\}$
4. $S = \{(4, 1), (8, 3), (11, 0), (8, 10)\}$
5. $A = \{(a, 0), (7, 1), (0, a)\}$
6. $Q = \{(m, m), (n, m), (4, m), (0, 0)\}$

7.

x	y
3	8
1	4
5	4
2	7

8.

x	y
1	2
2	1
3	4
4	1
5	1

9.

x	y
2	2/3
3	1/8
4	1/4
3	5/7
0	4/5

10.

x	y
8	4
8	1
8	10
8	12

11. When a fair die is rolled once, the probability 1/6 is assigned to each of the numbers 1, 2, 3, 4, 5, and 6.

12. For determining the base premium of auto insurance, states are divided into territories according to accident rate experience. Suppose that for a particular territory in a particular state, the following table associates the base premium with the amount of bodily injury insurance.

Base Premium	Amount of Coverage
$65	$10,000–$20,000
$70	$15,000–$30,000
$75	$25,000–$50,000
$80	$50,000–$100,000

13. Series E bonds issued by the Federal Government can be purchased for 75% of their maturity value. At maturity, the value will be paid to the holder. The following table illustrates the assignment of a maturity value to a purchase price.

Maturity Value	Purchase Price
$25	$18.75
$50	$37.50
$75	$56.25
$100	$75.00
$200	$150.00
$500	$375.00
$1000	$750.00

14. The amount of carbon monoxide $C(t)$ (in cubic centimeters) produced by a particular Chevrolet engine is related to the number of seconds t the engine idles. The equation describing the relationship is $C = 1000t$.

15. At a particular time of the day, the equation $L(h) = 0.3h$ relates the height of a person to the length of the shadow he or she casts.

For exercises 16–21, use the vertical line test to determine which relations are functions.

16.

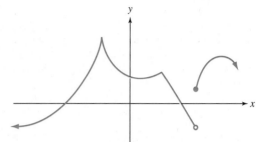

17.

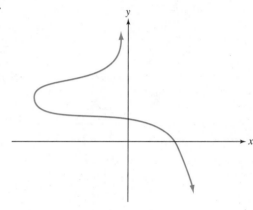

18.

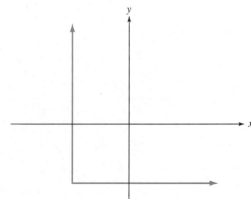

19.

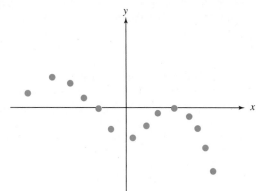

20.

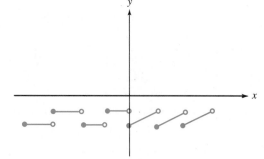

21.

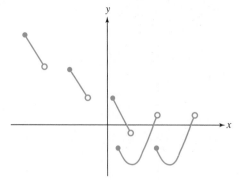

Use your graphing calculator to determine the domain and range of each function for exercises 22–30. Be sure to write down in paragraph form why the graph may not give you all the information you need, and the steps you take to get the correct answer. You may have to adjust the viewing window to get a better picture of

the graph. Use the RANGE key to reset the Xmin, Xmax, Ymin, and Ymax values in those situations.

22. $f(x) = 2x + 8$

23. $f(x) = \dfrac{7}{x - 8}$

24. $f(x) = \dfrac{x + 1}{2x + 3}$

25. $f(x) = -4x^2 + x - 1$

26. $f(x) = \sqrt{3x - 6}$

27. $f(x) = \dfrac{\sqrt{x + 2}}{x - 2}$

28. $f(x) = \dfrac{\sqrt{x + 1}}{x + 3}$

29. $f(x) = \dfrac{1}{\sqrt{x + 4}}$

30. $f(x) = \dfrac{\sqrt{x - 6}}{\sqrt{x + 5}}$

For exercises 31–36, use your graphing calculator to construct the complete graph of the function. Then draw a sketch of your graph on paper. Next, using the complete graph, estimate the range of the problem situation.

31. The marketing department of a plastic container manufacturer has estimated that the monthly demand $D(x)$ (in thousands of units) is related to the price x (in dollars) of a unit by the function $D(x) = \dfrac{48.26}{\sqrt{x}}$. Because of competition, labor, and capital, the company has to price each item at no less than \$1 and no more than \$9.

32. The concentration $C(t)$ (in milligrams per liter) of a particular drug in a person's bloodstream t hours after the drug has been injected into the bloodstream is approximated by the function $C(t) = \dfrac{80.06t}{25 + 0.08t^{3.2}}$. After 36 hours, the drug has essentially no effect on the patient.

33. Physicians use the function $W(t) = 0.033t^2 - 0.3974t + 7.3032$ to estimate

the average weight of infants during their first 14 days of life.

34. Manufacturers pay for components they order and their shipping. For a particular manufacturer, the cost $C(x)$ (in thousands of dollars) is related to the size of the order x (in units) by the function $C(x) = \dfrac{8.06}{x} + \dfrac{8.06x}{x + 2.54}$. The manufacturer must always order at least 1 unit, but, due to space restrictions, no more than 10 units.

35. A restaurant estimates that the temperature $T(t)$ of food placed in its main freezer is related to the time t (in hours) that the food has been in the freezer by the function $T(t) = \dfrac{580}{t^2 + 2.3t + 8.465}$.

36. When $10,000 is invested in an account paying $r\%$ annual interest compounded monthly, the amount of money accumulated at the end of 5 years is given by the function $A(r) = 10,000\left(1 + \dfrac{r}{1200}\right)^{60}$.

37. For two consecutive days a waitress working the 12 AM to 4 AM shift keeps a record of the number of customers she serves. Her record for these two days shows a first number representing the hour of night and a second number representing the number of customers during that hour. What are the domain set and the range set? Can this relation be a function? Give reasons for your answer.

38. A rectangular window being built must have an area of 24 square feet. If x denotes the length of the window, what is the perimeter function? Graph the function. Do you see any strange behavior near $x = 0$? What are the domain and range of the function? Is there any need to restrict the domain? Give reasons for your answer.

2.3   Evaluating Functions

Introduction

Evaluating Functions

Introduction

We know that functions are instruments for describing information. When an input value is known, it can be placed into the function, and the corresponding output value determined through computation. The process of computing the output value for a specified input value is called *evaluating a function*. In fact, the phrase *evaluate a function* means to compute the output value that corresponds to a specified input value. The input and output values can be numerical or symbolic.

   In this section we present a detailed look at the process of evaluating functions. The process is important because one of the duties of a function is to produce an output value for a specified input value not only for physical phenomena, but for theoretical purposes as well. In fact, one of your first important evaluations in calculus is theoretical and is demonstrated in Example Set B.

Evaluating Functions

Because the characteristic action of a function is to assign one output value to each input value, we can (and should) think of a function as a rule that provides us with instructions for assigning an output value to a chosen input value. If the input value is represented by x, the output value can be represented by $f(x)$, $g(x)$, $h(x)$, $\phi(x)$, or some letter or symbol other than f, g, h, or ϕ. Figure 2.15 and Figure 2.16 may be helpful in illustrating this concept.

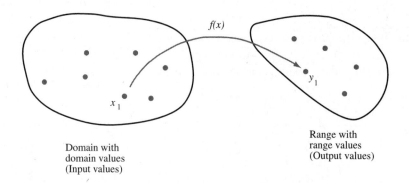

Domain with
domain values
(Input values)

Range with
range values
(Output values)

Figure 2.15

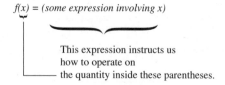

$f(x) = $ *(some expression involving x)*

This expression instructs us
how to operate on
the quantity inside these parentheses.

Figure 2.16

For example, the function $f(x) = x^3 + 8$ instructs us to take the input quantity (the quantity that appears inside the parentheses) and

$$\text{cube it} \quad \cdots \quad x^3, \quad \text{and}$$

$$\text{add 8} \quad \cdots \quad x^3 + 8$$

The variable inside the parentheses is just a placeholder for an input value that is to be inserted later. The input value may be represented by a number, a letter, a symbol, or even an expression such as $a + h$.

Evaluation of functions plays a big part in the study of much of mathematics from the calculus level up. So as to make the notation and computations as clear as possible, we illustrate the evaluation process with 6 examples. This technique is important. Study it carefully.

EXAMPLE SET A

1. For the function $f(x) = x^3 - x + 5$, find $f(2)$.

   Solution:
   This rule instructs us to take the input quantity (the quantity that appears inside the parentheses), 2, and

$$\text{cube it} \quad \cdots \quad 2^3 \quad \text{and}$$

$$\text{subtract 2} \quad \cdots \quad 2^3 - 2 \quad \text{and}$$

$$\text{add 5} \quad \cdots \quad 2^3 - 2 + 5$$

   So,

$$f(2) = 2^3 - 2 + 5$$

$$f(2) = 8 - 2 + 5$$

$$f(2) = 11$$

   Thus, when the input is 2, the output is 11.

2. For the function $f(x) = -2x + 10$, find $f(6)$.

   Solution:
   This rule instructs us to take the input quantity (the quantity that appears inside the parentheses), 6, and

$$\text{take the opposite of twice it} \quad \cdots \quad -2(6) \quad \text{and}$$

$$\text{add 10} \quad \cdots \quad -2(6) + 10$$

So,

$$f(6) = -2(6) + 10$$

$$f(6) = -12 + 10$$

$$f(6) = -2$$

Thus, when the input is 6, the output is -2.

3. For the function $g(x) = \dfrac{-x + 2a}{a}$, find $g(a)$, where $a \neq 0$.

Solution:

This rule instructs us to take the input quantity (the quantity that appears inside the parentheses), a, and

take the opposite of it $\cdots$ $-a$

and add $2a$ $\cdots$ $-a + 2a$

then divide the result by a $\cdots$ $\dfrac{-a + 2a}{a}$

So,

$$g(a) = \frac{-a + 2a}{a}$$

$$g(a) = \frac{a}{a}$$

$$g(a) = 1$$

Thus, when the input is a, the output is 1.

The next three examples illustrate an evaluation made frequently in calculus and in more advanced mathematics.

EXAMPLE SET B

1. For the function $f(x) = 5x - 7$, find $f(a + h)$.

Solution:

This rule instructs us to take the input quantity (the quantity that appears inside the parentheses), $a + h$, and

$$\text{multiply it by 5} \quad \cdots \quad 5(a + h) \quad \text{and}$$

$$\text{subtract 7} \quad \cdots \quad 5(a + h) - 7$$

So,

$$f(a + h) = 5(a + h) - 7$$

$$f(a + h) = 5a + 5h - 7$$

Thus, when the input is $a + h$, the output is $5a + 5h - 7$.

2. For the function $f(x) = x^2 + 4$, find $f(x + h) - f(x)$.

Solution:

We shall compute $f(x + h)$ first. This rule instructs us to first take the input quantity, $x + h$, and

$$\text{square it} \quad \cdots \quad (x + h)^2 \quad \text{and}$$

$$\text{add 4} \quad \cdots \quad (x + h)^2 + 4$$

so

$$f(x + h) = (x + h)^2 + 4$$

Next, we perform the subtraction $f(x + h) - f(x)$:

$$f(x + h) - f(x) = \overbrace{(x + h)^2 + 4}^{f(x+h)} - \overbrace{(x^2 + 4)}^{f(x)} \quad \leftarrow \text{Use ( ) here}$$

$$= x^2 + 2hx + h^2 + 4 - x^2 - 4$$

$$= 2hx + h^2$$

Thus, $f(x + h) - f(x) = 2hx + h^2$.

Notice that if this difference is computed correctly, all terms not involving h will drop out.

3. For the function $f(x) = 5x^2 - 8$, find $\dfrac{f(x + h) - f(x)}{h}$, $h \neq 0$.

Solution:

We shall compute $f(x+h)$ first. This rule instructs us to first take the input quantity, $x + h$, and

$$\text{square it} \quad \cdots \quad (x+h)^2 \quad \text{and}$$

$$\text{multiply by 5} \quad \cdots \quad 5(x+h)^2 \quad \text{and}$$

$$\text{subtract 8} \quad \cdots \quad 5(x+h)^2 - 8$$

Thus, $f(x+h) = 5(x+h)^2 - 8$.

Next, we compute the difference-quotient $\dfrac{f(x+h) - f(x)}{h}$:

$$\frac{f(x+h) - f(x)}{h} = \frac{\overbrace{5(x+h)^2 - 8}^{f(x+h)} - \overbrace{(5x^2 - 8)}^{f(x)}}{h}$$

$$= \frac{5(x^2 + 2hx + h^2) - 8 - 5x^2 + 8}{h}$$

$$= \frac{5x^2 + 10hx + 5h^2 - 8 - 5x^2 + 8}{h}$$

$$= \frac{10hx + 5h^2}{h}$$

$$= \frac{10hx + 5h^2}{h}$$

$$= \frac{5h(2x + h)}{h}$$

$$= \frac{5\cancel{h}(2x + h)}{\cancel{h}} \quad \text{since} \ \ h \neq 0$$

$$= 5(2x + h)$$

Thus, $\dfrac{f(x+h) - f(x)}{h} = 5(2x + h)$.

Notice that the expression no longer has a denominator.

Writing Activity   Write in paragraph form the steps you would take to find

$$\frac{f(x+h)-f(x)}{h}$$

for the function $2x^2 - 6x + 3$. Then work the problem and simplify your results.

EXERCISES

1. For $f(x) = 8x + 7$, find $f(3)$.
2. For $f(x) = -2x - 11$, find $f(8)$.
3. For $f(x) = -4x + 1$, find $f(-6)$.
4. For $f(x) = 2x + 10$, find $f(-n)$.
5. For $g(x) = 7x - 12$, find $g(-s)$.
6. For $H(r) = -5r^2 - r - 1$, find $H(1)$.
7. For $h(n) = n^2 + 3n + 8$, find $h(-4)$.
8. For $f(s) = s^3 - 27$, find $f(3)$.
9. For $f(z) = \dfrac{z^3 + 64}{z + 4}$, find $f(0)$.
10. For $g(a) = a^2 + 3a - 7$, find $g(\$)$.
11. For $P(n) = 4n^3 - 6n^2 + 5n - 8$, find $P(k)$.
12. For $f(x) = 4x^2 + 3x - 1$, find $f(h)$.
13. For $f(x) = 6x^2 - 8$, find $f(h)$.
14. For $f(x) = x + 7$, find $f(h)$.
15. For $f(x) = 5x - 8$, find $f(x + h)$.
16. For $f(x) = x^2 + x - 5$, find $f(x + h)$.
17. For $f(x) = 7x^2 - 3x - 1$, find $f(x + h)$.
18. For $f(x) = 3x - 7$, find $f(x + h)$.
19. For $f(x) = 10x^2 - 11x + 4$, find $-f(x)$.
20. For $f(x) = 5x + 6$, find $-f(x)$.
21. For $f(x) = 8x - 9$, find $-f(x)$.
22. For $f(x) = 4x + 1$, find $f(x + h) - f(x)$.
23. For $f(x) = 2x - 5$, find $f(x + h) - f(x)$.
24. For $f(x) = x^2 + x + 7$, find $f(x + h) - f(x)$.
25. For $f(x) = x^2 + 5x + 10$, find $f(x + h) - f(x)$.
26. For $f(x) = 8x^2 - 3x - 7$, find $f(x + h) - f(x)$.
27. For $f(x) = 7x^2 - 8x + 12$, find $f(x + h) - f(x)$.

For exercises 28–37, find $\dfrac{f(x+h)-f(x)}{h}$ for each of the following functions and simplify your results.

28. For $f(x) = 6x + 1$, find $\dfrac{f(x+h)-f(x)}{h}$

29. For $f(x) = 4x + 7$, find $\dfrac{f(x+h)-f(x)}{h}$

30. For $f(x) = x^2 + 2x + 4$, find $\dfrac{f(x+h)-f(x)}{h}$.

31. For $f(x) = x^2 - 6x - 1$, find $\dfrac{f(x+h)-f(x)}{h}$.

32. For $f(x) = 3x^2 + 7x + 1$, find $\dfrac{f(x+h)-f(x)}{h}$.

33. For $f(x) = 5x^2 + 9x - 4$, find $\dfrac{f(x+h)-f(x)}{h}$.

34. For $f(x) = -4x^2 + 3x + 2$, find $\dfrac{f(x+h)-f(x)}{h}$.

35. For $f(x) = -2x^2 + 8x + 7$, find $\dfrac{f(x+h)-f(x)}{h}$.

36. For $f(x) = \dfrac{2}{x}$, find $\dfrac{f(x+h)-f(x)}{h}$.

37. For $f(x) = \dfrac{x-3}{x+5}$, find $\dfrac{f(x+h)-f(x)}{h}$.

38. The percent P of concentration of a particular drug in the bloodstream t hours after the drug has been injected is given by the function $P(t) = \dfrac{4.06t}{0.64t^2 + 8.1}$. Use your graphing calculator to graph this function and then from the graph determine (a) how long it will take for the drug to reach its maximum percentage level and specify that level, and (b) the number of hours after injection that the percentage level of the drug is approximately

50%. (Hint: Use the TRACE key. Use the ZOOM key for part (b).)

39. The profit $P(x)$ (in thousands of dollars) a company realizes for selling x (in hundreds) compact disc players is given by the function $P(x) = 0.92x^2(x - 8.06)^2$, for $0 < x \le 8$. Use your graphing calculator to graph this function and then from the graph (a) approximate the number of compact disc players the company must sell to maximize its profit, (b) specify that profit, and (c) approximate the number of compact disc players the company must sell to make a profit of about $135,000.

40. The owner of an avocado orchard has determined that the function $y(x) = -284,700 + 1300x - x^2$ relates the number of trees, x, planted per acre and the yield of avocados, $y(x)$. Use your graphing calculator to graph this function. (a) Approximate the number of trees per acre that will produce the maximum yield, (b) specify that yield (do you really need the graph for this?), (c) approximate the number of trees per acre that will produce about 3500 avocados, and (d) approximate the number of trees per acre that will result in no avocado yield at all.

41. The function $h(t) = 150 + 64t - 16t^2$ approximates the height h of an object thrown nearly straight upward from the edge of the top of a building 150 feet high, and the number of seconds t since the object's launch. Use your graphing calculator to graph this function and then approximate (a) the amount of time required for the object to reach its maximum height, (b) specify that height, and

(c) approximate the amount of time required for the object to hit the ground below.

42. In the marketplace, the price p of an item is related to both the supply S of the item and the demand D for the item. That is, the price per item is a function of the number of items x supplied or demanded. Symbolically, $p = S(x)$ and $p = D(x)$. The function $S(x)$ is called the *supply* function and represents the price per unit for which the supplier will supply x units of the item. The function $D(x)$ is called the *demand* function and represents the price per unit for which the consumer will buy x units of the item. The point (x_e, p_x) at which the two functions intersect is called the *equilibrium point* and is the point at which the supply level equals the demand level. Supply functions are increasing functions since suppliers will provide more items as the price increases. Demand functions are decreasing functions since consumers will buy less items as the price increases.

Suppose for a company, the supply and demand functions are $S(x) = 12 + 0.16x^2$ and $D(x) = \dfrac{70}{1 + 0.088x}$, respectively. Use your graphing calculator to graph each of these functions on the same coordinate system and then find and interpret the coordinates of the equilibrium point.

43. The volume of a cylinder is kept fixed at $.756 \text{ cm}^3$. Find the height of the cylinder as a function of the radius. If the radius is 2.1 cm, what is the height? Graph the function and find the minimum height.

2.4   Graphs Derived From Function Graphs

Introduction

Translations

Reflection

Expansion and Contraction

Summary of Shifts, Expansions, Contractions, and Reflections

Introduction

Functions can be used to describe information. If the information being described changes, the function used for its description must also change. Particular types of changes in the symbolic form of a function have the effect of moving the graph of the function to a different location in the Cartesian coordinate system. Some changes have the effect of reflecting the graph about the input axis. In this section you see what effect the addition or subtraction of a constant has on the location of a function in the Cartesian coordinate system, and you also have the opportunity to change the symbolic form of a function to more properly describe the information it is to model.

Translations

The graph of a function in the Cartesian coordinate system can be translated (moved) horizontally and/or vertically to a different position. We examine the effect each type of translation has on the symbolic form of the original function, and then examine the effect of the combined translation on the form of the original function.

1. **Vertical Translations**

   Let's consider the symbolic form of a function, $f(x)$, and vary it by adding the nonzero constant c; that is, let's consider the function $f(x) + c$. Now we look at whether or not there is a relationship between the graph of $f(x)$ and its variation, $f(x) + c$. There is, indeed, a relationship and selecting $f(x) = x^2$, $c = 1$, and then $c = -3$ as examples should help us suggest what it is.

   The left graph in Figure 2.17 shows the graph of the function $f(x) = x^2$ while the right graph shows the graph of the function $f(x) = x^2 + 1$.

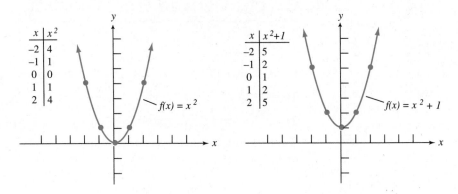

Figure 2.17

Figure 2.18 shows the graph of the function $f(x) = x^2 - 3$. Placing

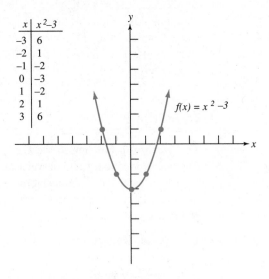

Figure 2.18

the three graphs on the same coordinate system makes the relationship clear. (See Figure 2.19.) The constant c produces a vertical translation

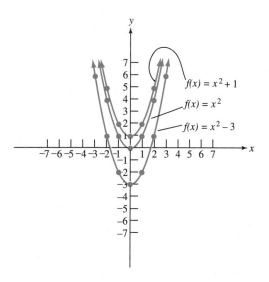

Figure 2.19

of the graph of $f(x)$. In fact, the graph of $f(x)$ is translated vertically $|c|$ units. In this case, the constant $c = 1$ translates the graph of $f(x) = x^2$ a distance $|1| = 1$ unit upward and $c = -3$ translates the graph of $f(x) = x^2$ a distance $|-3| = 3$ units downward. We summarize this result as follows.

Function	Vertical Shift
$f(x) + c$	Upward c units
$f(x) - c$	Downward c units

2. **Horizontal Translations**

To produce a vertical translation of the graph of $f(x)$ we need only evaluate $f(x)$ at x and then add or subtract c. Suppose we evaluate the function $f(x)$ at $x + c$ rather than at x. That is, rather than adding or subtracting c after we evaluate $f(x)$, let's add or subtract c to x and then evaluate. To see what effect evaluating at $x + c$ has on the graph of $f(x)$, let's again select $f(x) = x^2$ and $c = 1$ and then $c = -3$. The graphs of $f(x), f(x + 1)$, and $f(x - 3)$ appear in Figure 2.20. The graphs appear superimposed in Figure 2.21.

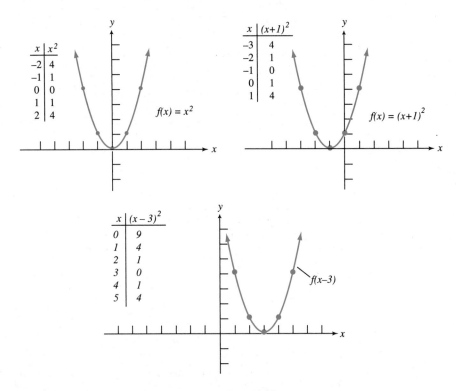

x	x^2
-2	4
-1	1
0	0
1	1
2	4

$f(x) = x^2$

x	$(x+1)^2$
-3	4
-2	1
-1	0
0	1
1	4

$f(x) = (x+1)^2$

x	$(x-3)^2$
0	9
1	4
2	1
3	0
4	1
5	4

$f(x-3)$

Figure 2.20

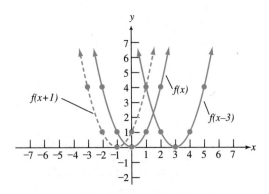

$f(x+1)$ $f(x)$ $f(x-3)$

Figure 2.21

Notice the effect of the constant. The left-right shifts may not be what you expected! The graph of $f(x+1)$ is shifted to the *left* of the graph of $f(x)$ rather than to the right. This makes sense when you analyze what each function is doing: because $x + 1$ is one larger than x, it must be made one smaller (a move to the left) to produce the same output-value as $f(x)$. Example Set E will help make this clearer.

We summarize this result as follows.

Function	Horizontal Shift
$f(x + c)$	Left c units
$f(x - c)$	Right c units

3. **Vertical and Horizontal Translations**

Vertical and horizontal translations can occur simultaneously. We summarize as follows.

Function	Shift
$f(x + h) + k$	h units left and k units up
$f(x + h) - k$	h units left and k units down
$f(x - h) + k$	h units right and k units up
$f(x - h) - k$	h units right and k units down

EXAMPLE SET A

Use the function $f(x)$ pictured in Figure 2.22 to graph the following variations:

1. $f(x) - 2$ and $f(x) + 1$
2. $f(x + 2)$ and $f(x - 1)$
3. $f(x + 1) - 2$.

Solution:

The basic function we are working with is $f(x)$. The other five functions we recognize as vertical and horizontal translations of $f(x)$.

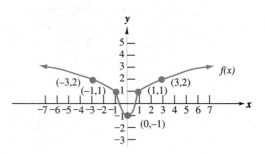

Figure 2.22

1. We obtain the graph of $f(x) - 2$ by translating the graph of $f(x)$ downward 2 units. The graph of $f(x) + 1$ is obtained by translating the graph of $f(x)$ upward 1 unit. (See the left graph in Figure 2.23.)

2. The graphs of $f(x + 2)$ and $f(x - 1)$ are obtained by translating the graph of $f(x)$ to the left 2 units and to the right 1 unit, respectively. (See the right graph in Figure 2.23.)

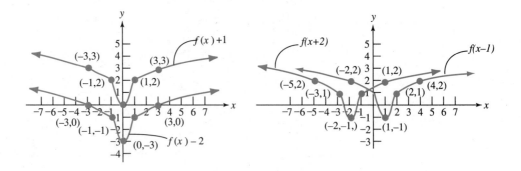

Figure 2.23

3. The graph of $f(x + 1) - 2$ is obtained by translating the graph of $f(x)$ downward 2 units and then to the left 1 unit (or to the left 1 unit and then downward 2 units). (See Figure 2.24.)

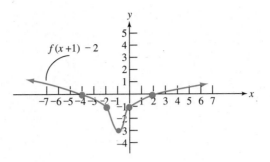

Figure 2.24

Reflection

Now let's consider the relationship between $f(x)$ and $-f(x)$. For any particular input value, x, $f(x)$ is the corresponding output value and $(x, f(x))$ is a
particular point on the graph of $f(x)$. Now $-f(x)$ is the negative of $f(x)$, and
hence $(x, -f(x))$ is precisely opposite that of $(x, f(x))$ with respect to the x-
axis. That is, the point $(x, -f(x))$ is the reflection of the point $(x, f(x))$ about
the x-axis. For this reason, $-f(x)$ is called the *reflection* of the graph of $f(x)$.
We summarize this fact as follows.

Function	Relationship to $f(x)$
$-f(x)$	Reflection of $f(x)$ across the x-axis

EXAMPLE SET B

Graph $f(x) = x^2$ and $g(x) = -x^2$.

   Notice that $g(x) = -f(x)$ and is therefore its reflection. (We verify this
solution by plotting several points.) Figure 2.25 shows the graphs of $f(x) = x^2$
and $g(x) = -x^2$.

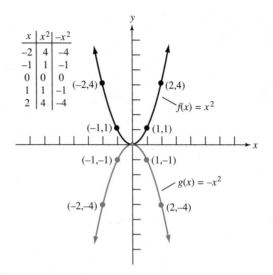

Figure 2.25

A reflection can be used simultaneously with a vertical and/or horizontal translation.

EXAMPLE SET C

Referring to the function pictured in Figure 2.26, construct the graph of $-f(x - 3) + 4$.

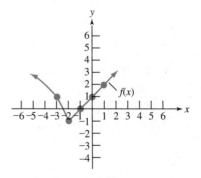

Figure 2.26

Solution:

Figure 2.27 and Figure 2.28 show how the graph of $-f(x-3)+4$ is obtained.

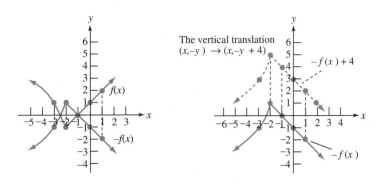

Figure 2.27

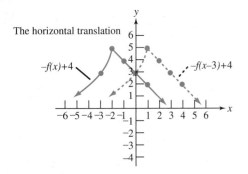

Figure 2.28

Expansion and Contraction

We now explore the relationship between the graphs of $f(x)$ and $cf(x)$. We select $f(x) = x^2$, $c = 2$, and then $c = \dfrac{1}{3}$ as values to help us suggest the relationship.

1. **Expansion:** $c > 1$

   The graphs in Figure 2.29 are, respectively, $f(x) = x^2$ and $g(x) = 2x^2$.

Notice that $g(x) = 2 \cdot f(x)$. Notice also that in the table of ordered pairs, each output-value of $g(x) = 2x^2$ is exactly twice that of $f(x) = x^2$. The output-values of $g(x) = 2x^2$ are larger than those of $f(x) = x^2$ and are, therefore, *expanding* away from the x-axis more quickly than those of $f(x) = x^2$, as $|x|$ increases. For this reason $cf(x)$ is said to produce an *expansion* of $f(x)$. Because of this expansion, the graph of $cf(x)$ is narrower than the graph of $f(x)$.

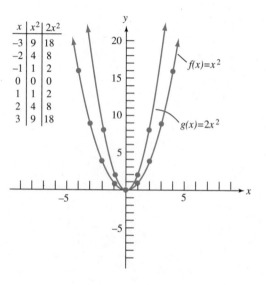

x	x^2	$2x^2$
-3	9	18
-2	4	8
-1	1	2
0	0	0
1	1	2
2	4	8
3	9	18

Figure 2.29

2. **Contraction:** $0 < c < 1$ The graphs in Figure 2.30 are, respectively, $f(x) = x^2$ and $g(x) = \frac{1}{3}x^2$. Notice that $g(x) = \frac{1}{3} \cdot f(x)$. Notice also that in the table of ordered pairs, each output-value of $g(x) = \frac{1}{3}x^2$ is exactly one-third that of $f(x) = x^2$. The output-values of $g(x) = \frac{1}{3}x^2$ are smaller than those of $f(x) = x^2$ and, therefore, are *contracting* toward the x-axis more quickly than those of $f(x) = x^2$, as $|x|$ decreases. For this reason $cf(x)$ is said to produce a *contraction* of $f(x)$. Because of this contraction, the graph of $cf(x)$ is wider than the graph of $f(x)$.

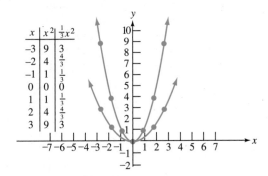

x	x^2	$\frac{1}{3}x^2$
-3	9	3
-2	4	$\frac{4}{3}$
-1	1	$\frac{1}{3}$
0	0	0
1	1	$\frac{1}{3}$
2	4	$\frac{4}{3}$
3	9	3

Figure 2.30

Figure 2.31 should make the concepts of reflection and contraction clearer. We

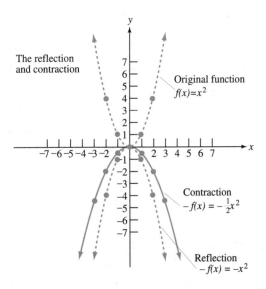

Figure 2.31

summarize the facts about expansion and contraction as follows.

Expansion and Contraction	
Function	**Relationship to $f(x)$**
$cf(x)$, $c > 1$	Expansion, narrower
$cf(x)$, $0 < c < 1$	Contraction, wider

Summary of Shifts, Expansions, Contractions, and Reflections

The shifting, expanding, contracting, and reflecting of functions we have just examined are summarized in Table 2.3. Each of these variations can occur simultaneously.

Table 2.3	
Function	**Shift/Relationship to $f(x)$**
$f(x + h) + k$	h units left and k units up
$f(x + h) - k$	h units left and k units down
$f(x - h) + k$	h units right and k units up
$f(x - h) - k$	h units right and k units down
$-f(x)$	Reflection of $f(x)$ across the x-axis
$cf(x)$, $c > 1$	Expansion, narrower
$cf(x)$, $0 < c < 1$	Contraction, wider

EXAMPLE SET D

If $f(x) = x^2$, construct the graph of $g(x) = -\frac{1}{2}f(x - 2) + 3$.

Solution:

Careful examination shows $g(x)$ to be a variation of $f(x)$. More precisely, it is a reflection, a contraction, a horizontal translation, and a vertical translation. The left and right graphs of Figure 2.32 show a progression of steps leading to the graph of $g(x) = -\frac{1}{2}f(x - 2) + 3$. The final result is graphed by itself in Figure 2.33 with labeled points.

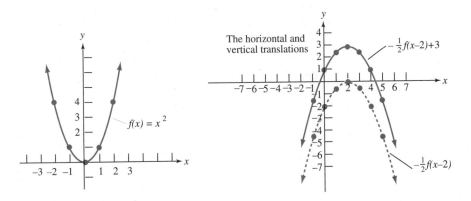

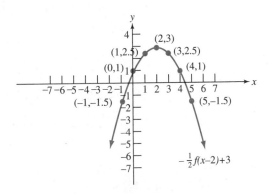

Figure 2.32

Figure 2.33

The next illuminator set and example set should help you understand why $f(x+c)$ translates $f(x)$ horizontally to the left and why $f(x-c)$ translates $f(x)$ horizontally to the right.

ILLUMINATOR SET A

A spherical balloon can be filled with only so much air before it pops. We'll call this critical volume the balloon's popping volume. The volume of a spherical balloon is related to its radius and is given by the function $V(x) = \frac{4}{3}\pi x^3$, where $V(x)$ represents the volume of the balloon and x represents the radius. Suppose that two people, A and B, inflate identical balloons by pumping air into them. Suppose further that person B starts the process before person A and that the process is such that the radius of person B's balloon is always 1 inch greater than person A's balloon. Then, if x represents the radius of person A's balloon at any particular time, $x + 1$ represents the radius of person B's balloon at that time, and the respective volume functions are $V(x) = \frac{4}{3}\pi x^3$ and $V(x + 1) = \frac{4}{3}\pi(x + 1)^3$. Suppose that the popping volume of these balloons is 60 cubic inches. Enter these functions on your calculator with $V(x)$ as Y1, $V(x + 1)$ as Y2, and 60 as Y3. Set Xmin = 0, Xmax = 5, Ymin = 0, and Ymax = 100. Keeping in mind that the calculator constructs graphs in numerical order, Y1, Y2, and Y3, press the GRAPH key to complete the graphs. Notice that the graph of $V(x + 1)$ appears *to the left* of the graph of $V(x)$. Window 2.1 shows this graph in the calculator viewing window. Since $V(x + 1)$ intercepts the horizontal popping

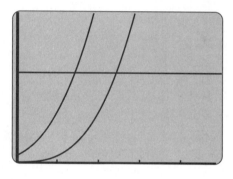

Window 2.1

volume line before that of $V(x)$, person B's balloon pops first.

EXAMPLE SET E

A nationwide trucking company determines that the alertness of an average driver is related to the number of hours of driving since the driver's last stop by the function $A(t) = 12 - \dfrac{1.3t}{17.5 - 0.8t}$, $t \leq 12$. Suppose that Jamal and Henry are two average drivers and that Jamal leaves Sue Ellen's Truck Stop $\frac{1}{2}$ hour ahead of Henry. If $A(t)$ represents Jamal's alertness, create a function that will represent Henry's alertness at time t.

Solution:
Figure 2.34 shows the graphs of $A(t)$ and a new function that we call $L(t)$, L representing *later*. Notice that at $t = 4\frac{1}{2}$ and $t = 5$, the output values

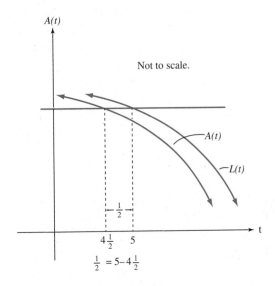

Figure 2.34

(alertness levels) of these functions are equal, that is, at $t = 4\frac{1}{2}$ and $t = 5$,

$$A(4\frac{1}{2}) = L(5)$$

$$= A(5 - \frac{1}{2})$$

This analysis indicates that for some time t, $A(t - \frac{1}{2})$ is precisely the graph of $A(t)$ shifted to the right one-half unit. Thus, $L(t) = A(t - \frac{1}{2})$.

EXERCISES

For exercises 1–12, write down in paragraph form the effect that the given expression has on the graph of $f(x) = x^2$. Then draw a sketch of each.

1. $f(x) + 2$

2. $f(x) - 4$

3. $f(x - 3)$

4. $f(x + 4)$

5. $f(x + 1) - 2$

6. $f(x - 3) - 3$

7. $-f(x + 2) + 3$

8. $-f(x + 3) + 5$

9. $\frac{1}{3}f(x)$

10. $\frac{1}{3}f(x) + 1$

11. $-\frac{1}{3}f(x - 2) + 4$

12. $-\frac{1}{3}f(x + 3) + 2$

For exercises 13–25, use functional notation to show the desired effect. Then draw a sketch of each.

13. Contract the function $f(x) = x^2$ by $\frac{1}{4}$.

14. Contract the function $f(x) = x^2$ by $\frac{1}{6}$.

15. Expand the function $f(x) = x^2$ by a factor of 4.

16. Expand the function $f(x) = x^2$ by a factor of 12.

17. Expand, by a factor of 5, and reflect about the x-axis, the function $f(x) = x^2$.

18. Expand, by a factor of 7, and reflect about the x-axis, the function $f(x) = x^2$.

19. Horizontally shifts $f(x) = x^2$ 5 units to the left and reflect it about the x-axis.

20. Horizontally shifts $f(x) = x^2$ 10 units to the right and reflect it about the x-axis.

21. Horizontally shifts $f(x) = x^2$ 15 units to the right, vertically shifts it 4 units downward, and reflect it about the x-axis.

22. Horizontally shifts $f(x) = x^2$ 1 unit to the left, vertically shifts it 10 units upward, and contract it to $1/5$.

23. Horizontally shifts $f(x) = x^2$ 8 units to the right, vertically shifts it 8 units upward, and contract it to $1/8$.

24. Horizontally shifts $f(x) = x^2$ 3 units to the right, vertically shifts it 2 units downward, and expand it by a factor of 4.

25. Horizontally shifts $f(x) = x^2$ 6 units to the right, vertically shifts it 20 units downward, and expand it by a factor of 2.

For exercises 26–31, use the illustrated function $f(x)$ to construct the graph of its specified variation.

26. $f(x + 2) - 4$

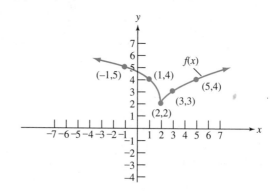

27. $f(x - 3) - 2$

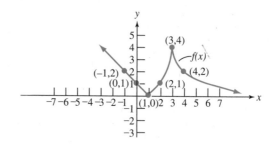

28. $-f(x - 1) - 5$

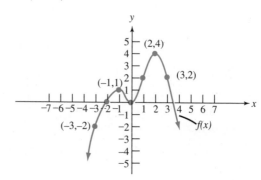

29. $-f(x - 2) - 4$

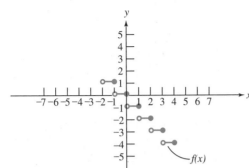

30. $-f(x + 3) - 2$

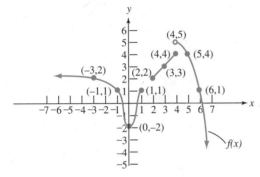

31. $-f(x + 1) - 2$

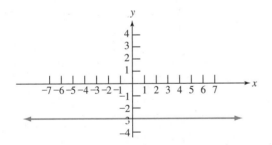

For exercises 32–36, use complete sentences to explain, in terms of vertical and horizontal shifts, expansions and contractions, and reflections, what has been done to the function $f(x)$.

32. $2f(x) + 6$

33. $\frac{1}{5}f(x - 2)$

34. $\frac{1}{2}f(x + 4) - 3$

35. $-\frac{3}{4}f(x + 1) + 5$

36. $-8f(x - 23) - 4$

Exercises 37–40 illustrate the concept of *mathematical modeling*. The idea behind mathematical modeling is to create a function that approximates a physical phenomenon, test it to see how accurate it is, adjust it to obtain a better approximation, test it again, adjust it again, and so on. Write in paragraph form the details of the steps you take to solve each of the following exercises.

37. A toxic chemical cleanup company originally assumes that the concentration of a particular pollutant in a lake t hours after it has been introduced into the lake is approximated by the function $C(t) = \dfrac{2.56t + 35}{1.8t + 45}$, where $t > 0$. Later analysis shows this not to be the case, and that the actual function that describes the situation is a slight modification of the original function. Create a new function that more closely approximates the concentration of the pollutant in the lake if (a) a particular concentration is attained 2 hours sooner than originally thought, and (b) the concentration is only two-thirds that originally thought.

38. An advertising company promises a manufacturer that t weeks after a new advertising campaign begins, the weekly demand for its product will be approximated by the function $D(t) = (4.2t + 95)^{3/2}$. Several months into the campaign however, the company finds that it must adjust the function slightly. Create a new function for each of the following that more closely approximates the weekly demand if new data show that (a) a particular weekly demand is attained 3 weeks earlier than originally thought, and (b) the weekly demand is actually twice what was originally projected.

39. A country's government approximates its consumption function to be

$$G(I) = \frac{5.5(2.5I + \sqrt{I + 30})}{\sqrt{I + 30}}$$

where I is in billions of dollars and represents the country's national income, and $G(I)$ is also in billions of dollars and represents the country's propensity to consume. After some time and after new data are available, the government realizes it must adjust the function $G(I)$ to better represent the actual situation. Create a new function that better approximates the country's propensity to consume if the new data show that the actual consumption is (a) 1.8 billion dollars more than originally thought, and (b) only $\frac{2}{3}$ that originally thought.

40. Tests for diabetes are made to determine how fast a person's body metabolizes glucose. One particular test involves ingesting a specified amount of glucose and then, as time passes, measuring the concentration of glucose in the bloodstream. Suppose that for this test, the concentration of glucose in the bloodstream, t hours after ingestion, of an average 40-year-old man is approximated by the function $C(t) = 5.2t^{-1/2}$. After some testing, a particular 40-year-old man's doctor realizes she must adjust the function to better approximate his glucose concentration. Create a new function that better approximates this man's glucose concentration if it is found that his glucose concentration (a) equals the concentration specified by $C(t)$ 15 minutes before, and (b) is 110% the concentration specified by $C(t)$ 30 minutes later.

41. The accountant of a company arrives at the following erroneous function for the cost of production and marketing x number of a certain product: $C(x) = 8.3x + 0.004x^2$. Graph this function. Which part of the cost has he neglected to consider? If the fixed cost is $7,500 and this is now included in the cost, how will the graph change?

42. The profit function for a product is estimated to be $P(x) = 250x + 0.03\sqrt{x}$ where x is the number of units manufactured. Draw the graph of the function. Now suppose that a new owner buys the company, which has 4,000 items in stock at the time of purchase. In the short run, if x is the number manufactured, what is the profit function? How does the graph change?

2.5 The Algebra of Functions

Introduction

Just as with numbers, it is possible to operate on functions. In this section you will see how to add, subtract, multiply, and divide functions, as well as use one function as the input quantity to another function. You will also investigate how algebraically combining functions effects the domains of each of the individual functions.

The Algebra of Functions

It is often important in the analysis of functions to be able to break down a particular function into a combination of simpler functions. For example, the function $f(x) = \dfrac{1}{x-1} + \sqrt{2x+3}$ might be best analyzed by considering $\dfrac{1}{x-1}$ and $\sqrt{2x+3}$ separately.

To aid in this regard we write the following definition, leaving the examination of domains until later.

Algebra of Functions

Algebra of Functions

Let $f(x)$ and $g(x)$ be two functions whose domains intersect (have some common elements). Then, we define the **sum** $f + g$, the **difference** $f - g$, the **product** $f \cdot g$, and the **quotient** $\dfrac{f}{g}$ as

1. $(f + g)(x) = f(x) + g(x)$ (Read this as *f plus g, of x*.)
2. $(f - g)(x) = f(x) - g(x)$ (Read this as *f minus g, of x*.)
3. $(f \cdot g)(x) = f(x) \cdot g(x)$ (Read this as *f times g, of x*.)
4. $\left(\dfrac{f}{g}\right)(x) = \dfrac{f(x)}{g(x)}, \ g(x) \neq 0$ (Read this as *f divided by g, of x*.)

These definitions are used because they describe the geometry of the combinations of actions of the functions $f(x)$ and $g(x)$. Recalling that $f(x)$ and $g(x)$ represent output values, $(f + g)(x) = f(x) + g(x)$ does exactly what it states: it adds the two output values together producing a third output value called $(f + g)(x)$. Suppose $f(x) = x^2$, and $g(x) = x + 3$, then the left portion of Figure 2.35 shows the functions and their sum. The right portion of Figure 2.35 shows $(f + g)(x) = f(x) + g(x)$ evaluated at $x = 2$. $f(2) = 4$, $g(2) = 5$, and $f(2) + g(2) = 9$, so that $(f + g)(2) = 9$.

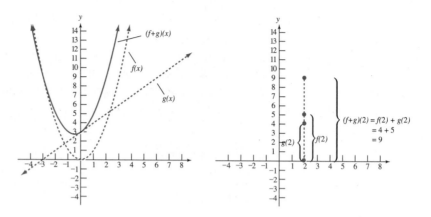

Figure 2.35

EXAMPLE SET A

If $f(x) = \dfrac{1}{x-1}$ and $g(x) = \sqrt{2x+3}$, evaluate

1. $(f+g)(3)$,
2. $(f-g)(3)$,
3. $(f \cdot g)(3)$,
4. $\left(\dfrac{f}{g}\right)(3)$.

Solution:

1. We find the sum by:

$$(f+g)(3) = f(3) + g(3)$$

$$= \frac{1}{3-1} + \sqrt{2 \cdot 3 + 3}$$

$$= \frac{1}{2} + 3$$

$$= \frac{7}{2}$$

2. We find the difference by:

$$(f-g)(3) = f(3) - g(3)$$

$$= \frac{1}{3-1} - \sqrt{2 \cdot 3 + 3}$$

$$= \frac{1}{2} - 3$$

$$= \frac{-5}{2}$$

3. We find the product by:

$$(f \cdot g)(3) = f(3) \cdot g(3)$$

$$= \frac{1}{3-1} \cdot \sqrt{2 \cdot 3 + 3}$$

$$= \frac{1}{2} \cdot 3$$

$$= \frac{3}{2}$$

4. We find the quotient by:

$$\left(\frac{f}{g}\right)(3) = \frac{f(3)}{g(3)}$$

$$= \frac{\frac{1}{3-1}}{\sqrt{2 \cdot 3 + 3}}$$

$$= \frac{\frac{1}{2}}{3}$$

$$= \frac{1}{6}$$

The Domain of Algebraically Combined Functions

The domain of the functions defined above is restricted to the intersection (the set of common elements) of the domain of $f(x)$ with that of $g(x)$. The domain of the quotient $\frac{f(x)}{g(x)}$ is additionally restricted to exclude values that result in a zero denominator.

EXAMPLE SET B

1. If $f(x) = \dfrac{1}{x-1}$ and $g(x) = \sqrt{2x+3}$, find $(f+g)(x)$, $(f-g)(x)$, $(f \cdot g)(x)$, and $\left(\dfrac{f}{g}\right)(x)$ and their domains.

   Solution:
   First we find the domains of $f(x)$ and $g(x)$ individually.

   The domain of $f(x) = \dfrac{1}{x-1}$ is $\{x|x \neq 1\}$. The domain of $g(x) = \sqrt{2x+3}$ is the set of real numbers for which $2x+3 \geq 0$, namely, $x \geq -\frac{3}{2}$. Hence, the domain of $g(x)$ is $\{x|x \geq -\frac{3}{2}\}$.

   a. $(f+g)(x) = \underbrace{\dfrac{1}{x-1}}_{f(x)} + \underbrace{\sqrt{2x+3}}_{g(x)}$,

   b. $(f-g)(x) = \dfrac{1}{x-1} - \sqrt{2x+3}$,

   c. $(f \cdot g)(x) = \dfrac{1}{x-1} \cdot \sqrt{2x+3} = \dfrac{\sqrt{2x+3}}{x-1}$.

   The domain of these functions is

   $$\left\{x|x \neq 1\right\} \cap \left\{x|x \geq -\frac{3}{2}\right\} = \left\{x|x \geq -\frac{3}{2} \text{ and } x \neq 1\right\}$$

   d. $\left(\dfrac{f}{g}\right)(x) = \dfrac{\dfrac{1}{x-1}}{\sqrt{2x+3}} = \dfrac{1}{(x-1)\sqrt{2x+3}}$

   and has domain $\left\{x|x > -\frac{3}{2} \text{ and } x \neq 1\right\}$. Notice that $-\frac{3}{2}$ has been excluded since $g(-\frac{3}{2}) = 0$, which produces a zero denominator.

2. If $f(x) = \sqrt{x}$ and $g(x) = \sqrt{4-x}$, find $(f+g)(x)$, $(f-g)(x)$, $(f \cdot g)(x)$, and $\left(\dfrac{f}{g}\right)(x)$ and their domains.

   Solution:
   The domain of $f(x) = \sqrt{x}$ is $\{x|x \geq 0\}$. The domain of $g(x) = \sqrt{4-x}$ is $\{x|x \leq 4\}$.

   a. We find the sum by:

   $$(f+g)(x) = \underbrace{\sqrt{x}}_{f(x)} + \underbrace{\sqrt{4-x}}_{g(x)}$$

   b. We find the difference by:

   $$(f-g)(x) = \sqrt{x} - \sqrt{4-x}$$

c. We find the product by:

$$(f \cdot g)(x) = \sqrt{x} \cdot \sqrt{4-x} = \sqrt{x(4-x)} \quad \text{or} \quad \sqrt{4x - x^2}$$

The sum, difference, and product have domain

$$\{x | x \geq 0\} \cap \{x | x \leq 4\} = \{x | 0 \leq x \leq 4\}$$

d. We find the quotient by:

$$\left(\frac{f}{g}\right)(x) = \frac{\sqrt{x}}{\sqrt{4-x}}$$

The quotient $\left(\dfrac{f}{g}\right)(x)$ has domain $\{x | 0 \leq x < 4\}$.

The number 4 has been excluded since $g(4) = 0$, which produces a zero denominator.

Composite Functions

Composite Function   A **composite function** is a function that results from a chain type process of composing one function with another. One function, say $g(x)$, is used as the input to another function, say $f(x)$. For example, consider the two functions $f(x) = x^2 + 4$ and $g(x) = 5x - 11$. Evaluate $g(x)$ at $x = 2$. That is, evaluate $g(2)$.

$$g(2) = 5(2) - 11$$

$$= 10 - 11$$

$$= -1$$

Now, evaluate $f(x)$ at this result. That is, evaluate $f(-1)$.

$$f(-1) = (-1)^2 + 4$$

$$= 1 + 4$$

$$= 5$$

Thus, $g(2) = -1$ and $f(-1) = 5$. Figure 2.36 shows this chain of events, which leads from 2 to 5. The output 5 is obtained first by applying $g(x)$

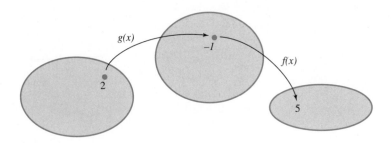

Figure 2.36

to 2, then by getting the intermediate result -1, and finally applying $f(x)$ to -1. In other words, 5 is obtained by composing (that is, making something by combining other things) the functions $f(x)$ and $g(x)$, and evaluating the composition at 2. By composing the functions $f(x)$ and $g(x)$ at the outset, it is possible to produce 5 directly from 2, bypassing the intermediate value -1.

To see how, first we develop a logical notation to describe the composition of two functions. Consider

$5 = f[-1]$.

But $-1 = g(2)$.

Replace -1 with $g(2)$.

So $5 = f[-1] = f[g(2)]$

This is the notation we're looking for, and we restate it as follows.

Composition of Functions

Composition of Functions

The **composition of the functions** $f(x)$ and $g(x)$ is denoted by $(f \circ g)(x)$ and is given by $(f \circ g)(x) = f[g(x)]$. $(f \circ g)(x)$ is read as *f circle g, of x* or *f composite g, of x*.

The composite function $f[g(x)]$ is evaluated by first evaluating $g(x)$ for some particular x-value, and then evaluating $f(x)$ at that resulting value. Also, although it is not entirely accurate, it is not uncommon to denote $(f \circ g)(x)$ by $f \circ g$. In fact, it is common to relax the output notation by writing simply f rather than the more accurate description $f(x)$. Mathematicians have traded some accuracy in notation for simplicity of notation. Figure 2.37 shows that the composite of $f(x)$ and $g(x)$ bypasses the intermediate value -1.

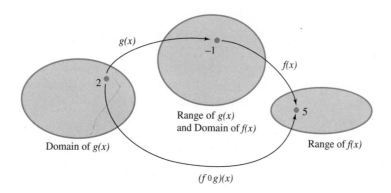

Figure 2.37

EXAMPLE SET C

If $f(x) = \sqrt{x-3}$ and $g(x) = 2x + 1$, find the algebraic expression for

1. $(f \circ g)(x)$,
2. $(g \circ f)(x)$.

Solution:

1. $(f \circ g)(x)$ indicates we should input $g(x)$ into $f(x)$.

$$(f \circ g)(x) = f[g(x)] \qquad \text{(definition of } f \circ g)$$

$$= f(2x + 1) \qquad \text{(definition of } g)$$

$$= \sqrt{(2x + 1) - 3} \qquad \text{(definition of } f)$$

$$= \sqrt{2x - 2}$$

2. $(g \circ f)(x)$ indicates we should input $f(x)$ into $g(x)$.

$$(g \circ f)(x) = g[f(x)] \qquad \text{(definition of } g \circ f)$$

$$= g(\sqrt{x - 3}) \qquad \text{(definition of } f)$$

$$= 2\sqrt{x - 3} + 1 \qquad \text{(definition of } g)$$

It is important to note that, in general, $f \circ g \neq g \circ f$. The problems of the previous example set illustrate the fact that the composition operation $\circ$ is not commutative.

The Domain of a Composite Function

A natural question is: How is the domain of a composite function related to the domains of the functions involved in the composition? To help answer this question, consider further the functions $f(x) = \sqrt{x-3}$ and $g(x) = 2x+1$ as defined in the previous example set.

$$f(x) = \sqrt{x-3} \text{ with domain } \{x|x \geq 3\} \text{ and range } \{f(x)|f(x) \geq 0\}$$

$$g(x) = 2x+1 \text{ with domain } \{x|x \in R\} \text{ and range } \{g(x)|g(x) \in R\}$$

We need to give extra concentration to the idea of determining the domain of the composition of two functions. This idea is a bit more involved than the algebraic combinations of functions, but may be more understandable by considering the following. Since we are looking at $f(g(x))$, we know that any replacement we make for $g(x)$ has to be an allowable domain value for $f(x)$. The domain of $f(x)$ is $\{x|x \geq 3\}$, so that the only values of $g(x)$ we can use must satisfy $g(x) \geq 3$. Now, what domain values of $g(x)$ produce range values satisfying $g(x) \geq 3$? This brings the problem to a more manageable position. We set $g(x) \geq 3$ as follows:

$$g(x) \geq 3$$

$$2x + 1 \geq 3$$

$$2x \geq 2$$

$$x \geq 1$$

Domain values of $g(x)$ that are ≥ 1 produce range values of $g(x)$ that are ≥ 3. We know that numbers that are ≥ 3 are allowable domain values for $f(x)$. Thus, the set $\{x|x \geq 1\}$ is the domain of the composite function $f \circ g$.

This discussion helps suggest the following theorem regarding the domain of a composite function:

The Domain of a Composite Function

The **domain of the composite function** $(f \circ g)(x)$ is the subset of numbers in the domain of $g(x)$ that produce values in the domain of $f(x)$.

The Domain of a Composite Function

We omit the proof of this theorem but illustrate its use in Example Set D.

EXAMPLE SET D

1. For the functions $f(x) = \dfrac{1}{x}$ and $g(x) = (x - \dfrac{1}{3})(x+1)(x+2)$, find an algebraic expression for $(f \circ g)(x)$ and then find the domain of $(f \circ g)(x)$.

   Solution:
   By its definition, the function $f(x)$ produces the reciprocal of the input quantity. Thus,

   $$(f \circ g)(x) = f[g(x)]$$

   $$= f[(x - \tfrac{1}{3})(x+1)(x+2)]$$

   $$= \frac{1}{(x - \tfrac{1}{3})(x+1)(x+2)}$$

   The function $g(x)$ has as its domain the set of all real numbers (R) while the function $f(x)$ is undefined for $x = 0$. The numbers $\dfrac{1}{3}, -1$, and -2 in the domain of $g(x)$ all produce 0 as outputs, and therefore cannot be in the domain of $f(x)$. Hence,

   The domain of $(f \circ g)(x) = \{x | x \in R \text{ and } x \neq \tfrac{1}{3}, -1, -2\}$

2. For the functions $f(x) = \sqrt{4 - x}$ and $g(x) = \sqrt{x}$, find an algebraic expression for $(f \circ g)(x)$ and find the domain of $(f \circ g)(x)$.

   Solution:
   We begin by evaluating $(f \circ g)(x)$:

$$(f \circ g)(x) = f[g(x)]$$

$$= f[\sqrt{x}]$$

$$= \sqrt{4 - \sqrt{x}}$$

The domain of $g(x) = \{x | x \geq 0\}$ and the domain of $f(x) = \{x | x \leq 4\}$. The numbers in the domain of $g(x)$ for which $g(x) \leq 4$ are all those numbers x such that $\sqrt{x} \leq 4$. Since $x \geq 0$, this implies that $0 \leq x \leq 16$. Hence,

The domain of $(f \circ g)(x) = \{x | x \in R \text{ and } 0 \leq x \leq 16\}$

Example Set E illustrates why we cannot simply examine $(f \circ g)(x)$ to determine the domain of $f \circ g$.

EXAMPLE SET E

For the functions $f(x) = x^2$ and $g(x) = \sqrt{x}$, find an expression for $(f \circ g)(x)$ and determine the domain of $(f \circ g)(x)$.

Solution:

$$(f \circ g)(x) = f[g(x)]$$

$$= f[\sqrt{x}]$$

$$= [\sqrt{x}]^2$$

$$= x$$

We might conclude, from looking only at $(f \circ g)(x) = x$, that the domain of $(f \circ g)(x)$ is the set of all real numbers. However, the domain of $g(x)$ is restricted to the set $\{x | x \geq 0\}$. The domain of $f(g(x))$ cannot be any bigger than the domain of $g(x)$. In addition, all of the values in the domain of $g(x)$ have output in the domain of $f(x)$. Thus, we conclude that the domain of $(f \circ g)(x) = \{x | x \geq 0\}$.

Composite Functions and the Graphing Calculator

The calculator is very useful for simplifying computations and graphing composite functions.

EXAMPLE SET F

A study of a northeastern community indicates that the average daily level of carbon monoxide in the air, in parts per million, is approximated by the function $C(p) = \sqrt{0.46p^2 + 15.5}$, $0 < p \leq 50$, where p represents the population of the community in thousands of people. In turn, the population of the community depends on the time t in years from now, and is approximated by the function $p(t) = 2.8 + 0.12t^2$, $0 < t \leq 25$.

1. Approximate, to two decimal places, the average daily level of carbon monoxide in the community 15 years from now.

2. With an error of at most 0.01 years, determine, in years from now, when the level of carbon monoxide will be 8.8 parts per million.

Solution:
Begin by entering the expression that defines $p(t)$ for Y1. Then enter the expression that defines the composition $C[p(t)] = (C \circ p)(t)$ for Y2. To do so we need to replace p in $C(p)$ with its defining expression, which was entered for Y1. So, for Y2 enter $\sqrt{(0.46Y_1\hat{\ }2 + 15.5)}$.

1. To approximate the average daily level of carbon monoxide in the community 15 years from now, return to the computation window and store 15 into X (by pressing 15 STO X) and compute Y2 using the 2nd Y-vars key. This computation results in 20.59. Thus, we conclude that 15 years from now, the level of carbon monoxide will be, to two decimal places, 20.59 parts per million.

2. To determine, with an error of at most 0.01, when the level of carbon monoxide will be 8.8 parts per million, we must recall the approximation techniques we studied in Section 1.3. Enter Xmax − Xmin for Y3, and 8.8 for Y4. Turn off the graphing capabilities of Y1 and Y3, and graph the composite function, Y2, and the horizontal output value, Y4. We can use the information contained in the definition of the individual functions to set Xmin, Xmax, Ymin, and Ymax. We will use the following settings: Xmin = 0, Xmax = 25, Ymin = 0, and Ymax = 50. Tracing and zooming several times until Y3 is less than 0.01, we conclude that $C[p(t)] = 8.8$ when x is approximately 8.5653. Thus, with an error of at most 0.01 years, the level of carbon monoxide in this community will be 8.8 parts per million 8.5653 years from now.

Writing Activity   Use complete sentences to write down the process of finding the composition of two functions f and g. Then apply your process to obtain $(f \circ g)(x)$ for the functions $f(x) = \frac{3}{5x}$ and $g(x) = \sqrt{x + 5}$. In addition, explain how the domain of a composite function is obtained in general and then give the domain of $(f \circ g)(x)$.

EXERCISES

For each of exercises 1–4, find the domain and rule of the functions $(f + g)(x)$, $(f - g)(x)$, $(f \cdot g)(x)$, and $\left(\dfrac{f}{g}\right)(x)$.

1. $f(x) = 3x + 2$, $g(x) = x^2 + 2x - 1$
2. $f(x) = x^2 - 2x$, $g(x) = 3x^2 + 2x - 1$
3. $f(x) = \sqrt{x^2 + 3x}$, $g(x) = \dfrac{1}{\sqrt{x^2 - x - 6}}$
4. $f(x) = \sqrt{x^2 - 4}$, $g(x) = \sqrt{x + 2}$

For each of exercises 5–6, suppose that $f(x) = \dfrac{x - 3}{4}$ and $g(x) = 2x - 1$. Find each of the following:

5. (a) $(f + g)(x)$.
   (b) $(f - g)(-2)$.
   (c) $(f \cdot g)(h + 1)$.
   (d) $\left(\dfrac{f}{g}\right)(4)$.

6. (a) $(f + g)(\sqrt{x})$, (b) $(f - g)(x + h)$, (c) $(f \cdot g)(-x)$, and (d) $\left(\dfrac{f}{g}\right)(-1)$

For each of exercises 7–19, find the domain and rule for $(f \circ g)(x)$, and the domain and rule for $(g \circ f)(x)$.

7. $f(x) = 2x + 5$ and $g(x) = x^2 - x$
8. $f(x) = -5x + 1$ and $g(x) = x^2 - 10$
9. $f(x) = x^2 + x - 4$ and $g(x) = 2x^2 - 5$
10. $f(x) = -(x + 1)^3$ and $g(x) = 4x + 2$
11. $f(x) = \dfrac{3}{5x}$ and $g(x) = x - 4$
12. $f(x) = \dfrac{x + 1}{x - 3}$ and $g(x) = 3x + 6$
13. $f(x) = \sqrt{x}$ and $g(x) = x + 1$
14. $f(x) = \dfrac{3}{5x}$ and $g(x) = \sqrt{x + 5}$
15. $f(x) = \dfrac{x^2 - 1}{x + 1}$ and $g(x) = \dfrac{5}{x}$

16. $f(x) = \dfrac{x + 3}{x - 4}$ and $g(x) = \dfrac{x + 2}{x - 4}$
17. $f(x) = \dfrac{2x - 1}{x + 2}$ and $g(x) = \dfrac{3x + 22}{2x - 6}$
18. $f(x) = \{(2, 5),\ (3, 8),\ (4, 11),\ (5, 12)\}$ and $g(x) = \{(2, 1),\ (3, 6),\ (4, 11),\ (5, 15)\}$
19. $f(x) = \{(-5, 0),\ (-1, -2),\ (0, 2),\ (4, 4)\}$ and $g(x) = \{(-5, 4),\ (-1, 3),\ (0, 2),\ (4, 1)\}$

For exercises 20–26, find the indicated composition function(s).

20. If $f(x) = 3x - 4$, $g(x) = 2x + 11$, and $h(x) = 5x + 6$, find the rule for $(f \circ g \circ h)(x)$ and then evaluate this composition at $x = 2$.
21. If $f(x) = x^2 - 1$, $g(x) = -5$, and $h(x) = x - 2$, find the rule for $(f \circ g \circ h)(x)$ and then evaluate this composition at $x = 0$.
22. If $f(x) = 2x + 7$ and $g(x) = \dfrac{x - 7}{2}$, find the rules for both $(f \circ g)(x)$ and $(g \circ f)(x)$.
23. If $f(x) = 5x + 4$ and $g(x) = \dfrac{x - 4}{5}$, find the rules for both $(f \circ g)(x)$ and $(g \circ f)(x)$.
24. If $f(x) = 2x + 9$, find the rule and the domain for $(f \circ f)(x)$.
25. If $f(x) = \dfrac{x}{x - 1}$, find the domain and the rule for $(f \circ f)(x)$.
26. If $f(x) = \dfrac{-x}{1 - x}$, find the domain and the rule for $(f \circ f)(x)$.

27. A manufacturer believes that the function $C(x) = 0.16x^2 + 0.92x + 850$ closely approximates the cost of a daily production run of x units. The manufacturer also knows that the number of units produced depends on the time of day t (in hours since the start of a shift) and is given by the function $x(t) = .95t^2 + 90t$. Find the approximate cost of producing x units (a) 5 hours after the start of a shift, and (b) 6 hours after the start of the shift. (c) With an error of at most 0.01

hours, how many hours after the beginning of a shift will the cost of a production run be $20,000?

28. The volume of a spherical cancer tumor is related to the diameter of the tumor by the function $V(x) = \dfrac{\pi x^3}{6} + 1.5$, where x is in millimeters. However, the diameter of the tumor is related to the time t (in days) since detection of the tumor by the function $x(t) = 0.00006t^2 - 0.002t + 0.005$, $0 < t \le 210$. Find the volume of the tumor (a) 60 days after detection, and (b) 90 days after detection. (c) With an error of at most 0.01 days, how many days after detection will the volume of the tumor be 6 cubic millimeters?

29. A government study estimates that consumers will buy a quantity $Q(p)$ of imported spices when the price per pound is p dollars. The price per pound depends on the time of year and is given by the function $p(t) = 0.03t^2 + .09t + 5.8$, $0 < t \le 26$, where t is given in weeks from the beginning of the year. The function $Q(p) = \dfrac{5286}{p^{1.8}}$ represents the quantity, in pounds, of spice that people are willing to buy at the price p. Estimate the number of pounds of spice that is bought (a) 10 weeks after the beginning of the year, and (b) 20 weeks after the beginning of the year. (c) With an error of at most 0.01 weeks, estimate the number of weeks after the beginning of the year that people will be buying 123 pounds of spice.

30. A salesman earns $25 per day and a 10% commission on his total daily sales.

   (a) How much money will he make in one day if he sells x dollars worth of goods?

   (b) He notices that there is a relationship between the amount of sales x and the number n of customers he encounters, and arrives at a formula that is generally valid for $n > 10$. The formula is

$x = 50n + 0.2\sqrt{n(n-1)}$. Express the commission as a function of the number of customers.

   (c) According to this formula, how many customers must he encounter in order to earn more than $500? (Graph and use TRACE.)

31. The curve on a decorative arch on top of the new City Hall is modeled by the equation $y = \sqrt{25 - x}$. The slope of this function is given by $f(y) = -\dfrac{1}{2y}$, $y \ne 0$. Write the slope as a function of x. Find the domain and range of the slope function. What happens as x approaches 25?

32. While on vacation Sue decides to rent a car. The car rental agency charges $40 per day plus $0.25 per mile. Write down the function expressing Sue's cost for one day if she drives x miles that day.

33. Referring to the previous exercise: Sue's vacation lasted ten days. On the first day she drove 30 miles. After that she increased her mileage each day by 25 miles. Write an expression for the number of miles she drove on the nth day. Combine the two functions in the previous exercise and this exercise to give an expression for Sue's cost on the nth day. How much was the total cost for the car rental and mileage?

34. Marcy is getting married this June and has chosen her bridesmaids' dresses. One bridesmaid's dress requires five yards of material. If a seamstress sews x dresses, how much material does she use? Suppose each yard of material requires 2.5 hours of labor to produce and each dress requires 28 hours of labor. How many hours of labor are required for x dresses?

35. Referring to the previous exercise, if the seamstress receives $7.50 an hour and the fabric designer receives $8.50 an hour, what is the cost of making one garment?

36. Ann is walking at the steady rate of 4 mph on a sloped road. For each 100 yards she walks up along the slope her elevation increases by 5 feet.
    (a) Find a function that expresses the distance covered in terms of time t. (Convert mph to ft./sec.)
    (b) Express the increase in elevation as a function of the distance traveled on a slope.
    (c) Can you express the increase in elevation as a function of time?

37. An auditorium holds 850 people. The general admission price is $7.50 while the senior admission price is $3.50. Assuming that x individuals buy general admission tickets, write a function describing the gate receipts if all tickets are sold. Graph the function and find the least amount and the most amount of money the auditorium owners can make.

38. Express the area of a square as a function of the edge. Express the volume of a cube as a function of the edge. Now, express the volume of the cube as a function of the area of a face. Do the reverse and express the area of the face of a cube as a function of the volume of the cube.

39. Express the area of a circle as a function of the radius. Express the volume of a sphere of the same radius as a function of the radius. Now find the area as a function of V and the volume as a function of A.

40. The demand d for a product is linearly dependent on the price as follows: $d = 5040 - \frac{4}{7}p$. The supply s for the product also depends on the price given by $s = 600p - 2300$. Express the supply as a function of the demand and the demand as a function of the supply.

41. An oil spill in the ocean is in the shape of a circle and its radius at time t seconds is given by $y = \sqrt{t+1}$ where y is expressed in feet. Find the circumference and the area of the oil spill as functions of t. What is the circumference at one minute? What was the circumference at $t = 0$?

2.6   Using Functions to Describe Variation

Introduction

Four Types of Variation

Examples of Direct Variation

Examples of Inverse Variation

Examples of Joint and Combined Variation

Introduction

Functions are instruments that describe relationships between quantities. It can be interesting and important to know how a change of a particular magnitude in one or more of the quantities effects the other quantities. In this section you see how to manipulate functions to investigate and answer such questions.

Four Types of Variation

In Chapter 1, we stated that the notation $f(x)$ represents the output value of a function when the input value is x. In this section, and because of convention only, we denote output values by a single letter. So, for example, the output value $f(x)$ may be denoted by the single letter y. We also let the letter k represent some nonzero constant.

There are four types of variation that occur so often in many fields (and particularly the sciences) that they are given special names: direct variation, inverse variation, joint variation, and combined variation.

Direct Variation

Direct Variation

In **direct variation** the quantities x and y change in such a way that their quotient is always the same; that is, $\frac{y}{x} = k$. Solving for y, we get $y = kx$. Since one x value produces precisely one y value, this statement is a function. Expressed in standard function form, $y = kx$ is $f(x) = kx$.

Inverse Variation

Inverse Variation

In **inverse variation** the quantities x and y change in such a way that their product is always the same; that is, $xy = k$. Solving for y, we get $y = \frac{k}{x}$. Since one x value produces precisely one y value, this statement is a function. Expressed in standard function form, $y = \frac{k}{x}$ is $f(x) = \frac{k}{x}$.

Joint Variation

Joint Variation

Joint variation involves more than two variables. One quantity varies directly as the product of the others.

Combined Variation

Combined Variation

Combined variation involves more than two variables. It is a combination of a direct variation and an inverse variation.

Constant of Variation

In each type of variation, the nonzero constant k is called the **constant of variation** or the constant of proportionality. The following example summarizes the variation equations and demonstrates the use of the corresponding terminology.

ILLUMINATOR SET A

1. The equation $y = kx$ is an example of direct variation and is read as either:

    y varies directly as x, or as

    y is directly proportional to x

2. The equation $y = kx^n$ is an example of an extension of direct variation and is read as either:

    y varies directly as the nth power x, or as

    y is directly proportional to the nth power of x

3. The equation $y = \dfrac{k}{x}$ is an example of inverse variation and is read as either:

    y varies inversely as x, or as

    y is inversely proportional to x

4. The equation $y = \dfrac{k}{x^n}$ is an example of an extension of inverse variation and is read as either:

    y varies inversely as the nth power x, or as

    y is inversely proportional to the nth power of x

5. The equation $y = kxz^n$ is an example of joint variation and is read as either:

    y varies jointly as the product of x and the nth power z, or as

    y is jointly proportional to the product of x and the nth power z

6. The equation $y = \dfrac{kxz^n}{w^m}$ is an example of combined variation and is read as either:

    y varies as the product of x and the nth power z, and inversely as the mth power of w, or as

y is proportional to the product of x and the nth power of z, divided by the mth power of w

Examples of Direct Variation

Table 2.4

x	y
0	0
1	3
2	6
3	9
4	12
10	30
15	45

ILLUMINATOR SET B

Suppose that data on some physical phenomenon are collected and summarized in Table 2.4. The input is represented by x and the output by y. The pattern in the table seems to indicate that the value of y is always 3 times that of x; that is; $y = 3x$. This is the form of direct variation, and we conclude that y varies directly as x. The constant of variation is 3. Notice also that each x produces precisely one y, so that this variation equation is a function. Using standard function notation, we could write this direct variation relationship as $f(x) = 3x$.

General Variation Functions
Particular Variation Functions

Variation functions that involve the constant of variation k are called **general variation functions** because they represent the relationship in the general case, that is, for any value of k. Variation relationships that involve a particular value of the constant of variation k are called **particular variation functions general variation functions** because they represent the relationship for a particular case only.

EXAMPLE SET A

1. Suppose it is known that y varies directly as x. Experimentation shows that $y = 40$ when $x = 5$. Find the general and particular variation functions and determine the value of y when $x = 4$.

   Solution:
   Since y varies directly as x, $y = kx$ is the general variation function.
   Since $y = 40$ when $x = 5$, we have, upon substitution, $40 = k \cdot 5$. Solving for k we get $k = \dfrac{40}{5} = 8$. Thus, $y = 8x$ is the particular variation function. Now we can use the particular variation function to find the output value y for the input value 4.

   $$y = 8 \cdot 4 = 32$$

   Thus, when $x = 4$, $y = 32$.

2. The force required to compress a spring x units varies directly as x. It is determined that a force of 9 lb compresses a particular spring 3 inches. How many inches will a force of 18 lb compress this spring?

Solution:
Since force is a function of x, we represent it by $F(x)$. Then, since $F(x)$ varies directly as x, $F(x) = kx$ is the general variation function.
Since $F(x) = 9$ when $x = 3$, we get $9 = k \cdot 3$. Solving for k we get $k = \dfrac{9}{3} = 3$. Thus, $F(x) = 3k$ is the particular variation function.
Now, when $F(x) = 18$, $18 = 3x$, which implies that $x = 6$.
Thus, a force of 18 lb will compress this spring 6 inches.

The next example illustrates how we can use variation to compare a new output to an original output when changes are made to one or more of the quantities in the variation function.

EXAMPLE SET B

The length L, in feet, of skid marks made by a car varies directly as the square of the speed s, in miles per hour (mph), of the car. Experimentation has shown that, under ideal conditions, a car traveling 30 mph will leave a skid mark 40-feet long. What is the effect on the length of the skid mark if the speed is tripled?

Solution:
Since the length L is a function of the speed s, we denote it by $L(s)$. Then, since length varies directly as the square of the speed,
   $L(s) = ks^2$ is the general variation function.
   The idea is to compare two quantities. We need to express one quantity in terms of the other. Specifically, we want to express the new length in terms of the original length.
   We'll let $L(s)_{orig} =$ original length of the skid mark. Then we have $L(s)_{orig} = ks^2$.
   We can describe tripling s as $3s$.
   We'll let $L(s)_{new} =$ new length of the skid mark.

Then $L(s)_{new} = k(3s)^2$ $\leftarrow$ manipulate and search for ks^2 $\left(L_{orig} \right)$

$$= k \cdot 3^2 \cdot s^2$$

$$= 9ks^2$$

$$= 9 \underbrace{\left(ks^2\right)}_{L(s)_{\text{orig}}}$$

$$L(s)_{\text{new}} = 9 \cdot L(s)_{\text{orig}}$$

We interpret this equation as saying "The new length is nine times the original length."

Examples of Inverse Variation

EXAMPLE SET C

1. Suppose that y varies inversely as x and it is determined, by experimentation, that $y = 2$ when $x = 8$. Find the value of y when $x = 1$.

   Solution:

   Since y varies inversely as x, $y = \dfrac{k}{x}$ is the general variation function.

   Since $y = 2$ when $x = 8$, we have, $2 = \dfrac{k}{8}$. Solving for k we get $k = 16$.

   Thus, $y = \dfrac{16}{x}$ is the particular variation function.

   When $x = 1, y = \dfrac{16}{1} = 16$.

   Thus, $y = \dfrac{16}{x}$ and $y = 16$ when $x = 1$.

2. The intensity of illumination of a light on an object varies inversely as the square of the distance between the light source and the object. It is determined that at a distance of 10 feet, the intensity of illumination is 5 foot-candles. What is the intensity at 4 feet?

   Solution:

   Since the intensity is a function of distance, we denote it by $I(d)$, where d represents distance. Then, since intensity varies inversely as the square of the distance, we can write

   $I(d) = \dfrac{k}{d^2}$ is the general variation function. Since $I(d) = 5$ when $d = 10$,

   we have $5 = \dfrac{k}{10^2}$. Solving for k we get $k = 5 \cdot 10^2 = 500$.

   Thus, $I(d) = \dfrac{500}{d^2}$ is the particular variation function. When $d = 4$,

   $$I(d) = \frac{500}{4^2}$$

$$I(d) = \frac{500}{16}$$

$$I(d) = 31.25 \text{ foot-candles}$$

Thus, $I(d) = \dfrac{500}{d^2}$, and the intensity at 4 feet is 31.25 foot-candles.

Examples of Joint and Combined Variation

The next example set illustrates the need and use of a function of several variables.

EXAMPLE SET D

1. The pressure on a sail varies jointly as the area of the sail and the square of the speed of the wind. It is known that for a wind speed of 20 mph, a 25-square-foot sail experiences a pressure of 1 pound per square foot. What is the pressure on the sail if the wind's speed is 30 mph?

   Solution:
   Since the pressure (say P) depends on both the area (say A) of the sail and the speed (say s) of the wind, it is a function of these two quantities and we denote it by $P(A, s)$. Then, since the pressure varies jointly as the area and the square of the wind's speed, we can write $P(A, s) = kAs^2$ as the general variation function.
   Since $P(A, s) = 1$ when $A = 25$ and $s = 20$, $1 = k \cdot 25 \cdot 20^2$. Solving for k we get $1 = k \cdot 10,000$ or $k = 0.0001$.
   Thus, $P(A, s) = 0.0001As^2$ is the particular variation function.
   When $A = 25$ and $s = 30$,

   $$P(A, s) = (0.0001) \cdot 25 \cdot 30^2$$

   $$= (0.0001) \cdot 25 \cdot 900$$

   $$= 2.25$$

   Thus, $P(A, s) = 0.0001As^2$, and the pressure is 2.25 pounds per square foot when the wind's speed is 30 mph.

2. The load at which a circular pillar will crush varies directly as the fourth power of the diameter and inversely as the square of the height of the pillar. What effect will there be on the crushing load if the diameter is cut in half and the height is doubled?

Solution:

Since the crushing load (say C) at which a pillar will crush depends on both the diameter (say d) and the height (say h) of the pillar, it is a function of these two variables, and we denote it by $C(d, h)$. Then, since the crushing load varies directly as the fourth power of the diameter and inversely as the square of the height, we can write $C(d, h) = \dfrac{kd^4}{h^2}$ is the general variation function.

We'll let $C(d, h)_{\text{orig}} = $ original crushing load.

Then we have $C(d, h)_{\text{orig}} = \dfrac{kd^4}{h^2}$.

We'll let $C(d, h)_{\text{new}} = $ new crushing load, and compare $C(d, h)_{\text{new}}$ to $C(d, h)_{\text{orig}}$.

We can describe halving d as $\dfrac{d}{2}$.

We can describe doubling h as $2h$. Then

$$C(d, h)_{\text{new}} = \frac{k(d/2)^4}{(2h)^2} \quad \leftarrow \quad \text{manipulate and search for } \frac{kd^4}{h^2} \quad \left(C(d, h)_{\text{orig}} \right)$$

$$= \frac{k\dfrac{d^4}{2^4}}{2^2 h^2}$$

$$= \frac{\dfrac{kd^4}{16}}{4h^2}$$

$$= \frac{kd^4}{64h^2}$$

$$= \frac{1}{64} \cdot \underbrace{\frac{kd^4}{h^2}}_{C(d,h)_{\text{orig}}}$$

$$= \frac{1}{64} C(d, h)_{\text{orig}}$$

We can interpret this equation as saying "the new crushing load is one sixty-fourth the original crushing load."

Writing Activity   Write down in paragraph form the different types of variation that were covered in this section. Include the function rule that is associated with each variation you discuss.

EXERCISES

For each of exercises 1–10, express each statement as a general variation function.

1. m varies directly as n.

2. p varies directly as w.

3. u varies directly as the fifth power of p.

4. t varies directly as the fourth power of s.

5. y varies inversely as the square of x.

6. n varies inversely as the cube root of m.

7. t varies directly as the square of s and inversely as the cube of q.

8. s varies jointly as the square of t and square root of p.

9. y varies jointly as the cube of x and the square of z, and inversely as the product of w and the square root of z.

10. u varies jointly as the fourth power of m and inversely as the product of the square root of x and the cube of z.

11. If y varies directly as x, and $y = 30$ when $x = 6$, find the general and particular variation functions and also the value of y when $x = 2$.

12. If L varies directly as t, and $L = -28$ when $t = 14$, find the general and particular variation functions and also the value of L when $t = 8$.

13. If y varies inversely as x, and $y = 20$ when $x = 20$, find the general and particular variation functions and also the value of y when $x = 40$.

14. If y varies inversely as x, and $y = 0.1$ when $x = 0.1$, find the general and particular variation functions and also the value of y when $x = 1$.

15. If y varies jointly as x and the square of w, and $y = 6$ when $x = 4$, and $w = 5$, find the general and particular variation functions and also the value of y when $x = 2$ and $w = 7$.

16. If y varies jointly as the square of x and inversely as the product of w and the cube of z, and $y = 8$ when $x = 2$, $w = 5$, and $z = 3$, find the general and particular variation functions and also the value of y when $x = 6$, $w = 9$, and $z = 3$.

17. The lift L of an airplane wing of fixed design varies directly as the square of its width. For a particular wing design, $L = 35$ pounds when $w = 5$ inches. Find L when $w = 20$.

18. The volume V of water under a constant pressure delivered through a pipe varies directly as the square of the diameter d of the pipe. If a $1\frac{1}{2}$-inch pipe delivers 270 gallons of water each minute, how many gallons per minute would a $2\frac{1}{2}$-inch pipe deliver?

19. The distance d traveled by a rolling ball on an inclined plane varies directly as the square of the time t it has been in motion. If the ball has traveled a distance of 80 inches after 4 seconds, how far will it travel in 5 seconds?

20. For a particular load L attached to a wire of diameter d, the amount the wire stretches varies inversely as the square of the diameter of the wire. If a wire with diameter 0.8 inches is stretched 0.008 inches by supporting a particular load, by how much will a wire of the same type but having a diameter of 0.4 inches be stretched?

21. The frequency f of an electromagnetic wave varies inversely as the length L of the wave. If a 400-foot wave has a frequency of 400 kilocycles, what will be the frequency of a 200-meter-long wave?

22. The electrical resistance R of a wire varies directly as the length L of the wire and inversely as the square of its diameter d. If a 36-inch-long wire having diameter 0.03 inches has a resistance of 4 ohms, what will

be the resistance of the same type of wire that is 64-inches long with diameter 0.04 inches?

23. The weight W that a horizontal beam of length L can support without breaking varies inversely as the length of the beam. If a 9-foot beam can support a maximum of 600 pounds, how many pounds can a 30-foot beam safely support?

24. In chemistry, Boyle's Law states that the pressure P of a gas varies directly as the temperature T of the gas and inversely as the volume V of the gas. If a gas that is heated to $210°K$ occupies 30 cubic centimeters of space when it is under a pressure of 14.7 kilograms per square centimeter, what will be the pressure of this type of gas when it is heated to a temperature of $240°K$ and occupies only 8 cubic centimeters?

25. The number N of genes that are mutated when exposed to a dosage d of x-rays varies directly as the amount of the dosage. What is the effect on the number of mutated genes if the dosage of x-rays is tripled?

26. Poiseuill's Law in physiology states that the blood flow in a particular arteriole (a small artery) varies directly as the fourth power of the radius of the arteriole. It is possible for the body to adjust blood flow by changing the radius of arterioles. How does the blood flow through an arteriole that has been reduced to 30% of its original radius compare to the blood flow before the reduction?

27. The force of attraction between two objects varies jointly as their masses, M_1 and M_2, and inversely as the square of the distance d between them. How is the force of attraction between two bodies affected if one mass is doubled, the other is tripled, and the distance between them halved?

28. Naval architects know that the thrust of a particular type of propeller varies jointly as the fourth power of its diameter and the square of the number of revolutions per minute that it is turning. What is the effect on the thrust if the diameter of the propeller is cut in half and the number of revolutions it turns per minute is doubled?

29. The velocity of a satellite in a circular orbit about the earth varies inversely as the square root of radius of its orbit. What effect would there be on the velocity if the radius of the orbit is doubled? By what factor would we have to change the orbital radius if we want to double the velocity?

30. The frequency of pitch of a musical string varies directly as the square root of the tension of the string and inversely as the product of the length and diameter of the string. How is the pitch affected if the diameter of the string is left unchanged, but the tension is quadrupled and the length is doubled?

Decide what type of variation is represented by the data in the tables below. Then use complete sentences to describe how you reached your decision.

31.

x	y
-2	-8
-1	-4
0	0
1	4
2	8
3	12
4	16

32.

x	y
-3	15
-2	10
-1	5
0	0
1	-5
2	-10
3	-15
4	-20

33.

x	y
0	0
1	1
2	8
3	27
4	64
5	125

34.

x	y
0	0
1	2
2	8
3	18
4	32
5	50

35.

x	y
1	20
2	10
4	5
5	4
10	2
20	1
40	1/2

36.

x	y
1	100
2	25
5	4
10	1
20	0.25

2.7   Data-Generated Functions

The Three Parts of the Creation Process

The Scatter Diagram

The Coefficient of Correlation

The Method of Least Squares

Calculator Mechanics

Accuracy

The Three Parts of the Creation Process

As we have seen, a function can provide information about a physical phenomenon. Function output values can be used to predict behavior of the phenomenon for specific input values. For example, if $d(t)$ represents the distance a person walks and is approximated by the function $d(t) = 3.2 \cdot t$, then we can predict that the walker will travel about 16 miles in 5 hours ($d(t) = 3.2 \cdot 5 = 16$). A natural question is "How are functions that describe physical phenomena created?" There are three parts to the creation of such functions.

1. The first part occurs when you wonder if two quantities are related.

2. The second part involves actually determining if a relationship exists. One

Table 2.5		
P	x	y
1	1.1	4.5
2	1.3	6.9
3	1.8	7.4
4	2.2	8.9
5	2.6	8.3
6	2.6	10.
7	2.9	10.0
8	3.0	11.5
9	3.3	11.0
10	3.7	10.4
11	3.9	11.2
12	4.0	12.0
13	4.5	13.6
14	5.2	14.0
15	5.3	16.1
16	5.5	18.7
17	7.1	21.6
18	7.8	20.0
19	9.1	25.8
20	9.7	26.3

Scatter Diagram

way to accomplish this is to use the statistical techniques of *correlation analysis*. Correlation analysis tells us if a relationship exists and if so, how strong or weak it is. (We will not concern ourselves with these techniques, since the examples and exercises presented in this text deal with quantities that are strongly related.)

3. The third part specifically deals with creation. If there is a relationship between the two quantities, it *may* be possible to construct a function that nicely approximates that relationship using the statistical techniques of *regression analysis*. (The graphing calculator is very efficient with both correlation and regression techniques.) In this section, we restrict our attention to *linear regression*, that is, regression that produces linear functions. We consider nonlinear regression in subsequent sections.

The Scatter Diagram

Let's begin with an example in which a psychiatrist wonders if there is a relationship between the amount of time a patient spends in therapy and the severity of the patient's disorders. Part (1) is complete. It seems reasonable to want to predict *time in therapy* from *severity of disorders*. Hence, we introduce
   the input variable x to represent *severity of disorders*, and
   the output variable y to represent *time in therapy*.
   Part (2) begins by collecting data. The psychiatrist might devise a rating system, from 1-10, in which mild disorders are given lower numbers and more severe disorders are given higher numbers. At the time of the original diagnosis, each patient is assigned an x value. When therapy is complete, that patient is then assigned a y value. Thus, after some time, an ordered pair, (x, y) has been assigned to each patient. Suppose that the psychiatrist collects the data shown in Table 2.5 on twenty of her patients. (The Patient's number is in the column headed P.)
   Now, having collected the data, the psychiatrist plots each point on a Cartesian coordinate system and looks for trends in behavior. The graph of these points is illustrated in Figure 2.38 and is called a **scatter diagram**. For this set of data, it appears that as the severity of the disorders (input) increases, the length of time in therapy (output) increases. Moreover, the data appear to exhibit a linear trend. That is, it appears that a straight line could be drawn through the data points so that for a particular input value, the corresponding

output value on the line would serve as a reasonable approximation to the actual output value.

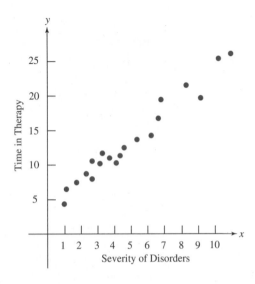

Figure 2.38

The Coefficient of Correlation

Coefficient of Correlation

This analysis partially answers question 2, "Does a relationship exist between the two quantities?" The question is more fully answered by computing the **coefficient of correlation**, r, a quantity that measures the strength of the relationship. The value of the coefficient of correlation always lies between -1 and 1.

$$-1 \le r \le +1$$

Values of r near 0 indicate a weak relationship or no relationship at all. The weaker the relationship, the worse the approximation of the line provides. Values of r near -1 or 1 indicate a strong relationship. The stronger the relationship, the better the approximation the line provides. Positive values of r indicate a rising line; the input and output values increase together. Negative values of r indicate a declining line; output values decrease as input values increase. Figure 2.39 illustrates some of these cases. Scatter diagrams may

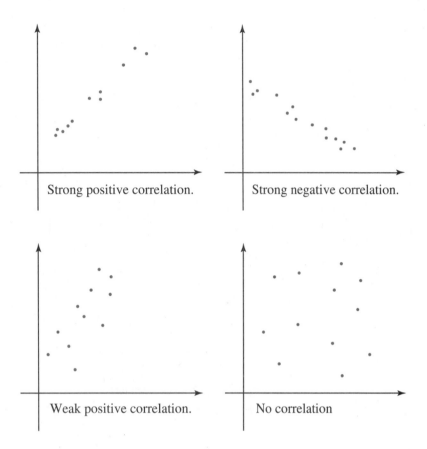

Figure 2.39

point out relations other than linear relationships. Figure 2.40 illustrates some nonlinear relationships. We investigate various nonlinear relations (such as power, exponential, and logarithmic) in subsequent chapters.

The psychiatrist, now convinced of the existence of a strong linear relationship, proceeds to part (3), creating the linear function from which time in therapy can be predicted from severity of disorders.

The Method of Least Squares

Least Squares   The creation of the linear function is based on the process of **least squares**. The idea is to place a line through all the data points so that it comes as close as possible to each of them. Figure 2.41 illuminates the idea. The least-

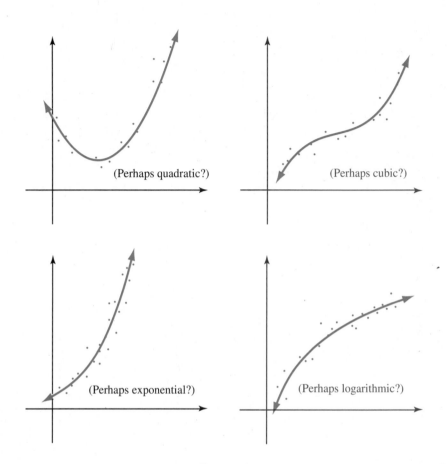

(Perhaps quadratic?)

(Perhaps cubic?)

(Perhaps exponential?)

(Perhaps logarithmic?)

Figure 2.40

squares process is built into the graphing calculator, and by using it we can generate the value of the coefficient of correlation, r, and the linear function itself. First, the calculator specifies the slope of the line with the letter b, and the y−intercept with the letter a. Thus, the form of the linear function is $f(x) = bx + a$.

You can use your calculator to:

1. Enter data.

2. Compute the regression constant and the coefficient of correlation.

3. Enter the regression expression as a function definition.

4. Plot the scatter diagram along with the regression line.

5. Make a prediction.

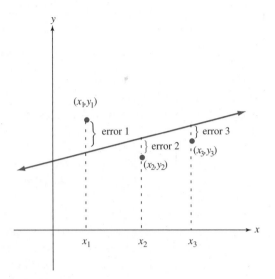

Figure 2.41

Calculator Mechanics

TI-82   We begin the process of creating the linear function by entering the data in the x- and y-columns of Table 2.5 as lists. In the STAT menu, select EDIT to display the lists labeled L1 and L2. If data are already in these lists, clear them by highlighting a label (L1, for example) using the arrow keys, pressing CLEAR and then ENTER. You can now enter the x-column in list L1 by moving to the first entry position in that column, typing the number and then pressing ENTER (or the down-arrow key) to move to the next position in the column. When you have finished entering the first column, use the arrow keys to move to the first entry position in list L2 to enter the y-column.

You are almost ready to produce a scatter diagram. Press 2nd and then Y= to access the STAT PLOT menu. From this menu, you can set up how you will display your data. Press 1 to set up Plot1. Press ENTER to turn on Plot1. Highlight the first graph icon in the Type row and press ENTER to select a scatter diagram presentation of the data. Then select L1 in the Xlist row and L2 in the Ylist row so that L1 will be used for the x-coodinates and L2 for the y-coordinates. You have three choices for the mark to be used in the display of the data. Select the one you prefer.

Before plotting the scatter diagram, you need to be sure your graphing mode and graphing window are appropriate. You should be in Func mode and the Y= functions should be cleared or deselected. You can deselect a function by highlighting its = sign and pressing ENTER. Press WINDOW to set the Xmin, Xmax, Ymin, and Ymax to values appropriate for your data. For our case,

$$0 \le x \le 10 \quad \text{and} \quad 0 \le y \le 30 \qquad (2.1)$$

would be a reasonable choice. You can now display the scatter diagram by pressing GRAPH.

To compute the coefficients of the line that will model these data, return to the STAT menu, select CALC and then choose 3 to set up the lists to be used in our computations. Use the arrow keys to highlight L1 on the Xlist row toward the bottom of the screen below 2-Var Stats. Press ENTER to select list L1. Now move down another row and select list L2 in the row labeled Ylist in the same way.

To complete the computation of the modeling or prediction line, first obtain the calculation window and then enter the statistics mode again by pressing STAT and selecting CALC. Use the arrow keys to move down to number 9, which is the second linear regression choice, select it, and press ENTER. The values of a, b, and r will be displayed on the viewing screen. Window 2.2 shows the information in this case. You can have your calculator display the

```
LinReg
y=ax+b
a =2.836656023
b=2.441880826
r=0.983104784
```

Window 2.2

prediction line along with the scatter diagram in the following way. Press the Y= key and enter the expression 2.836656023+2.441880826X in Y1=. Another method is to let your calculator paste the function there. To do this, first press Y=, clear the expression in Y1=, and leave the cursor there. Then press VARS, select Statistics, highlight the EQ menu and select RegEQ. The expression that defines the prediction linear function is now entered for Y1. When you press GRAPH, the linear regression line and the data points will appear in the viewing window.

TI-81   The procedure for obtaining the scatter diagram and prediction line expression for this data is much different on the TI-81. You enter the data using the STAT menu as (x, y) pairs. To do this on the TI-81, press 2nd STAT, highlight DATA and press ENTER to obtain a data entry window. In this case, you enter the data in Table 2.5 as (x, y) pairs rather than as lists.

After entering the data, return to the computation window. You can display the data as a scatter diagram by selecting DRAW in the STAT menu and pressing 2. This causes the command Scatter to appear in the computation window. Pressing ENTER will display the data as a scatter diagram provided that you have set the range (see equation 2.1) and cleared the Y= menu.

To determine the prediction line, first return to the computation window and then select LinReg in the STAT menu with CALC highlighted. This places the command LinReg in the computation window. Pressing ENTER gives the a, b, and r values of the prediction line. You can graph this by entering the a and b values in the form $a + bx$ for Y1 in the Y= menu.

Accuracy

The values for the slope b and the y-intercept a of the prediction line are produced using data values and cannot, in the end, produce more accurate predictions than justified by the original data values. For example, the data values in our psychiatric example are accurate to only one decimal place. To maintain as much accuracy as possible, it is best to round off only at the last step. We shall adhere to the following rule. We will construct the linear prediction function using the values of a and b displayed in the calculator viewing window. When reporting predicted values, we will report them to the accuracy prescribed in the original data. If rounding is necessary, we will round to an even digit when a 5 occurs in the following position.

Prediction Part (3) of the process is now complete and the linear function is created! To indicate that the output $f(x)$ is an approximation, it is common to place a hat over it and express the function as

$$\hat{f}(x) = bx + a$$

For our psychiatric example, the calculator displays $a = 2.836656023$, $b = 2.441880826$, and $r = 0.983104784$, so that

$$\hat{f}(x) = 2.441880826x + 2.836656023$$

The function can now be used to predict the number of months in therapy from the severity of the disorders. For example, a psychiatrist who assigns a disorders rating of 2.8 to a patient can expect that patient to spend approximately 9.7 months in therapy:

$$\hat{f}(x) = 2.441880826(2.8) + 2.836656023$$

$$= 9.6739223358 \approx 9.7 \qquad \text{Rounded to one decimal place}$$

Your calculator can make this computation. Return to the calculation window by pressing 2nd and QUIT. Type in the desired value of the input variable, in this case, 2.8. Press STO and X|T to store this value in the variable x. Press ENTER to finish the storing process. To make the computation, press 2nd, Y-VARS, 1, ENTER. The value of the output variable will be displayed on the screen. In this case, Y1=9.6739223358.

Notice the information contained in the slope of the line. The slope of this line is 2.441880826, or as a fraction, $\dfrac{2.441880826}{1}$. This can be interpreted as follows: If the severity of disorders is increased by 1 point, the number of months in therapy will increase by approximately 2.4 months.

Writing Activity   Table 2.6 shows the net profits (P, in thousands of dollars) for a small, independent business during the first eight years of operation. Construct a scatter diagram for the data. If the data show an apparent linear relationship between the input and output values, compute the values of a, b, and r. Create the linear function that describes the relationship if the correlation (as measured by r) is sufficiently strong. If you create a linear function, make a statement about the expected change in the value of the output variable when the input value is increased by one unit. If you judge it nonlinear, so state.

Table 2.6

Years	P
1	87.9
2	120.3
3	162.2
4	212.7
5	286.3
6	403.8
7	532.0
8	717.3

EXERCISES

For exercises 1–7, examine the scatter diagram in each figure and determine if a strong or weak linear relationship exists between the variables, a nonlinear relationship exists, or no relationship exists.

1.

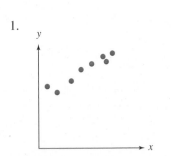

2.

3.

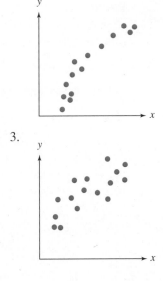

4.

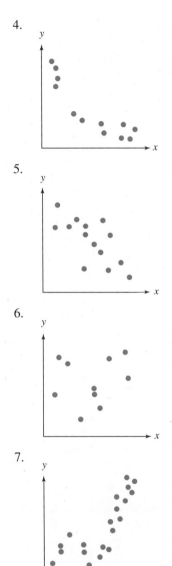

5.

6.

7.

For exercises 8–19, use your calculator to construct a scatter diagram for the given data. If your scatter diagram shows an apparent linear relationship between the input and output values, compute the values of a, b, and r. Create the linear function that describes the relationship. If the scatter diagram shows an apparent nonlinear relationship between the variables, state so, and

proceed to the next exercise. In each case where you create a linear function, make a statement about the expected change in the value of the output variable when the input value is increased by one unit. Describe in paragraph form the details of your solutions.

8. A person riding in a car that traveled at 45 miles per hour for 6 hours recorded the following data in the hope of establishing a function that describes the relation between time and distance.

Data Number	x	y
1	0	0
2	1	44.6
3	2	97.7
4	3	135.2
5	4	180.4
6	5	270.7

9. In the January 1986 issue of *Transactions of the ASME*, the results of an experiment that studied the stress corrosion cracking of Type 304 stainless steel in a simulated boiling water reactor environment were reported. The following table displays the maximum load x and the resulting crack growth rate y (in meters per second) for the six specimens that were tested.

Data Number	x	y
1	30.0	1.0
2	35.6	2.2
3	41.5	3.9
4	50.2	5.8
5	55.5	5.0
6	61.1	14.0

10. The following data are from the *United Nations Statistical Yearbook*, and display the average life expectancy x and the population-physician ratio y for 10 developing African countries.

Data Number	x	y
1	63.00	1,907
2	32.00	47,889
3	48.30	26,447
4	52.70	815
5	53.50	6,411
6	49.05	10,136
7	38.30	7,306
8	50.00	22,291
9	47.35	18,657
10	52.50	7,378

11. In their article "The Impact of Cable Penetration on Network Viewing" in the *Journal of Marketing Research*, Vol. 27, No. 9, Oct/Nov 1987, pp. 9–12, D. M. Drugman and R. T. Rust reported the following data for the years 1980 – 1985. The x values represent the percentage of U.S. cable TV households and y represents the network share of the TV revenues.

Data Number	x	y
1	21.1	98.9
2	23.7	97.9
3	25.8	96.5
4	30.0	95.2
5	35.7	94.0
6	41.1	91.9

12. The following data record the relationship between the radius r of six circles, as measured by 20-year-old college students, and the area A of that circle.

Data Number	r	A
1	0	0
2	1.02	3.17
3	2.02	12.56
4	3.04	28.26
5	4.01	50.24
6	4.45	61.9

13. The following table displays the order of the known planets and the asteroid belt x (counting outward from the sun) and the distance D (in units of 1/10 times the earth's distance from the sun) from the sun.

Data Number	x	D
1	1	3.87
2	2	7.23
3	3	10.00
4	4	15.24
5	5	29.00
6	6	52.03
7	7	96.46
8	8	192.0
9	9	300.9
10	10	395.0

14. A psychobiology student, trying to specify a relationship between anger level x and activity level y of college football players, collected the following data.

Data Number	x	y
1	2	16
2	4	21
3	9	35
4	5	26
5	12	50
6	3	17
7	4	22
8	10	402

15. The following table represents the correspondence between the carbon content of coal x and the hydrogen content y (in parts per million) of moon rocks collected by the crews of Apollo 14 and 15. (Source: *Journal of Research*, U.S. Geological Survey, Vol. 2, 1974.)

Data Number	x	y
1	110	82
2	105	120
3	99	90
4	74	66
5	51	85
6	50	38
7	50	20
8	45	20
9	22	8
10	7.7	2.0
11	7.3	2.8

16. The following data represent the correspondence between the average number of pages of notes taken by students in a college cultural diversity class x and the number of weeks y since the beginning of the quarter.

Data Number	x	y
1	2	3.3
2	3	8.7
3	4	4.1
4	10	3.0
5	8	5.1
6	11	2.1

17. The following table shows the net profits x (in thousands of dollars) for a small independent business during the first eight years of business.

Data Number	x	y
1	1	87.9
2	2	120.3
3	3	162.2
4	4	212.7
5	5	286.3
6	6	340.5
7	7	372.1
8	8	424.5

18. A clothing store owner wonders if there is a relationship between the number n of employees on duty and the number of dollars d lost per week due to shoplifting. The following table shows the data she collected over a twelve-week period.

Data Number	n	d
1	3	380
2	4	360
3	5	360
4	6	320
5	7	340
6	8	290
7	9	270
8	10	280
9	11	230
10	12	170
11	13	190
12	14	180

Data Number	n	d
1 (1960)	2,725	2.9
2 (1965)	3,769	3.5
3 (1970)	4,845	4.3
4 (1975)	6,282	6.5
5 (1977)	6,551	8.0
6 (1978)	6,699	8.6
7 (1979)	6,787	8.8
8 (1980)	6,856	9.9
9 (1981)	6,944	10.8
10 (1982)	7,059	11.5
11 (1983)	7,125	11.6
12 (1984)	7,230	12.3

19. You might wonder if there is a relationship between the number n of golf courses in the U.S. with at least eighteen holes and the divorce rate r in the U.S. The following data are taken from the *U.S. Bureau of the Census, Statistical Abstract of the United States* (1986 Washington, D.C.: Government Printing Office, 1986), pp. 35, 230. The divorce rate is given in millions.

20. The quantity of cars sold by a manufacturer is a function of the price. Suppose that the quantity sold doubles each time the price is halved. Complete the following table of values and then write down a defining expression for this function.

Price	Quantity
5,000	100,000
	50,000
	25,000
	12,500
	6,250

Summary and Projects

Summary

The Distance Formula The distance formula is a major idea in the development of mathematical relationships. The distance between two points $P_1(x_1, y_1)$ and $P_2(x_2, y_2)$ in a plane is given as

$$d = \sqrt{(x_2 - x_1)^2 + (y_2 - y_1)^2}$$

The less important but useful midpoint formula is also developed and is given by

$$\left(\frac{x_1 + x_2}{2}, \frac{y_1 + y_2}{2} \right)$$

This is the formula for determining the coordinates of the point that is exactly half-way between the two points (x_1, y_1) and (x_2, y_2).

Relations and Functions A relation is a rule that associates one or more output values with each allowable input value. A function, on the other hand, has a bit more restriction placed on it. A function is a relation in which each input value is associated with exactly one output value. To know a function means knowing the rule of association, the domain, and the range. The domain of a relation or function is the set of allowable input values for the function. The range is the corresponding set of values produced by the input values using the given rule. For our purposes, this means obtaning real numbers as outputs for the allowable real numbers that are input. For example, consider the function

$$f(x) = \frac{1}{\sqrt{x}}$$

Note that the denominator should not become 0. This would cause you to restrict the input values so that $x \neq 0$. Further, realize that the square root could return complex numbers for the range. If you desire to have real numbers as output, you need to further restrict the input values so that $x > 0$. Graphs of functions and relations are extremely helpful in determining their domains and ranges.

It is important to know that a function can be described by tables, by ordered pairs, by symbolic rules, and by graphs. The vertical line test is useful in determining if the graph of a relation is actually the graph of a function. Imagine a vertical line moving from left to right across the graph. The graph will represent a function if all the vertical lines intersect the graph only once. This is just a visual restatement of the definition of a function.

Shifting, Reflecting, Expanding, and Contracting   If you start with a function $f(x)$, then adding (subtracting) a constant to the function or the argument of the function will shift its graph. Adding (subtracting) a constant to the function will shift the graph of $f(x)$ vertically. For example, $f(x) + k$ will shift the graph of $f(x)$ up k units if k is positive and down k units if k is negative. In addition, $f(x + k)$ will shift the graph of $f(x)$ left k units if k is positive and right k units if k is negative. The graph of $f(x)$ will be reflected about the x-axis by taking the opposite of the function as in $-f(x)$. Finally, the effect that $kf(x)$ has on the graph of $f(x)$ is to expand (narrow) the graph if $k > 1$ and contract (widen) the graph if $0 < k < 1$. Knowledge of these effects expands your depth of understanding regarding the behavior of a much larger number of functions and is also a powerful resource for analyzing and solving problems.

Combining Functions   Whenever you add, subtract, multiply, or divide two functions, the result is a function. In addition, the domain of the resulting function is the intersection of the domains of the two original functions. There is an additional restriction on the domain of the function resulting from the quotient of two functions. In this case, you need to further restrict the domain to exclude any input values that cause the denominator function to become 0. The composition of two functions is also covered and is an operation that produces a function. In this case, one function serves as the input quantity of another function. The domain of the composition of two functions is slightly more difficult. For example, if the composition function is $(f \circ g)(x)$, then its domain is constructed as follows: start with the domain of g, since these input values must be allowable for g. Next consider the range of g, since these will be the input values for f. If any of the resulting range values of g are not allowable input values for f, then the x values that produced those range values of g must also be excluded.

Variation   There are two major types of variation: direct variation given by $f(x) = kx$ and inverse variation given by $f(x) = \dfrac{k}{x}$. The other variations are fairly straightforward once you understand direct and inverse variation.

Constructing Functions   The graphing calculator gives you the resource needed to construct a fairly wide variety of functions from data you collect. The inclusion of this important topic is new to courses like this one and is a direct result of the technology available. This particular section focuses on producing linear functions from data that you or someone else has collected. The process of constructing a linear function from a set of data is called linear regression. The coefficient of correlation, r, is a measure of the strength of the relationship between two quantities and its value can be obtained using the statistics capabilities of your graphing calculator. The value of r is between -1

and 1. Its value can help you determine if the corresponding linear function constructed from the data will or will not act as a good predictor. The closer r is to 1 or -1, the stronger the relationship between the two quantities.

Projects

1. The following data were collected using a computer and a motion detector as a ball was dropped toward the motion detector. The data show the distance between the ball and the motion detector. Determine the height of the ball above the motion detector when the ball was dropped, and when the ball struck the motion detector. Write an explanation of your solution for a classmate.

Data Number	Time	Distance
1	0.2	5.33
2	0.3	4.54
3	0.4	3.46
4	0.5	2.01

2. In this project you will need to use your graphing calculator *User's Guide* as a reference on programming and programming commands. Modify the program given below to display the the quotient of each pair of (x, y)-coordinates of data points you have stored using the STAT Edit menu. Use this program to investigate the data in the following table. Describe

TI-81	**TI-82**
Prgm1:TABLECHK	PROGRAM:TABLECHK
1⟶J	For(J,1,dim L1)
Lbl 1	L2(J)+L1(J) ⟶Q
{ y} (J)+{ x} (J) ⟶Q	Disp Q
Disp Q	Pause
Pause	End
IS > (J,Dim{ x})	Stop
Goto 1	
Stop	

the pattern in the quotients that are displayed by the program. Write down a function that approximately describes this pattern. What kind of variation does this function exhibit?

Data Number	x	y
1	0.2	3.9092
2	0.3	4.1247
3	0.4	4.3402
4	0.5	4.5556
5	0.6	4.7711
6	0.7	4.9866

Modify the program again to display the product of each pair of (x, y)-coordinates of the data points. Use this modified program to investigate the data in the table that follows. Describe the pattern in the products that are displayed. Write down a function that approximately describes this pattern. What kind of variation does this function exhibit?

Data Number	x	y
1	0.2	11.36
2	0.3	7.87
3	0.4	5.84
4	0.5	4.62
5	0.6	3.86
6	0.7	3.29

CHAPTER 3

Polynomials

3.1 Polynomial Functions

Introduction

Polynomial Functions and Related Terminology

Odd and Even Functions

Dominating Terms

Shapes of Polynomial Graphs

Introduction

Along with the introduction to polynomials in both your elementary and inter-mediate algebra courses came the related terminology. Much of this section serves as a review of that terminology. We also examine some concepts that are not usually presented in elementary and intermediate algebra courses. These concepts relate to the shapes of polynomial functions and are the concepts of odd and even functions, dominating terms of polynomials functions, turning points of polynomial functions, and continuous polynomial functions.

Polynomial Functions and Related Terminology

Algebraic Expression

An **algebraic expression** is a number, a letter, or a finite collection of numbers and letters along with meaningful signs of the arithmetic operations: addition, subtraction, multiplication, and division. For example, $x + 8$, 7, and $\dfrac{x - 3}{2x + 9}$ are algebraic expressions. However, $\dfrac{5x + 8}{0}$ is not an algebraic expression since division by 0 is not a meaningful operation.

Terms and Coefficients

The product of a number and one or more variables is called a **term**. For example, $3x^5$ is a term with one variable, $-2x^2y^3$ is a term with two variables, and $5ab^2c^8$ is a term with three variables. The **coefficient** of a particular group of factors in a term is the remaining group of factors. For example, $3x^5$ and $6x^{2/3}$ are terms with one variable. The coefficient of x^5 is 3. In fact, 3 is referred to as the *numerical coefficient* of x^5. The coefficient of $x^{2/3}$ is 6. In the term $-2x^2y^3$, the coefficient of y^3 is $-2x^2$, and the coefficient of $-2x$ is xy^3.

Degree of a Term

The **degree of a term** with one variable is the value of the exponent on that variable. For example, the degree of the term $3x^5$ is 5, the value of the exponent on the variable. The degree of a term with more than one variable is the value of the *sum* of the exponents on the variables. For example, the degree of the term $5ab^2c^8$ is $1 + 2 + 8 = 11$ (the exponent of a is 1).

Monomial, Binomial, and Trinomial

A single term involving variables that have whole number exponents only is called a **monomial**. For example, the term $3x^5$ is a monomial but the term $6x^{2/3}$ is not, because the exponent is not a whole number. The sum of two monomials is called a **binomial**, and the sum of three monomials is called a **trinomial**. Generally, the sum of a finite number of monomials is called a **polynomial**.

Degree of a Polynomial

The degree of a polynomial is the degree of the term of highest degree. For example, the degree of the polynomial $4x^3 - 7x^2 + 8x + 1$ is 3, the degree of the first term, and the degree of the polynomial $8x^3y^2 - 3x^2y^4 - 4y$ is 6, the degree of the second term.

Form of a Polynomial

Symbolically, a polynomial of degree n in the variable x can be expressed in the form

$$a_n x^n + a_{n-1} x^{n-1} + a_{n-2} x^{n-2} + \cdots + a_2 x^2 + a_1 x + a_0,$$

Leading Term

where a_n, a_{n-1}, a_{n-2}, ..., a_2, a_1, and a_0 are constants, $a_n \neq 0$, and n is a natural number. Notice that the terms are written in descending order of degree. The term $a_n x^n$ is called the **leading term** of the polynomial and the numerical coefficient of $a_n x^n$ is called the **leading coefficient** of the polynomial. For example, $6x^4 - 8x^3 + 2x + 10$ is a fourth-degree polynomial in the variable x with leading term $6x^4$ and leading coefficient 6.

Polynomial Function

A **polynomial function** in the variable x is a function in which the defining expression is a polynomial in x. Thus, a polynomial function in x is a function of the form

$$P(x) = a_n x^n + a_{n-1} x^{n-1} + a_{n-2} x^{n-2} + \cdots + a_2 x^2 + a_1 x + a_0.$$

To completely describe the function, the domain and range must be presented. Since a polynomial expression is defined for all values of its variables, the domain of a polynomial function is all real numbers. The range of a polynomial function depends on the defining polynomial expression. For example, $P(x) = 6x^4 - 8x^3 + 2x + 10$ describes a polynomial function in x. The description is only partial since only two of the three descriptive parts are presented (the rule and domain are present, the domain being all real numbers, but the range is not).

Similarly, a polynomial function in the variables x and y is a function in which the defining expression is a polynomial in x and y. For example, $P(x, y) = 5x^2y + 2xy^2 - 7y - 2$ describes a polynomial function in the variables x and y. Again, the description is only partial. The function $P(x, y) = 3x^2 + 5x - 4$ is a polynomial function of the two variables x and y (as indicated by $P(x, y)$), but the output does not depend on the value of y (just as the output of $f(x) = 4$ does not depend on the value of x). The value of y has no effect on the output value. For example,

$$P(2, 4) = 3 \cdot 2^2 + 5 \cdot 2 - 4 = 18$$

The y value 4 is not needed in the computation. The shapes of the graphs of polynomial functions can be used to describe the function's behavior. We will focus on how to identify the shapes of graphs of polynomial functions of one variable from a knowledge of the degree of the defining polynomial expression.

Odd and Even Functions

Consider the leftmost function pictured in Figure 3.1. Notice that each pair of opposites, x and $-x$, produce the same output value $f(x)$. Symbolically, $f(-x) = f(x)$. Functions for which this is true are symmetric about the output axis. (A curve is symmetric about a line if on one side of the line, the curve is the mirror image of the curve on the other side of the line.) Functions that are symmetric about the y-axis are called *even* functions.

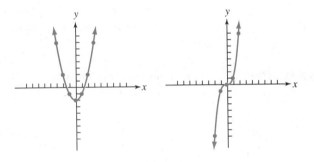

Figure 3.1

Now consider the rightmost function pictured in Figure 3.1. Notice that each pair of opposites, x and $-x$, produce opposite output values, $f(x)$ and $-f(x)$. Symbolically, $f(-x) = -f(x)$. Functions for which this is true are symmetric about the origin. Graphs symmetric about the origin are called *odd* functions.

Even and Odd Functions

Even and Odd Functions

A function $f(x)$ is called **even** if $f(-x) = f(x)$, and **odd** if $f(-x) = -f(x)$, for every x in the domain of $f(x)$.

The definition says that a function is *even* if opposite input values, x and $-x$, produce the same output values, $f(x)$. A function is *odd* if opposite input values, x and $-x$, produce opposite output values, $f(x)$ and $-f(x)$, respectively. We know now that opposite input values produce the same output value for even functions. Hence, we can state that those portions of the graph of an even function that correspond to input values that are extreme (values that are far out to the left or right) will lie on the same side of the input axis. Similarly, those portions of the graph of an odd function that correspond to input values that are extreme will lie on opposite sides of the input axis. The leftmost function and rightmost function in Figure 3.2 illustrate these facts for an even and odd function, respectively.

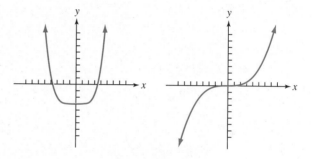

Figure 3.2

ILLUMINATOR SET A

1. The polynomial function $P(x) = x^4$ is an even function since, as the argument below will show, both x and its opposite, $-x$, produce the same output x^4, that is, $P(-x) = P(x) = x^4$.

$$P(-x) = (-x)^4$$

$$= (-x)(-x)(-x)(-x)$$

$$= x^4$$

$$= P(x)$$

Notice the even number of "$-$" signs in $P(-x)$. Do you see why it is reasonable to call this type of function an even function?

2. The polynomial function $P(x) = 3x^6 - 5x^2 + 8$ is an even function since, as the argument below will show, both x and its opposite, $-x$, produce the same output, namely, $3x^6 - 5x^2 + 8$.

$$P(-x) = 3(-x)^6 - 5(-x)^2 + 8$$

$$= 3x^6 - 5x^2 + 8$$

$$= P(x)$$

3. The polynomial function $P(x) = x^3$ is an odd function since, as the argument below will show, x and its opposite, $-x$, produce the opposite outputs, x^3 and $-x^3$, that is, $P(-x) = -P(x)$.

$$P(-x) = (-x)^3$$

$$= (-x)(-x)(-x)$$

$$= -x^3$$

$$= -P(x)$$

Notice the odd number of "$-$" signs in $P(-x)$. Do you see why it is reasonable to call this type of function an odd function?

4. The polynomial function $P(x) = -6x^5 + 2x^3 + x$ is an odd function since, as the argument below will show, x and its opposite, $-x$, produce the opposite outputs, x^3 and $-x^3$, that is, $P(-x) = -P(x)$.

$$P(-x) = -6(-x)^5 + 2(-x)^3 + (-x)$$

$$= 6x^5 - 2x^3 - x$$

$$= -\underbrace{(-6x^5 + 2x^3 + x)}_{P(x)}$$

$$= -P(x)$$

5. The polynomial function $P(x) = x^2 + 3x - 8$ is neither an odd nor an even function since, as the argument below will show, x and its opposite, $-x$, produce neither the same nor opposite outputs.

$$P(-x) = (-x)^2 + 3(-x) - 8$$

$$= x^2 - 3x - 8$$

But $P(x) = x^2 + 3x - 8 \neq P(-x)$ and $-P(x) = -x^2 + 3x + 8 \neq P(-x)$. Thus, $P(x)$ is neither an odd nor an even function.

Dominating Terms

Consider the two polynomial functions $P(x) = x^2$ and $Q(x) = x^2 + 25x + 35$. Notice that $Q(x)$ is a variation of $P(x)$. Specifically, $Q(x) = P(x) + 25x + 35$. We wish to compare the values of these polynomial functions for extreme values of x, that is, for very large and very small values of x. The following table shows the output values for several input values of these two functions.

Comparison of $P(x)$ and $Q(x)$ Values		
x	$P(x)$	$Q(x)$
50	2, 500	3, 785
500	250, 000	262, 535
500, 000	250, 000, 000, 000	250, 012, 500, 035
5, 000, 000	25, 000, 000, 000, 000	25, 000, 125, 000, 035
50, 000, 000	2, 500, 000, 000, 000, 000	2, 500, 001, 250, 000, 035
500, 000, 000	250, 000, 000, 000, 000, 000	250, 000, 012, 500, 000, 035
500, 000, 000, 000	250, 000, 000, 000, 000, 000, 000, 000	250, 000, 000, 012, 500, 000, 000, 035
−50	2, 500	1, 285
−500	250, 000	237, 535
−500, 000	250, 000, 000, 000	249, 987, 500, 035
−5, 000, 000	25, 000, 000, 000, 000	24, 999, 875, 000, 035
−50, 000, 000	2, 500, 000, 000, 000, 000	2, 499, 998, 750, 000, 035
−500, 000, 000	250, 000, 000, 000, 000, 000	249, 999, 987, 500, 000, 035
−500, 000, 000, 000	250, 000, 000, 000, 000, 000, 000, 000	249, 999, 999, 987, 500, 000, 000, 035

The table shows that for some smaller values of x, such as 500 or even $500, 000, 000$, $Q(x) = x^2 + 25x + 35$ differs in value from $P(x) = x^2$. As the values of x become larger, the differences between $Q(x)$ and $P(x)$ become smaller relative to the sizes of the function values. ($Q(x)$ and $P(x)$ will appear to have the same value on your graphing calculator as the values of x become extreme because of calculator roundoff.) Apparently, as the values of x become larger, the two terms, $25x$ and 35 of $Q(x)$, contribute relatively less and less to the value of $Q(x)$. This means that for extreme values of x, the term of primary importance in either polynomial is the leading term x^2. It is the leading term that, for extreme values of x, dominates the contribution to the output value.

Dominating Term

Dominating Term
The **dominating term** of a polynomial function is the leading term of the polynomial expression that defines the function.

This fact implies that the leading term of a polynomial function determines the behavior of the function at extreme input values. Also, both odd and even polynomial functions will behave as prescribed by their leading terms at extreme input values.

Shapes of Polynomial Graphs

Graphs of polynomial functions fall into two major categories: those that, for extreme values of the input variable, have branches that lie on the same side of the input axis, and those that, for extreme values of the input variable, have branches that lie on opposite sides of the input axis. Figure 3.3 shows these two categories.

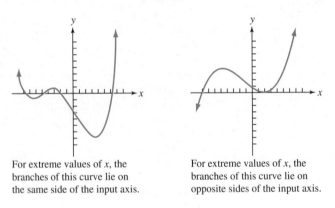

For extreme values of x, the branches of this curve lie on the same side of the input axis.

For extreme values of x, the branches of this curve lie on opposite sides of the input axis.

Figure 3.3

The category that a particular function falls into can be determined by the dominating term of the defining polynomial expression. If the dominating term is of even degree and has a positive coefficient, then extreme values of the input variable will generate positive output values and a curve with portions above the horizontal axis. Figure 3.4 shows the graph of the polynomial function $P(x) = 0.0002x^6 + x^3 + x^2 - 2x - 5$. Notice that when x is extreme

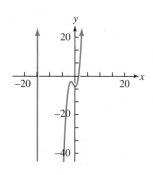

Figure 3.4

Dominating Terms and Category of a Graph

enough (larger than 0 and smaller than about -15), both branches lie above the x-axis. The positive extreme values of the input variable will generate positive output values for a polynomial function if the dominating term is of odd degree and has a positive coefficient. The portions of the curve for these input values will lie above the horizontal axis. Likewise, for such functions, the negative values of the input variable will generate negative output values and a curve that lies below the horizontal axis. Figure 3.5 shows a graph of the polynomial function $P(x) = 0.0002x^3 + 0.05x^2 - 0.009x - 3$. Notice that when x is extreme enough (larger than 10 and smaller than about -240), the branches lie on opposite sides of the x-axis.

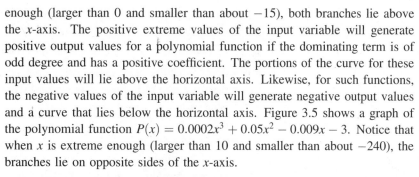

Dominating Terms and Category of a Graph

1. If the dominating term of a polynomial function is of even degree, the graph of the function will have branches that, for extreme values of the input variable, lie on the same side of the input axis.

2. If the dominating term of a polynomial function is of odd degree, the graph of the function will have branches that, for extreme values of the input variable, lie on opposite sides of the input axis.

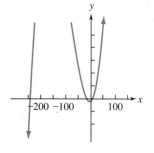

Figure 3.5

At nonextreme values of the input variable, polynomial functions can have turning points, that is, points at which the function changes from increasing to decreasing or from decreasing to increasing. For example, Figure 3.4 shows a polynomial function with three turning points, two that occur near the origin and one apparent turning point that occurs at an x value between -5 and -15. Figure 3.5 shows a polynomial function with two turning points, one at the origin and one apparently between -100 and -200. Once the input values become extreme enough, however, there will be no more turning points.

We now state several important facts about the graphs of polynomial functions that are commonly proven in calculus.

1. A polynomial function will, at extreme values of the input variable, behave as its leading term.

2. The output values of a polynomial function will, for extreme input values, be extreme values themselves. Graphically, this means that, as the input values move far enough out on their axis and then continue to move even farther out, the curve will move away (either upward or downward) from

the input axis. The curve will never turn around and move back toward the input axis. Figure 3.3 illustrates this fact.

3. A polynomial function of degree n will have at most $n - 1$ *turning points*.

4. Polynomial functions are *continuous*. That is, graphs of polynomial functions have no holes, breaks, or gaps. Figure 3.6 shows the graph of a function with a hole at $x = a$, a break at $x = b$, and a gap between $x = c$ and $x = d$. This graph is not continuous and does not describe a polynomial function.

5. Polynomial functions are *smooth*. That is, graphs of polynomial functions have no corners or cusps. Figure 3.6 shows the graph of a function with a cusp at $x = e$ and a corner at $x = f$.

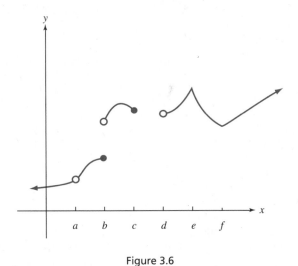

Figure 3.6

It is worth noting that just because a function is continuous and smooth does not mean it has to be a polynomial function. A continuous and smooth graph indicates only that the function could be a polynomial function. The graphs in Figure 3.7 are not graphs of polynomial functions, but rather an exponential function, a logarithmic function, and a trigonometric function, respectively. They are all smooth graphs that we study in detail later in the text.

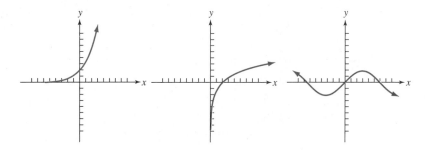

Figure 3.7

One last fact about polynomial functions is that they follow all the rules of translations, reflection, expansion, and contraction that we discussed in Section 2.4. Window 3.1 shows the polynomial function $P(x) = 0.002x^3 + 1.03x$ and its horizontal translation $P(x-6) = 0.002(x-6)^3 + 1.03(x-6)$ on the graphing calculator viewing window $[-25, 25]$ for x and $[-50, 50]$ for y. (You should enter $P(x)$ for Y1 and $P(x-6)$ for Y2 and construct these graphs to see that $P(x-6)$ actually lies 6 units to the right of $P(x)$. Remember that Y1 is graphed first and Y2 is graphed second.)

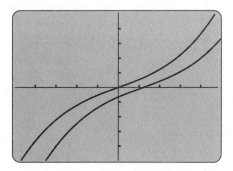

Window 3.1

EXAMPLE SET A

The population of a town in northwest Arkansas has been growing for the last fifteen years because several new industries have located there. Town managers have hired a company to look at the situation and develop a mathematical model that can be used to predict the future population of the town. The town is given the polynomial function $P(t) = 0.16t^2 + 8.05$ as the model where $P(t)$ is the population (in thousands of people) and t is the time since 1975.

1. What was the population of the town in 1975?

2. What was the population of the town in 1982?

3. With an error of at most 0.01, how many years from 1975 was the population of the town 14,500?

4. Describe this function completely by writing the defining rule, the domain, and the range.

5. Do you think this function provides a reasonable description of the town's population trend for a 35 year period?

Solution:

1. In 1975, $t = 0$ so that $P(0) = 0.16(0)^2 + 8.05 = 8.05$. That is, in 1975, the population of the town was 8.05 thousand, or 8,050.

2. In 1982, $t = 1982 - 1975 = 7$ so that $P(7) = 0.16(7)^2 + 8.05 = 15.89$. That is, in 1982, the population of the town was 15.89 thousand, or 15,890.

3. Entering $0.16x^2 + 8.05$ for Y1 and 14.5 for Y2 on the graphing calculator and using the methods presented in Section 1.3 (The Calculator and Approximation), we conclude, with an error of at most 0.01 years, that 6.34 years from 1975 the population of the town was 14,500.

4. Since the given rule runs over a 15-year period of time, from 1975 to 1990, the domain of the function is $0 \leq t \leq 15$. The graph obtained in part (3) shows an increasing curve and no turning points between $t = 0$ and $t = 15$. Therefore we know that the minimum output value is $P(0) = 8.05$ and the maximum output value is $P(15) = 44.05$. Thus, this population function is described completely as

$$P(t) = 0.16t^2 + 8.05, \quad 0 \leq t \leq 15, \quad 8.05 \leq P(t) \leq 44.05$$

5. Since $P(35) = 204.05$, it is unlikely that this function still provides a reasonable approximation to the town's population. A population of 204,050 people is more like some large cities. So we would conclude that the function would not provide a reasonable description of the town's population trend for a 35-year period.

Writing Activity   Study the following problem and write down the solution using complete sentences. The total cost $C(x)$ (in dollars) of producing x hundred pounds of crystalline cleaning compound is approximated by the function $C(x) = 2.08x^4 - 26.82x^3 + 10.95x^2 + 20.06x$. What amount (other than none), to the nearest hundredth of a pound, of cleaning compound should be produced to minimize the cost per pound? (Hint: The function $C(x)$ is the *total* cost function. You do not want to minimize this function, but rather a slightly different version of it. If $C(x)$ represents the total cost of producing x

hundred pounds, what function represents the cost per pound? Remember, x represents the number of hundreds of pounds.)

EXERCISES

For exercises 1–14, determine if the function could or could not be a polynomial function. If the function could not be a polynomial function, state *not a polynomial function*. If it could be, state *possible polynomial function* and then state if its degree would be odd or even and if its leading coefficient is positive or negative.

1.

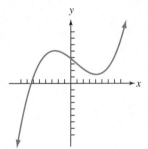

2.

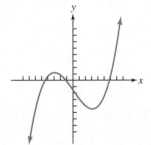

3.

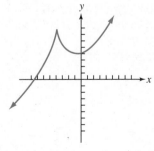

4.

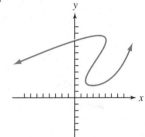

5.

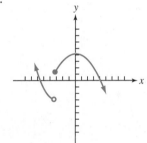

6.

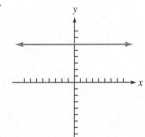

7.

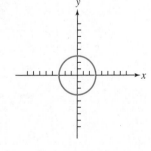

8.

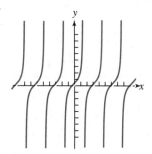

9.

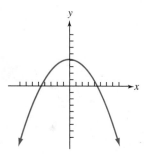

10.

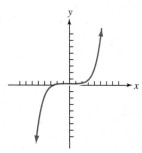

11.

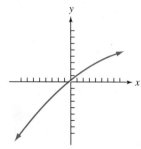

12.

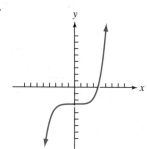

13.

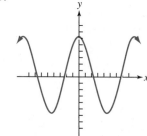

14.

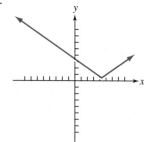

For exercises 15–21, determine if the function is or is not a polynomial function. If the function is not a polynomial function, state *not a polynomial function*. If it is, state *polynomial function*. Then state the maximum number of turning points it can have.

15. $f(x) = 3x^2 - 6x + 5$

16. $f(x) = x - 2$

17. $f(x) = 5x^3 - 2x^{-2} + x + 4$

18. $f(x) = -3x^2 - \dfrac{5}{2}x$

19. $f(x) = \sqrt{7}x^4 + 2x^3 - 6x^2 + 2$

20. $f(x) = 16 + 4x - 3x^2$

21. $f(x) = \dfrac{3x + 1}{2x + 5}$

For exercises 22–32, state if the given polynomial is odd, even, or neither.

22.

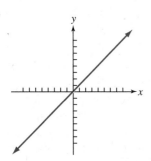

23.

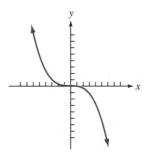

24.

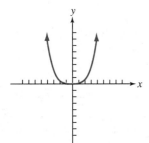

25.

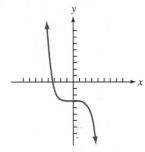

26.

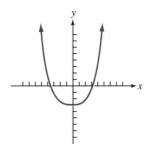

27.

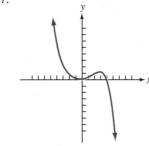

28. $f(x) = x^5 - x^3 + 5x$

29. $f(x) = x^5 - x^4 + x^3$

30. $f(x) = 8x^4 - 2x^2 + 6$

31. $f(x) = x^3 + 5$

32. $f(x) = x^2 - x$

For exercises 33–38, use the fact that polynomial functions of degree n have at most $n - 1$ turning points to determine the minimum and maximum number of real zeros the given polynomial function can have. Describe, using complete sentences and a sketch, how you arrived at your answer. (See Section 1.4 if you feel you need a review of what a zero of a function is.)

33. $f(x)$ is a third degree polynomial function.

34. $f(x)$ is a fifth degree polynomial function.

35. $f(x)$ is a second degree polynomial function.

36. $f(x)$ is a fourth degree polynomial function.

37. $f(x) = ax^6 + bx^4 + cx^3 + dx + e$

38. $f(x) = ax^7 + bx^6 + cx^5 + dx^4 + ex^3 + fx^2 + gx + h$

For exercises 39–48, use your graphing calculator, if necessary, to approximate, with an error of at most 0.01, the zeros, should any exist, of each polynomial function.

39. $f(x) = (x + 4)(x - 6)$

40. $f(x) = (x - 5)(x - 2)$

41. $f(x) = (4x + 1)(2x - 7)$

42. $f(x) = (3x - 2)(3x - 5)(2x - 1)$

43. $f(x) = 0.3x^2 - x - 12.5$

44. $f(x) = x^2 + 0.6x + 0.12$

45. $f(x) = x^2 - 8.2x + 16.81$

46. $f(x) = x^2 + 11.2x + 31.36$

47. $f(x) = x^2 + 5x + 3$

48. $f(x) = x^2 - 5x - 20$

For exercises 49–52, write down in paragraph form the approach(es) you might consider to solve the given problem. Then carry out the details of your solution. Use your graphing calculator to answer each question.

49. A propane tank is to be constructed in the shape of a right circular cylinder with hemispheres at each end. The figure shows this tank.

The total length of the tank is to be 15 feet. (a) Express the volume of the tank as a function of the radius of a hemisphere only, (b) Find the volume of the tank if the radius of a hemisphere is 2.6 feet, and (c) Find, with an error of at most 0.01 feet, the radius of the tank if the volume of the tank is to be 261.67 cubic feet.

50. A stockbroker studies the behavior of a stock over a period of time and feels that the cubic function $P(n) = 0.03n^3 - 0.64n^2 + 1.97n + 35.80$ approximates the price per share $P(n)$ as a function of the nth day from now over a period of 30 days. To the nearest tenth of a day, determine (a) when (in days from now), according to this stockbroker's prediction function, the price of a stock will be $26.50 per share, (b) When a person should buy the stock, and (c) When a person should sell the stock.

51. A drug manufacturer believes that the effectiveness $E(t)$ (on a scale of 0 to 1) of a new drug t hours from the time it is administered is approximated by the function $E(t) = -0.03t^3 + 0.1t^2 + 0.29t$, $0 \le t \le 6$. To the nearest tenth of an hour, determine (a) when (in hours after administration) the drug will be at its maximum effectiveness, and (b) When the drug will no longer have any perceivable effect.

52. When an object is thrown (nearly straight upward) from a building 206 feet high, the height $h(t)$ of the object from the ground t seconds from when it is thrown is approximated by the function $h(t) = -16t^2 + 46.8t + 206$. To the nearest hundredth of a second: (a) When does the object reach its maximum height above the ground? (b) When does the object hit the ground?

53. A designer creates a pattern by repeating the same motif over and over again (see the accompanying figure). Write down the function representing the piecewise curve. (Hint: The first segment has the equation $f(x) = 1 - (x - 1)^2$.)

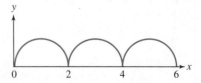

54. Sadia can rent a car at $25 a day and $0.30 per mile or at $30 a day and $0.20 a mile.

Write the cost of renting a car and driving for x miles. Of course Sadia wants to get the best deal possible and she needs to choose which car is the better option for her. Graph the equations. At how many miles traveled will the cars cost the same? Which car should she choose if she is driving less than that amount or more than that amount?

55. If two polynomials $f(x)$ and $g(x)$ have the same degree and exactly the same zeros, can you conclude that the two polynomials are equal [i.e. $f(x) = g(x)$]? Give reasons for your answer.

56. The bottom of a swimming pool gently slopes away from the shallow end to the deep end as shown in the diagram. If x denotes the horizontal distance from the shallow end to the deep end and y represents the depth of the bottom of the pool at any point, write down equations for y as a function of x. ← piecewise

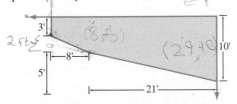

57. The polynomial function $P(x) = 3000 + 2.25x^2 - 0.34x^3$ gives the approximate count of an animal population during a ten-year period, where x represents the number of years from the initial count. The first term in the polynomial is the initial population. A lack of predators and sufficient food supplies give rise to an increase of population (represented by the second term). However, too rapid a rise in population creates food scarcity and decreases the population (the third term). What is the animal population at the end of the ten years? When during

this period was the population at a maximum? Graph the function to find your answer.

58. Achamma has an 8' x 4' balcony which is not draining rainwater properly. To solve this problem she decides to slope the surface of the balcony by pouring cement (see the accompanying figure). The maximum height of the cement is 2″ at the house side tapering to 0″ on the opposite side. Find, in cubic feet, the amount of cement required.

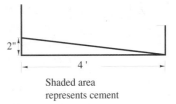

Shaded area
represents cement

If cement bags come in 3/4-cubic-feet boxes, how many will be needed? If the maximum height is x inches instead of 2″, what is the volume? How many bags of cement will be needed?

59. The Ly family is adding a new room to their house. They find a special window trim that is 26 feet long and they want to use all of the trim around the perimeter of a rectangular window. Find the dimensions of a rectangular window that will provide the most light (that is, window with the largest area).

60. (Refer to the previous exercise.) The Ly's are also considering installing a window shaped as a semicircle on top of a rectangle (a norman window). The same 26-foot-long trim is to be used around the perimeter of the window. Find the dimensions of the window (radius of the semicircle and the length and the width of the rectangle) that will maximize the light. Which of these two windows will provide more light? Solve by graphing.

3.2 Complex Numbers

Introduction

Complex Numbers

Operations with Complex Numbers

Introduction

In the next three sections, we will be studying polynomials and their zeros. We will find both polynomial division and factoring useful in this study. Since not all zeros of polynomials are real numbers, it is wise to begin our study with a discussion of complex numbers. Since the traditional calculus does not operate on complex numbers, we will present only a brief review of complex numbers and the operations on them. Complex numbers may arise in certain operations within a calculus operation, but they will not be found at the beginning or the end of those calculus operations.

Complex Numbers

Let's begin the study of complex numbers by using the graphing calculator to graph the functions $f(x) = x^2 + 10x + 21$ using Y1 and $g(x) = x^2 + 2$ using Y2 in the standard viewing window, -10 to 10 for x and -10 to 10 for y. As you have the calculator construct the graphs, notice that the first function to be graphed, $f(x) = x^2 + 10x + 21$, intercepts the x-axis at two points, once at $x = -7$ and once at $x = -3$, and therefore has two real number zeros, $x = -7$ and $x = -3$. (Remember, zeros of a function are input values that produce zero as the output value, and they occur where the curve intercepts the input axis. See Section 1.4 for a review of these ideas.) The second function to be graphed, $g(x) = x^2 + 2$, does not intercept the x-axis at all, and therefore has no real number zeros in $-10 \leq x \leq 10$. Since each function is quadratic, neither have any more turn-around points and will, therefore, not intercept the x-axis at any other points. This means that the function $g(x) = x^2 + 2$ has no real number zeros. The graphs of $f(x) = x^2 + 10x + 21$ and $g(x) = x^2 + 2$ appear in Window 3.2.

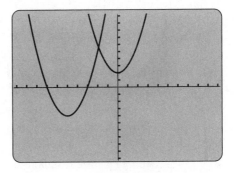

Window 3.2

However, notice that if we input $\sqrt{-2}$ into $g(x) = x^2 + 2$, and use the fact that $(\sqrt{x})^2 = x$, we get,

$$(\sqrt{-2})^2 + 2 = -2 + 2$$

$$= 0$$

Since $g(\sqrt{-2}) = 0$, $\sqrt{-2}$ must be a zero of the function $g(x)$. Numbers such as $\sqrt{-2}$ are not real numbers, but are *imaginary numbers*.

We now introduce the set of numbers that includes both real and pure imaginary numbers.

Complex Numbers

Complex Numbers

A **complex number** is a number of the form $a + bi$, where a and b are real numbers and i is the imaginary number $\sqrt{-1}$. The real number a is called the **real part** of the complex number, and the real number b is called the **imaginary part** of the complex number. The set of all such numbers is called the set of complex numbers and is denoted C.

Complex numbers are of the form
$a + bi$

Notice that the imaginary part of the complex number $a + bi$ is b and not bi. This is simply by definition. The proper format is

$$\underbrace{a}_{real\ part} + \underbrace{b}_{imaginary\ part} i$$

Notice also that if $b = 0$, $a + bi$ simply becomes the real number a. This means that every real number can be expressed in the form $a + bi$ as $a + 0i$.

Therefore, every real number is also a complex number. Also, if $a = 0$ and $b \neq 0$, then $a + bi$ simply becomes the pure imaginary number bi.

If a is a positive number, then $-a$ is a negative number and $\sqrt{-a}$ can be expressed as

$$\sqrt{-a} = \sqrt{-1 \cdot a}$$

$$= \sqrt{-1} \cdot \sqrt{a}$$

Imaginary Unit

The symbol i is used to denote $\sqrt{-1}$ and it is called the **imaginary unit**. Thus, $\sqrt{-a}$ can be expressed as $i\sqrt{a}$. Also, since $i = \sqrt{-1}$, $i^2 = -1$.

Figure 3.8 shows the relationship between the sets of natural numbers N, the whole numbers W, the integers Z, the rational numbers Q, the irrational numbers Ir, the real numbers R, and the complex numbers C.

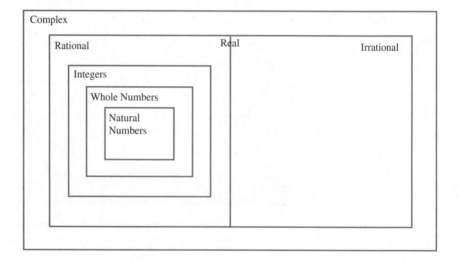

Figure 3.8

Operations with Complex Numbers

Now that we have introduced complex numbers, it would be useful to have methods for performing arithmetic on them. However, first we define equality for two complex numbers.

Equality of Complex Numbers

Equality of Complex Numbers

Two complex numbers $a + bi$ and $c + di$ are equal if and only if the real parts are equal and the imaginary parts are equal. Symbolically,

$$a + bi = c + di \quad \text{if and only if} \quad a = c \quad \text{and} \quad b = d$$

For example, the complex numbers $-8 + 6i$ and $a + 6i$ will be equal only when $a = -8$. The complex numbers $3 - 10i$ and $3 + 7i$ are not equal since their imaginary parts, -10 and 7, are not equal.

Addition of Complex Numbers

Addition of Complex Numbers

Addition of complex numbers is accomplished by adding the real parts together and adding the imaginary parts together. Symbolically,

$$(a + bi) + (c + di) = (a + c) + (b + d)i$$

We perform subtraction of complex numbers in a similar way.

EXAMPLE SET A

1. Find the sum of $2 + 3i$ and $5 + 9i$.

   Solution:
   $$(2 + 3i) + (5 + 9i) = (2 + 5) + (3 + 9)i$$
   $$= 7 + 12i$$

2. Find the difference of $6 - 14i$ and $4 + 3i$.

   Solution:
   $$(6 - 14i) - (4 + 3i) = (6 - 4) + (-14 - 3)i$$
   $$= 2 - 17i$$

Multiplication of Complex Numbers

> ### Multiplication of Complex Numbers
>
> Multiplication of complex numbers is accomplished by treating each as a binomial and multiplying them using the method you studied in elementary and intermediate algebra, that is, multiplying each term of one number by each term of the other. The multiplication is completed by converting i^2 to -1.
>
> $$(a + bi)(c + di) = (ac - bd) + (ad + bc)i$$

You can derive this result by performing the multiplication.

$$(a + bi)(c + di) = a(c + di) + bi(c + di)$$
$$= ac + adi + bic + bidi$$
$$= ac + adi + bci + bdi^2$$
$$= ac + adi + bci + bd(-1)$$

which gives

$$(a + bi)(c + di) = (ac - bd) + (ad + bc)i$$

EXAMPLE SET B

Find the product of $2 + 3i$ and $5 + 9i$.

Solution:

$$(2 + 3i) \cdot (5 + 9i) = 2 \cdot 5 + 2 \cdot 9i + 3i \cdot 5 + 3i \cdot 9i$$
$$= 10 + 18i + 15i + 27i^2$$
$$= 10 + 33i + 27(-1)$$
$$= 10 + 33i - 27$$
$$= -17 + 33i$$

Division of complex numbers requires the use of multiplication of complex numbers as well as the concept of complex conjugates.

Complex Conjugates

Complex Conjugates
The complex numbers $a + bi$ and $a - bi$ are **complex conjugates**.

Before we introduce an important property of complex conjugates, we introduce a common notation used for complex numbers. It is common to represent a complex number $a + bi$ with the letter z, and its conjugate $a - bi$ with $\bar{z}$, so that $z = a + bi$ and $\bar{z} = a - bi$.

Complex conjugates, z and $\bar{z}$, have the property that their product is a real number. In fact, the product is the sum of the squares of the real and imaginary parts. We illustrate this as follows:

$$z \cdot \bar{z} = (a + bi) \cdot (a - bi)$$

$$= a^2 - abi + abi - b^2 i^2$$

$$= a^2 - b^2(-1)$$

$$= a^2 + b^2$$

Thus, $z \cdot \bar{z} = a^2 + b^2$. Note that $a^2 + b^2$ is the sum of the squares of the real and imaginary parts of $a + bi$.

EXAMPLE SET C

Find the product of $2 + 3i$ and its conjugate $2 - 3i$.

Solution:

$$(2 + 3i)(2 - 3i) = 2^2 - 2 \cdot 3i + 3i \cdot 2 - 3^2 i^2$$

$$= 4 - 6i + 6i - 9(-1)$$

$$= 4 + 9$$

$$= 13$$

Note that 13 is the sum of the squares of 2 and 3.

Division of Complex Numbers

Division of Complex Numbers

Division of complex numbers is accomplished by multiplying the complex expression by 1 in the form $\dfrac{\text{conjugate of denominator}}{\text{conjugate of denominator}}$.

We illustrate the general process of division of complex numbers:

$$\left.\frac{a+bi}{c+di} = \frac{a+bi}{c+di} \cdot \frac{c-di}{c-di}\right\} \quad \longrightarrow \quad \frac{\text{conjugate of } \ c+di}{\text{conjugate of } \ c+di}$$

$$= \frac{(ac+bd) + (bc-ad)i}{c^2+d^2}$$

$$= \frac{ac+bd}{c^2+d^2} + \frac{bc-ad}{c^2+d^2}i$$

EXAMPLE SET D

Divide $5 + 9i$ by $2 + 3i$.

Solution:

$$\frac{5+9i}{2+3i} = \frac{5+9i}{2+3i} \cdot \frac{2-3i}{2-3i}$$

$$= \frac{(5+9i)(2-3i)}{13}$$

$$= \frac{10 - 15i + 18i - 27i^2}{13}$$

$$= \frac{10 + 3i - 27(-1)}{13}$$

$$= \frac{37 + 3i}{13}$$

$$= \frac{37}{13} + \frac{3}{13}i$$

The result is given in standard $a + bi$ format.

We end this section by demonstrating a mathematical proof.

EXAMPLE SET E

Prove that for any two complex numbers z_1 and z_2, $\overline{\overline{z_1} + \overline{z_2}} = z_1 + z_2$. That is, prove that the conjugate of the sum of the conjugates of two complex numbers is the same as the sum of the two original complex numbers.

Solution:

Assume that z_1 and z_2 are complex numbers. Show that $\overline{\overline{z_1} + \overline{z_2}} = z_1 + z_2$.

$$\overline{\overline{z_1} + \overline{z_2}} = \overline{\overline{a + bi} + \overline{c + di}}$$

$$= \overline{(a - bi) + (c - di)} \quad \text{by the definition of conjugate}$$

$$= \overline{(a + c) + (-b - d)i} \quad \text{by the definition of complex addition}$$

$$= (a + c) - (-b - d)i \quad \text{by the definition of conjugate}$$

$$= (a - (-b)i) + (c - (-d)i) \quad \text{by the definition of complex addition}$$

$$= (a + bi) + (c + di)$$

$$= z_1 + z_2$$

EXERCISES

For exercises 1–26, find the indicated sum, difference, product, or quotient as indicated. Express the result in the form $a + bi$.

1. $(2 - 3i) + (-1 + 7i)$

2. $(-1 - i) + (-4 - 4i)$

3. $(12 + 8i) - (4 + 2i)$

4. $(-5 - 9i) - (-2 - 6i)$

5. $-3 - (-5 + 2i)$

6. $8 + (2 - 9i)$

7. $-3i + (5 - 3i)$

8. $4i - (12 + 2i)$

9. $(1 - 8i)(3 + 2i)$

10. $(-3 - 5i)(-2 + 7i)$

11. $(4 - 7i)^2$

12. $(3 - 8i)^2$

13. $(5 + 2i)(5 - 2i)$

14. $(6 + 7i)(6 - 7i)$

15. $(2 + \sqrt{-3})(5 - \sqrt{-3})$

16. $(1 + \sqrt{-8})(6 - \sqrt{-8})$

17. $\dfrac{5}{5 + 2i}$

18. $\dfrac{4}{3 - 5i}$

19. $\dfrac{2 + 4i}{1 - 6i}$

20. $\dfrac{1 - 5i}{3 - 9i}$

21. $\dfrac{-6+i}{2-i}$

22. $\dfrac{4i}{6i}$

23. $(3+4i)^{-1}$

24. $(1-3i)^{-1}$

25. i^{i^2}

26. i^{-i^2}

For exercises 27–32, let z represent the complex number $a+bi$, and $\bar{z}$ its conjugate, $a-bi$.

27. Prove that the conjugate of the sum of two complex numbers is the sum of the conjugates of the numbers. That is, prove that if z_1 and z_2 are complex numbers, then

$$\overline{z_1+z_2} = \overline{z_1} + \overline{z_2}$$

28. Prove that for any two complex numbers z_1 and z_2, and any two real constants p and q,

$$\overline{pz_1+qz_2} = p\overline{z_1} + q\overline{z_2}$$

29. Prove that the conjugate of the product of two complex numbers is the product of the conjugates of the numbers. That is, prove that if z_1 and z_2 are complex numbers, then

$$\overline{z_1 \cdot z_2} = \overline{z_1} \cdot \overline{z_2}$$

30. Prove that the difference of a complex number and its conjugate is a pure imaginary number, that is, a number of the form bi.

31. Prove that the sum of a complex number and its conjugate is a real number.

32. Prove that if x and y both represent negative numbers, $\sqrt{x}\sqrt{y} = -\sqrt{xy}$.

33. Find the values of i, i^2, i^3, i^4, i^5, i^6, i^7, i^8, where $i = \sqrt{-1}$. Do you see a pattern? Define a function $f(n) = i^n$ where n is an integer. From the pattern you observed, simplify the value of $f(n)$.

In Section 3.5, you will see that a polynomial function of degree n has exactly n zeros, provided that some may occur more than once. Using this fact and what you know about the relationship between the degree of a polynomial and the shape of its graph, use your graphing calculator to construct the graph of each of the following functions. State how many real number zeros each has and how many nonreal complex zeros each has. Once you complete exercises 34–42, see if you can make a conjecture regarding the number of nonreal complex zeros a polynomial function has, if it has any at all.

34. $f(x) = x^2 + 0.88x + 0.704$

35. $f(x) = 0.56x^2 - 2.88x + 3.72$

36. $f(x) = 0.2x^3 + 1.5x^2 + 1.3x + 0.6$

37. $f(x) = 0.189x^4 - 2.244x^3 + 5.114x^2 + 8.676x + 6.945$

38. $f(x) = 2.24x^4 + 7.552x^3 - 29.8336x^2 - 17.6064x + 60.8552$

39. $f(x) = x^6 + 2x^5 - 18x^4 - 20x^3 + 89x^2 + 18x + 1$

40. $f(x) = x^7 + 3x^6 - 53x^5 + 111x^4 + 844x^3 + 900x^2 - 3600x + 2605$

41. $f(x) = x^4 + x^3 - 7x^2 - x + 19.2$

42. $f(x) = x^4 - 7x^3 + 8.75x^2 + 18.25x + 6.50$

3.3   The Division Algorithm

Introduction

A Review of Polynomial Division

The Division Algorithm

Synthetic Division

Introduction

Polynomial division and factoring are useful tools in the study of polynomial functions. Both are used in determining zeros and factors of polynomial functions. Although it provides a way to find zeros and factors, it is only convenient to use when the coefficients of the defining expression are integers. When the coefficients are noninteger rational numbers, the computations involved in polynomial division become cumbersome, even on a calculator, and the educated guessing involved in factoring becomes very hard, if not impossible. You should already be familiar with, and have a working knowledge of, the topics in this section and this discussion is intended only as a review.

A Review of Polynomial Division

The process of dividing polynomials is the same as the process used with whole numbers: divide, multiply, subtract, and bring down, divide, multiply, subtract, and bring down, ... The division, multiplication, and subtraction take place one term at a time. The division is done with only the *leading term* of the divisor polynomial. The process is concluded when the polynomial remainder has degree less than the polynomial divisor.

EXAMPLE SET A

Divide $3x^3 + 2x^2 - x + 4$ by the linear binomial $x - 3$.

Solution:
Notice first that 3 is not a zero of $P(x) = 3x^3 + 2x^2 - x + 4$ since

$$P(3) = 3(3)^3 + 2(3)^2 - 3 + 4$$

$$= 81 + 18 - 3 + 4$$

$$= 100$$

$$\neq 0$$

$$3x^2 + 11x + 32 \quad \longrightarrow \text{quotient}$$

$$\text{divisor} \longrightarrow x - 3 \, \overline{)\, 3x^3 + 2x^2 - x + 4} \quad \longrightarrow \text{dividend}$$

$$\underline{3x^3 - 9x^2}$$

$$11x^2 - x$$

$$\underline{11x^2 - 33x}$$

$$32x + 4$$

$$\underline{32x - 96}$$

$$100 \longrightarrow \text{remainder}$$

Thus, $\dfrac{3x^3 + 2x^2 - x + 4}{x - 3} = 3x^2 + 11x + 32 + \dfrac{100}{x - 3}$

The Division Algorithm

When the polynomial $2x^4 - 7x^3 + 10x + 4$ is divided by the linear polynomial $x - 3$, the result is $2x^3 - x^2 - 3x + 1 + \dfrac{7}{x - 3}$, that is,

$$\frac{2x^4 - 7x^3 + 10x + 4}{x - 3} = 2x^3 - x^2 - 3x + 1 + \frac{7}{x - 3}$$

If we multiply each side by $x - 3$, we obtain

$$(x - 3) \cdot \frac{2x^4 - 7x^3 + 10x + 4}{x - 3} = (x - 3) \cdot \left[(2x^3 - x^2 - 3x + 1) + \frac{7}{x - 3} \right]$$

$$2x^4 - 7x^3 + 10x + 4 = (x - 3)(2x^3 - x^2 - 3x + 1) + 7$$

To help us visualize more easily, we make the following representations:

$r = 3$,   the constant being subtracted from x.

$P(x) = 2x^4 - 7x^3 + 10x + 4$,   the polynomial to be divided.

$Q(x) = 2x^3 - x^2 - 3x + 1$,   the quotient without the remainder.

$R = 7$,   the remainder.

Since the divisor is of degree one, the remainder will always be a constant so we use R rather than $R(x)$.

With these representations, the equation

$$2x^2 - 7x^3 + 10x + 4 = (x - 3)(2x^3 - x^2 - 3x + 1) + 7$$

becomes

$$P(x) = (x - r)Q(x) + R$$

Algorithms

This example suggests a standardized procedure for division problems. Such standardized procedures are called **algorithms**, and we now introduce the *division algorithm* (which we state without proof).

Division Algorithm

Division Algorithm
If a polynomial $P(x)$ of degree one or more is divided by the linear polynomial $x - r$, there exists a unique polynomial quotient $Q(x)$, of degree one less than that of $P(x)$, and a constant remainder R such that for all values of x, $P(x) = (x - r)Q(x) + R$.

EXAMPLE SET B

If $2x^4 - 7x^3 + 10x + 4$ is divided by $x - 3$, then, by the division algorithm,

$$2x^4 - 7x^3 + 10x + 4 = (x - 3)(2x^3 - x^2 - 3x + 1) + 7$$

In this case, $2x^4 - 7x^3 + 10x + 4$ corresponds to $P(x)$, the polynomial to be divided, $x - 3$ corresponds to $x - r$, the linear polynomial doing the dividing, $2x^3 - x^2 - 3x + 1$ corresponds to $Q(x)$, the quotient, and 7 corresponds to R, the remainder.

Synthetic Division

The method that is commonly used to perform the division of a polynomial by $x - r$ is quick and accurate. Called **synthetic division**, it is an abbreviated form of the usual long division, and it is most easily understood through an example.

Divide $3x^3 + 2x^2 - x + 4$ by the linear binomial $x - 3$. From our work in Example Set A, we have

$$
\begin{array}{r}
3x^2 + 11x + 32 \qquad \longrightarrow \text{quotient} \\
\text{divisor} \longrightarrow x - 3 \,\overline{\big)\, 3x^3 + 2x^2 - x + 4} \quad \longrightarrow \text{dividend} \\
\underline{3x^3 - 9x^2} \qquad\qquad\quad \\
11x^2 - x \qquad\qquad \\
\underline{11x^2 - 33x} \qquad\quad \\
32x + 4 \qquad \\
\underline{32x - 96} \quad \\
100 \longrightarrow \text{remainder}
\end{array}
$$

If we eliminate the x's from the above work, the problem has a simpler appearance.

$$
\begin{array}{l}
\phantom{\text{divisor} \longrightarrow 1-3\,|\,}3 + 11 + 32 \qquad \longrightarrow \text{quotient} \\
\text{divisor} \longrightarrow 1 - 3 \,\big|\, \overline{3 + 2 - 1 + 4} \qquad \longrightarrow \text{dividend} \\
\phantom{\text{divisor} \longrightarrow 1-3\,|\,}\underline{3 - 9} \\
\phantom{\text{divisor} \longrightarrow 1-3\,|\,33}11 - 1 \\
\phantom{\text{divisor} \longrightarrow 1-3\,|\,33}\underline{11 - 33} \\
\phantom{\text{divisor} \longrightarrow 1-3\,|\,33333}32 + 4 \\
\phantom{\text{divisor} \longrightarrow 1-3\,|\,33333}\underline{32 - 96} \\
\phantom{\text{divisor} \longrightarrow 1-3\,|\,3333333}100 \longrightarrow \text{remainder}
\end{array}
$$

Observe that in the computations below the dividend, the second term in every first line (-1 and 4) and the first term in every second line (3, 11, 32) are duplications and can be omitted.

$$
\begin{array}{l}
\phantom{\text{divisor} \longrightarrow 1-3\,|\,}3 + 11 + 32 \qquad \longrightarrow \text{quotient} \\
\text{divisor} \longrightarrow 1 - 3 \,\big|\, \overline{3 + 2 - 1 + 4} \qquad \longrightarrow \text{dividend} \\
\phantom{\text{divisor} \longrightarrow 1-3\,|\,}\underline{-9} \\
\phantom{\text{divisor} \longrightarrow 1-3\,|\,3}11 \\
\phantom{\text{divisor} \longrightarrow 1-3\,|\,333}\underline{-33} \\
\phantom{\text{divisor} \longrightarrow 1-3\,|\,33333}32 \\
\phantom{\text{divisor} \longrightarrow 1-3\,|\,333333}\underline{-96} \\
\phantom{\text{divisor} \longrightarrow 1-3\,|\,3333333}100 \longrightarrow \text{remainder}
\end{array}
$$

The work can be condensed even more by moving the terms above the lines into a single row and doing the same for the terms below the lines.

$$
\begin{array}{l}
\overline{3 + 11 + 32} \\
1 - 3 \,\big|\, 3 + 2 - 1 + 4 \\
\underline{-9 - 33 - 96} \\
11\ \ 32\ \ 100
\end{array}
$$

Notice that the bottom row lacks only a 3 to contain all the coefficients of the quotient. If 3 is inserted, the top row can be omitted. Also, the 1 from the coefficients of the divisor can be omitted. Hence,

$$-3 \overline{\left) \, 3 + 2 - 1 + 4 \right.}$$
$$\underline{\quad -9 - 33 - 96 \quad}$$
$$3 \quad 11 \quad 32 \quad 100$$

The terms in the third row are simply the result of subtracting the terms in the second row from those in the first row. The terms in the second row are obtained by multiplying the preceding term in the third row by -3. Since it is generally easier to add than subtract, changing the signs of the terms in the second row and of the remaining divisor coefficients gives us

r from $x-r \longrightarrow$ $\underline{3}$ $\quad 3 \qquad 2 \qquad -1 \qquad 4 \quad$ coefficients of the dividend
$\qquad\qquad\qquad\qquad\qquad 9 \qquad 33 \qquad 96$
$\qquad\qquad\qquad \underbrace{3 \quad 11 \qquad 32} \qquad 100 \longleftrightarrow$ remainder

coefficients of $\nearrow$
the quotient

Now the third row is the sum of the first and second rows. The steps taken in using synthetic division are as follows.

Synthetic Division Outline

Synthetic Division Outline

1. Synthetic division can be used when the divisor is of the form $x - r$.

2. Arrange the coefficients of the dividend in order of decreasing powers, using 0 as the coefficient of each missing power.

3. Bring down the leading coefficient.

4. Multiply the leading coefficient by r and add the product to the next coefficient. Multiply this sum by r and add the product to the next coefficient. Continue in this manner until there are no more terms in the dividend to add to the product.

5. The last number in the third row is the remainder. The other numbers in the third row are the coefficients of the quotient. The quotient will be of degree one less than the degree of the dividend.

The following examples illustrate how to use synthetic division and how the last row, the remainder, and the quotient as a polynomial in x are related.

EXAMPLE SET C

1. Divide $2x^4 - 7x^3 + 10x + 4$ by $x - 3$.

   Solution:

   In this case, $r = 3$.

   $$
   \begin{array}{r|rrrrr}
   3 & 2 & -7 & 0 & 10 & 4 \\
     &   & 6 & -3 & -9 & 3 \\
   \hline
     & 2 & -1 & -3 & 1 & 7
   \end{array}
   $$

   Quotient $= 2x^3 - x^2 - 3x + 1$
   Remainder $= 7$
   Therefore, $\dfrac{2x^4 - 7x^3 + 10x + 4}{x - 3} = 2x^3 - x^2 - 3x + 1 + \dfrac{7}{x - 3}$.

2. Find the remainder in the division

   $$\frac{x^3 - 8x^2 + 12x + 10}{x + 2}.$$

   Solution:

   In this case, $x + 2 = x - (-2)$ so that $r = -2$.

   $$
   \begin{array}{r|rrrr}
   -2 & 1 & -8 & 12 & 10 \\
      &   & -2 & 20 & -64 \\
   \hline
      & 1 & -10 & 32 & -54
   \end{array}
   $$

   The remainder is -54.

3. This example illustrates that synthetic division can work well even when the number r is complex. Divide $2x^3 - x^2 + 4x + 15$ by $x - (1 + 2i)$.

   Solution:

   $$
   \begin{array}{r|rrrr}
   1 + 2i & 2 & -1 & 4 & 15 \\
          &   & 2 + 4i & -7 + 6i & -15 \\
   \hline
          & 2 & 1 + 4i & -3 + 6i & 0
   \end{array}
   $$

4. This example illustrates the convenience of synthetic division when multiple divisions are necessary. Multiple divisions are performed by using the last row of a division as the new first row (the dividend row). Divide $x^5 - 9x^4 + 34x^3 - 58x^2 + 45x - 13$ by $x - 1$ three times.

Solution:

$$
\begin{array}{r|rrrrrr}
\underline{1} & 1 & -9 & 34 & -58 & 45 & -13 \\
  &   & 1 & -8 & 26 & -32 & 13 \\
\hline
\underline{1} & 1 & -8 & 26 & -32 & 13 & 0 \\
  &   & 1 & -7 & 19 & -13 & \\
\hline
\underline{1} & 1 & -7 & 19 & -13 & 0 & \\
  &   & 1 & -6 & 13 & & \\
\hline
  & 1 & -6 & 13 & 0 & & \\
\end{array}
$$

EXERCISES

For exercises 1–19, use synthetic division to (a) write each quotient, and (b) write each division in the form specified by the division algorithm, that is, in the form $P(x) = Q(x)(x - r) + R$.

1. $\dfrac{x^2 - x + 3}{x + 1}$

2. $\dfrac{x^2 + 5x + 5}{x + 5}$

3. $\dfrac{a^2 + 4}{a + 2}$

4. $\dfrac{y^3 - 1}{y + 1}$

5. $\dfrac{x^3 + x + 6}{x - 1}$

6. $\dfrac{x^3 + 2x + 1}{x - 3}$

7. $\dfrac{x^3 + 3x^2 + x - 2}{x - 2}$

8. $\dfrac{y^3 + 2y^2 - y + 1}{y - 3}$

9. $\dfrac{x^3 - x^2 - 11x + 15}{x - 2}$

10. $\dfrac{2x^2 + 4x + 1}{x - 3}$

11. $\dfrac{2x^3 - 16x^2 + 38x - 24}{x - 4}$

12. $\dfrac{3x^4 - 5x^3 + 2x^2 - 12x + 10}{x - 5}$

13. $\dfrac{x^5 + 2x^4 - 3x^3 - 6x^2 + x + 2}{x + 1}$

14. $\dfrac{x^2 + 2x + 1}{x + 3}$

15. $\dfrac{x^5 + 1}{x + 1}$

16. $\dfrac{3x^3 + 2x^2 + x - 5}{x - (1 + 3i)}$

17. $\dfrac{2x^3 + x^2 - 5x + 10}{x - (1 - 2i)}$

18. Divide $x^5 - 13x^4 + 66x^3 - 164x^2 + 200x - 96$ by $x - 2$ three times.

19. Divide $3x^5 + 31x^4 + 86x^3 - 24x^2 - 224x + 128$ by $x + 4$ three times.

20. Tim is shown the graph of the accompanying function and he is challenged to come up with an equation describing it.

(a) He notices that the curve increases in the positive direction when x attains large positive or negative values. What does this indicate?

(b) The graph crosses the x-axis at $x = 1$, $x = 2$, $x = 3$, and $x = 4$. This indicates that $(x - 1)(x - 2)(x - 3)(x - 4)$ is a factor of the polynomial. Is this statement true or false?

(c) What is the possible minimum degree of the polynomial?

(d) Are there any other possibilities (i.e., can there be a 6th-degree polynomial that will be similar to this graph)?

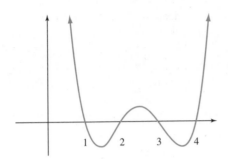

21. Suppose that you want to generate a function where $f(x)$ approaches $-\infty$ as x goes to $+\infty$ and $f(x)$ approaches $+\infty$ as x goes to $-\infty$. In addition, the function must intersect the x-axis at the points 1, 2, and 7.

    (a) What kind of polynomial do you start with?

    (b) Are there any other choices?

3.4 The Factor Theorem

Introduction

The Factored Form of a Polynomial Function

Relating Zeros to the Factored Form

The Factor Theorem

A Review of Factoring Techniques

Introduction

Because physical phenomena usually require functions more complex than polynomial functions in order to model their behavior, polynomial functions are rarely used to describe physical phenomena. Studying polynomial functions is of value, however, because they are relatively simple in form and can be discussed without getting bogged down in computational details, and some of what we discover can be extended to other, more sophisticated functions.

Recall that a polynomial function of degree n, as we saw in Section 3.1, is a function of the form

$$P(x) = a_n x^n + a_{n-1} x^{n-1} + \cdots + a_2 x^2 + a_1 x + a_0,$$

where $a_n \neq 0$. Then, as we noted in Section 1.4, a zero of a polynomial function is a number r such that $P(r) = 0$. That is, a zero of a polynomial function is a number that when used as an input value, produces the output value zero.

Real zeros (that is, zeros that are real numbers) of polynomial functions of *any* degree can be approximated by the calculator using the approximation techniques of Section 1.4. Nonreal complex zeros (that is, zeros that are nonreal complex numbers like $3 - 4i$) cannot be found using the graphing calculator unless we use the programming functionality of the calculator. In certain special cases, real zeros and nonreal complex zeros can be found exactly using analytic (pencil-and-paper) means. We review some of those cases you studied in elementary and intermediate algebra later in this section. First we examine several examples that illustrate an important relationship between the factored form of a polynomial function, the zeros of the polynomial function, and the input-intercepts of the graph form of the function.

The Factored Form of a Polynomial Function

From our study of functions in Chapters 1 and 2, it should be clear that for the same input value, two functions are equal when the output values are equal. For example, Figure 3.9 shows two functions, $f(x)$ and $g(x)$, that are equal when $x = a$. At this input value, both functions have the same output value, b.

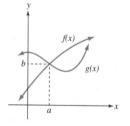

Figure 3.9

Now, using the standard viewing window, graph the functions $f(x) = x^2 - 5x + 4$ (as Y1) and $g(x) = (x - 1)(x - 4)$ (as Y2). Remember that Y1 is graphed first, Y2 is graphed second. It is useful to note that when the highlighted cursor in the top right corner of the viewing window is visible, the functions are still being graphed. When the cursor disappears, the graphing is complete. Think about what you see displayed when the graphs become complete. Window 3.3 shows these graphs. (Don't just look at these pictures, but actually construct these graphs yourself.) It might seem that only one curve is displayed in the viewing window. But in fact, both curves are displayed. They are simply the same curve. That is, both functions, $f(x)$ and $g(x)$, produce

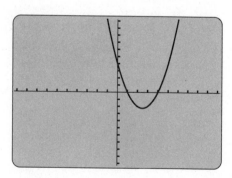

Window 3.3

the same output value at each x-value. This means that these functions are equal for every x-value. When this occurs, we say that the two functions are equal, or equivalent, or identical, and denote this fact symbolically by writing $f(x) = g(x)$. But this means that the defining expressions for these functions must be equivalent. That is, that $x^2 - 5x + 4 = (x-1)(x-4)$. Hence, $f(x)$ can be expressed as either $f(x) = x^2 - 5x + 4$ or as $f(x) = (x-1)(x-4)$. The form

Factored Form $f(x) = (x-1)(x-4)$ is called the **factored form** of $f(x)$, and $(x-1)(x-4)$ is called the factored form of $x^2 - 5x + 4$.

We are reminded of the factoring process studied in elementary and intermediate algebra. In fact, the expression $x^2 - 5x + 4$ can be factored as $(x-1)(x-4)$ using the techniques presented in those courses. But the expressions we worked with then were special cases (quadratic polynomials and cubic polynomials), and in those special cases, we were able to find exact factorizations for those polynomials. We did not attempt to factor polynomial expressions such as $x^4 + 0.3x^3 - 10.78x^2 + 2.208x + 10.2816$. But with our current level of calculator experience, we have the ability to approximate the factorization of such polynomials.

Relating Zeros to the Factored Form

The figure on the left of Window 3.4 shows the graph of the polynomial function $f(x) = (x-4)^2$, or multiplied out, $f(x) = x^2 - 8x + 16$, in the standard viewing window. The graph shows that 4 is a zero of $f(x) = (x-4)^2$. Notice

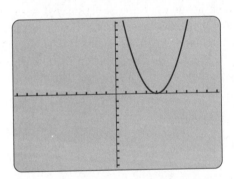 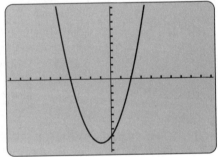

Window 3.4

the relationship between the zero and the factored form of the function,

$$f(x) = \left(x - \text{ the zero}\right)^2$$

The expression $x - \underbrace{\text{(the zero)}}_{x \ minus \ a \ zero}$ appears as a factor of $f(x)$.

The figure on the right of Window 3.4 shows the graph of the polynomial function $f(x) = (x + 4)(x - 2)$, or multiplied out, $f(x) = x^2 + 2x - 8$, in the standard viewing window.

The graph shows that both -4 and 2 are zeros of $f(x) = (x + 4)(x - 2)$. Notice the relationship between the zeros and the factored form of the function,

$$f(x) = (x - \text{ a zero})(x - \text{ a zero})$$

The expression $x - \underbrace{\text{(a zero)}}_{x \ minus \ a \ zero}$ appears as a factor of $f(x)$.

Window 3.5 shows the graph, in the standard viewing window, of the polynomial function $f(x) = (x + 2)(x - 1)(x - 4)$, or multiplied out, $f(x) = x^3 - 3x^2 - 6x + 8$. You can see from the graph that -2, 1, and 4 are zeros of

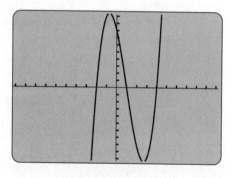

Window 3.5

$f(x) = (x + 2)(x - 1)(x - 4)$. Notice the relationship between the zeros and the factored form of the function,

$$f(x) = (x - \text{ a zero})(x - \text{ a zero})(x - \text{ a zero})$$

In this case, notice that the expression $(x + 2)$ can be expressed as $(x - (-2))$, which results in the form $x - \underbrace{\text{(a zero)}}_{x \ minus \ a \ zero}$.

The Factor Theorem

The zeros of these three functions lead us to conjecture that if the number r is a zero of the polynomial function $f(x)$, then the expression $x - r$ is a factor of $f(x)$. Notice that we have only conjectured that $x - r$ is a factor of $f(x)$. Three (or any number) of examples do not prove a conjecture. To prove that this conjecture is true, we must demonstrate that it is true for every possible choice of r (of which there are infinitely many). This is the work of the mathematician. Mathematicians create and prove conjectures, or theorems, and we have created a theorem through experimentation, and will, after we state it, prove it. The theorem is called the Factor Theorem.

Factor Theorem

Factor Theorem

If the number r is a zero of the polynomial function $f(x)$, then $x - r$ is a factor of $f(x)$. Furthermore, if $x - r$ is a factor of the polynomial function $f(x)$, then the number r is a zero of $f(x)$.

To prove the Factor Theorem, we must prove both the following statements.

1. If the number r is a zero of the polynomial function $f(x)$, then $x - r$ is a factor of $f(x)$.
2. If $x - r$ is a factor of the polynomial function $f(x)$, then the number r is a zero of $f(x)$.

Proof of Statement 1: Assume that r is a zero of the polynomial function $f(x)$ and show that $x - r$ is a factor of $f(x)$.

Since r is a zero of $f(x)$, $f(r) = 0$. Then, from the Division Algorithm,

$$f(x) = (x - r)Q(x) + R$$

$$f(r) = (r - r)Q(x) + R$$

$$0 = 0 \cdot Q(x) + R$$

$$0 = 0 + R$$

$$0 = R$$

Thus, when r is a zero of $f(x)$, $f(x) = (x - r)Q(x)$. But this means that $x - r$ is a factor of $f(x)$, which is what we wished to show.

Proof of Statement 2:   Assume that $x - r$ is a factor of $f(x)$ and show that $f(r) = 0$.

Since $x - r$ is a factor of $f(x)$, we can express $f(x)$ as $f(x) = (x - r)Q(x)$, where $Q(x)$ is some polynomial function.

Letting $x = r$ gives,

$$f(r) = (r - r)Q(r)$$

$$= 0 \cdot Q(r)$$

$$= 0$$

Thus, if $x - r$ is a factor of $f(x)$, then $f(r) = 0$, as we wished to show.

The following serve to demonstrate the Factor Theorem.

ILLUMINATOR SET A

1. If -6, 2, and 8 are the only zeros of the polynomial function $f(x)$ of degree 3, then $f(x) = (x - (-6))(x - 2)(x - 8)$, or $f(x) = (x + 6)(x - 2)(x - 8)$. Multiplied out, $f(x) = x^3 - 4x^2 - 44x + 96$. Since the zeros are evident in the factored form, it is usually more convenient to express polynomial functions in their factored form.

2. If $f(x) = (x - 5)(x + 7)$, or more conveniently $f(x) = (x - 5)(x - (-7))$, then the zeros of $f(x)$ are 5 and -7.

EXAMPLE SET A

Determine if $x - 6$ is a factor of the polynomial function $f(x) = x^4 - 6x^3 - 3x^2 + 20x - 12$.

Solution:

If $x - 6$ is a factor of $f(x)$, then 6 is a zero of $f(x)$. We'll use the calculator to evaluate $f(6)$.

Using Y1, enter the defining expression $x^4 - 6x^3 - 3x^2 + 20x - 12$. Press 2nd and QUIT to get back to the calculation window. Press 6 STO XIT and ENTER to store 6 for x. Use 2nd Y-VARS and select Y1 from the appropriate menu. Press ENTER to complete the evaluation. You should see that $f(6) = 0$, and then conclude that $x - 6$ is a factor of $f(x)$. (Can you think of another way of demonstrating that 6 is a zero of $f(x)$?)

A Review of Factoring Techniques

The factoring techniques you studied in your previous algebra courses have been summarized in the following table. Review the table, then continue on to some examples of factoring. As you study the examples, keep in mind that the factored form of a polynomial function is meaningful because it not only defines how the input and output quantities are related, but it displays the zeros of the function.

Polynomial Factorization Forms		
Type	**Expression**	**Factored Form**
Common Factor	$ax + ay$	$a(x + y)$
Grouping	$ax + ay + bx + by$	$(a + b)(x + y)$
Difference of Two Squares	$x^2 - y^2$	$(x + y)(x - y)$
Difference of Two Cubes	$x^3 - y^3$	$(x - y)(x^2 + xy + y^2)$
Sum of Two Cubes	$x^3 + y^3$	$(x + y)(x^2 - xy + y^2)$
Perfect Square Trinomial	$x^2 + 2x + y^2$	$(x + y)^2$
Perfect Square Trinomial	$x^2 - 2x + y^2$	$(x - y)^2$
Trinomial	$acx^2 + (ad + bc)xy + bdy^2$	$(ax + by)(cx + dy)$

Before we begin looking at some examples, it is important to know that some polynomial expressions cannot be factored if there are restrictions on the types of numbers that can be used. Such polynomial expressions are called *irreducible* or *prime* relative to those restrictions.

Irreducible or Prime Polynomials

Irreducible or Prime Polynomials

If a polynomial is not factorable for a particular subset of the complex numbers, it is said to be **irreducible** or **prime** on that set.

For example, $x^2 - 5$ is prime on the set of rationals since it cannot be expressed as a product in which only rational constants occur. It is, however, factorable over the real numbers since $x^2 - 5 = (x + \sqrt{5})(x - \sqrt{5})$.

Factoring a Common Factor   In some polynomials, a particular factor will appear in every term. If such a factor appears, factor it out.

ILLUMINATOR SET B

1. $\underbrace{ax + ab}_{a \text{ is common}} = \underbrace{a(x + b)}_{a \text{ is factored out}}$

2. $\underbrace{6a^4 + 10a^7}_{2a^4 \text{ is common}} = \underbrace{2a^4(3 + 5a^3)}_{2a^4 \text{ is factored out}}$

3. $\underbrace{x(a + y) + 3(a + y)}_{(a+y) \text{ is common}} = \underbrace{(a + y)(x + 3)}_{(a+y) \text{ is factored out}}$

4. $\underbrace{x^{2n+3} - 5x^{n+1}}_{x^{n+1} \text{ is common}} = \underbrace{x^{n+1}(x^{n+2} - 5)}_{x^{n+1} \text{ is factored out}}, \quad n$ is a natural number

Examples 2 and 4 above illustrate a concept that occurs frequently in calculus. When a factor is common to several terms, it should be factored out using the *smallest* exponent. The smallest exponent must be used since it is that exponent that has recorded the least number of times the common factor has appeared. In Example 2, the smallest exponent is 4, indicating that the common factor a appears in *both* the terms *at most* 4 times. This type of factorization occurs often enough in calculus that we will illustrate it with a detailed example.

EXAMPLE SET B

Factor $(2x + 1)^3(3x - 5)^5 + (2x + 1)^2(3x - 5)^8$.

Solution:
Notice that both $2x + 1$ and $3x - 5$ are common to both terms. The smallest number of times $2x + 1$ appears in both terms is 2, which is recorded by the exponent 2, and the smallest number of times $3x - 5$ appears in both terms is 5, which is recorded by the exponent 5. Therefore, we factor out $(2x + 1)^2$ and $(3x - 5)^5$. If we remove 2 factors of $2x + 1$ from the first term, there are $3 - 2 = 1$ of those factors left. If we remove 5 factors of $3x - 5$ from the first term, there are $5 - 5 = 0$ of those factors left. Thus, the first factor in the parentheses will be $(3x - 5)^5$. Similarly, if we remove 2 factors of $2x + 1$ from the second term, there are $2 - 2 = 0$ of those factors left. If we remove 5 factors of $3x - 5$ from the second term, there are $8 - 5 = 3$ of those factors left. Thus, the second factor in the parentheses will be $(2x + 1)^2$. Then,

$$(2x+1)^3(3x-5)^5 + (2x+1)^2(3x-5)^8 = (2x+1)^2(3x-5)^5\left[(2x + 1) + (3x - 5)^3\right]$$

Factoring Perfect Square Trinomials   Perfect square trinomials always factor as the square of a binomial. To recognize a perfect square trinomial, look for the following features.

1. The first and last terms are perfect squares.

2. The middle term is divisible by 2, and if you divide the middle term in half (the opposite of doubling it), you will get the product of the terms, which when squared, produce the first and last terms.

ILLUMINATOR SET C

1. $x^2 + 8x + 16$ is a perfect square trinomial.
   x^2 is the square of x.
   16 is the square of 4.
   The middle term, $8x$ is divisible by 2 and is twice the product of x and 4, the terms which when squared, produce the first and last terms.
   $x^2 + 8x + 16 = (x + 4)^2$
   The zero of the corresponding polynomial function $f(x) = (x + 4)^2$ is now evident as -4.

2. $a^6 - 12a^3b^2 + 36b^4$ is a perfect square trinomial.
   a^6 is the square of a^3.
   $36b^4$ is the square of $6b^2$.
   The middle term, $-12a^3b^2$ is divisible by 2 and is twice the product of a^3 and $-6b^2$, the terms which when squared, produce the first and last terms.
   $a^6 - 12a^3b^2 + 36b^4 = (a^3 - 6b^2)^2$.

Factoring Trinomials   We look to the pattern of the product of two binomials when attempting to factor a trinomial.

ILLUMINATOR SET D

1. $x^2 - 34x - 72 = (x + 2)(x - 36)$
   The zeros of the corresponding polynomial function $f(x) = (x + 2)(x - 36)$ are now evident. They are -2 and 36.

2. $20x^2 - 23x - 21 = (4x - 7)(5x + 3)$
   The zeros of the corresponding polynomial function $f(x) = (4x - 7)(5x + 3)$ are almost evident. The best we can get from this form is that $4x = 7$ and $5x = -3$. Solving these equations produces $x = \dfrac{7}{4}$ and $x = \dfrac{-3}{5}$.
   Notice the relationship between the zeros and their corresponding factors. The numerator of the zero is the constant of the factor, but with the opposite sign. The denominator is precisely the coefficient of the variable

in the factor. In the factor $4x - 7$, the constant appears with the opposite sign in the numerator of the zero, and the coefficient of x appears in the denominator.

3. Using the information provided about the relationship between zeros and factors in the previous item, if the only zeros of a second degree polynomial are $\frac{4}{3}$ and $\frac{-8}{5}$, the factors of the polynomial expressions must be $(3x - 4)$ and $(5x + 8)$. Thus, a corresponding polynomial function is $f(x) = (3x - 4)(5x + 8)$. Also, by the Factor Theorem, we could write the equivalent form, $f(x) = (x - \frac{4}{3})(x - \frac{-8}{5})$. These two factorization forms differ only in appearance. The first uses only the integers, whereas the second uses rational numbers.

Factoring by Grouping   This technique is useful when there is no one factor common to each term in the polynomial expression.

ILLUMINATOR SET E

Factor the polynomial $2a^2 + 3a - 2ab - 3b$ by grouping together terms that have common factors.

$$2a^2 + 3a - 2ab - 3b = \underbrace{2a^2 + 3a}_{a \text{ is common}} \quad \underbrace{-2ab - 3b}_{-b \text{ is common}}$$

$$= (2a^2 + 3a) - (2ab + 3b)$$

$$= \underbrace{a(2a + 3) - b(2a + 3)}_{\text{now } 2a+3 \text{ is common}}$$

$$= (2a + 3)(a - b)$$

Factoring the Sum and Difference of Two Cubes   These factorization forms are easier to remember if we note the following features.

1. The factorization consists of a binomial and a trinomial.

$$a^3x^3 - b^3 = \underbrace{(ax - b)}_{\text{binomial}} \underbrace{(a^2x^2 + abx + b^2)}_{\text{trinomial}}$$

$$a^3x^3 + b^3 = \underbrace{(ax + b)}_{\text{binomial}} \underbrace{(a^2x^2 - abx + b^2)}_{\text{trinomial}}$$

2. The terms of the binomial are the cube roots of the respective terms of the original expression.

3. The operation sign in the binomial is identical to that of the original expression.

4. The trinomial can be formed using the terms of the binomial. In the trinomial,

   (a) the first term is the square of the first term of the binomial;

   (b) the second term is the product of the terms of the binomial with the sign changed;

   (c) the third term is the square of the second term of the binomial.

When the corresponding polynomial function has real number coefficients, the real zeros will be evident from the factored form as they will always come from the binomial. The trinomial will always produce complex conjugate zeros. The next example illustrates this fact.

EXAMPLE SET C

Factor $27x^3 - 125$.

Solution:

$$\sqrt[3]{27x^3} = 3x \quad and \quad \sqrt[3]{125} = 5$$

First term squared is $(3x)^2 = 9x^2$.
   Second term squared is $5^2 = 25$.
   Product of both terms (with opposite sign) is $3x \cdot 5 = +15x$.
   Thus,

$$27x^3 - 125 = (3x)^3 - 5^3$$

$$= (3x - 5)((3x)^2 + 3x \cdot 5 + 5^2)$$

$$= (3x - 5)(9x^2 + 15x + 25)$$

The polynomial function $f(x) = 27x^3 - 125$ will then factor as $f(x) = (3x - 5)(9x^2 + 15x + 25)$. The real zero is evident from the binomial factor and is $\dfrac{5}{3}$. To get any other zeros, we will have to set the trinomial factor equal to zero and use the quadratic formula to solve for x. Identifying $a = 9$, $b = 15$, and $c = 25$, we get

$$x = \frac{-b \pm \sqrt{b^2 - 4ac}}{2a}$$

$$= \frac{-15 \pm \sqrt{15^2 - 4(9)(25)}}{2(9)}$$

$$= \frac{-15 \pm \sqrt{225 - 900}}{18}$$

$$= \frac{-15 \pm \sqrt{-675}}{18}$$

$$= \frac{-15 \pm 15i\sqrt{3}}{18}$$

$$= \frac{-5 \pm 5i\sqrt{3}}{6}$$

Thus, $x = \dfrac{-5}{6} + \dfrac{5\sqrt{3}}{6}i, \dfrac{-5}{6} - \dfrac{5\sqrt{3}}{6}i$.

Thus, this polynomial function has one real number zero, $\dfrac{5}{3}$, and two complex conjugate zeros, $\dfrac{-5}{6} + \dfrac{5\sqrt{3}}{6}i$ and $\dfrac{-5}{6} - \dfrac{5\sqrt{3}}{6}i$.

Writing Activity   Write down in paragraph form the method of factoring expressions that are of the following type: difference of two squares, difference of two cubes, sum of two cubes, perfect square trinomial, and general trinomial. Then factor the expression $8x^3 + 27y^3$.

EXERCISES

For exercises 1–4, specify the zeros of each polynomial function.

1. $f(x) = (x - 4)(x + 6)$
2. $f(x) = (x + 11)(x - 5)$
3. $f(x) = (x + 3.2)(x + 9.8)(x - 5.3)$
4. $f(x) = (x - 10.1)(x + 1.3)(x - +4.7)$

For exercises 5–8, use the Factor Theorem to determine if the first polynomial is a factor of the second.

5. $x + 2$, $x^3 - 4x^2 - 3x + 18$
6. $x - 6$, $x^3 + x^2 - 32x - 60$
7. $x + 1$, $x^3 - 6x^2 + 9x$
8. $x + 1$, $x^3 + 8x^2 + 21x + 18$

For exercises 9–14, graph each pair of functions in the same viewing window. On the basis of the graphs, state if any factor of the function $g(x)$ is a factor of the function $f(x)$. If there are no common factors, write *no common factors*. If there are one or more common factors, specify them.

9. $f(x) = x^3 - 3x^2 - 10x + 24$, $g(x) = x^2 + 4x + 3$

10. $f(x) = x^4 + 10x^3 + 23x^2 - 10x - 24$, $g(x) = x^2 + x - 30$

11. $f(x) = x^4 - 1.5x^3 - 11.25x^2 + 7.375x + 26.250$, $g(x) = x^3 - 12.25x - 15$

12. $f(x) = x^5 + x^4 - 29x^3 - 45x^2 + 216x + 432$, $g(x) = x^3 - 11x^2 + 7x + 147$

13. $f(x) = x^4 - 10x^3 + 33x^2 - 40x + 16$, $g(x) = -x^3 + 13x^2 - 56x + 80$

14. $f(x) = x^5 + 24x^4 + 182x^3 + 196x^2 - 3087x - 9604$, $g(x) = x^4 + 12x^3 - 432x - 1296$

For exercises 15–20, use your calculator to determine if each factorization statement is *true* or *false*.

15. $x^2 - 0.8x - 4.68 = (x + 1.8)(x - 2.6)$

16. $x^2 + 3.9x - 2.20 = (x - 0.5)(x + 4.2)$

17. $x^3 - 13x - 12 = (x - 4)(x + 1)(x + 3)$

18. $x^4 - 5x^3 - 25x^2 + 65x + 84 = (x + 4)(x + 1)(x - 7)(x - 2)$

19. $x^7 - 3.2x^6 - 7.20x^5 + 51.840x^4 - 107.1360x^3 + 109.48608x^2 - 56.733696x + 11.943936 = (x - 1.2)^6(x + 4)$

20. $x^2 + 7.6x + 11.88 = (x + 2.2)(x + 5.4)$

For exercises 21–34, construct a polynomial function having real number coefficients that meets all the given conditions. If the degree is two or three, express the function in both factored and expanded form. If the degree is four or more, express the function only in factored form.

21. Degree is 2 with zeros -5 and 2.

22. Degree is 2 with zeros 4 and -7.

23. Degree is 3 with zeros -3, 1, and 3.

24. Degree is 3 with zeros -4, -3, and 2.

25. Degree is 2 with zeros $-\sqrt{7}$ and $\sqrt{7}$.

26. Degree is 2 with zeros $-\sqrt{10}$ and $\sqrt{10}$.

27. Degree is 4 with zeros -8, -4, 2, and 5.

28. Degree is 5 with zeros -10, -6, -2, 3 and 4.

29. Degree is 2 with zeros $3 + 2i$ and $3 - 2i$.

30. Degree is 2 with zeros $1 - 6i$ and $1 + 6i$.

31. Degree is 3 with zeros 3, $2 - i$, and $2 + i$.

32. Degree is 3 with zeros -4, $1 + i$, and $1 - i$.

33. Degree is 3 with zeros -2, 0, and 5.

34. Degree is 3 with zeros -1, 0, and 1.

For exercises 35–49, use your calculator to express each polynomial function in factored form over the real numbers. Specify all the real number zeros of each function.

35. $f(x) = x^4 - 2x^3 - 33x^2 + 50x + 200$

36. $f(x) = x^6 + x^5 - 95x^4 - 89x^3 + 2134x^2 + 1600x - 9600$

37. $f(x) = x^6 + 15x^5 - 6x^4 - 692x^3 - 888x^2 + 4320x$

38. $f(x) = x^7 + x^6 - 529x^5 - 1865x^4 + 71,576x^3 + 339,682x^2 - 2,391,680x - 11,827,200$

39. $f(x) = x^2 + 2x - 24$

40. $f(x) = 6x^2 - 5x - 4$

41. $f(x) = 24x^2 + 14x - 3$

42. $f(x) = 24x^2 - 7x - 6$

43. $f(x) = 6x^2 - 150$

44. $f(x) = 14x^2 - 126$

45. $f(x) = x^3 - 216$

46. $f(x) = 3x^4 + 192x$

47. $f(x) = x^5 + 4x^3 + 3x^2 + 12$

48. $f(x) = x^3 + 6x^2 - 5x - 30$

49. $f(x) = x^6 - 9x^4 - 16x^2 + 144$

50. Use the Factor Theorem to prove that if a and b are real numbers, and if n is a natural number, then $a - b$ is a factor of $a^n - b^n$. (Hint: Begin by showing that b is a zero of the polynomial function $f(x) = x^n - b^n$, and then let $x = a$.)

51. The graph of a function is given in the accompanying figure. What does the point $x = -1$ signify? If this is the graph of a polynomial, what factors does the function have? What is the degree of the polynomial? What does 2 represent? Write a possible equation for the function.

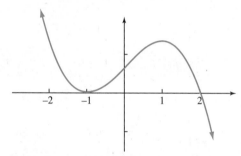

52. A sculptor from the School of Mathematical Art is creating a sculpture in the form of a big cube with a smaller cubic shape cut out at one corner (see the accompanying figure). The larger cube is 5 units in length and the length of the smaller one is x units. The artist then writes an expression for the volume of his sculpture. Since this is a cubic polynomial he feels he can factor the expression. Can you factor the polynomial? What are the zeros of this polynomial? Are they all real?

3.5 Zeros, Roots, Factors, and Intercepts of Polynomial Equations and Functions

Introduction

Relating Zeros, Roots, Factors, and Intercepts of Polynomial Equations and Functions

Multiple Zeros and Intercepts

The Number of Zeros of a Polynomial Function

Introduction

We are now in a good position to examine four important functional relationships. These are: the relationship between the factored form of a polynomial function, the zeros of the polynomial function, the roots of the corresponding polynomial equation, and the input-intercepts of the graph form of the function. Understanding these relationships will give you a good foundation for the study of functions in calculus.

Relating Zeros, Roots, Factors, and Intercepts of Polynomial Equations and Functions

The following is a description of the relationship between zeros, roots and factors.

Zeros, Roots, Factors, Intercepts

Zeros, Roots, Factors, Intercepts

The following four statements are equivalent in the sense that if any one of them is true, the other three are also true.

1. The number r is a zero of the function $f(x)$.
2. The number r is a root of the polynomial equation $f(x) = 0$.
3. The expression $x - r$ is a factor of the function $f(x)$.
4. The graph of the function $f(x)$ intercepts the x-axis at r.

The next example will help you to visualize this relationship.

EXAMPLE SET A

Construct the graph of the function $f(x) = (x - 3)(x^2 + x + 4)$, or multiplied out, $f(x) = x^3 - 2x^2 + x - 12$, in the viewing window -5 to 5 for x, and -20 to 20 for y. Before you press the GRAPH key, think about where you expect this cubic function to intercept the x-axis.

Solution:
Window 3.6 shows the graph of $f(x) = (x - 3)(x^2 + x + 4)$.

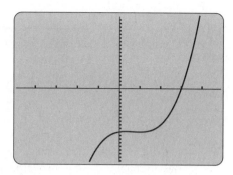

Window 3.6

As is clear from the defining expression, $x - 3$ is a factor of $f(x)$. Since this is true, the graph of $f(x)$ intercepts the x-axis at 3, so that 3 is a zero of $f(x)$. The graph shows that the curve intercepts the x-axis at 3, so that statement is verified. We can verify that 3 is a zero by inputting it into the function and computing.

$$f(3) = (3 - 3)(3^2 + 3 + 4)$$

$$= (0)(16)$$

$$= 0$$

Since $f(x) = 0$ when $x = 3$, 3 is a root of the polynomial equation $(x - 3)(x^2 + x + 4)$.

Thus, for the function $f(x) = (x - 3)(x^2 + x + 4)$, or in its expanded form, $f(x) = x^3 - 2x^2 + x - 12$, each of the following statements is true:

1. 3 is a zero of $f(x)$,

2. 3 is a root of the polynomial equation $f(x) = 0$,

3. $x - 3$ is a factor of $f(x)$, and

4. the graph of $f(x)$ intercepts the x-axis at 3.

Multiple Zeros and Intercepts

Multiple Zero

The next example illustrates what the graph of a function with multiple zeros looks like. A number a is called a **multiple zero** of a function if $(x - a)$ is a multiple factor of the function. Specifically, a is said to be a zero of multiplicity n if the factor $(x - a)$ occurs exactly n times in the function.

EXAMPLE SET B

Construct the graph of the function $f(x) = (x + 5)^2$.

Solution:
Notice first that $x + 5$ can be written in the more descriptive form $x - (-5)$. This form makes it more apparent that -5 is a zero of the function. In fact, -5 is a zero of multiplicity two since $(x + 5)^2$ is actually $(x + 5)(x + 5)$. Does this mean the curve will intercept the x-axis at -5 twice? Hopefully not, for if it did, $f(x)$ would not be a function. (The curve would have to pass through -5, then turn around and come back to get to -5 again. This would produce more than one output value for some input value, perhaps -5. Draw some curves that do this. You'll see!) Make two attempts at the graph. Use the standard viewing window on the first attempt. The curve intercepts the x-axis at -5, but it looks as if it intercepts the axis at many other points between -6 and -4. For your second graphing attempt, change the output scale to Ymin$= -1$ and Ymax$= 1$. Window 3.7 shows this view of the graph.

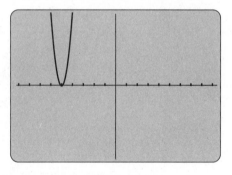

Window 3.7

This view of the graph more clearly shows that the curve intercepts the x-axis only at -5. In fact, it intercepts the x-axis only once.

The function in the previous example illustrates that even though a number may be a multiple zero of a function, the graph of the function intercepts the input axis only once at that value.

Multiple Zeros and Intercepts

Multiple Zeros and Intercepts

1. If a real number is a zero of **even** multiplicity of a polynomial function, the graph of the function intercepts the input axis at that value exactly once and lies only on one side of the input axis near that point. Conversely, if a curve lies on only one side of the input axis near a point and intercepts the input axis at that point (tangentially), then that input value is a zero of even multiplicity of the polynomial function.

2. If a real number is a zero of **odd** multiplicity of a polynomial function, the graph of the function will intercept the input axis at that value exactly once, will lie on both sides of the input axis near that point, and will be somewhat flat near that point.

Notice that we cannot make as strong a statement about zeros of odd multiplicity as we can about zeros of even multiplicity. If a curve intercepts the input axis at a point in such a way that it lies on both sides of the input axis near that point and so that it is somewhat flat at that point, we *cannot* conclude that that input value is a zero of odd multiplicity. Figure 3.10 illustrates this idea. In the figure, the graph of $f(x)$ satisfies the requirements that a zero of even multiplicity exists at $x = a$. There is also a zero at $x = b$, but is it a multiple zero? Since the graph is not very flat at $x = b$, it probably does not have a zero of odd multiplicity there. We don't know this for sure, but it is a reasonable guess based on the shape of the graph. Since the curve appears somewhat flat near $x = b$, the number b *may* be a zero of odd multiplicity. Again, we don't know this for sure.

As a zero increases in multiplicity, the curve becomes flatter near the input-intercept, but will intercept the axis only once. The higher the multiplicity, the flatter the curve. Try this for the function $f(x) = (x + 5)^4$, then increase the exponent (and hence the multiplicity of the zero -5) to 6, 8, and 10. (As you increase the exponent, you will get a better view of the curve near the axis if you also decrease the Ymin and Ymax values. For the exponent of 10, you might try Ymin$= -0.0001$ and Ymax$= 0.0001$.) Each curve should exhibit the same shape (since they are of even degree), but each curve will become flatter near the x-axis as the exponent increases. (You might want to try changing to odd exponents. That should alter the shape of the curve.)

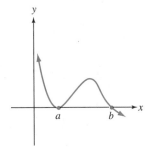

Figure 3.10

Some functions have no real number zeros. The function in the next example is just such a function.

EXAMPLE SET C

Determine the zeros of the function $f(x) = x^2 + 3x + 4$.

Solution:
Constructing the graph of this function (and you should do so now) in the standard viewing window shows that it does not intercept the x-axis between -10 and 10. Since the function is of degree two, it has only one turnaround point, and since this point is visible in the viewing window, the function will not turn around again outside this window. Thus, this function never intercepts the x-axis, and therefore, has no real number zeros. It does, however, have two nonreal complex zeros as the quadratic formula, with $a = 1$, $b = 3$, and $c = 4$, shows:

$$x \quad = \frac{-b \pm \sqrt{b^2 - 4ac}}{2a}$$

$$= \frac{-3 \pm \sqrt{(-3)^2 - 4(1)(4)}}{2(1)}$$

$$= \frac{-3 \pm \sqrt{-7}}{2}$$

$$= \frac{-3 \pm \sqrt{7}i}{2}$$

$$= \frac{-3}{2} \pm \frac{\sqrt{7}}{2}i$$

Thus, this function has two nonreal complex zeros,

$$\frac{-3}{2} + \frac{\sqrt{7}}{2}i$$

and

$$\frac{-3}{2} - \frac{\sqrt{7}}{2}i$$

Notice that these two complex zeros are conjugates.

The Number of Zeros of a Polynomial Function

When the graph of a function does not intercept the input axis, so that it has no real number zeros, how would one know that it has nonreal complex zeros, or any zeros at all? The knowledge about the existence of zeros of polynomial functions is provided in a theorem that was proved by the German mathematician Karl Friedrich Gauss in 1797, when he was 20 years old. The theorem is important because it tells us that every polynomial function of degree one or higher has at least one zero. That zero may be a real number or it may be a nonreal complex number. The theorem is so important that it is called the Fundamental Theorem of Algebra. However, the theorem only guarantees the existence* of a zero. It does not indicate how many real or nonreal complex zeros the function has, nor does it provide a method of finding them. Since its proof relies on some calculus concepts, we'll state the theorem without proof.

The Fundamental Theorem of Algebra

The Fundamental Theorem of Algebra

Every polynomial function of degree one or higher has at least one zero (be it real or nonreal complex).

You should reflect back on each of the examples we have examined and note that each one has at *least one* zero.

The next theorem is a corollary[†] to the Fundamental Theorem of Algebra and relates the degree of a polynomial function to the number of zeros it has.

Corollary to the Fundamental Theorem

Corollary to the Fundamental Theorem

Every polynomial of degree n, where $n \geq 1$, has exactly n zeros, provided a zero of multiplicity k is counted as k zeros.

ILLUMINATOR SET A

1. The polynomial function $P(x) = 2x^3 + 2x^2 - 4x + 5$ is of degree 3 and therefore has exactly 3 zeros. The zeros may be real numbers or nonreal

---

*The Fundamental Theorem of Algebra is an example of an *existence* theorem.
[†]A corollary of a theorem is a theorem, the proof of which follows directly from that theorem.

complex numbers. We cannot tell the number of real and nonreal complex zeros by looking at the function in its current form.

2. The polynomial function $f(x) = 4x^5 - 2x + 7$ is of degree 5 and therefore has exactly 5 zeros. The zeros may be real numbers or nonreal complex numbers. Again, from this form, we do not know how many of each there are.

3. The polynomial function $f(x) = (x - 4)^7(x + 2)^3$ is of degree 10 and therefore has exactly 10 zeros. With the function expressed in this form, we know the zeros. The number 4 is a zero of multiplicity 7, and the number -2 is a zero of multiplicity 3.

4. The polynomial function $f(x) = (x - 2)(x^2 + 3x + 1)$ is of degree 3 (you'll see if you multiply it out) and therefore has exactly 3 zeros. In this case, we know one of the zeros. From the first factor, one zero is 2. There are two other zeros. In this case, they happen to be nonreal complex zeros. In fact, they turn out to be complex conjugates. (You can convince yourself of this by setting $x^2 + 3x + 1$ equal to zero and using the quadratic formula to solve for x.)

A second corollary to the Fundamental Theorem involves the factorization of a polynomial function, and relates to the first corollary.

Corollary to the Fundamental Theorem

Corollary to the
Fundamental
Theorem

If $P(x)$ is a polynomial function of degree n, with leading coefficient a_n, then it can be factored into n linear factors (that are not necessarily distinct) where the replacement set is the set of complex numbers. That is, $P(x)$ can be expressed as

$$P(x) = a_n(x - r_1)(x - r_2) \cdots (x - r_n)$$

where $n \geq 1$, n is a positive integer, $r_1, r_2, \ldots, r_n$ are the n zeros of $P(x)$.

Proof: Let

$$P(x) = a_n x^n + a_{n-1}x^{n-1} + a_{n-2}x^{n-2} + \cdots + a_1 x + a_0$$

be a polynomial function of degree n, and let r_1 be one of the zeros of $P(x)$ that are guaranteed by the Fundamental Theorem. Then, by the Factor Theorem, $x - r_1$ is a factor of $P(x)$, and $P(x)$ can then be expressed in factored form as

$$P(x) = (x - r_1)P_{n-1}(x)$$

where $P_{n-1}(x)$ is a polynomial function of degree $n - 1$. The Fundamental Theorem guarantees us that $P_{n-1}(x)$ has at least one zero. Assume it to be r_2. Then, $(x - r_2)$ is a factor of $P_{n-1}(x)$, and, again by the Factor Theorem, $P_{n-1}(x)$ can be factored as

$$P_{n-1}(x) = (x - r_2)P_{n-2}(x)$$

where $P_{n-2}(x)$ is a polynomial function of degree $n - 2$. Now, $P(x)$ can be expressed as

$$P(x) = (x - r_1)(x - r_2)P_{n-2}(x)$$

Continuing in this way, we can show that

$$P(x) = (x - r_1)(x - r_2) \cdots (x - r_n)P_0(x)$$

where $P_0(x)$ is a polynomial function of degree 0. That is, $P_0(x)$ is a constant. It remains to be shown that $P_0(x) = a_n$. We leave that as an exercise.

It is important to note that the factors $(x - r_1)$, $(x - r_2)$, ..., $(x - r_n)$ may not all be distinct. It is possible that two or more of the r's (the zeros) may be identical. This would just indicate multiple zeros.

It is also important to note that this factorization accounts for *all* the zeros. There cannot be any more. Suppose that $P(x)$ is a polynomial function of degree n with the n zeros $r_1, r_2, \ldots, r_n$. Then $(x - r_1)$, $(x - r_2)$, ..., $(x - r_n)$ are factors of $P(x)$. Thus, as we have just shown,

$$P(x) = a_n(x - r_1)(x - r_2) \cdots (x - r_n)$$

Now suppose there is some other zero, say r_{n+1}, of $P(x)$. If this were true, then $(x - r_{n+1})$ would be a factor of $P(x)$ and then

$$P(x) = a_n(x - r_1)(x - r_2) \cdots (x - r_n)(x - r_{n+1})$$

But now $P(x)$ is an $(n + 1)$ st-degree polynomial function, which contradicts the fact that $P(x)$ is an n th-degree polynomial function. Thus, r_{n+1} cannot be a zero of $P(x)$.

ILLUMINATOR SET B

The polynomial function

$$f(x) = (x + 3)(x + 1)^2(x - 4)^3(x - 5)$$

is a 7th-degree polynomial with 7 zeros. One of the zeros is -3, two are -1, three are 4, and one is 5.

EXAMPLE SET D

By constructing its graph and using the approximation techniques of Section 1.4, approximate, with an accuracy of 0.01, the factorization of the function $f(x) = x^3 + 1.6x^2 - 9.6x - 18.432$.

Solution:

This is a 3rd-degree polynomial function so we expect three factors (perhaps not all distinct). Window 3.8 shows the graph of this function in the standard viewing window.

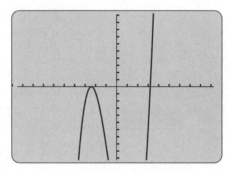

Window 3.8

The leftmost intercept occurs at approximately $x = -2.4$. The rightmost intercept occurs at approximately 3.2. In the standard viewing window, these are the only two zeros.

Do we need to explore a bigger window to search for more zeros? No. This is a third-degree function and therefore has only two turnaround points. Since the function will not turn around again, it will not approach and intercept the x-axis again. Thus, these are the only two real number zeros of this function. But this function is of the third degree so it must have exactly three zeros. We have two real zeros. We also know that the third zero cannot be a nonreal complex number because nonreal complex zeros, as you may recall from intermediate algebra, occur in conjugate pairs. If there is one nonreal complex zero, there would automatically be two. Then a third-degree function would have four zeros. This contradicts the corollary to the Fundamental Theorem of Algebra. The only thing we can conclude is that one of the real number zeros must be a zero of multiplicity two. We even know which one it is. Since the curve only skims the axis at -2.40, -2.40 must be a zero of multiplicity two. Thus, an approximate factored form of $f(x) = x^3 + 1.6x^2 - 9.6x - 18.432$ is

$$f(x) = (x - 3.2)(x + 2.4)^2$$

To investigate what a small change in the form of the function can do, investigate what happens to the zeros of this function if the coefficient of the third term is changed from $9.6x$ to $9.61x$, a change of one hundredth of a unit.

EXERCISES

For exercises 1–7, use the Factor Theorem to determine if the first polynomial is a factor of the second.

1. $x + 4$; $3x^4 + 13x^3 + 7x^2 + 5x + 4$

2. $x - 2$; $x^4 - x^3 + x^2 - 4$

3. $x - 1$; $x^3 + 7x^2 - 4x - 3$

4. $x + 2$; $5x^3 + 3x^2 - 8x + 12$

5. $x - 5$; $3x^4 - 10x^3 - 23x^2 - 12x + 1$

6. $x + r$; $x^n + r^n$,   if n is a positive odd integer.

7. $x - r$; $x^n + r^n$,   if n is a positive even integer.

For exercises 8–17, find all the factors of the second polynomial given that the first polynomial is a factor of the second.

8. $x + 3$; $x^3 + 3x^2 - 4x - 12$

9. $x - 1$; $x^3 - 10x^2 + 23x - 14$

10. $x + 5$; $x^3 - x^2 - 21x + 45$

11. $x - 2$; $x^3 + 6x^2 - 32$

12. $x - 4$; $x^4 - 4x^3 - 7x^2 + 22x + 24$

13. $x + 4$; $x^4 + 2x^3 - 13x^2 - 14x + 24$

14. $x - 1$; $x^5 - 20x^3 + 30x^2 + 19x - 30$

15. $x - 1$; $x^5 - x^4 - 13x^3 + 13x^2 + 36x - 36$

16. $x - (2 + 3i)$; $x^3 - 5x^2 + 17x - 13$

17. $x - (1 - 5i)$; $x^3 - 4x^2 + 30x - 52$

Describe an approach you might use to solve exercises 18–26. Then write out the details of your solutions.

18. Show that $x - 1$ is a factor of $x^{40} + 15x^{26} - 2x^{16} - 15x + 1$.

19. Show that $x + 1$ is a factor of $x^{68} - 14x^{55} + 23x^{54} + 9x^{11} - 29$.

20. Show that 4 is a zero of the function $f(x) = x^4 - 5x^3 + 5x^2 - 5x + 4$. (There are several ways to do this. Choose any method.)

21. Show that -5 is a zero of the function $f(x) = x^4 + 10x^3 + 22x^2 - 30x - 75$. (There are several ways to do this. Choose any method.)

22. Show that $2 + i$ is a zero of the function $f(x) = x^2 - 4x + 5$. (There are several ways to do this. Choose any method.)

23. Show that $4 - 3i$ is a zero of the function $f(x) = x^2 - 8x + 25$. (There are several ways to do this. Choose any method.)

24. Show that i is a zero of the function
$f(x) = x^3 + 2x^2 - 3x - x^2i - 2xi + 3i$.
(There are several ways to do this. Choose any method.)

25. Show that $-i$ is not a zero of the same function as in exercise 24. (There are several ways to do this. Choose any method.)

26. Reflect back on exercises 24 and 25. Explain why the fact that i and its conjugate $-i$ are not both zeros of the function $f(x) = x^3 + 2x^2 - 3x - x^2i - 2xi + 3i$ does not violate the Complex Conjugate Zero Theorem presented in this section.

For exercises 27–34, construct a polynomial function having real coefficients that meet all the given conditions.

27. Degree is 2 with zeros 2 and -2

28. Degree is 3 with zeros 0, 3, and 5

29. Degree is 4 with zeros -2, -1, 0, and 4

30. Degree is 2 with zeros $2 + 2i$ and $2 - 2i$

31. Degree is 2 with zeros $4 - i$ and $4 + i$

32. Degree is 4 with zeros -3, $1 - i$, 0, and $1 + i$

33. Degree is 2 with zeros 1.6 and -2.5

34. Degree is 3 with zeros 0.04, -0.14, and -0.42

35. A university atmospheric sciences department has arranged to drop a weather device from an altitude of 14,000 feet. A particular device is activated when it is 200 feet above the ground. At what time (in seconds) after the drop should the device be triggered? The height of the device above the ground is given by $y(t) = -16t^2 + 14t + 14000$. (Hint: Graph $y(t)$ as well as $y = 200$ and find the point of intersection.)

36. A mountain climber drops a pebble from the edge of a vertical cliff to a lake below. She hears the pebble splash into the water 5.2 seconds after dropping it. What is the height of the cliff if $y = 16t^2$ gives the depth of the free-falling object and the speed of sound is 1080 feet per second?

3.6　Some Important Functions

Introduction

The Constant Function

The Identity Function

The Absolute Value Function

Piecewise Functions

The Greatest Integer Function

Introduction

Some functions occur so frequently in mathematical analysis that they are worth special attention. In this section we will examine the constant function, the identity function, and a class of functions known as piecewise functions, which include the absolute value function and the greatest integer function.

The Constant Function

If, in the polynomial function

$$f(x) = a_n x^n + a_{n-1} x^{n-1} + \cdots + a_1 x + a_0$$

all the coefficients a_n, a_{n-1}, ..., and a_1 are 0, then we are left with the polynomial function

$$f(x) = a_0$$

Constant Function

This function is called the **constant function**. Notice that the defining expression a_0 is independent of the input variable. This means that regardless of the input value, the output value is a_0. That is, as the input variable assumes values throughout its domain, the output remains *constant*. The graph of a constant function is therefore a horizontal line, and unless $a_0 = 0$, it will have no zeros. Window 3.9 shows the graph of the constant function $f(x) = -4$. You can obtain this graph on your calculator by entering -4 for Y1.

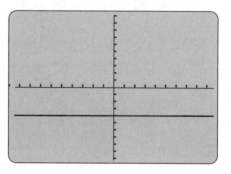

Window 3.9

The Identity Function

If, in the polynomial function $f(x) = a_n x^n + a_{n-1} x^{n-1} + \cdots + a_1 x + a_0$, all the coefficients a_n, a_{n-1}, ..., and a_2 are 0, the coefficient a_1 is 1, and the constant term a_0 is 0, then we are left with the polynomial function

$$f(x) = x$$

Identity Function

This function is called the **identity function**. In the identity function, each output value is identical to the input value. The graph of the identity function appears in Window 3.10.

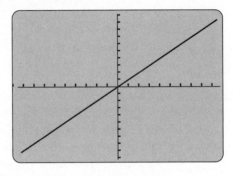

Window 3.10

The Absolute Value Function

The absolute value function is not a polynomial function, but can be formed using polynomial functions.

One useful function with which you are already familiar is the absolute value function, and its defining expression is commonly denoted using two vertical bars around some expression. If the expression is represented by x, then this is called the standard absolute value function and is commonly denoted by $f(x) = |x|$. Before we illustrate the absolute value function, let's review the definition of absolute value.

Absolute Value

Absolute Value

The **absolute value** of a real number a is

$$|a| = \begin{cases} a, & \text{if } a \geq 0; \\ -a, & \text{if } a < 0. \end{cases}$$

This definition takes into account the fact that the number a could be either positive or zero ($a \geq 0$) or negative ($a < 0$).

1. If the number a is positive or zero ($a \geq 0$), the upper part of the definition applies. The upper part of the definition tells us that if the number enclosed within the absolute value bars is a nonnegative number, then the absolute value of the number is the number itself.

2. The lower part of the definition tells us that if the number enclosed within absolute value bars is negative, then the absolute value of the number is the opposite (negative) of the number. The opposite of a negative number is a positive number.

ILLUMINATOR SET A

Find each absolute value using the definition.

1. $|8|$. The number enclosed within the absolute value bars is a positive number, so the upper part of the definition applies. The upper part of the definition says that the absolute value of 8 is 8 itself. Thus, $|8| = 8$.

2. $|-6|$. The number enclosed within the absolute bars is a negative number, so the lower part of the definition applies. The lower part of the definition says that the absolute value of -6 is $-(-6) = 6$. Thus, $|-6| = 6$.

3. $|a|$ if it is known that $a < -10$. Since a is less than -10, it must be a negative number. Since it is a negative number, the lower piece of the definition applies. The lower piece of the definition says that the absolute value of a is $-a$. Thus, $|a| = -a$. If you're thinking that absolute value is always positive, your're right, and this makes sense because if a is negative (which it is, $a < -10$), then $-a$ is positive!

4. Compare $|9 - 5|$ and $|5 - 9|$.
   (i) $|9 - 5| = |4| = 4$.   (Using the upper part of the definition.)
   (ii) $|5 - 9| = |-4| = 4$.   (Using the lower part of the definition.)
   Thus, $|9 - 5| = |5 - 9|$.

Part 4 in the previous Illuminator Set illustrates an important fact concerning computations with absolute values.

Distances on the
Number Line

Distances on the Number Line
If a and b are the coordinates of any two points on the number line, then $|a - b|$ represents the distance between the points a and b and $$|b - a| = |a - b|$$

The next example illustrates the computation of lengths of several line segments on the number line. The notation $\overline{AE}$ indicates the *length of the line segment AE*. The first letter indicates one endpoint of the line segment and the second letter, the other endpoint. Thus, AE represents the line segment from A to E, or equivalently, from E to A, the direction making no difference. The points on the line are indicated in Figure 3.11

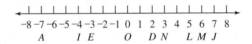

Figure 3.11

ILLUMINATOR SET B

1. $|\overline{DI}| = |2 - (-4)| = |2 + 4| = |6| = 6.$ Equivalently, $|\overline{DI}| = |-4 - 2| = |-6| = 6.$

2. $|\overline{MA}| = |6 - (-7)| = |6 + 7| = |13| = 13.$ Equivalently, $|\overline{MA}| = |-7 - 6| = |-13| = 13.$

3. $|\overline{JN}| = |7 - 3| = |4| = 4.$ Equivalently, $|\overline{JN}| = |3 - 7| = |-4| = 4.$

4. $|\overline{LO}| = |5 - 0| = |5| = 5.$ Equivalently, $|\overline{LO}| = |0 - 5| = |-5| = 5.$

5. $|\overline{EO}| = |-3 - 0| = |-3| = 3.$ Equivalently, $|\overline{EO}| = |0 - (-3)| = |0 + 3| = |3| = 3.$

Whenever the line segment is determined by the origin and another point on the number line, say with coordinate x, the notation $|x - 0|$ reduces to $|x|$. For example, and as problems 4 and 5 of the previous example verify, $|\overline{LO}| = |5 - 0| = |5|$, and $|\overline{EO}| = |-3 - 0| = |-3|$.

But this suggests a relationship between the length of a line segment determined by the origin and another point and the definition of absolute value. Indeed, the absolute value of a real number x, $|x|$, can be interpreted as the length of the line segment determined by the origin and the point with coordinate x.

The Absolute Value Function

The Absolute Value Function

The **absolute value function** is defined as

$$f(x) = \begin{cases} x, & \text{if } x \geq 0; \\ -x, & \text{if } x < 0. \end{cases}$$

To graph the absolute value function, we must consider both pieces of its defining expression.

Since $f(x)$ can be either x or $-x$, its graph is the combination (union) of the graphs of $f(x) = x$ for $x \geq 0$, and $f(x) = -x$ for $x < 0$. Notice the domain of each piece of the graph. For $f(x) = x$, the domain is the set of all nonnegative real numbers. This means that this piece of the graph has an endpoint at $x = 0$ and does not exist for $x < 0$. The domain of $f(x) = -x$ is $x < 0$. This means that this piece of the graph has an endpoint at $x = 0$ but does not include it and does not exist for $x \geq 0$. Graph $f(x) = |x|$ by entering the expression ABS(X) in the Y= menu.

Each piece of $f(x) = |x| = \begin{cases} x, & \text{if } x \geq 0; \\ -x, & \text{if } x < 0. \end{cases}$ is graphed separately according to its domain. Window 3.11 shows the graph of $f(x) = |x|$ in the standard viewing window. Notice that it always lies above the input axis, indicating that $f(x) \geq 0$, which is what we expect. The absolute value function

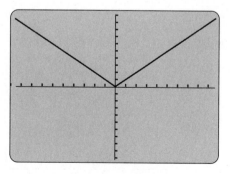

Window 3.11

is important, so we illustrate it further by constructing the graph of one of its many versions.

EXAMPLE SET A

Graph $f(x) = |2x - 1|$ first using your calculator, and then using pencil and paper.

Solution:
To construct the graph using the calculator, enter ABS (2x - 1). Once the graph is complete, notice that it has the same basic shape as that of $f(x) = |x|$, but that it is a bit narrower and is offset slightly to the right. Press ZOOM 2 and ENTER to zoom in around the zero of this function. The zero appears to be at $x = \frac{1}{2}$. Indeed it is. All graphs of absolute value functions will have this basic shape. Also, recall from our discussion of translations of functions in Section 2.4 that coefficients greater than one expand (narrow) the graph, and subtracting a positive number from the variable translates the graph to the right. Thus, if we rewrite the function as

$$|2x - 1| = \left|2\left(x - \frac{1}{2}\right)\right| = |2|\left|x - \frac{1}{2}\right| = 2\left|x - \frac{1}{2}\right|$$

we can see that this graph is an expansion (narrowing) and translation by one-half unit to the right of $f(x) = |x|$.

To construct the graph using pencil and paper, we begin by writing $f(x) = |2x - 1|$ so we may clearly see its pieces and determine domains.

$$f(x) = \begin{cases} 2x - 1, & \text{if } 2x - 1 \geq 0; \\ -(2x - 1), & \text{if } 2x - 1 < 0. \end{cases}$$

Now, $2x - 1 \geq 0$ when $x \geq \frac{1}{2}$. Also, $2x - 1 < 0$ when $x < \frac{1}{2}$. We now rewrite $f(x)$ in the even more informative form

$$f(x) = \begin{cases} 2x - 1, & \text{if } x \geq 1/2; \\ -2x + 1, & \text{if } x < 1/2. \end{cases}$$

We construct the graph of $f(x) = |2x - 1|$ by graphing each piece separately (paying close attention to domains) on the same coordinate axes. With no surprise, the graph is the same as that obtained using the calculator.

Piecewise Functions

Piecewise Function   The absolute value function is an example of a more general class of functions called **piecewise functions**. A piecewise function is a function in which the expression that produces the ordered pairs can change as the independent variable assumes various subsets of the domain. In other words, the ordered-pair producing expression is broken into pieces that are defined for various pieces of the domain. The function $f(x)$ is the union of these various pieces. The following example will help to illuminate this idea.

ILLUMINATOR SET C

1. The function

$$f(x) = \begin{cases} -2x - 3, & \text{if } x \leq -1; \\ 3, & \text{if } -1 < x \leq 2; \\ x + 1, & \text{if } x > 2 \end{cases}$$

is a piecewise function. The domain has been separated into three pieces: $x \leq -1, -1 < x \leq 2$, and $x > 2$. In this function, each piece of the domain is acted on by a different defining expression. The function is the union of these three pieces. Figure 3.12 illustrates this concept.

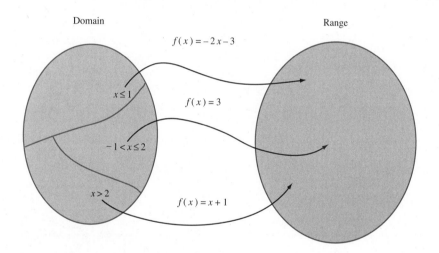

Figure 3.12

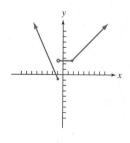

Figure 3.13

The union of these three graphs results in the graph of $f(x)$, shown in Figure 3.13.

2. You can use your graphing calculator to construct the graph of

$$f(x) = \begin{cases} x^2 + 2, & \text{if } x \leq -1; \\ 1, & \text{if } -1 < x \leq 2; \\ x, & \text{if } x > 2. \end{cases}$$

First, change the mode from Connected to Dot by pressing the MODE key, then using the arrow to move to Dot. When you press ENTER, Connected becomes unhighlighted and Dot becomes highlighted. The Dot mode makes the function appear discontinuous in that it appears as a collection of individual dots. This is a distortion of the reality that the function is describing, but graphing piecewise functions in the Connected mode produces an even greater distortion. (Try it.)

In the first piece of the function, the defining expression $x^2 + 2$ is quadratic and will result when the input values are less than or equal to -1 ($x \leq -1$). Using Y1, enter the defining expression in parentheses. Immediately following the closing right parenthesis, and within another set of parentheses, enter the inequality that describes the appropriate piece of the domain, in this case $x \leq -1$. You can find the $\leq$ symbol in the TEST menu (2nd MATH).

Using Y2, enter, within parentheses, the defining expression for the second piece of the function, 1. This piece of the function is active only when the input value is between -1 and 2, that is, when $x > -1$ and $x \leq 2$. Unfortunately, you cannot enter $-1 < x \leq 2$. You must enter this interval as $x > -1$ and $x \leq 2$ by entering (X> −1)(X≤2), using the TEST key as with the first piece.

Now, using, say Y3, enter, within parentheses, the defining expression for the third piece of the function, x. Immediately following the closing right parenthesis, enter, in its own set of parentheses, the inequality for this piece.

The Y= viewing window should appear as in Window 3.12. Window 3.13 shows the graph of $f(x)$ in the standard viewing window.

```
:Y1=(X^2+2)(X≤ −1)
:Y2=(1)(X>-1)(X≤ 2
)
:Y3=(X)(X>2)
:Y4=
```

Window 3.12

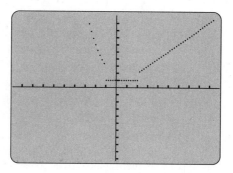

Window 3.13

The Greatest Integer Function

An important class of functions that occur in many applications is the class of *step functions*. There are many types of step functions, but we examine only the *greatest integer function*.

ILLUMINATOR SET D

The greatest integer function is denoted by $f(x) = [\![x]\!]$ and is the largest integer that is less than or equal to x. Window 3.14 shows the graph (in Dot mode) of $f(x) = [\![x]\!]$. The greatest integer function can be accessed by selecting MATH and then choosing NUM from the menu, and selecting INT. (Be sure to change the graphing mode from Connected to Dot.)

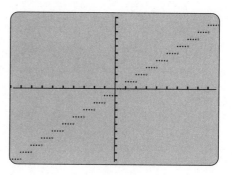

Window 3.14

To understand more fully what this function does, zoom in once. Now press TRACE, and move the cursor onto the first step below and to the left of the origin. Move the cursor to some x-value, say -0.7631579. The greatest (largest) integer less than or equal to -0.7631579 is -1, which is precisely what the y-coordinate shows. Now, use the arrow key to move the cursor to the next step up, the interval between 0 and 1. Move the cursor to some x-value, say 0.3410526. The corresponding y-value is 0, which is the greatest (largest) integer less than or equal to 0.3410526. Move the cursor slowly to the right endpoint of this step and then watch the output value change from 0 to 1 as the cursor moves to the next higher step. Indeed, for every x-value in this interval, 1 is the greatest integer less than or equal to it.

What happens at a left endpoint of an interval? Use 3 for example. What is the greatest (largest) integer less than or equal to 3? It's 3, of course. As you move along the interval from 3 toward 4, the greatest integer less than or equal to any of these numbers is still 3. This means that as the input moves from left to right, there is a point at $(3, 3)$, and a horizontal line from 3 moving toward 4. What happens at the right endpoint of this interval, that is, what happens at 4? At 4, the greatest integer less than or equal to 4 is 4. The output has changed from 3 to 4. So for input values just less than 4, the output is 3. But at 4, the output jumps to 4. So when the input is 4, the point on the graph is at $(4, 4)$. This means that at the left endpoint of an interval, there is a point, and we commonly denote this fact with a closed circle. But at the right endpoint, the graph jumps up to the next step and to a closed circle. So right endpoints must be denoted with open circles, to indicate that that input value has no corresponding output value on that part of the curve. Figure 3.14 shows this more clearly than can be displayed on the calculator.

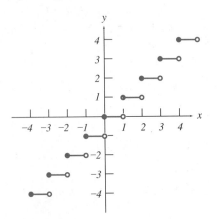

Figure 3.14

You should use some other negative domain values and try to convince yourself this fact is still true. You'll see that it is.

Writing Activity   Use complete sentences to describe the following types of functions (without referring to the book): absolute, identity, greatest integer, and piecewise. Be sure to include a discussion of the domain and range of each.

Writing Activity   The function $f(x) = mx + b$, where m and b are real numbers, is called the linear function. The function $g(x) = ax^2 + bx + c$, where a, b, and c are real numbers, is called the quadratic function. Describe all the different types of graphs they would produce with various choices of the constants. Also, describe what the domains and ranges of these various cases would be and why.

EXERCISES

For exercises 1–12, the function $g(x)$ is the absolute value of the function $f(x)$. Graph each pair of functions and explain what effect the absolute value function has on the function $f(x)$, if any at all. Do so in terms of how the behavior of $g(x)$ may differ from the behavior of $f(x)$.

1. $f(x) = -5$, and $g(x) = |f(x)|$.

2. $f(x) = -3$, and $g(x) = |f(x)|$.

3. $f(x) = x - 4$, and $g(x) = |x - 4|$.

4. $f(x) = x - 7$, and $g(x) = |x - 7|$.

5. $f(x) = 2x - 4$, and $g(x) = |2x - 4|$.

6. $f(x) = 3x - 7$, and $g(x) = |3x - 7|$.

7. $f(x) = x^2 - 6x + 7$, and $g(x) = |f(x)|$.

8. $f(x) = x^2 - 8x + 13$, and $g(x) = |f(x)|$.

9. $f(x) = x^2 + 8x + 18$, and $g(x) = |f(x)|$.

10. $f(x) = x^2 - 2x + 4$, and $g(x) = |f(x)|$.

11. $f(x) = -|x^2 - 8x + 13|$, and $g(x) = |f(x)|$.

12. $f(x) = -|x^2 - 6x + 7|$, and $g(x) = |f(x)|$.

For exercises 13–17, construct each graph using the standard viewing window, -10 to 10 for x and -10 to 10 for y.

13. $f(x) = \begin{cases} 3x - 1, & \text{if } x < 1; \\ x^2 - 2x + 5, & \text{if } x > 1. \end{cases}$

14. $f(x) = \begin{cases} x^2 - 4, & \text{if } x \leq 0; \\ \dfrac{3}{4}, & \text{if } x > 1. \end{cases}$

15. $f(x) = \begin{cases} |x - 3|, & \text{if } x < 3; \\ -x^2 + 6x + 8, & \text{if } x \geq 3. \end{cases}$

16. $f(x) = \begin{cases} -x^2 - 2x, & \text{if } x < -1; \\ 3, & \text{if } x = -1; \\ x^2 + 2x + 2, & \text{if } x > -1. \end{cases}$

17. $f(x) = \begin{cases} x^2 + 8x + 13, & \text{if } x < -4; \\ 0, & \text{if } x = -4; \\ -3, & \text{if } x > -4. \end{cases}$

For exercises 18–21, construct the graph of each function and then describe its behavior in terms of the given situation.

18. The monthly charge to a household, in dollars, for water in a particular city is given by the function

$$C(x) = \begin{cases} 0.3x, & \text{if } 0 \leq x \leq 15; \\ 0.9x - 8, & \text{if } x > 15, \end{cases}$$

where x is the number of cubic feet of water used by the household.

19. The monthly charge to a household, in dollars, for x kilowatt hours of electricity used by a residential customer in a particular city is given by the function

$$C(x) = \begin{cases} 0.05x, & \text{if } 0 \le x < 210; \\ 0.08x, & \text{if } 210 \le x < 340; \\ 0.1x - 6.8, & \text{if } 340 \le x < 450; \\ 0.12x - 14, & \text{if } x > 450. \end{cases}$$

20. In a particular county, income tax in x dollars earned during the year is determined by the function

$$C(x) = \begin{cases} 0x, & 0 \le x < 10,000; \\ 0.15x - 1,500, & 10,000 \le x < 25,000; \\ 0.2x - 2,750, & 25,000 \le x < 40,000; \\ 0.25x - 4,750, & x \ge 40,000. \end{cases}$$

21. Suppose that the production of N items of a new line of products is given by the function

$$N(t) = \begin{cases} 300 \left[(t + 10) - \dfrac{600}{t + 60} \right], & 0 \le t < 20; \\ 500 \left[(t - 7) - \dfrac{1}{t + 18} \right], & 20 \le t \le 52, \end{cases}$$

where t is the number of weeks the new line has been in production.

22. The diagram shown below represents the roof of a house. Because the side view consists of four lines that are joined at three points, the outline of the roof can be called a piecewise function. Find the equations of the roof if the axes are chosen as shown in the diagram.

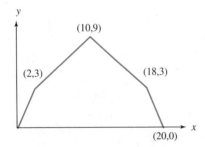

23. A boy is a passenger in a car and wants to throw his baseball at a tree up ahead. The tree is on his side of the freeway. He waits until his car (going 50 mph) is just crossing in front of the tree, aims his ball directly at it and throws. Is he successful? If not, which way did the ball fall and why? (Note: Everything inside a moving car is traveling at the same speed as the car.) What would be a better strategy for the boy?

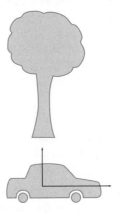

24. John earns $10.33 per hour at his job. However, if he works more than 40 hours in a week, he gets 1.6 times his usual rate for every hour worked over 40 hours. If x denotes the number of hours worked per week, write down the wage function $w(x)$ (wage earned each week) covering all possibilities.

25. A motorcycle stuntwoman is planning to jump across a 352-foot-wide river and sets up ramps on both sides of the river. Her vertical speed on takeoff is 50 feet/second, so that $y =$ height of the stuntwoman above the ramp $= 50t - 16t^2$. She figures that if her horizontal speed at the time of takeoff is 124 feet per second, she will land exactly on top of the ramp on the other side. Is she correct? If not, will she drop into the water or overshoot

her target? Give reasons for your answer.

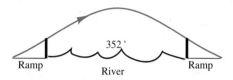

26. (Refer to the previous exercise) In her next attempt the stuntwoman changes her horizontal speed to 130 feet per second. What is the result this time?

27. (Refer to the previous two exercises) In her final attempt, the stuntwoman does some calculations. She solves the equation $y = 50t - 16t^2 = 0$ for t. (Why?). How does she adjust the horizontal speed with this information?

3.7   Inequalities Involving Polynomial Functions

Introduction

Graphical Methods

Analytic Methods

Applications of Inequalities

Rational Inequalities

Introduction

The graphing calculator can be used to approximate the input values for which the output value of a function is not just zero, as was described previously, but for which it also is greater than or less than some specified value. Noting that $f(x)$ represents the output of a function, the output will be positive when $f(x) > 0$ and the graph of $f(x)$ lies *above* the x-(input) axis. The output will be negative when $f(x) < 0$ and the graph of $f(x)$ lies *below* the x-axis. Figure 3.15 illustrates these relationships. The graph shows that the function $f(x)$ is positive, that is, $f(x) > 0$, on the intervals $(-\infty, a)$ and $(b, +\infty)$. It is

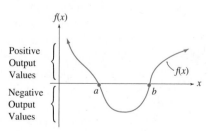

Figure 3.15

negative, that is, $f(x) < 0$, on the interval (a, b). The function $f(x)$ has zeros at $x = a$ and $x = b$.

Graphical Methods

We can use the graphing computation to approximate such intervals by graphing the function and using the approximation techniques described in Section 1.4.

EXAMPLE SET A

1. Construct the graph of $f(x) = 0.14x^4 - 0.59x^3 + 0.2x - 6$ and determine with an error of at most 0.01 (a) the zeros of $f(x)$, and (b) the intervals upon which $f(x)$ is positive and the intervals upon which it is negative.

   Solution:
   The graph of $f(x)$ is displayed in Window 3.15.

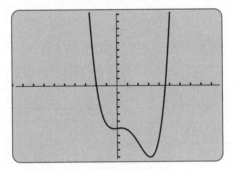

Window 3.15

   a. The zeros of $f(x)$ are, with an error of at most 0.01, -1.95 and 4.59.

   b. Since this function is a 4th degree polynomial function, it will have at most three turnaround points. The one at $x = 3$ is very evident from the graph. Let's zoom in (by creating boxes) around the region that looks somewhat flat. This activity shows two more turnaround points. Thus, all three turnaround points are visible in the standard viewing window, and we can conclude that we have a complete graph. Then, $f(x) > 0$ on approximately $(-\infty, -1.95)$ and $(4.59, +\infty)$, and $f(x) < 0$ on approximately $(-1.95, 4.59)$.

2. Determine by three methods, and in each case with an error of at most 0.01, the values of x for which $f(x) \leq g(x)$, where $f(x) = 0.14x^4 - 0.59x^3 + 0.2x - 6$ and $g(x) = -0.46x^2 + 5.24x - 10.65$.

Solution:

a. One way of using the calculator to approximate the desired input values is to enter the defining expression for $f(x)$ using Y1, the defining expression for $g(x)$ using Y2, and then Y1−Y2 using Y3. Notice that Y1−Y2 is equivalent to $f(x) - g(x)$. Now, $f(x) \leq g(x)$ when $f(x) - g(x) \leq 0$. Use the method presented in the previous example to determine the x-values for which Y3 lies below the x-axis. In this case, with an error of at most 0.01, $f(x) \leq g(x)$ on the interval $(0.93, 4.79)$. (When graphing Y3=Y1−Y2, it is convenient to deactivate Y1 and Y2. Do this by placing the cursor over the corresponding $=$ sign and pressing the ENTER key. The function defined in this location is still active, but the graphing capability is not.)

b. A second way to determine the x-values for which $f(x) \leq g(x)$ is to graph $f(x)$ and $g(x)$ separately to see where the graph of $f(x)$ lies below the graph of $g(x)$. Because it may be difficult at times to see which function is which, we can have the calculator shade the region for which $f(x) \leq g(x)$. We do this by using the Shade feature of the calculator found on the DRAW menu with $f(x)$ as Y1 and $g(x)$ as Y2. The Shade function has the form

Shade(lower function, upper function)

In this case, you would place Shade(Y1,Y2) in the calculator window and press ENTER.

c. A third way to determine the x-values for which $f(x) \leq g(x)$ is to enter $f(x)$ using Y1, $g(x)$ using Y2, and for Y3, Y1−Y2≤ 0, using the TEST menu to access the $\leq$ symbol.

This activity defines Y3 as Y1−Y2≤ 0, which corresponds to $f(x) - g(x) \leq 0$. It will be true when $f(x) \leq g(x)$. The TEST feature tests for the truth of the statement Y1−Y2≤ 0. When it is true, a 1 results, and when it is false, a 0 results. The calculator should be placed in the Dot mode. When Y3 is graphed, a horizontal line at $y = 1$ indicates the x-values that produce a true statement. Since the output will be 0 or 1, you might want to set the range at, say $-10 \leq x \leq 10$ and $-1 \leq y \leq 2$ to get a better view. The functions Y1 and Y2 need to be deactivated by unhighlighting their equal signs. Window 3.16 shows this viewing window. Again, with an error of at most 0.01, $f(x) \leq g(x)$ on approximately $(0.93, 4.79)$.

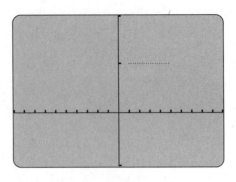

Window 3.16

Analytic Methods

As with finding zeros of polynomial functions, there are special cases for which inequalities involving polynomial functions can be solved exactly. These are inequalities involving linear and quadratic polynomial functions. Because these methods were studied extensively in previous algebra courses, we present only two examples that are intended to serve as memory joggers.

EXAMPLE SET B

1. Find the values of x for which $f(x) > g(x)$, where $f(x) = 5x + 4$ and $g(x) = 7x - 8$.

   Solution:

$$f(x) > g(x)$$

$$5x + 4 > 7x - 8 \quad \text{Solve for } x$$

$$-2x + 4 > -8$$

$$-2x > -12 \quad \text{Divide each side by } -2$$

$$x < 6 \quad \text{Notice the symbol reversal}$$

   Remember, when both sides of an inequality are multiplied or divided by a negative number, the inequality symbol reverses direction. Thus, $5x + 4$ will be strictly greater than $7x - 8$ when x is strictly less than 6. (You might try a graphical solution using the methods of the previous example.)

2. Find the values of x for which $f(x) > 0$, where $f(x) = x^2 - 2x - 24$.

Critical values

Solution:

We begin by finding the **critical values** of the function defined by $x^2 - 2x - 24$. Critical values are values for which the function is either zero or undefined. Since $f(x)$ is a polynomial function, it is never undefined (its domain is the set of all real numbers). Critical values occur only where $f(x) = 0$. We'll set $f(x)$ equal to 0 and solve for x.

$$f(x) = 0$$

$$x^2 - 2x - 24 = 0 \quad \text{Factor the quadratic expression}$$

$$(x - 6)(x + 4) = 0 \quad \text{Solve for } x$$

$$x = 6, -4$$

Thus, the critical values are -4 and 6, and we plot them on a number line. We'll use open circles to indicate that they are *not* included as members of the solution set. (They are not included because of the strict inequality of the original problem.) Figure 3.16 shows a number line with these two critical values sketched in.

Figure 3.16

Now, in each interval, choose a *test value* (any number in the interval will do). If the test value produces a true (or false) statement when substituted into the inequality, the inequality is true (or false) for all values in that interval. We'll use $(x - 6)(x + 4) > 0$ as an equivalent form to $x^2 - 2x - 24 > 0$.

In the leftmost interval, choose as a test point any number to the left of -4, say -5. Then

$$(-5 - 6)(-5 + 4) = (\text{negative})(\text{negative}) = (\text{positive}) > 0$$

which is true. Thus, all values in this interval are solutions. That is, solutions exist in the interval $(-\infty, -4)$.

In the middle interval, choose and test a value, say 0.

$$(0 - 6)(0 + 4) = (\textit{negative})(\textit{positive}) = (\textit{negative}) > 0$$

which is false. Thus, no values in this interval are solutions.

In the rightmost interval, choose and test, say 7.

$$(7 - 6)(7 + 4) = (\textit{positive})(\textit{positive}) = (\textit{positive}) > 0$$

which is true. Thus, all values in this interval are solutions. That is, solutions exist in the interval $(6, +\infty)$.

Thus, $x^2 - 2x - 24 > 0$ on the intervals $(-\infty, 4)$ and $(6, +\infty)$.

(Are you convinced, or do you want to construct the graph?) Figure 3.17 shows the sign chart for this inequality.

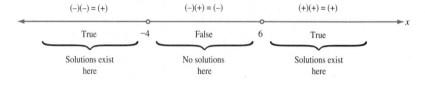

Figure 3.17

Applications of Inequalities

EXAMPLE SET C

The revenue $R(x)$ (in thousands of dollars) realized by a company on the sale of x units of a product has been researched and modeled (approximated) by the formula $R(x) = 446x - 0.02x^2$, $0 < x < 25,000$. The cost $C(x)$ (in thousands of dollars) of producing x units of the product is modeled by $C(x) = 155x + 243,000$, $0 < x < 25,000$. (1) Determine the number of units produced for which there is a profit, and (2) approximate the number of units that, when produced, yields the maximum profit and approximate that profit.

Solution:

1. A profit will occur when the revenue is greater than the cost, that is when $R(x) > C(x)$. Thus, we need to solve the inequality $R(x) > C(x)$, or equivalently, $446x - 0.02x^2 > 155x + 243,000$. Since the functions themselves are only approximations to the revenue and the cost, we will use the calculator to approximate the solution. (After all, what good is an exact solution to a function that generates only approximate values.) Window 3.17 shows the graphs of these functions in the viewing window $0 < x < 30,000$ and $0 < y < 3,000,000$. The interval of x-values can be taken from the designated domain, with a little room for a margin. The interval of y-values must be found by some trial and error and some good use of the TRACE key. Rough estimates (estimates using the trace key and without any zooming) for the number of units that must be produced in order to effect a profit are a minimum of approximately 790 and a maximum of about 13,680.

2. A rough estimate for the number of units that must be produced in order to maximize the profit is 11,000, which produces a maximum profit of about $2,486,000.

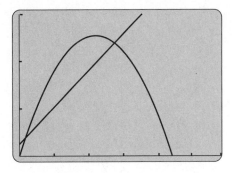

Window 3.17

Rational Inequalities

Rational functions are functions that are quotients of polynomial functions. We will study them in some detail in Chapter 4. At this point, however, we wish to consider inequalities of the form

$$\frac{f(x)}{g(x)} < 0, \quad g(x) \neq 0$$

and $f(x)$ and $g(x)$ are polynomial functions. The next example illustrates how to solve, analytically, a rational inequality.

EXAMPLE SET D

Find the values of x for which $h(x) \geq p(x)$, where $h(x) = \dfrac{2x - 1}{x + 4}$ and $p(x) = 1$.

Solution:
We cannot simply multiply both sides of $\dfrac{2x - 1}{x + 4} \geq 0$ by $x + 4$ since we don't know whether or not to change the direction of the $\geq$ symbol. Is $x + 4 > 0$ or is $x + 4 < 0$? If $x + 4 < 0$, the inequality symbol changes direction upon multiplication. If $x + 4 > 0$, the inequality symbol does not change direction upon multiplication. Since we don't know if $x + 4$ is positive or negative, we

cannot use it as a multiplier. We begin by setting the inequality equal to zero in order to find the critical values.

$$\frac{2x-1}{x+4} - 1 \geq 0$$

Now, write the left-side expression as a single fraction.

$$\frac{2x-1}{x+4} - 1 \geq 0$$

$$\frac{2x-1-1(x+4)}{x+4} \geq 0$$

$$\frac{x-5}{x+4} \geq 0$$

1. The expression $\dfrac{x-5}{x+4}$ is zero when $x = 5$. (Fractions are zero only when the numerator is zero.) Thus, 5 is a critical value.

2. The expression $\dfrac{x-5}{x+4}$ is undefined when $x = -4$. (Polynomial fractions are undefined only when the denominator is zero.) Thus, -4 is a critical value.

   This function has two critical values, -4 and 5. Plot these on the number line, using a closed circle at 5 (since the original inequality states that equality is possible), and an open circle at -4 (since -4 produces zero in the denominator and therefore must be excluded.) Figure 3.18 shows a number line and these values breaking the line into the three distinct intervals $(-\infty, -4)$, $(-4, 5)$ and $(5, +\infty)$. Choose test points in each interval to locate solutions.

Figure 3.18

3. Select a test number in the interval $(-\infty, -4)$. We'll choose $x = -5$.

$$\text{Is} \quad \frac{-5-5}{-5+4} \overset{?}{\geq} 1$$

$$10 \overset{\checkmark}{\geq} 0.$$

All numbers in this interval satisfy the inequality and are, therefore, solutions.

4. Select a test number in the interval $(-4, 5)$. We'll choose $x = 0$.

$$\text{Is} \quad \frac{0-5}{0+4} \overset{?}{\geq} 1$$

$$\frac{-5}{4} \geq 0.$$

No numbers in this interval satisfy the inequality and, therefore, there are no solutions in this interval.

5. Select a test number in the interval $(5, +\infty)$. We'll choose $x = 6$.

$$\text{Is} \quad \frac{6-5}{6+4} \geq 1 \quad \text{true or false?}$$

$$\frac{1}{10} \overset{\checkmark}{\geq} 0. \quad \text{True}$$

All numbers in this interval satisfy the inequality and are, therefore, solutions.

Thus, all numbers in the intervals $(-\infty, -4)$ and $(5, +\infty)$ are solutions. Figure 3.19 shows the solutions on a number line.

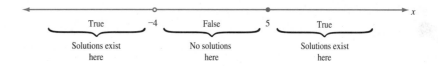

Figure 3.19

Writing Activity   Write down in paragraph form any methods you can think of for solving inequalities involving polynomial and rational functions.

EXERCISES

For exercises 1–22, use analytic methods to obtain exact solutions to the given inequality. Check each solution with your graphing calculator.

1. $x^2 - 4x - 21 \leq 0$.

2. $x^2 + 6x + 8 < 0$.

3. $x^3 + 3x - 4 > 0$.

4. $x^2 - 3x - 10 \geq 0$.

5. $(x-4)(x-1)(x+2) \geq 0$.

6. $(x-5)(x+2)(x+5) \leq 0$.

7. $(x^2 - 4)(x^2 - 9) \geq 0$.

8. $(x^2 - 1)(x^2 - 36) < 0$.

9. $x^3 - x^2 - 6x \leq 0$.

10. $x^3 + 2x^2 - 8x > 0$.

11. $x^2 - 8x + 16 > 0$. $(x-4)^2 > 0$

12. $x^2 + 10x + 25 < 0$.

$x \in \mathbb{R}$
$x \neq 4$

Calculator does not show correct solution

13. $x^2 + 7x < -10$.

14. $x^2 - x - 17 \geq 25$.

15. $6x^2 + 5x - 4 < 0$.

16. $8x^2 + 18x - 5 \geq 0$.

17. $x^3 - 27 \geq 0$.

18. $x^3 + 64 < 0$.

19. $x^4 + x^3 - 24x^2 - 4x + 80 \leq 0$, given that 4 and -2 are zeros.

20. $x^4 - 4x^3 - 23x^2 + 54x + 72 \leq 0$, given that 6 and -4 are zeros.

21. $x^5 + 5x^4 - 35x^3 - 125x^2 + 194x + 280 > 0$, given that -7, -4, and 2 are zeros.

22. $x^5 + 2x^4 - 16x^3 - 2x^2 + 15x \geq 0$, given that -1, 1, and 3 are zeros.

Write down in paragraph form an approach you might use to solve exercises 23–30. Then carry out the details of your solution.

23. If we drop an object from 686 feet above the ground, its height $h(t)$ above the ground t seconds after it starts its descent is given by

$$h(t) = -16t^2 + 686$$

For how many seconds after the object starts to fall is it still 200 or more feet above the ground?

24. The number $N(t)$ of bacteria in a colony is a function of time t in hours and is given by

$$N(t) = 1.45t^2 + 23t + 1026$$

After how many hours will there be more than 1500 bacteria in the colony?

25. The number $N_A(t)$ of individuals in colony A of bacteria t hours from now is given by the function

$$N_A(t) = 1.45t^2 + 23t + 1026$$

The number $N_B(t)$ of individuals in colony B of bacteria t hours from now is given by the function

$$N_B(t) = 2.88t^2 + 14t + 875$$

For what hours, over the next 24-hour period, will the population of the colony A be greater than the population of colony B?

26. A colony, A, of bacteria has an initial population of 1026 individuals. The number $N_A(t)$ of bacteria in colony A t hours from now is given by the function

$$N_A(t) = 1.5t^2 + 36t + 1026$$

Another colony, B, of bacteria also has an initial population of 1026 individuals. Because this colony is treated with a mild toxin, the number $N_B(t)$ of bacteria in colony B t hours from now is given by the function

$$N_B(t) = -7.6t^2 + 98t + 1026$$

For what hours, over the next 18-hour period, will the population of the colony treated with the toxin be smaller in number than the population that is not treated?

27. The management of a company has decided to test two different sales strategies, strategies A and B. The number $N_A(x)$ of sales using strategy A depends on x, the number of days from the beginning of the month, and is given by the function

$$N_A(x) = -0.03x^2 + 0.55x + 213$$

The number $N_B(x)$ of sales using strategy B depends on x, the number of days from the beginning of the month, and is given by the function

$$N_B(x) = -0.01x^2 + 0.12x + 213$$

For what days from the beginning of the month is strategy A better than strategy B, if it is at all?

28. A company realizes a profit of

$$P(x) = -0.01x^2 + 62.5x - 625$$

dollars on the sale of x units of one of its products. For how many units sold will the profit be greater than 10,000?

29. The area $A(x)$ of a concrete patio x feet in length is given by the function

$$A(x) = -2.6x^2 + 124x$$

For what lengths will the patio have an area greater than 1000 square feet?

30. The cost $C(x)$ to a company to produce x units of an item is given by the function

$$C(x) = -0.02x^2 + 25.7x + 315$$

The revenue $R(x)$ realized from the sale of x of those units is given by

$$R(x) = 0.002x^2 + 6.4x$$

For what number of items sold will the revenue be greater than the cost?

Summary and Projects

Summary

Behavior of Polynomials The dominating term of a polynomial function is the term of highest degree. When the input values are very large in absolute value, the entire polynomial behaves as the dominating term. Because of this, polynomial functions fall into two major categories. At large positive and small negative input values, the curve approaches infinity in opposite directions for odd-degree leading terms. It approaches infinity in the same direction for even-degree leading terms. In addition, a polynomial function of degree n will have at most $n - 1$ turning points and they are smooth and continuous.

Division of Polynomials The division algorithm shows you that when you divide a polynomial of degree two or higher by a linear polynomial, you obtain a polynomial of degree one less than the original polynomial, and that the remainder is a constant.

Factors of Polynomials The Factor Theorem shows you that if the number r is a zero of the polynomial function $f(x)$, then $x-r$ is a factor of $f(x)$ and vice-versa. You have quite a bit more information available when a polynomial is in factored form. The zeros are apparent and locations of possible local maximum and minimum values are readily available.

Multiple Zeros Polynomial functions behave differently at zeros of even multiplicity than they do at zeros of odd multiplicity. You can identify possible multiple zeros by examining the graph of the function. An additional important fact is that a polynomial function of degree n will have exactly n zeros if we allow complex number zeros.

Some Special Functions   Having a resource of numerous basic functions is valuable. The constant function $f(x) = k$ where k is a real number is a function that remains flat over its entire domain. The function $f(x) = x$ is called the identity function and has the feature that every output value is the same as its corresponding input value. The absolute value function $f(x) = |x|$ is a V-shaped graph with its corner at the origin. Piecewise functions are those that have two or more rules over separate domain intervals. They are made up of pieces of functions. The greatest integer function is the basis of the step function, which is like a flight of stairs without the vertical risers.

Solving Inequalities   The graphing calculator has revamped the way you approach solving inequalities. A graph of the expression(s) in the inequality gives you quite a bit of information. For example, the graph immediately shows where a function is positive and negative. While the standard algebraic techniques remain important, they can be approached with much more confidence and knowledge through the use of the graphing calculator.

Projects

1. The polynomial function

$$f(x) = (x - 2)(x - 4)(x - a)$$

has zeros at 2, 4, and a. If $a > 4$, describe the changes in the graph of $f(x)$ as a gets close to 4 $(5, 4.1, 4.01, \ldots)$. Describe what happens to the graph of $f(x)$ as the value of a gets large, say, values like $a = 5, 10, 20, 50, 100, 1000, \ldots$.

2. Describe what happens to the zeros of the quadratic polynomial

$$f(x) = 2x^2 - 3x - c$$

as c increases through the integer values from -5 to 5. Suppose a quadratic polynomial has integer coefficients and two positive *rational* roots. Discuss the possibilities that may occur for the leading coefficient of the polynomial if the two zeros are less than one-thousandth of a unit apart. Use examples to support your conclusions. If the restriction that the zeros are rational is removed, do your conclusions still hold? Discuss your findings in light of your description of

$$f(x) = 2x^2 - 3x - c \quad \text{with } -5 \leq c \leq 5$$

CHAPTER 4

Rational and Parametric Functions

4.1   Introduction to Rational Functions

Applications of Rational Functions

Describing the Behavior of f(x) = 1/x

f(x) =1/x as a Decreasing Function

f(x)=1/x as a Reciprocal Function

Translations

Powers of f(x) = 1/x

Reflection of f(x) = 1/ x

Applications of Rational Functions

The intensity of light and sound varies inversely as the square of the distance. Thus, light and sound intensity is a function of distance from the light or sound source. This relationship can be represented by the function

$$intensity(distance) = \frac{k}{distance^2}, \text{where } k \text{ depends on the units of measure}$$

or more simply,

$$I(d) = \frac{k}{d^2}$$

We can dilute an acid solution with pure water. When 10 liters of a 25% acid solution are diluted with pure water, the concentration of the resulting acid solution varies as the inverse of 10 + (the number of liters of pure water added). This relationship can be represented by the function

$$concentration(liters\ of\ water) = \frac{10\ liters \times 0.25}{10 + (liters\ of\ water)}$$

or

$$C(w) = \frac{2.5}{10 + w}.$$

The functions I and C are examples of rational functions. A rational function is the quotient of two polynomial functions. Our study of these functions begins with the standard function $f(x) = \dfrac{1}{x}$.

Describing the Behavior of f(x) = 1/x

The function $f(x) = \dfrac{1}{x}$ is a simple rational function. The graph of this function appears in the calculator screen on the left of Window 4.1. We can describe

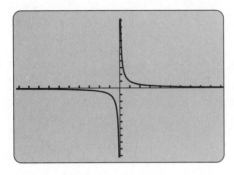
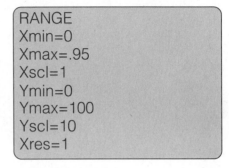

Window 4.1

the behavior of this function by carefully considering this graph. The function is decreasing starting at the far left $(-\infty)$ through negative values up to 0. It also is decreasing starting at values of x larger than 0 and moving to the right. How can we describe the behavior near 0? If we use the values in the RANGE menu in the calculator screen on the right of Window 4.1, we can use TRACE to complete Table 4.1. The table shows that the values of the function are large for x values close to 0. We can also see from the function that as we divide a fixed length by larger and larger values, the value of $1/x$ gets smaller and smaller. This occurs because as we divide a fixed length into more and more parts, each part is smaller. For example, if a 1 foot board is divided into 10 equal lengths, each length is 0.1 feet. If the same board is divided into 100 equal lengths, then each length is 0.01 feet. Thus, the function values decrease and are closer and closer to zero as the values of x get larger and larger (positively).

From this discussion, we have that the function $f(x) = \dfrac{1}{x}$ is a decreasing function on the interval $(0, \infty)$. We can verify this in Window 4.1. We prove this statement next.

Table 4.1

x	y
0.01	?
0.05	20
0.1	10
0.5	2
1	1

f(x) =1/x as a Decreasing Function

To show that a function is decreasing on an interval, we must show that for $x_1 < x_2$ we have that $f(x_1) > f(x_2)$. Suppose x_1 and x_2 are positive and

$x_1 < x_2$. Then dividing each side by the positive number x_1x_2, we have

$$\frac{x_1}{x_1x_2} < \frac{x_2}{x_1x_2}$$

Reducing produces

$$\frac{1}{x_2} < \frac{1}{x_1}$$

or, equivalently

$$\frac{1}{x_1} > \frac{1}{x_2}$$

but since

$$\frac{1}{x_1} = f(x_1) \quad \text{and} \quad \frac{1}{x_2} = f(x_2)$$

then

$$f(x_1) > f(x_2)$$

and, thus, $f(x_1) > f(x_2)$. The proof convinces us that our idea about dividing a fixed length into pieces actually works.

The same proof works if x_1 and x_2 are both negative since again their product is positive. The approach fails if x_1 is negative and x_2 is positive. Since $x_1 < 0 < x_2$ and since the reciprocal of a negative number is negative and the reciprocal of a positive number is a positive number,

$$\frac{1}{x_1} < 0 < \frac{1}{x_2}$$

From what we have done so far, we can quickly sketch the graph of $f(x) = \dfrac{1}{x}$ using paper and pencil. Our graph should look like the graph in Figure 4.1. Thus, $f(x)$ is decreasing on the intervals $(-\infty, 0)$ and $(0, \infty)$ and on any subintervals of these intervals. However, it is not decreasing on any interval that contains 0. Hence, it is not decreasing on the entire real line.

f(x)=1/x as a Reciprocal Function

Our old friend $I(x) = x$ can be multiplied by $f(x) = \dfrac{1}{x}$ in the following way.

$$I(x) \cdot f(x) = x \cdot \frac{1}{x} = 1$$

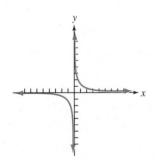

Figure 4.1

where 1 is the multiplicative identity for real numbers. These two functions are said to be reciprocals of one another. Thus,

$$f(x) = \frac{1}{I(x)}$$

is the reciprocal of $I(x)$.

Translations

Let's look at translations of this function.

$$f(x - 1) = \frac{1}{x - 1}$$

This is a translation to the right as we saw in Chapter 2. A graph of this function appears in Window 4.2. The table in Window 4.2 shows exactly how the values change.

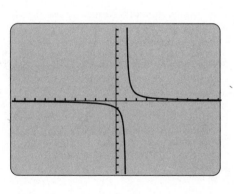

x	$\frac{1}{x}$	$\frac{1}{x-1}$
−1	1/−1	1/−2
0	1/0 undefined	1/−1
1	1/1	1/0 or undefined
2	1/2	1/1
3	1/3	1/2
4	1/4	1/3
5	1/5	1/4
6	1/6	1/5
7	1/7	1/6
8	1/8	1/7

Window 4.2

EXAMPLE SET A

Sketch the graph of the function

$$g(x) = \frac{1}{x + 3}$$

Table 4.2		
x	$\frac{1}{x}$	$\frac{1}{x+3}$
1	1	1/4
2	1/2	1/5
3	1/3	1/6
4	1/4	1/7
5	1/5	1/8
6	1/6	1/9
7	1/7	1/10
8	1/8	1/11

Solution:

This will be a translation of the standard function $f(x) = \dfrac{1}{x}$ to the left three units. This function appears in Window 4.3. Table 4.2 shows how the values change.

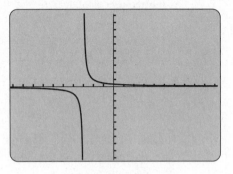

Window 4.3

Can you see how the same function values of $\dfrac{1}{x}$ appear 3 units earlier in the column for the function values of $\dfrac{1}{x+3}$? The line $x = -3$ is called

Asymptote

an **asymptote**. The word asymptote comes from the Greek word *symptote* which means together and the prefix *a* which means nearly, or not so. Thus, asymptote means not quite together.

Powers of f(x) = 1/x

The function

$$g(x) = [f(x)]^2 = \frac{1}{x^2}$$

is always positive and increasing on $(-\infty, 0)$ and decreasing on $(0, \infty)$ as illustrated in Window 4.4.

EXAMPLE SET B

Let $f(x) = \dfrac{1}{x}$. Sketch the graph of $h(x) = [f(x)]^3 = \dfrac{1}{x^3}$.

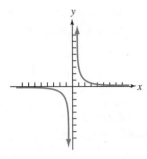

Figure 4.2

Solution:
This graph will be similar to $f(x)$ since an odd power of a negative number is negative. So this graph will be positive when $f(x)$ is positive, and negative when $f(x)$ is negative. The graph of this function appears in Figure 4.2.

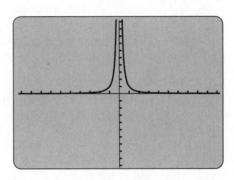

Window 4.4

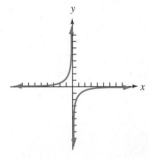

Figure 4.3

Reflection of f(x) = 1/ x

The function $g(x) = -\dfrac{1}{x}$ is graphed in Figure 4.3. Recall from Chapter 1 that the graph of the negative of a function is a reflection of the original graph about the x−axis. We should be able to see that is the case here.

EXAMPLE SET C

Sketch the graph of

$$g(x) = \frac{1}{2-x}$$

Solution:
As we have seen before, we can use an algebraic transformation to help us with the sketch of this graph.

$$g(x) = \frac{1}{2-x} = \frac{1}{-x+2} = \frac{1}{-(x-2)} = -\frac{1}{(x-2)}$$

Thus, $g(x)$ is both a reflection and a translation of $f(x) = \dfrac{1}{x}$, respectively about the x-axis and 2 units to the right. The graph appears in Figure 4.4.

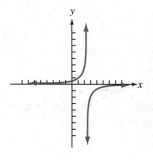

Figure 4.4

As we saw at the beginning of this section, rational functions appear in applied situations. We take a functional approach to a mixture problem in the next example.

EXAMPLE SET D

Pure water is added to 10 mL of a 40% acid solution. Find the concentration of acid as a function of the amount of pure water added and graph the function. Describe the behavior of the function.

Solution:

The concentration of acid is the original (unchanging) amount of acid divided by the total volume of the solution. Thus,

$$C(x) = \frac{0.40 \cdot 10}{10 + x} \quad \text{where } x \text{ is the amount of pure water added.}$$

The graph of the function C with x positive appears in Window 4.5. As the amount of pure water that is added increases, the concentration of acid in the solution decreases but is always positive. It decreases at a slower and slower rate.

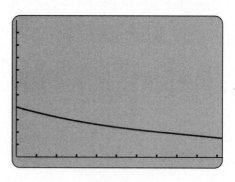

Window 4.5

EXERCISES

In exercises 1–8, draw a rough paper-and-pencil sketch of the given functions on the interval $[0, 4]$ on the same axes.

1. $f(x) = \dfrac{a}{x}$ for $a = 1, 2, 3, 4, 5.$

2. $f(x) = \dfrac{a}{x^2}$ for $a = 1, 2, 3, 4, 5.$

3. $f(x) = -\dfrac{a}{x}$ for $a = 1, 2, 3, 4, 5.$

4. $f(x) = -\dfrac{a}{x^2}$ for $a = 1, 2, 3, 4, 5.$

5. $f(x) = \dfrac{1}{x^a}$ for $a = 1, 2, 3, 4, 5.$

6. $f(x) = -\dfrac{1}{x^a}$ for $a = 1, 2, 3, 4, 5.$

7. $f(x) = \dfrac{a}{x}$ and $g(x) = ax$ for $a = 1, 2, 3.$

8. $f(x) = -\dfrac{a}{x}$ and $g(x) = -ax$
   for $a = 1, 2, 3.$

In exercises 9–16, make a paper-and-pencil sketch of the graph of the indicated function. Be sure to carry out the paper-and-pencil thought process before checking your results with your calculator. You will find that this will greatly increase your ability to analyze the behavior of functions.

9. $f(x) = \dfrac{1}{x - 3}$

10. $f(x) = \dfrac{1}{x - 4}$

11. $f(x) = \dfrac{1}{x + 2}$

12. $f(x) = \dfrac{1}{x + 4}$

13. $f(x) = \dfrac{2}{x - 3}$

14. $f(x) = \dfrac{5}{x - 3}$

15. $f(x) = \dfrac{-2}{x - 3}$

16. $f(x) = \dfrac{-3}{x - 3}$

In exercises 17–20, use your graphing calculator to help you sketch a graph of the following functions.

17. $f(x) = \dfrac{1}{2x^2}$

18. $f(x) = -\dfrac{1}{(x - 2)^2}$

19. $f(x) = -\dfrac{1}{3(x - 2)^4}$

20. $f(x) = -\dfrac{3}{(x - 2)^4}$

In exercises 21–24, describe in paragraph form the behavior of the given function.

21. $g(x) = \dfrac{1}{x - a}$

22. $g(x) = \dfrac{-1}{x - b}$

23. $g(x) = \dfrac{1}{x - 3} + 4$

24. $g(x) = \dfrac{1}{x - 3} - 2$

In exercises 25–27, complete the given table and answer the questions.

25.

x	$1/x^2$
10^1	10
10^2	100
10^3	1000
10^4	10000
10^5	
10^6	
10^7	
10^8	
10^9	

As the x values increase, what happens to the function values?

26.

x	$1/x$
10^1	
10^2	
10^3	
10^4	
10^5	
10^6	
10^7	
10^8	
10^9	

As the x values increase, what happens to the function values?

27.

x	$1/x$
10^0	
10^{-1}	
10^{-2}	
10^{-3}	
10^{-4}	
10^{-5}	
10^{-6}	
10^{-7}	
10^{-8}	
10^{-9}	

Are the numbers in the left (domain) column increasing or decreasing? As the values of x decrease through positive values toward 0, what happens to the values of the function?

28.

x	$1/x^2$
10^0	
10^{-1}	
10^{-2}	
10^{-3}	
10^{-4}	
10^{-5}	
10^{-6}	
10^{-7}	
10^{-8}	
10^{-9}	

Are the numbers in the left (domain) column increasing or decreasing? As the values of x decrease through positive values toward 0, what happens to the values of the function?

29. Bea builds a decorative pond that can hold 160 cubic feet of water. She starts to fill the pond using a hose that supplies 2 cubic feet of water per minute. She expects the pond to be filled in 80 minutes. However, it takes exactly two hours for the pond to be completely filled. Since no leaks in the hose are found, she concludes that the pond must have sprung a leak. Let x denote the number of minutes it would take the pond to completely empty if the pond were full and the hose were off. Find an expression for the rate at which the pond fills when the hose is on. Next, deduce the amount of water added in 120 minutes. Graph the function setting Xmin in the RANGE equal to 1. (Why?) Also graph $y = 160$. What does the x value of the point of intersection indicate?

30. Bea fixes the leak and fills the pond using two different hoses at the same time. One of the hoses is the one used before in exercise 29; she does not know the rate for the second hose. It requires 50 minutes to complete the process. Find how long it would take if Bea uses the second hose alone to completely fill the pond (volume of the pond is 160 cubic feet; rate for the first hose is 2 cubic feet/minute). Describe how you could solve the problem graphically.

4.2 Rational Functions I

Introduction

Function Multiples

Reciprocals of Factored Quadratic Polynomials

Introduction

In this section, we consider two of the easier types of rational functions: quotients of linear polynomials and reciprocals of polynomials. Such polynomials sometimes occur in the real world. For example, what happens when a 10% acid solution is added to 10 mL of a 40% acid solution? We can create a function of concentration with respect to the amount of 10% acid solution added. This function will be the amount of acid divided by the total amount of solution. Thus,

$$C(x) = \frac{0.40 \cdot 10 + 0.10x}{10 + x} = \frac{4 + 0.1x}{10 + x}$$

where x is the amount of 10% acid solution added. Here we have a more complex rational function (the quotient of two polynomials).

How can we analyze and graph such functions?

EXAMPLE SET A

Analyze the function

$$f(x) = \frac{x + 3}{x + 4}$$

Solution:

A standard window calculator graph of this function is misleading. We see from the symbolic rule for the function that it is defined for all $x \neq -4$. When x is near -4, the numerator is close to -2 and the denominator is close to 0. Table 4.3 gives a glimpse of the behavior of the function near -4. A vertical asymptote seems to occur. Also, as x gets large the values of the function get closer and closer to 1. Enter Y1=X+3, Y2=X+4 and then Y3=Y1Y2>0 by selecting the > symbol from the TEST menu. Use $[-7, 7]$ for the x and y RANGE or WINDOW values. Window 4.6 shows that the function is positive to the left of -4 and negative to the right of -4. The asymptotic behavior of the function is shown in Figure 4.5.

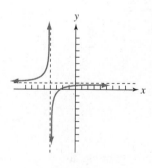

Figure 4.5

Table 4.3	
x	*y*
−1000	1.001004
−100	1.010417
−10	1.166667
−5	2
−4.5	3
−4.1	11
−4.01	101
−4.001	1001
−4	undefined
−3.999	−999
−3.99	−99
−3.9	−9
−3	0
10	0.928571
100	0.990385
1000	0.999004

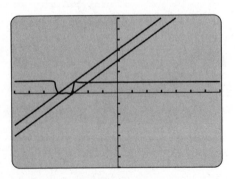

Window 4.6

We recall that the TI graphing calculator is simply plotting the points as they appear in the current graphing window and joining them with line segments to the best of its ability. Select Dot in the MODE menu and graph the function. We get the representation of the function graph displayed in Window 4.7. Notice that we can imagine the asymptote at $x = -4$ and the

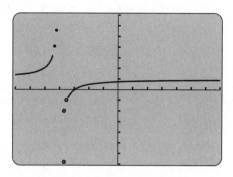

Window 4.7

$$x + 4 \overline{)\begin{array}{l} 1 \\ x + 3 \\ \underline{x + 4} \\ -1 \end{array}}$$

asymptote at $y = 1$. Using long division as a tool (displayed in the margin), we have the following symbolic transformation of the function expression:

$$f(x) = \frac{x+3}{x+4} = 1 + \frac{-1}{x+4} = -\frac{1}{x+4} + 1$$

Thus, this function is a horizontal translation of 4 units to the left, a reflection, and then a vertical translation of 1. We can now observe that our seeming horizontal asymptote at $y = 1$ is indeed an asymptote as it is a simple displacement of the original x-axis asymptote of $h(x) = \frac{1}{x}$.

Function Multiples

The graph of $f(x) = \frac{2}{x}$ is displayed in Window 4.8 along with the graph of $h(x) = \frac{1}{x}$ in a $[-5, 5]$ by $[-5, 5]$ window. You can see both from Table 4.4

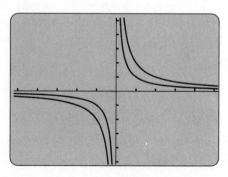

Window 4.8

and the TRACE feature that $f(x) = \dfrac{2}{x}$ has the same general shape as $h(x) = \dfrac{1}{x}$. But each y-value of f is twice the y-value of h, which is an expansion of h.

Table 4.4

x	$h(x) = 1/x$	$f(x) = 2/x$
1/8	8	16
1/4	4	8
1/2	2	4
1	1	2
2	1/2	1
4	1/4	1/2
8	1/8	1/4

EXAMPLE SET B

Analyze the graph of

$$f(x) = \frac{x+2}{x-5}$$

Solution:

By long division, we have

$$f(x) = 1 + \frac{7}{x-5}$$

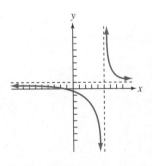

Figure 4.6

Here we have a horizontal translation, an expansion followed by a vertical translation. The graph is given in Figure 4.6. The asymptotes are indicated by dashed lines. This function has a vertical asymptote at $x = 5$. The function is negative immediately to the left of 5 and positive immediately to the right of 5. The second term, $\dfrac{7}{x-5}$, in the function $f(x) = 1 + \dfrac{7}{x-5}$ has a horizontal asymptote at $y = 0$ as we know from our previous work. The graph of $f(x)$ is the graph of a function whose rule is $\dfrac{7}{x-5}$, which has been translated up 1 unit. Thus, f has a horizontal asymptote at $y = 1$. We can describe the behavior of this function as follows: For x values from $-\infty$ to 5, the function is decreasing. There is a vertical asymptote at $x = 5$ where the function is undefined. For x values from 5 to $+\infty$, the function is decreasing. The graph of the function approaches the horizontal asymptote $y = 1$ from below as x gets large negatively (moving to the left). The graph of the function also approaches the horizontal asymptote $y = 1$ from above as x gets large positively (moving to the right).

Reciprocals of Factored Quadratic Polynomials

We were able to analyze the quotient of two linear polynomials through a transformation involving long division. How can we deal with a rational function that is the reciprocal of a factored polynomial?

Here we can analyze the reciprocal of a polynomial by first investigating the graph of the polynomial itself. Consider the function

$$f(x) = \frac{1}{(x + 2)(x - 3)}$$

The polynomial $(x + 2)(x - 3)$ is graphed in Window 4.9. This function is

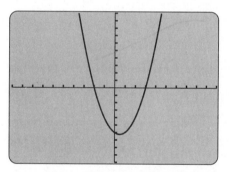

Window 4.9

a parabola with zeros at -2 and 3, vertex at the midpoint of the zeros or $\left(\frac{1}{2}, -\frac{25}{4}\right)$, and is large positively as x gets large both positively and negatively. This information allows us to see that the reciprocal of this polynomial has vertical asymptotes at $x = -2$ and $x = 3$. The function will be negative between -2 and 3 since the polynomial is negative there. Also, between -2 and 3 the polynomial moves down and away from the x-axis and then moves up toward the x-axis. The reciprocal function, therefore, moves up toward the x-axis and then away from it. We call the point where the direction changes in this way a local maximum. It occurs at the vertex where the polynomial has a minimum. Thus, the local maximum occurs at $\frac{1}{2}$ and is $-\frac{4}{25}$. Finally, the reciprocal function will approach 0 as x gets large since the polynomial is large positively when x is large both positively and negatively. Thus, the reciprocal function has a horizontal asymptote at $y = 0$.

We can sketch the graph of the reciprocal function from the polynomial function as is done in Figure 4.7.

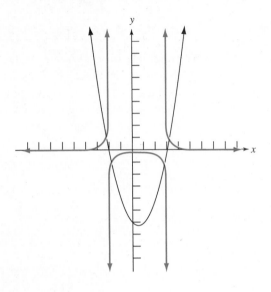

Figure 4.7

EXAMPLE SET C

Describe the behavior of the function $f(x) = \dfrac{1}{(x-4)(x+3)}$.

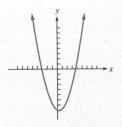

Figure 4.8

Solution:
The quadratic polynomial $(x-4)(x+3)$ is graphed in Figure 4.8. The zeros of this polynomial are the asymptotes of f. The polynomial is large positively as x gets large both positively and negatively. Thus, f has a horizontal asymptote at $y = 0$. The graph of f is given in Window 4.10 with the MODE set to Dot in a graphing window with $-1 \le y \le 1$. The graph of the polynomial $(x - 4)(x + 3)$ can be used to deduce the behavior of f. The function f is positive immediately to the left of the $x = -3$ asymptote and negative immediately to the right; the function is negative just to the left of the $x = 4$ asymptote and positive just to the right as seen in Window 4.10. There is a horizontal asymptote at $y = 0$. The minimum value of the polynomial occurs at the vertex. Thus, the local maximum of the function f on the interval where it is negative is the reciprocal of the minimum value of the quadratic polynomial. Figure 4.9 is a sketch of the graph of f with the vertical

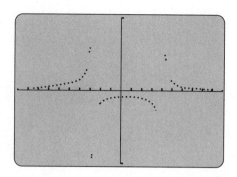

Window 4.10

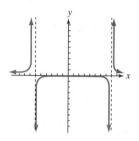

Figure 4.9

asymptotes appearing as dashed lines. The function f can be described as follows: The function f is an increasing function on the interval $(-\infty, -3)$. There is a vertical asymptote at $x = -3$. The function is increasing on the interval $(-3, \frac{1}{2})$ and decreasing on the interval $(\frac{1}{2}, 4)$. There is a vertical asymptote at $x = 4$. The function is decreasing on the interval $(4, +\infty)$. There is a horizontal asymptote at $y = 0$ and a local maximum of $-\frac{4}{49}$ at $x = \frac{1}{2}$.

What do we do when the polynomial whose reciprocal function we wish to investigate does not factor over the real numbers?

EXAMPLE SET D

Describe the behavior of the function defined by

$$f(x) = \frac{1}{x^2 + 4}$$

Solution:

The polynomial $x^2 + 4$ does not factor over the reals. The graph of this polynomial function is displayed in Window 4.11. The parabola opens upward because the lead coefficient is positive. We also note that the graph of the polynomial is above the x-axis, has no real zeros, and so the function is always positive. There are no asymptotes because there are no zeros of the polynomial. The reciprocal function is always positive and has a local maximum for the same value of x that the polynomial has a local minimum, namely, at $x = 0$. Finally, the reciprocal function has $y = 0$ as an asymptote since the polynomial gets large positively as x gets large both positively and negatively. The graph of the function is displayed in Figure 4.10 with

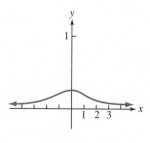

Figure 4.10

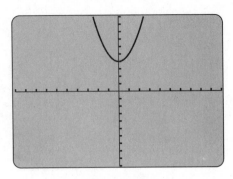

Window 4.11

$-5 \leq x \leq 5$ and $-1 \leq y \leq 1$. We can describe the function's behavior as follows:

The function is increasing on the interval $(-\infty, 0)$ and decreasing on the interval $(0, +\infty)$. The function has a local maximum at $x = 0$, which is also the largest value of the function. The maximum of the function is $\frac{1}{4}$ since the minimum of the polynomial is 4. The horizontal asymptote is $y = 0$.

We may encounter reciprocals of polynomials of degree higher than 2.

EXAMPLE SET E

Sketch the graph of the function defined by

$$f(x) = \frac{1}{(x-3)(x-5)(x+5)}$$

by analyzing the polynomial in the denominator.

Solution:

The polynomial has zeros at $x = 3,\ 5,$ and -5. Its graph is displayed in Window 4.12 with $-10 \leq x \leq 10$, $-120 \leq y \leq 120$, and tick marks every 10 units. There are vertical asymptotes at the zeros of the polynomial and $y = 0$ is a horizontal asymptote. The reciprocal function is negative where the polynomial is negative and positive where the polynomial is positive. There is a local minimum for the reciprocal function where there is a local maximum for the polynomial and there is a local maximum for the function

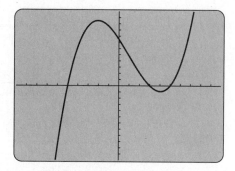

Window 4.12

where there is a local minimum for the polynomial. We can sketch the graph of the reciprocal function with this information and it should look something like Figure 4.11 with $-0.5 \leq y \leq 0.5$. You may wish to check this result

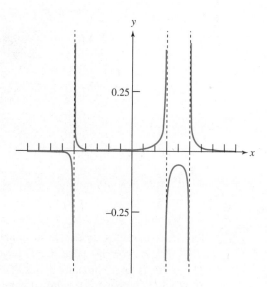

Figure 4.11

using your graphing calculator. You will find that it is very hard to get a reasonable display of the entire function using a single graphing window.

We now return to the problem situation posed at the beginning of the section.

EXAMPLE SET F

How many milliliters of a 10% acid solution must be added to 10 mL of a 40% acid solution to obtain a 30% acid solution?

Solution:

Recall that the concentration of the solution is given by

$$C(x) = \frac{0.40 \cdot 10 + 0.10x}{10 + x} = \frac{4 + 0.1x}{10 + x}$$

where x is the number of milliliters of 10% acid solution that is added to the 40% solution. The variable x is greater than zero in this problem situation since we are adding the 10% acid solution and we can add as much 10% as we like. The concentration, $C(x)$, is a fraction between 0 and 1. A graph of $C(x)$ is displayed in Window 4.13, a graphing window with $0 \le x \le 8$ and $0 \le y \le 1$. The tick marks in the y-direction are every 1/4 unit apart.

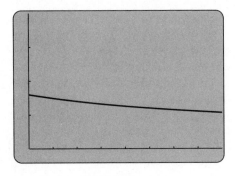

Window 4.13

We notice that the graph is decreasing and always above the line $y = \frac{1}{10}$ for positive values of x. The original acid solution is to be diluted to a concentration of 30%. Use the Trace cursor to determine where this occurs. The result is approximately 5 mL. Thus, we should add about 5 mL of the 10% acid solution to the original solution to obtain a 30% acid solution.

EXERCISES

In exercises 1–6, determine the horizontal and vertical asymptotes.

1. $f(x) = \dfrac{x + 5}{x - 6}$

2. $f(x) = \dfrac{2x + 3}{x - 5}$

3. $f(x) = \dfrac{6x + 3}{2x - 6}$

4. $f(x) = \dfrac{4x - 1}{2x - 6}$

5. $f(x) = \dfrac{1}{x^2 + 3}$

6. $f(x) = \dfrac{2}{x-7}$

In exercises 7–12, the graph of a polynomial is given. Sketch the graph of the reciprocal of each polynomial.

7.

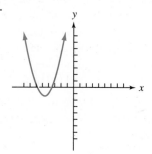

8.

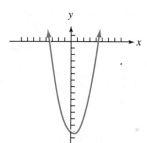

9.

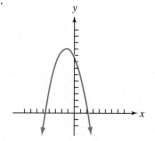

10.

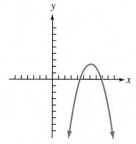

11.

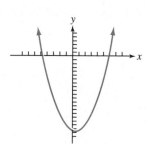

12.

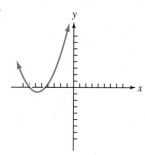

In exercises 13–22, sketch the graph of the rational function, indicating asymptotes if they exist.

13. $f(x) = \dfrac{x+2}{x-2}$

14. $f(x) = \dfrac{x+3}{x-4}$

15. $f(x) = \dfrac{4}{x-2}$

16. $f(x) = \dfrac{3}{x+3}$

17. $f(x) = \dfrac{5x+4}{x-1}$

18. $f(x) = \dfrac{32}{x^2+16}$

19. $f(x) = \dfrac{10}{x^2+5}$

20. $f(x) = \dfrac{1}{(x+3)(x-2)}$

21. $f(x) = \dfrac{\frac{1}{3}}{(x-2)(x-5)}$

22. $f(x) = \dfrac{\frac{1}{20}}{(x+3)(x+11)}$

In exercises 23–32, describe the behavior of the function.

23. $f(x) = \dfrac{6}{x^2 + 3}$

24. $f(x) = \dfrac{1}{5x^2 - 4x - 1}$

25. $f(x) = \dfrac{1}{x - 3}$

26. $f(x) = \dfrac{1}{x^2 - 2x - 8}$

27. $f(x) = \dfrac{1}{4 - 2x}$

28. $f(x) = \dfrac{4x - 1}{2x + 5}$

29. $f(x) = \dfrac{-3}{2x + 5}$

30. $f(x) = x^2 + 2x - 3$

31. $f(x) = 3x^2 - x - 5$

32. $f(x) = \dfrac{1}{x^2 - 3x - 7}$

33. A chemist needs five liters of a 32% acid solution. He mixes pure concentrated acid with water to get his solution.

(a) How much of each ingredient is needed to obtain the required solution?

(b) The chemist inadvertently adds too much water so that he has five liters of a 24% solution. He must add more pure acid to get the 32% solution he desires. How much pure acid must he add to the five liters of the 24% solution to make it a 32% solution? What is the volume of the resulting solution?

(c) Later the chemist mixes some of the 32% solution with a 45% solution to get eight liters of a 40% solution. How much of each ingredient is mixed?

34. A soda manufacturer wants to minimize her costs by using the least amount of sheet metal to make the cylindrical containers. Each can should contain 355 mL of soda. What is the optimum radius and height of the cylinder in centimeters? (1 liter = 1000 cubic centimeters.) Solve by graphing.

4.3 Rational Functions II

Introduction

Equal Degrees

Unequal Degrees

Introduction

In this section, we consider a variety of rational functions and how to analyze their behavior. A key concept in understanding the behavior of rational functions is the degree of the polynomial function in the numerator versus the degree of the polynomial function in the denominator. There are three possibilities.

$$\text{degree of the numerator} = \text{degree of the denominator}$$

$$\text{degree of the numerator} < \text{degree of the denominator}$$

$$\text{degree of the numerator} > \text{degree of the denominator}$$

We deal with each of these cases in turn.

Equal Degrees

If the numerator has the same degree as the denominator, then, as the value of x increases, the function value of the rational expression approaches the quotient of the dominant terms in the numerator and denominator.

EXAMPLE SET A

Analyze the function

$$f(x) = \frac{3x^2 - 5x - 6}{2x^2 + 4x - 3}$$

Solution:
As x gets large we can see from Table 4.5 that the value of $f(x)$ gets closer and closer to $\frac{3}{2}$. The same is true as x gets large negatively.

Table 4.5	
x	y
1	-2.666667
10	1.029536
100	1.445997
1,000	1.494510
10,000	1.499450
100,000	1.499945
1,000,000	1.499995

Thus, there is a horizontal asymptote at $y = \frac{3}{2}$. This function will have zeros at the zeros of the numerator that are not common to the denominator. The zeros of the numerator are $x \approx -0.8$ and $x \approx 2.5$ while the zeros of the denominator are $x \approx -2.5$ and $x \approx 0.6$. Thus, there are zeros at $x \approx -0.8$ and $x \approx 2.5$. Graphing each of the polynomials that make up this rational function, we see that the overall function changes signs at $x \approx -2.5$, $x \approx -0.8$, $x \approx 0.6$, and $x \approx 2.5$. From the graph of the denominator, we see that there are vertical asymptotes at $x \approx -2.5$ and $x \approx 0.6$. The graph of this function is displayed in Window 4.14.

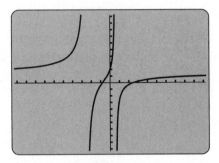

Window 4.14

We can verify the horizontal asymptote by using long division to divide the denominator into the numerator. The long division computation is shown below. Thus, the function can be represented symbolically by

$$2x^2 + 4x - 3 \overline{\smash{\big)}\, 3x^2 - 5x - 6} \quad \begin{array}{c} \frac{3}{2} \end{array}$$

$$3x^2 + 6x - \frac{9}{2}$$

$$\overline{-11x - \frac{3}{2}}$$

$$f(x) = \frac{3x^2 - 5x - 6}{2x^2 + 4x - 3} = \frac{3}{2} + \frac{-11x - 1\frac{1}{2}}{2x^2 + 4x - 3}$$

Notice that the remainder in this division is a function whose value gets close to zero as x increases toward $+\infty$.

Unequal Degrees

EXAMPLE SET B

Analyze the function

$$f(x) = \frac{x^2 - x - 6}{x^3 - 19x + 30}$$

Solution:

Here the degree of the numerator is less than the degree of the denominator. Thus, the process of long division we have used so successfully in previous analyses will fail to yield a nonzero polynomial quotient. (Why?) Still, we can determine the approximate zeros of the numerator and the denominator by graphing each of them. The zeros of the numerator are $x = 3$ and $x = -2$ while the zeros of the denominator appear to be at $x = 3$, $x = 2$, and $x = -5$. We can verify the zeros of the denominator by substitution. We see that the numerator and the denominator share a zero. Thus, because the denominator has a zero at $x = 3$, the rational function will not be defined at this value. When the numerator and denominator have a single factor of $x - 3$, there will not be a vertical asymptote at that point. Both the numerator and denominator approach zero as x approaches 3, unlike the situation for the expression $\frac{1}{x-3}$. Thus, this rational function is equivalent to

$$f(x) = \frac{x + 2}{x^2 + 3x - 10}, \quad x \neq 3$$

From this we can see that there will be two vertical asymptotes at $x = -5$ and $x = 2$. There will be a zero at $x = -2$, and the function is undefined at $x = 3$ and there will be a "hole" in the graph of the function there. The calculator graph of the function and a "paper-and-pencil" sketch of the graph appear in Figure 4.12. Notice that the x-axis is a horizontal asymptote. Since the degree of the denominator is larger than the degree of the numerator, the denominator will greatly exceed the numerator as x increases toward positive or negative ∞ and the function value will approach zero. This will always happen when the degree of the denominator is larger than the degree of the numerator.

Writing Activity Describe when a rational function will have a horizontal asymptote at $y = 0$. Generalize by describing when a rational function will have a horizontal asymptote at $y = k$, where k is a real number.

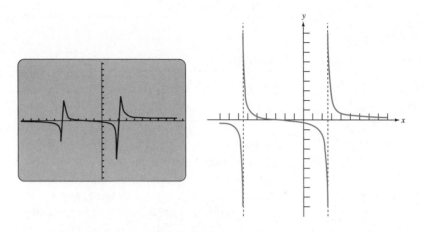

Figure 4.12

EXAMPLE SET C

Analyze the following function:

$$f(x) = \frac{3x^4 - 2x^2 - 2x - 5}{x^2 + 5x - 3}$$

Solution:

The numerator has degree greater than the denominator. We can, once again, transform the expression for f by long division. We find that

$$f(x) = 3x^2 - 15x + 82 + \frac{241 - 457x}{x^2 + 5x - 3}$$

The last term of this expression is a rational function, the degree of whose denominator exceeds the degree of its numerator. Thus, as x gets large positively or negatively, the value of this term approaches zero. Hence, the value of f approaches the value of the expression $3x^2 - 15x + 82$ as x gets large positively or negatively. We can see that this is true using our calculators with the range or window settings given in Window 4.15. The graphs of the function and the quadratic function it approaches are given in Window 4.16. Thus, we can visualize the end behavior of the function as approaching a quadratic polynomial.

We know the end behavior of the function, but what about zeros and vertical asymptotes? For the zeros of f, we determine the set of real zeros of the numerator of the defining expression for f and check that none of them

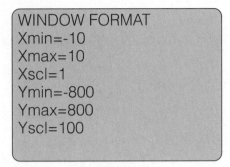

Window 4.15

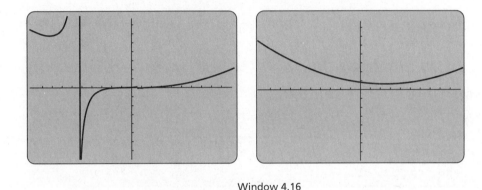

Window 4.16

are zeros of the denominator of f. In this case, approximate values for the zeros of the numerator and denominator are

$$x \approx -1.2 \quad \text{and} \quad 1.4 \quad \text{for the numerator}$$

$$x \approx -5.5 \quad \text{and} \quad 0.6 \quad \text{for the denominator}$$

Since there are no common zeros in these sets, the zeros of the function f occur at $x \approx -1.2$ and $x \approx 1.4$. A nice additional benefit of this analysis is that we now also know that the vertical asymptotes occur at $x \approx -5.5$ and $x \approx 0.6$ where the denominator of f is zero.

Here we have a situation where the calculator graph of the function is not satisfactory. When we are able to see the end behavior of the function, we are not able to see clearly the zeros of the function and the behavior of the function at its vertical asymptotes. A paper-and-pencil sketch of the graph of this function is given in Figure 4.13.

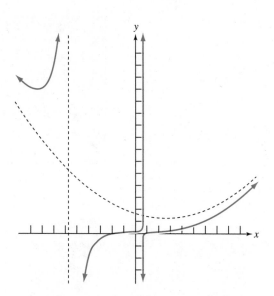

Figure 4.13

Very Important

EXERCISES

In exercises 1–6, determine the function that $f(x)$ is asymptotic to as x gets large positively and negatively.

1. $f(x) = \dfrac{3x^2 + 2x + 4}{x - 3}$

2. $g(x) = \dfrac{x^2 + 2x - 3}{x + 3}$

3. $h(x) = \dfrac{x^3 + 2x^2 + 3x - 4}{x - 2}$

4. $f(x) = \dfrac{5x^4 - 2x^2 + 3x - 2}{2x^4 - 3x^2 + 2x - 1}$

5. $g(x) = \dfrac{x^2 - 3x - 4}{x - 2}$

6. $h(x) = \dfrac{x^2 - 3x + 2}{x - 5}$

In exercises 7–12, find the vertical asymptotes.

7. $f(x) = \dfrac{x^3 + 5x^2 - 8x - 48}{x + 4}$

8. $g(x) = \dfrac{x^3 + 8x^2 + 5x - 50}{x^2 - x - 2}$

9. $h(x) = \dfrac{x^2 + x - 6}{(x - 2)(x + 3)}$

10. $f(x) = \dfrac{x^3 - 3x - 2}{x^2 - 5x + 6}$

11. $g(x) = \dfrac{x^2 + x - 12}{x^3 - 2x^2 - 15x + 36}$

12. $h(x) = \dfrac{x^2 - x - 6}{x^3 - 4x^2 - 3x + 18}$

In exercises 13–24, sketch a graph of the function indicating asymptotes.

13. $f(x) = \dfrac{3x^2 - 2x + 4}{x^2 - x - 3}$

14. $g(x) = \dfrac{2x^2 - 2x - 5}{x - 4}$

15. $h(x) = \dfrac{x + 2}{x - 3}$

16. $f(x) = \dfrac{x-2}{x+3}$

17. $g(x) = \dfrac{(x-3)^2}{x-2}$

18. $h(x) = \dfrac{2}{x^2-3}$

19. $f(x) = \dfrac{x-1}{x^2-3x+2}$

20. $g(x) = \dfrac{x-6}{x^2-8x+15}$

21. $h(x) = \dfrac{3x^3-2x^2+4}{x^2-3x+1}$

22. $f(x) = \dfrac{3x^3-2}{3x}$

23. $g(x) = \dfrac{4x^2-2x-1}{4-x}$

24. $h(x) = \dfrac{9x^2-3}{3x^2-1}$

In exercises 25–28, describe the behavior of the function.

25. $f(x) = \dfrac{1}{1-x}$

26. $g(x) = \dfrac{x^2+3x-1}{x+2}$

27. $h(x) = \dfrac{x^2-3x-4}{x-4}$

28. $f(x) = \dfrac{x-4}{x^3-4x^2+3x-12}$

29. Jane wants to have a vegetable garden in her backyard. She feels that a 100-square-foot garden would be adequate, and wants it to be rectangular in shape. Suppose x is the length of the rectangle. What is the width as a function of x? What restriction do you place on x? What is the perimeter as a function of x? If she puts up fencing along the perimeter and it costs $6.60 per foot, what is the cost function for the fencing? Graph the function and find the minimum cost and the corresponding x value.

4.4 The Parametric Representation

Introduction

Other Uses of Parametric Representations

Introduction

Figure 4.14

You take off from a regional airport in a small propeller-driven airplane. The plane lifts off the ground and begins to gradually climb into the sky at a certain point on the runway. The plane ascends along the flight path pictured in Figure 4.14. If you fly the airplane with a ground speed of 80 feet per second (horizontally), how far will the airplane rise in half a minute? We create Table 4.6 that indicates the position of the plane each second both along the ground (x) and perpendicular to the ground (y) measuring from the point of lift-off. You see from this table that both the horizontal $x(t)$ position and vertical $y(t)$ position are a function of time (t). In particular,

$$\begin{cases} x(t) = 80t \\ y(t) = 20t \end{cases}$$

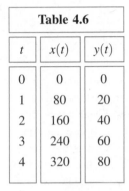

Table 4.6

t	$x(t)$	$y(t)$
0	0	0
1	80	20
2	160	40
3	240	60
4	320	80

The airplane rises so that it moves 80 feet per second horizontally (ground speed) and 20 feet per second vertically.

We can use the parametric mode of our graphing calculators to simulate this motion. Here's how we do it:

1. Using the MODE menu, switch to parametric mode.

2. In the Y= menu, use the X|T key to enter X1T as 80T and Y1T as 20T.

3. Use the RANGE key to set

    Tmin = 0
    Tmax = 60
    Tstep = 1
    Xmin = 0
    Xmax = 4800
    Xscl = 1000
    Ymin = 0
    Ymax = 1200
    Yscl = 100

This will be a simulation of the airplane's first 60 seconds in the air. The plane moves horizontally from 0 to 4800 feet and vertically rises from 0 to 1200 feet.

4. Press GRAPH.

The calculator graph will look something like the graph in Window 4.17. Now press the TRACE key. The Trace cursor will appear along with three numbers in the lower left corner of the screen. These numbers give the T value and the corresponding X(T) and Y(T) values. The variable t is called a

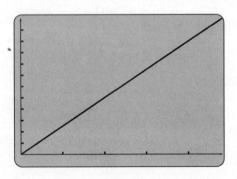

Window 4.17

Parameter

parameter while the functions $x(t)$ and $y(t)$ are called the parametric representations or parametric equations of the curve. The function values $x(t)$ and $y(t)$ are the coordinates of the point on the graph for the given value of t.

This graphing process gives an illuminating way to describe the movement of the plane with respect to time.

Let's look at the table again. Since $y(t)$ changes at $\frac{1}{4}$ the rate of $x(t)$, we can rewrite $x(t)$ and $y(t)$ in the following way:

$$\begin{cases} x(t) = 80t \\ y(t) = (1/4)(80t) \end{cases}$$

With this in mind, if we change the horizontal speed of the airplane to 120 feet per second, the equations change as follows:

$$\begin{cases} x(t) = 120t \\ y(t) = (1/4)120t \end{cases}$$

Enter these equations in the calculator and graph the simulation again. We get the same graph, but it is traced out at a different rate on the calculator since the rate of change of $x(t)$ is larger (120 versus 100).

Table 4.7

t	$x(t)$	$y(t)$
0	0	0
1	80	20
2	160	40
3	240	60
4	320	80

EXAMPLE SET A

1. Determine the parametric equations for an airplane that is moving horizontally at 100 feet per second on a flight path that rises 5 feet for evey 40 feet of horizontal motion when it lifts off the ground.

   Solution:
   The horizontal motion is in the x direction and the vertical motion is at the rate of 5/40 times the horizontal motion. Thus,

   $$\begin{cases} x(t) = 100t \\ y(t) = (5/40)100t \end{cases}$$

2. What happens if we begin with the airplane already 800 feet horizontally from the lift off point and 100 feet in the air? (See Table 4.7.)

   Solution:
   This sounds difficult; but the changes in the formulae are minimal because we are dealing with the situation parametrically.

Other Uses of Parametric Representations

Parametric graphing allows us to simulate a variety of different physical events. For example, if we wish to show what happens when an arrow is shot from

Table 4.8		
t	$x(t)$	$y(t)$
0	800	100
t	$800 + 100t$	$100 + (5/40)100t$

the top of an 80-foot building at a 45° angle with a horizontal velocity of 100 feet per second, we can proceed as follows from Figure 4.15. Assuming that

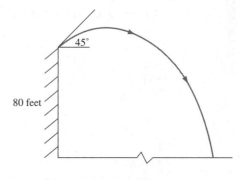

Figure 4.15

we can neglect wind resistance, the horizontal velocity will remain constant at 100 feet per second, which gives $x(t) = 100t$. The force of gravity pulls the arrow downward thus affecting the vertical distance. The -16 value is due to gravitational acceleration ($\frac{1}{2}gt^2$). Since the initial height of the arrow is 80 feet, we arrive at the following parametric description of the arrow's path:

$$\begin{cases} x(t) = 100t \\ y(t) = 80 + 100t - 16t^2 \end{cases}$$

We create the simulation using the following Y= window settings.

    X1T = 100T
    Y1T = 80 + 100T − 16T^2

When we graph this parametric representation of the situation, we get a feel for how the arrow would fly through the air.

A natural question is how far would the arrow go before it hits the ground? We can use the TRACE feature to determine the value of $x(t)$ that has 0 for the corresponding $y(t)$ value.

Table 4.9

t	x(t)	y(t)
0	0	0
1	1	1
2	2	4
3	3	9

A companion question is how long will it take the arrow to hit the ground? Again, we can use the TRACE feature to determine the value of t when $y(t)$ is 0.

How are graphs of functions represented as $y = f(x)$ displayed using parametric methods? Let's look at Table 4.9. What simple function is represented by the $x(t)$ and $y(t)$ columns? It would appear that the function $y = f(x) = x^2$ contains these pairs of values. Thus, we would be tempted to graph this function parametrically using the following Y= entries.

X1T = T
Y1T = T^2

If we graph the function represented in this way using the appropriate values for Tmin, Tmax, and Tstep, we get at least a portion of the graph of $y = f(x)$.

Set the values in the RANGE or WINDOW as follows:

Tmin=0
Tmax=3
Tstep=.1
Xmin=-10
Xmax=10
Xscl=1
Ymin=-10
Ymax=10
Yscl=1

Now graph the function. Your graph should look something like the graph in Window 4.18. The resulting graph is only a portion of the standard $y = x^2$

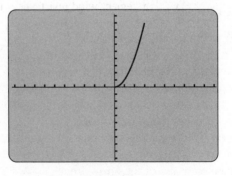

Window 4.18

Table 4.10

t	x(t)	y(t)
0	0	$f(0)$
1	1	$f(1)$
2	2	$f(2)$
3	3	$f(3)$

graph. This is because the values of $x(t)$ are restricted to the interval $(0, 3)$ since t runs from 0 to 3. We can graph functions that are represented in the standard way by $y = f(x)$ using a parametric approach. To do this, let's look at Table 4.10. Here, $x(t)$ and $y(t)$ together give points on the graph of $y = f(x)$.

Thus, we can use parametric techniques in graphing functions represented in the standard symbolic way.

EXERCISES

In exercises 1–6, graph the function parametrically.

1. $y = f(x) = x^2 + 2, \quad 0 \le x \le 7$
2. $y = g(x) = x - 3, \quad 0 \le x \le 7$
3. $y = h(x) = 2x, \quad 0 \le x \le 3$
4. $y = f(x) = 2 - x, \quad -4 \le x \le 4$
5. $y = g(x) = 5 - x^2, \quad 0 \le x \le 7$
6. $y = h(x) = x^2 - 3x, \quad -3 \le x \le 8$

In exercises 7–10, graph the function $y = f(x)$ on the indicated interval using parametric methods.

7. $y = f(x) = x - 5 \quad$ on $[-1, 3]$
8. $y = g(x) = x^2 + 3x - 4 \quad$ on $[-4, 5]$
9. $y = h(x) = x^2 + 5x \quad$ on $[-2, 3]$
10. $y = f(x) = 2 - x^2 \quad$ on $[2, 15]$

11. Suppose the plane in Example Set A of this section takes off in such a way that it rises 3 feet for every 50 feet of horizontal movement. Give a parametric representation of the flight path of the aircraft with initial time being the time of liftoff.

12. Suppose the arrow considered in this section is shot from the same building as in Figure 4.15 at the same angle with an initial velocity of 120 feet per second. How far will it go before it hits the ground? How long will it take to hit the ground? How does this length of time compare with that of the arrow from the example? Can you explain why these time lengths are related in this way?

13. Graph the function $y = f(x) = \ln(x)$ in Function mode.

14. Graph the function $y = f(x) = \log(x)$ in Function mode.

15. Sketch a graph of the parametric representation given by

$$\begin{cases} x(t) = \cos(t) \\ y(t) = \sin(t) \end{cases}$$

where SIN and COS are the names of these functions on your calculator. Was the figure drawn in the clockwise or counterclockwise direction? What is the shape of this graph? What is the shape of this graph after you have invoked the Square feature of the Zoom menu?

16. Sketch a graph of the parametric representation given by

$$\begin{cases} x(t) = 2\cos(t) \\ y(t) = \sin(t) \end{cases}$$

What is the shape of this graph? What is the shape of this graph after you have invoked the Square feature of the Zoom menu? How does this graph compare with the graph in exercise 15?

17. Are the graphs in exercises 15 and 16 the graphs of functions or just relations? Explain.

18. Both the demand and the supply of a commodity depend on the price. For a particular commodity the following was observed:

$$d(T) = 2000 - T$$

$$S(T) = 200 + 3T \quad T > 1$$

where T represents the price. Graph this parametric equation. What are the range settings? What is your conclusion from the graph?

19. The coordinates of a particle are observed as
$$x(t) = 2t$$

$$y(t) = \sqrt{2t + 1}$$

   Graph the equation. What is the resulting curve? Write the equation of the path.

20. The parametric equations of a curve are given by $x(t) = \sqrt{1 - t}$ and $y(t) = \sqrt{1 + 2t}$. What is the range for the parameter t? Eliminate t from the two equations to get an equation of the graph. [Hint: Square both sides of each equation. Solve for t in terms of x and substitute the result in the expression for y.] What kind of curve is it? What portion of the curve is given by these equations?

21. The parametric equation of the path of a ball thrown at an angle is given by
$$x(t) = 20t$$

$$y(t) = 160t - 16t^2$$

   Eliminate t and write (a) y as an expression involving x and (b) x as an expression involving y using the method from the previous problem. Which of these two expressions is a function? Give reasons for your answer.

22. An airplane pilot's mission is to drop relief provisions to an entrenched group of rebels. He is flying horizontally at an altitude of 1,600 feet and a speed of 352 feet per second when he drops his load.

   (a). Why wouldn't he drop the supplies when he is vertically over the target?

   (b). If he hits his mark, what is his horizontal distance from the indicated spot? (Use the following parametric equations.)

$$x = 352t$$

$$y = 16t^2$$

   where x is the horizontal distance from the drop-off point and y is the vertical distance from the drop-off point. (Hint: Every object in the moving plane has the same velocity as the plane.)

23. (Refer to the previous exercise.) The pilot needs to make two drops of provisions to complete his assignment. His first drop is successful, but when he begins to launch his second pass he is fired upon by some snipers and he quickly chooses to climb along a slanted path. His calculations are correct and his second load arrives on target. What is his horizontal distance from the target if at the time of drop his vertical speed is 300 feet per second, his horizontal speed is the same as in the previous exercise, and his altitude is 1800 feet? (Assume the equation for x remains unchanged.) Use $y = 300t - 16t^2 + 1800$. Graph this equation and solve for t (i.e., find t when $y = 0$.)

4.5 Inverse Functions

Introduction

A Simple Function

Inverses as Relations

Inverses as Functions

Inverse Function Notation

Introduction

As we know, a function consists of a domain set, a range set, and a rule that associates elements of the domain with elements of the range. In this section, we investigate what happens when we reverse the association of elements so that the range becomes the domain and the domain becomes the range.

A Simple Function

Table 4.11	
x	y
1	1
2	4
3	9
4	16

Given the function $f(x) = x^2$, for $x = 1,\ 2\ ,3\ ,4$, we can construct a table of function values that displays the domain and range of the function along with the association of elements from these sets as in Table 4.11. We can switch or interchange the columns of this table as in Table 4.12. We would like to consider this switching of columns in the table visually. We can affect this switching using the parametric representation of functions.

First, we can input our original function parametrically as shown on the left of Window 4.19.

The graph on the right of Window 4.19 displays (in a magnified version) the points from our original table of values with the calculator set in Dot mode and with $0 \leq t \leq 4$ and Tstep = 1 and

```
Xmin=−24
Xmax=24
Xscl=4
Ymin=−16
Ymax=16
Yscl=4
```

Table 4.12	
x	y
1	1
4	2
9	3
16	4

We wish to switch the columns of the original table. We can do this symbolically by interchanging the formulas for $x(t)$ and $y(t)$ in the second parametric representation in the Y= menu. The interchange of formulas is shown in Window 4.20.

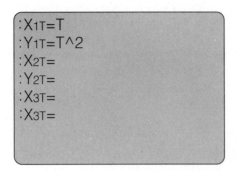

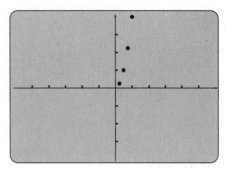

Window 4.19

$$:X_{1T}=T$$
$$:Y_{1T}=T\char`^2$$
$$:X_{2T}=$$
$$:Y_{2T}=$$
$$:X_{3T}=$$
$$:X_{3T}=$$

$$X_{1T}=T$$
$$Y_{1T}=T\char`^2$$
$$X_{2T}=T\char`^2$$
$$Y_{2T}=T$$
$$X_{3T}=$$
$$X_{3T}=$$

Window 4.20

We can see both functions graphed with the original RANGE or WINDOW settings and the graphing window X-values and Y-values between −10 and 10 on our graphing calculators. A more informative graph can be obtained using the ZOOM/Square feature of your calculator. The Y-values are now between −15 and 15 as shown in Window 4.21.

In this graphical representation of the function, we see that the points appear to show a certain symmetry. They are symmetric about the line $y = x$ although we will not prove this fact. You can convince yourself that this is actually the situation by using the TRACE feature and selecting the point $(2, 4)$ on the first function and then switching to the second function with the up-arrow key. The coordinates on the second function of the point with $t = 1$ are $(4, 2)$. The midpoint of the line segment between these two points $(2, 4)$ and $(4, 2)$ is $(3, 3)$, which is on the graph of the line $y = x$. If you had chosen any point (a, b) on the graph of the first function and switched the coordinates, you would have obtained the point (b, a). The midpoint of the line segment between (a, b) and (b, a) is always a point on the line $y = x$. (What are the coordinates of this midpoint?)

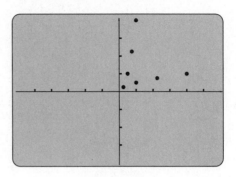

Window 4.21

Table 4.13

x	y
−4	16
−3	9
−2	4
−1	1
0	0
1	1
2	4
3	9
4	16

Inverses as Relations

We can change our function, $x(t) = t$ and $y(t) = t^2$, by expanding our t values so that negative values of t are used. Some function values are listed in Table 4.13. The new graphs of f and its paired function are displayed in Window 4.22. The second graph turns out not to be a graph of a function since both $(4, 2)$ and $(4, -2)$ are on the graph defying the vertical line test and the definition of a function. The inverse or switched graph is, however, a relation since it is the graph of ordered pairs. If you connect the dots in your mind, you might recognize this graph as the parabola with equation $x = y^2$.

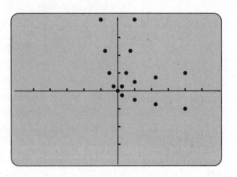

Window 4.22

EXAMPLE SET A

Determine whether the graph of the inverse of the function $f(x) = x^2 - 4x - 5$ is a function or only satisfies the less restrictive conditions for a relation. If the inverse is a relation, determine an appropriate restriction of the domain of f so that the inverse will be a function.

Solution:

We enter t for $x(t)$ and the expression for f as the $y(t)$ function in the first parametric representation. We then enter the expression for f (or Y1T) for $x(t)$ and t for $y(t)$ in the second parametric representation. Next, press ZOOM 6 and change the RANGE or WINDOW settings as follows:

    Xmin=-15
    Xmax=15

We can see that the inverse graph is the graph of a relation (not a function) in the graph on the left of Window 4.23.

In the next section, we find methods for determining the symbolic representation of this relation. If you study the display in the graph on the left of Window 4.23, you can see that the two graphs are symmetric about the line $y = x$.

How can we change the values of the independent variable (x) for f so that the inverse will be a function? We can change the domain of the function $x(t)$ by changing the Tmin-value so that the inverse is a function. There are many sets of values that will do this. The largest sets are those with $t \geq 2$ or $t \leq 2$. The graph with the first of these choices is shown in the graph on the right of Window 4.23. Now the inverse is a function as well.

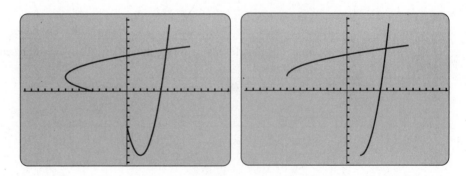

Window 4.23

Inverses as Functions

If the inverse of a function is a function, the inverse must pass the vertical line test. A function and its inverse are reflections of each other about the line $y = x$. The reflection of a vertical line about the line $y = x$ is a horizontal line.

**One-to-One
Functions**

This means that a function whose inverse is a function must pass a horizontal line test. Figure 4.16 illustrates this idea. Functions that pass a horizontal line test are called **one-to-one functions**. Pairs of functions that are "switches"

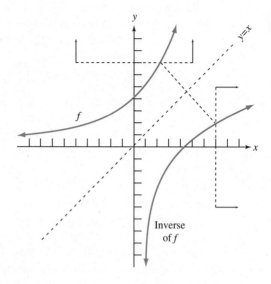

Figure 4.16

of each other are called *inverses* of each other. In Example A, we have two functions that are inverses of one another. They are represented parametrically as follows:

$$f : \begin{cases} x(t) = & t \\ y(t) = & t^2 - 4t - 5 \end{cases} \quad t \geq 2$$

and

$$g : \begin{cases} x(t) = & t^2 - 4t - 5 \\ y(t) = & t \end{cases} \quad t \geq 2$$

Inverse Function Notation

The inverse of a function is given a special function notation based on the notation used for the multiplicative inverse of a real number. The inverse of a nonzero number a with respect to multiplication is defined to be the unique number b that satisfies the equation:

$$a \cdot b = 1 \quad \text{where 1 is the multiplicative identity} \qquad (4.1)$$

Recall that for a nonzero real number a, its multiplicative inverse is denoted by a^{-1}. This is because the number a times a^{-1} gives

$$a \cdot a^{-1} = 1$$

Thus, a^{-1} is the multiplicative inverse of a since it satisfies 4.1.

If (c, d) is on the graph of f, (d, c) is on the graph of the inverse function g. We then have that $g[f(c)] = g(d) = c$ and $f[g(d)] = f(c) = d$. Or, we have

$$f[g(d)] = d$$

$$g[f(c)] = c$$

Functions that are inverses of each other undo each other's effects.

Composition Property of Inverses

Composition Property of Inverses

The function f and its inverse g satisfy the condition

$$(f \circ g)(x) = x \quad \text{and} \quad (g \circ f)(x) = x$$

Recalling that the identity function (for function composition operation) is defined to be $I(x) = x$, we can rewrite this condition as:

$$f \circ g = I$$

Thus, as with real number multiplication, we may denote g as f^{-1} for the process of function composition. The symbol f^{-1} or $f^{-1}(x)$ means the inverse function of f. This symbol is often confused with the reciprocal of the function, written $\dfrac{1}{f}$ or $\dfrac{1}{f(x)}$. The symbol f^{-1} is very seldom used in this way. The context in which it is used will indicate clearly when it means the reciprocal and not the function composition inverse.

EXAMPLE SET B

1. Find the inverse relation for

$$f(x) = 3$$

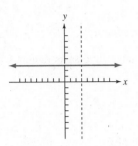

Figure 4.17

Figure 4.18

Solution:
The function $f(x) = 3$ can be represented as $y = 3$. Interchanging x and y means to change y to x in this case. Thus, the function f has the relation $x = 3$ as its inverse. The graph of both f and its inverse (dashed line) is given in Figure 4.17.

2. Find the inverse relation for

$$f(x) = \frac{1}{x^2 - 4}$$

Solution:
The graph of f in Figure 4.18 is not the graph of a one-to-one function. The graph of the inverse of f is, therefore, not a function. Thus, we must restrict the domain of f if its inverse is to be a function. There are several possible ways to do this. In Figure 4.19, we have graphed variations of f with two different restrictions of the domain. We will study how to find

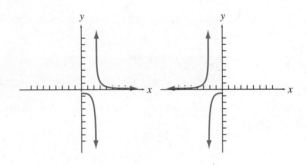

Figure 4.19

the symbolic expression for the inverse of f in the next section.

EXERCISES

In exercises 1–4, write down a table for the inverse of the given function and indicate if this inverse is a function.

1. This table is for f.

x	y
1	4
2	5
4	4

2. This table is for g.

x	y
−1	5
−2	6
−3	7
−4	9

3. This table is for *h*.

x	*y*
−1	2
−2	10
−3	10
−4	2

4. This table is for *k*.

x	*y*
1	4
2	21
3	5
4	11

In exercises 5–10, sketch the inverse of the given function. Indicate if the inverse relation is a function.

5.

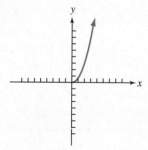

6.

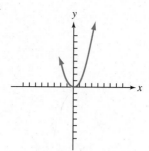

7.

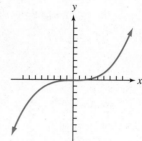

8.

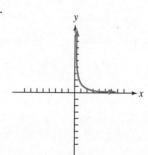

9.

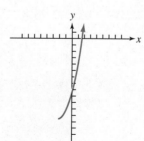

10.

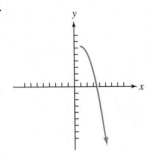

In exercises 11–20, sketch the graph of the inverse of the given function.

11. $f(x) = \dfrac{1}{x - 3}$ on $[3, 10]$

12. $g(x) = x^2 - 4x + 2$ on $[-3, 2]$

13. $h(x) = x^2 - 4x + 2$ on $[2, 7]$

14. $f(x) = x^2 - 4x - 3$ on $[-2, 7]$

15. $f(x) = x^2 - 4x - 3$ on $[-6, -2]$

16. $f(x) = 2x + 1$ on $(-\infty, \infty)$

17. $f(x) = 3x + 1$ on $(-\infty, \infty)$

18. $f(x) = -3x + 1$ on $(-\infty, \infty)$

19. $f(x) = -4x + 5$ on $(-\infty, \infty)$

20. $f(x) = 2x - 1$ on $(-\infty, \infty)$

In exercises 21–28, restrict the domain of the given function to create a new function whose inverse is a function. Sketch a graph of the inverse function of your new function with its domain.

21. $f(x) = x^2 + 4x - 11$

22. $g(x) = 2x^2 - 8x - 5$

23. $h(x) = x^2 - 3x - 6$

24. $f(x) = (1/3)x - 5$

25. $g(x) = -3x + 4$

26. $h(x) = -x^2 + 3x - 7$

27. $f(x) = \dfrac{1}{x^2 - x - 6}$

28. $g(x) = \dfrac{1}{x^2 + 2}$

In exercises 29–34, show that $(f \circ g)(x) = x$ and $(g \circ f)(x) = x$ by computing both composition functions.

29. $f(x) = 2x + 5$ and $g(x) = \frac{1}{2}(x - 5)$.

30. $f(x) = 3x - 2$ and $g(x) = \frac{1}{3}(x + 2)$.

31. $f(x) = \dfrac{1}{4x + 5}$ and $g(x) = \dfrac{1 - 5x}{4x}$.

32. $f(x) = \sqrt[3]{\dfrac{x + 3}{2}}$ and $g(x) = 2x^3 - 3$.

33. $f(x) = \sqrt[3]{\dfrac{1 - x}{3x}}$ and $g(x) = \dfrac{1}{3x^3 + 1}$.

34. $f(x) = -x + 3$ and $g(x) = 3 - x$. What can you say about $f(x)$?

4.6 A Symbolic Approach to Inverse Functions

Introduction

Switching Domain and Range

A Parametric Approach to Inverse Functions

Introduction

In the last section, we studied inverse relations and functions from a graphical and numeric point of view. In this section, we develop a symbolic approach to inverse functions.

Switching Domain and Range

In Table 4.14, we have some ordered pairs of the function f defined by $y = f(x) = x^2$, $x \geq 0$, and one for its inverse, g. The domain variable column is labeled x and the range column is labeled y. With these tables in mind, we

Table 4.14			
$f:$		$g:$	
x	y	x	y
1	1	1	1
2	4	4	2
3	9	9	3
4	16	16	4

see that, for f, the domain value squared is the associated range value; and, for g, the range value squared is the associated domain value. Of course, this is expected since g is the inverse (switch) of f.

We would like to determine the rule for g (you can easily guess the rule) using a systematic symbolic approach, if possible. We need to keep in mind that the two functions are different. The two statements $y = f(x)$ and $y = g(x)$ indicate what is the independent variable and what is the dependent variable in each function table. These two statements in no way imply that $f(x) = g(x)$, a fact that should be clear from the two tables for f and g above.

Concentrating on the table for g, the range value (y) comes from the domain value (x). For example, 3^2 in the x or domain column for g can be thought of as coming from 3 in the y or range column. Thus, the domain value for any range value y is (as a general rule) y^2. The ordered pair (y^2, y) that arises from this general rule is a kind of rule for the function g and can be interpreted for g as $x = y^2$. This is the hard part. Once we have that $x = y^2$, we can couple it with the symbolic expression $y = g(x)$ to see that $y = g(x) = \pm\sqrt{x}$.

You may see a difficulty in the expression for g: $g(x)$ can have only one value for each x value. If $x = y^2$, then surely y is equal to $\pm\sqrt{x}$ but there are additional constraints in this case. The y values are always positive since the domain values for f are always positive. Thus, $y = \sqrt{x}$ is correct and the actual rule for g is $y = g(x) = \sqrt{x}$.

Once we understand this rule we can shorten it considerably. The function f can be represented symbolically by $y = f(x) = x^2$ or by $y = x^2$. If we switch x for y in this statement, we obtain the statement $x = y^2$ immediately. Here, however, we must keep in mind that we have a new defining statement for a new function. In determining the defining rule for this inverse function, we must remember that the constraints on y for g are the same as the constraints on x for f. This occurs because the domain and range have been interchanged.

Let's look at several examples.

EXAMPLE SET A

1. Find the symbolic expression for g, the inverse function of f, where $y = f(x) = 3x + 5$.

   Solution:
   We begin with the statement defining f.

   $$f: \quad y = 3x + 5$$

   We switch x and y to obtain a defining statement for g.

   $$g: \quad x = 3y + 5$$

   We solve for y (the dependent variable) in terms of x (the independent variable).

   $$x = 3y + 5$$

   $$x - 5 = 3y$$

   $$\frac{x - 5}{3} = y$$

   $$y = \frac{x - 5}{3}$$

   The solution process for y seems not to take into account any constraints on y growing out of the definition of f because the domain of f is the entire set of real numbers. Thus, g, the inverse of f, is defined by

   $$g(x) = f^{-1}(x) = \frac{x - 5}{3}$$

2. Find the symbolic expression for g, the inverse function of f where $y = f(x) = \frac{1}{x} + 3$.

Solution:
We begin with the defining statement for f.

$$f: \quad y = \frac{1}{x} + 3, \quad x \neq 0$$

We interchange x and y to obtain a defining statement for g.

$$g: \quad x = \frac{1}{y} + 3$$

We then solve for y in terms of x to find the rule for the inverse function g recalling the need for constraints of y in the rule for g.

$$x = \frac{1}{y} + 3$$

$$x - 3 = \frac{1}{y}$$

$$y(x - 3) = 1$$

$$y = \frac{1}{x - 3}$$

The constraint on y is the same constraint that we had on the domain of f, namely, y cannot be equal to 0. We know from our work with rational functions that $\frac{1}{x-3}$ is never 0. Thus, we have satisfied the constraint. Using the standard inverse function designation, we have that

$$g(x) = f^{-1}(x) = \frac{1}{x - 3}, \quad x \neq 3$$

Does the range of f match the domain of g?

In the next example, it is slightly more difficult to find the rule for f^{-1} after we have made the interchange of x and y in the defining rule for f.

EXAMPLE SET B

Find f^{-1}, the inverse function for f, where

$$y = f(x) = x^2 - 2x - 3$$

Solution:

We begin with

$$f: \quad y = x^2 - 2x - 3$$

and interchange the variables

$$f^{-1}: \quad x = y^2 - 2y - 3$$

Then, as before, we solve for the dependent variable y in terms of the independent variable x in this defining statement for f^{-1}.

$$x = y^2 - 2y - 3$$

We write this equation in the standard form for a quadratic equation, namely,

$$y^2 - 2y - 3 - x = 0$$

We now use the quadratic formula to solve the equation for y in terms of its coefficients where the constant term is $-3 - x$. We can consider x to be a constant in this surprising way because the quadratic formula is based on the completing-the-square process, which requires only that the constant term $-3 - x$ be a number, variable or otherwise.

$$y = \frac{2 \pm \sqrt{(-2)^2 - 4(1)(-3-x)}}{2(1)}$$

$$y = \frac{2 \pm \sqrt{4 + 12 + 4x}}{2(1)}$$

$$y = 1 \pm \sqrt{4 + x}$$

Belatedly, we recognize that the original function was not one-to-one as shown in Window 4.24. From Window 4.24, we see that if we restrict the domain of f to $x \geq 1$, we can make f into a one-to-one function and its inverse into this new function's inverse.

However, a choice has to be made. Is $+$ or $-$ to be used in the defining expression

$$y = 1 \pm \sqrt{4 + x}$$

for f^{-1}? We will choose the $+$ sign, since the domain values for f and thus the range values for g are never less than 1. (Why?). You may wish to graph the function and its inverse using parametric mode to check this result.

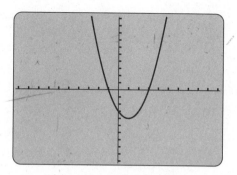

Window 4.24

A Parametric Approach to Inverse Functions

In the previous section, we used parametric representations of functions to graph a function and its inverse. The parametric representations of the function defined by $f(x) = x^2$, $x \geq 0$ and its inverse are given in 4.1 and 4.2.

$$f: \quad \begin{cases} x(t) = & t \\ y(t) = & t^2 \end{cases} \quad t \geq 0 \tag{4.1}$$

$$g: \quad \begin{cases} x(t) = & t^2 \\ y(t) = & t \end{cases} \quad t \geq 0 \tag{4.2}$$

By substituting $x(t)$ for t in the parametric representation of f we obtain

$$y(t) = [x(t)]^2$$

or

$$y = x^2$$

This is, of course, the same equation we've seen before. Making the same substitution in the parametric representation of g, we have

$$x = y^2$$

We can also see that if we apply a function to a domain value and then the inverse of the function to the resulting function value we return to the original domain value from this representation. Figure 4.20 illustrates this idea, which can also

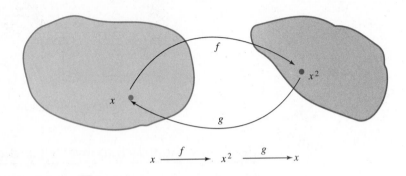

Figure 4.20

be illustrated using the parameter at, say $t = 3$, as shown below.

$$f : \begin{cases} x(3) = & 3 \\ y(3) = & 3^2 \end{cases}$$

$$g : \begin{cases} x(3) = & 3^2 \\ y(3) = & 3 \end{cases}$$

If we start at 3, we end at 3. In general, $g[f(x)] = x = I(x)$. From these ideas, we can see that the inverse function undoes the effect of the function and vice-versa.

EXAMPLE SET C

Use the following parametric representations of a function and its inverse to determine the standard rule for the inverse of the function.

$$f : \begin{cases} x(t) = & t \\ y(t) = & t^2 - 4t - 5 \end{cases} \quad t \leq 2$$

$$g : \begin{cases} x(t) = & t^2 - 4t - 5 \\ y(t) = & t \end{cases} \quad t \leq 2$$

Solution:

The function g has $y = t$ and thus the expression for g is $x = y^2 - 4y - 5$. We can use the constraints of the parametric representation of g to determine that $y \leq 2$. The quadratic formula gives us that

$$y = \frac{4 \pm \sqrt{(-4)^2 - 4(1)(-5 - x)}}{2(1)}$$

$$= 2 \pm 2\sqrt{9 + x}$$

Here, we choose the $-$ sign since $y \leq 5$ and square roots are always positive.

EXERCISES

In exercise 1–6, determine if the given function has an inverse that is a function.

1. $f(x) = \frac{1}{2}x + 2, \quad -1 \leq x \leq 5$

2. $g(x) = x^2 + 2x - 5, \quad 0 \leq x \leq 5$

3. $h(x) = \dfrac{1}{x - 3}, \quad -5 \leq x \leq 5$

4. $f(x) = \dfrac{1}{x} + 7, \quad 0 \leq x \leq 8$

5. $g(x) = \dfrac{1}{x^2 - x - 6}, \quad 3 \leq x \leq 11$

6. $h(x) = x^2 - 5x + 3, \quad -5 \leq x \leq 0$

In exercises 7–20, find an expression for the inverse relation. Indicate the range of the inverse function.

7. $f(x) = 2x + 3$

8. $g(x) = -3x + 5$

9. $h(x) = x^2 - 3, \quad 0 \leq x \leq +\infty$

10. $f(x) = x^2 + 2, \quad 0 \leq x \leq +\infty$

11. $g(x) = x^2 - 3, \quad -\infty \leq x \leq 0$

12. $h(x) = \dfrac{1}{x} + 4, \quad x \neq 0$

13. $f(x) = \dfrac{2}{x} - 3, \quad x \neq 0$

14. $g(x) = \dfrac{2}{x - 5}, \quad x \neq 5$

15. $h(x) = x^2 + 3x - 5, \quad x \leq -\frac{3}{2}$

16. $f(x) = x^2 - 4x - 6, \quad x \leq 2$

17. $g(x) = x^2 - 8x + 2, \quad x \leq 4$

18. $h(x) = x^2 - 8x + 7, \quad x \leq 4$

19. $f(x) = 5x^2 - 4x - 1, \quad x > 1$

20. $g(x) = \dfrac{3}{x - 1} + 4, \quad x \neq 1$

In exercises 21–24, sketch the graph of some inverse function for the given function based on a restriction of the domain of the original function.

21. $f(x) = \dfrac{1}{x^2 - 16}$

22. $g(x) = \dfrac{1}{25 - x^2}$

23. $f(x) = \dfrac{1}{x^2 - 6x + 8}$

24. $g(x) = \dfrac{1}{x^2 - 2x - 15}$

In exercises 25 and 26, sketch the graph of the inverse function and the reciprocal function for the given function.

25. $f(x) = x^2 - 3x - 4, \quad x > 1.5$

26. $g(x) = \dfrac{1}{x - 3}, \quad x \neq 3$

4.7 Radical Functions and Equations

Introduction

Radical Functions as Inverse Functions

Graphing Radical Functions

Solving Equations Involving Radicals

Introduction

In this section, we study functions that are the inverses of the power functions, functions of the form $f(x) = x^n$. Such inverse functions are called radical functions. We also study equations that involve radical functions.

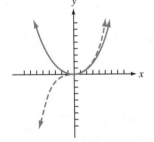

Figure 4.21

Radical Functions as Inverse Functions

We are familiar with $f(x) = x^2$ and $f(x) = x^3$ as displayed in Figure 4.21. We display several power functions in Figure 4.22.

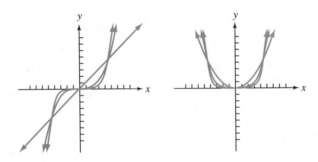

Figure 4.22

On the left, we display examples of functions with n an odd number; on the right, functions with n an even number. The functions on the left are all one-to-one; those on the right are not. Thus, power functions with odd exponents have inverses that are functions. Power functions with even exponent must have their domains restricted to obtain inverse functions that are also functions.

The standard domain restriction for power functions $f(x) = x^n$ with n an even number is $x \geq 0$, in order for their inverses to be functions. We can easily use parametric representations of functions to display these inverse functions as in Figure 4.23. How can we display these inverse functions symbolically

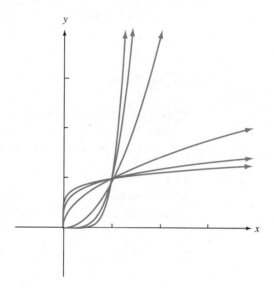

Figure 4.23

in the form

$$y = f^{-1}(x) = \text{an expression in } x$$

We begin with

$$y = f(x) = x^n$$

or

$$y = x^n$$

We wish to determine an expression for $f^{-1}(x)$ and so we interchange x and y and solve the resulting equation for the dependent variable of the equation defining f^{-1} or y.

$$\begin{cases} f: & y = x^n \\ f^{-1}: & x = y^n \end{cases} \quad \text{switching } x \text{ for } y$$

We know that

$$(a^n)^{1/n} = a$$

and so, by raising $x = y^n$ to the power $1/n$, we obtain $y = x^{1/n}$. We must be careful to ensure that x, the domain variable of f^{-1}, is nonnegative if n is even. But the restriction $x \geq 0$ made on the domain of f causes the domain of f^{-1} to be nonnegative when n is even. Thus, we have that the inverse functions of power functions are of the form $f(x) = x^{1/n}$.

Graphing Radical Functions

Entering Fractional Exponents   For the domain of the function $f(x) = x^{1/n}$, TI graphing calculators use real numbers when n is odd, and nonnegative real numbers ($x \geq 0$) when n is even. Problems arise, however, for functions like $f(x) = x^{5/3}$ because the algorithm used by the calculators assumes that the domain is $x \geq 0$ for all functions of the form $f(x) = x^{m/n}$ where m and n are nonzero integers with $n \neq 1$. Thus, the number (-8)^(5/3) gives rise to an error when it is entered on your calculator. You can obtain a value for this expression by representing it using the exponent 1/3 as follows:

((-8)^(1/3))^5

The correct integer value -32 will be displayed.

ILLUMINATOR SET A

We attempt to graph the function

$$f(x) = \sqrt[3]{x^2}$$

by entering the function rule as x^(2/3). The partial graph in Window 4.25 is displayed.

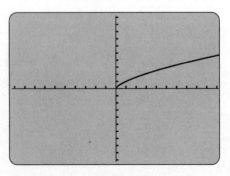

Window 4.25

Notice that the graph has no values for $x < 0$. Using the TRACE feature, we can see that the calculator has calculated no output values for input values to the left of the origin. We know that this is incorrect since this function is defined for input values that are negative from the original mathematical expression.

We should enter this function rule as $((x)^{(1/3)})^2$ because of our knowledge of the way the calculator evaluates expressions with fractional exponents. The graph is displayed in Window 4.26.

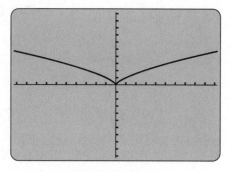

Window 4.26

Its domain is the positive and negative real numbers and zero, as it should be.

EXAMPLE SET A

Graph each of the following radical functions.

1. $f(x) = \sqrt[3]{(x-3)^5}$
2. $g(x) = 2 + \sqrt[5]{(x+3)^2}$
3. $h(x) = \sqrt[4]{1 + (x-3)^3}$

Solution:

1. We recognize this as a translation to the right by three units of the function $f(x) = x^{3/5}$. Since the denominator of the exponent is odd, we have a function that is defined for all real numbers. We enter this function as

$$((x\text{-}3)^{\wedge}(1/3))^{\wedge}5$$

The graph of the function appears in Window 4.27.

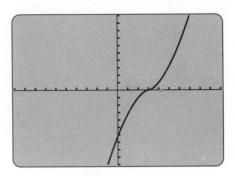

Window 4.27

2. We enter the function $g(x) = 2 + \sqrt[5]{(x+3)^2}$ as

$$2 + ((x+3)^\wedge(1/5))^\wedge 2$$

The graph of the function appears in Window 4.28.

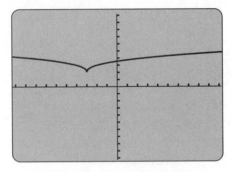

Window 4.28

3. The function $h(x) = \sqrt[4]{1 + (x-3)^3}$ should be entered as the function rule

$$((1 + (x-3)^3)^{1/4}$$

This function is an even root of a number that might be negative. To find the domain for this function, we need to determine when $1 + (x-3)^3$ is nonnegative. Thus, we need to solve the inequality

$$1 + (x-3)^3 \geq 0$$

A graph $1 + (x-3)^3$ will show that it is nonnegative when $x \geq 2$. The graph of the original function, $h(x)$, appears in Window 4.29.

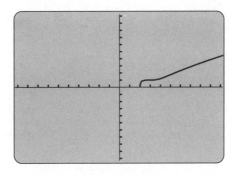

Window 4.29

Solving Equations Involving Radicals

Squaring Both Sides   In the equation $x = 2$, if we square both sides, we obtain the equation $x^2 = 4$, which we know has two solutions. Thus, when we square both sides, we may introduce solutions that do not satisfy the original equation. Such solutions of the squared equation are called *extraneous solutions*.

Two Methods of Solving Radical Equations   The real number solutions to radical equations can be found by determining the intersection points of the graphs of the functions whose rules are given by the left and right sides of the equation. Symbolic methods can sometimes be used to determine roots of a radical equation.

EXAMPLE SET B

Find the solution to the equation

$$\sqrt{2x + 9} - \sqrt{x + 16} = -1$$

Solution:
The method used to solve such equations is to square both sides, as you may recall from previous mathematics courses. In this case, if we square both sides of the original equation, the situation gets worse rather than better, for we introduce a more complicated expression, $2\sqrt{(2x + 9)(x + 16)}$, which is the middle term when we square the binomial $\sqrt{2x + 9} - \sqrt{x + 16}$. To make the computations simpler, we first take one of the radicals to the right side

and then square both sides once (and then once again) to clear the radicals as shown below.

$$\sqrt{2x+9} - \sqrt{x+16} = -1$$
$$\sqrt{2x+9} = \sqrt{x+16} - 1$$
$$2x+9 = x+16 - 2\sqrt{x+16} + 1$$
$$x - 8 = -2\sqrt{x+16}$$
$$x^2 - 16x + 64 = 4(x+16)$$
$$x^2 - 16x + 64 = 4x + 64$$
$$x^2 - 20x = 0$$
$$x(x - 20) = 0$$
$$x = 0, 20$$

Thus, the solution to this equation seems to be the two numbers 0 and 20.

The graph of the two functions that represent the right and left sides of this equation appear in Window 4.30 with $-5 \leq x \leq 30$ and $-4 \leq y \leq 4$.

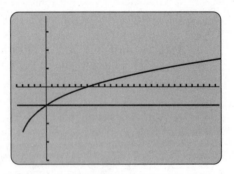

Window 4.30

These two functions seem to intersect only when $x = 0$. Why do the results differ?

We have squared both sides, perhaps introducing an extraneous solution. When we check these two solutions, we find that 0 is a true solution and that 20 is an extraneous solution.

In dealing with radical equations, we should check the solutions.

EXERCISES

For exercises 1–10, convert each of the following expressions into fractional exponent form.

1. $\sqrt[3]{(x-7)^2}$

2. $2 + \sqrt[4]{(x+2)^3}$

3. $\sqrt[6]{(x^2+3)^5} + 3$

4. $-4 + \sqrt[4]{(t-5)(t+2)^3}$

5. $\dfrac{1}{\sqrt[3]{(2x+a)^2}}$

6. $\dfrac{3}{\sqrt[9]{(x+1)^{11}}}$

7. $-4\sqrt[3]{(3-x)^2}$

8. $\sqrt[3]{(5u-3)^7}$

9. $\sqrt[7]{x^2 y^3 z^4}$

10. $\sqrt[3]{a^2 b^2 c^5}$

Graph the following functions in exercises 11–20.

11. $f(x) = \sqrt{(x-4)^2}$

12. $g(x) = \dfrac{1}{\sqrt{x+2}}$

13. $h(x) = \sqrt[3]{3x-5}$

14. $f(x) = \sqrt[4]{(5-x)^3}$

15. $g(x) = \sqrt[4]{(x-2)(x+4)}$

16. $h(x) = \dfrac{\sqrt[4]{(x-1)^7}}{10}$

17. $f(x) = \sqrt[2]{(2x-3)^5}$

18. $g(x) = \dfrac{x-3}{\sqrt[3]{(x+20)^2}}$

19. $h(x) = \dfrac{x-2}{\sqrt[3]{(4-x)}}$

20. $f(x) = \dfrac{1}{\sqrt[4]{x^5}}$

For exercises 21–32, find the real-number solutions to each of the following equations using symbolic methods. Check your solutions by graphing on your graphing calculator.

21. $\sqrt{x+1} = 5$

22. $\sqrt{2x-5} = 3$

23. $\sqrt{x+3} = -2$

24. $\sqrt{x-5} = -1$

25. $\sqrt{5-x^2} - 1 = x$

26. $\sqrt{3-4x^2} - 2 = 1 - 2x$

27. $\sqrt{3x-1} - \sqrt{4x+5} = 0$

28. $\sqrt{2x-6} = \sqrt{4x+5} + 1$

29. $\sqrt[3]{2x-5} = 3$

30. $\sqrt[3]{5-x} = 8$

31. $2 + \sqrt[5]{2-x} = 7$

32. $3 - \sqrt[5]{(x+2)^2} = 5$

Summary and Projects

Summary

The Behavior of $f(x) = 1/x$ The function $f(x) = 1/x$ might be called the standard rational function. Knowing the behavior of this function and powers of it allow you to study and understand the behavior of a wide class of rational functions. Figure 4.24 is a graph of $f(x) = 1/x$ with vertical asymptote $x = 0$ and horizontal asymptote $y = 0$. Vertical and horizontal shifts, reflection about

Figure 4.24

the x-axis, and expansions and contractions apply to this function as to every function. For example, the function

$$g(x) = \frac{-1}{x-3} - 2$$

will reflect the graph of $f(x) = 1/x$ about the x-axis, shift it to the right 3 units, and shift it down 2 units. This means the vertical asymptote will become $x = 3$ and the horizontal asymptote will become $y = -2$.

If we square $f(x) = 1/x$, we obtain $1/x^2$. The asymptotes remain the same but the graph is now entirely above the x-axis and the curve now approaches the asymptotes sooner. This is because the squaring causes the function values to be pushed away from the x-axis for input values betwen -1 and 1 but are pulled toward the x-axis for input values outside 1 and -1. Figure 4.25 shows the graph of $h(x) = 1/x^2$.

Higher powers of $f(x) = 1/x$ will alternate between looking like Figure **??** and Figure 4.25 for odd and even powers, respectively. In addition, the higher the power the more the expansion and contraction between and outside the interval $(-1, 1)$.

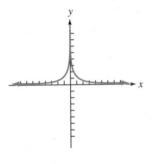

Figure 4.25

Vertical Asymptotes   Rational functions, which are quotients of polynomials, can be considered in a more global way with the help of your graphing calculator. For example, you can discover that a linear factor of the form $ax - b$ in the denominator will result in a vertical asymptote occurring at $x = \frac{b}{a}$ provided $ax - b$ is not a factor of the numerator also. In fact, a hole will occur at $x = \frac{b}{a}$ when $ax - b$ is a factor of both the numerator and denominator. A linear factor of the form $cx + d$ in the numerator of a rational function will give us the x-intercepts of the function. Again, this is true provided the same linear factor does not appear in the denominator.

Horizontal Asymptotes   You can examine horizontal asymptotes of rational functions by looking at function values corresponding to input values far away from the origin. When the degree of the numerator is the same as the degree of the denominator, then

$$y = \frac{\text{the coefficient of the leading term of the numerator}}{\text{the coefficient of the leading term of the denominator}}$$

will be a horizontal asymptote. You can see this by dividing the polynomials or by considering, for example

$$\frac{5x^2 - 2x + 3}{2x^2 + 7x - 1}$$

You can see that for very large values of x the linear and constant terms become less and less important. If you neglect these terms temporarily, the function looks like

$$\frac{5x^2}{2x^2} = \frac{5}{2}$$

so that the function is approaching the line $y = 5/2$ as a horizontal asymptote.

   This same approach can be applied to every rational function. For example, if the degree of the numerator is bigger than the degree of the denominator, the function will become infinite for input values far from the origin. Whereas, if the degree of the numerator is less than the degree of the denominator, the function values will approach 0 and $y = 0$ will be a horizontal asymptote.

Inverse Functions   The parametric representation of a function is introduced in this chapter because, combined with the graphing calculator, it allows for a very nice way of examining functions and their inverses. You can use the parametric mode of your graphing calculator to graph a function and its inverse and examine their behavior visually. First of all, the inverse of a function is obtained by interchanging the domain and range values of the original function. The result of doing this graphically through the use of the parametric mode of our graphing calclulator soon indicates that a function and its inverse are reflections of each other about the line $y = x$. This examination can also show us how to obtain the equation of the inverse of a function if we know the equation of the function. We simply interchange the variables x and y. It is useful to solve for y after the interchange when possible. The notation $f^{-1}(x)$ is used for the inverse of the function $f(x)$. One final observation about a function and its inverse is

$$f[f^{-1}(x)] = f^{-1}[f(x)] = x$$

which tells us that the inverse of a function undoes what the function did and vice-versa.

Projects

1. Jan is conducting an experiment under a very controlled situation, which requires her to measure the intensity of light at different points in an experimental tube. The tube is 10 meters long. These are her observations: The intensity becomes zero at $x = 1$ and $x = 2$ and as x approaches the beginning of the tube the intensity shows a sudden large increase. In between $x = 1$ and $x = 2$ the intensity rises slightly and then falls off. After $x = 2$, it rises sharply again, although not as fast as at the other end. Jan needs to come up with a function that will fit the situation. Can

you help? Are there other possibilities? If so, how would you check the validity of your formula?

Jan changes some of the conditions of the experiment and makes a very different kind of observation. This time, conditions near $x = 0$ still remain the same (i.e., the intensity increases very rapidly). However, as x increases to 10m, the intensity falls off almost towards zero. Give a function that may describe the situation.

CHAPTER 5

Exponential and Logarithmic Functions

5.1 Exponential Functions

Introduction

Exponential Functions

Properties of Exponential Functions

Applications of Exponential Functions

The Natural Exponential Function

Graphs of the Natural Exponential Function

Constructing Modeling Functions Using the Calculator

Introduction

The functions we have studied to this point have been *algebraic functions*, that is, functions in which the defining expressions are algebraic expressions. As important and useful as they are, algebraic functions do not model all applied and theoretical phenomena. In this section we introduce a new class of function, the *transcendental functions*. The functions in this class are nonalgebraic and are called transcendental because they transcend algebraic principles. In this section we study the *exponential function*, and in the next section, its inverse, the *logarithm function*. Transcendental functions are important and useful as they model many applied and theoretical phenomena. In both these first two sections, we investigate the nature and behavior of the exponential and logarithm functions, as well as examine examples of how they are used. We will end this section with a description of how to construct an exponential function that models phenomena that exhibit exponential behavior.

Exponential Functions

In algebraic functions, the variable always appears in the *base* of any power, and the exponent on any base is always a constant. For example, in the quadratic function $f(x) = x^2$, the variable x is positioned in the base, and the constant, 2, in the exponent. In the exponential function, we interchange the position of the variable and constant, making the constant the base of a power, and the variable the exponent. We begin by formally defining the exponential function. Following the definition, we discuss some of its properties.

The Exponential Function

The Exponential
Function

If $b > 0$ and $b \neq 1$, then the **exponential function** with base b is

$$f(x) = b^x$$

For example, the function $f(x) = 2^x$ is an exponential function. The base 2 satisfies the condition that the base is positive and not equal to 1.

Let's examine the domain of this function. Up to this point in your study of algebra, the only numbers allowable as exponents have been rational numbers. The definition seems to allow for irrational exponents. Before we construct the graph of an exponential function, we will loosely define numbers such as $4^{\sqrt{3}}$, where $\sqrt{3}$ is an irrational number. A formal definition of such numbers is covered in the calculus course. For now, we will define numbers such as $4^{\sqrt{3}}$ as the decimal approximation we obtain on our graphing calculator.

Now, with the information that, for any real number x, the expression b^x defines a real number, we can construct the graph of $f(x) = b^x$. The figure on the left of Window 5.1 shows the graph of $f(x) = 2^x$, and the figure on the right of Window 5.1 shows the graph of $f(x) = \left(\dfrac{1}{2}\right)^x$. (Use your calculator to graph $f(x) = 2^x$ two different ways, first by entering 2^x as the defining expression for, say, Y1, and then parametrically by selecting the Param mode and entering t and 2^t for X1T and Y1T, respectively.)

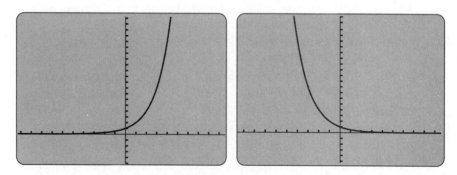

Window 5.1

All exponential functions have graphs similar to these two graphs. If $b > 1$, as in $f(x) = 2^x$, $f(x)$ is an increasing function. That is, the output values increase as the input values increase. These functions are useful for modeling growth behavior and are commonly called *growth functions*. If $0 < b < 1$, as in $f(x) = \left(\dfrac{1}{2}\right)^x$, $f(x)$ is a decreasing function. That is, the output values decrease as the input values increase. These functions are useful for modeling decay (negative growth) behavior and are commonly called *decay functions*. By constructing several graphs yourself, you will be able to see that for growth functions, the larger the base, the faster the long-term growth, and for decay functions, the smaller the base, the faster the initial decay.

The definition of the exponential function $f(x) = b^x$, $b > 0$, $b \neq 1$, specifies that the base b can be any positive real number except 1. These exceptions are made so that b^x models growth or decay for all real input values. If $b = 1$, $f(x) = 1^x = 1$, and is a constant function for all values of x. The graphs of constant functions are horizontal lines indicating that this behavior differs from the expected increasing or decreasing behavior of growth and decay functions. If $b < 0$, the graph of $f(x) = b^x$ is in no way similar to the growth or decay graph. Some points on the graph of $f(x) = (-2)^x$ appear in Figure 5.1. Notice that the points on the graph of $f(x) = (-2)^x$ have not been joined with a smooth curve. Since the defining expression $(-2)^x$ is not defined for exponents such as $1/2$, $1/4$, $1/6, \ldots$ (why not?), the graph has infinitely many breaks in it and is, therefore, discontinuous. It is also negative for various values of x. When $b < 0$, the domain does not include all real numbers as specified in the definition of the exponential function, and the graph does not reflect the growth or decay behavior that is modeled when the base b is positive and different from 1.

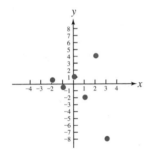

Figure 5.1

Properties of Exponential Functions

We note the following general facts about the exponential function:

1. The power function $f(x) = x^b$ and the exponential function $f(x) = b^x$ are two entirely different functions.

    The power function has the form $f(x) = (\text{variable})^{(\text{constant})}$.

    The exponential function has the form $f(x) = (\text{constant})^{(\text{variable})}$.

2. The domain of $f(x) = b^x$ is all real numbers, that is, the exponential function is computable for all real values of x.

3. The range of $f(x) = b^x$ is all positive real numbers. The exponential function is never zero or negative.

4. The graph of the exponential function is a smooth, continuous curve in that it has no holes, breaks, gaps, or corners.

5. The graph of the exponential function always opens upward. This means that growth functions always increase at an increasing rate, and that decay

functions always decrease at a decreasing rate. The graph also tells us that the exponential function is a one-to-one function.

6. Both the exponential growth and decay functions pass through the point $(0, 1)$. This means that when the input value is 0, the output value is 1. If either function is modified to the form $f(x) = A_0 b^x$, where A_0 is a constant, then when the input value is 0, the output value is A_0 (an expansion or contraction of the function $g(x) = b^x$ by the factor A_0). This is important when the exponential function models some physical phenomenon. The input value may be time, in which case, $t = 0$ indicates the initial time and the value A_0 indicates the initial output.

7. The exponential decay function $f(x) = b^x$, $0 < b < 1$ is often expressed with a negative exponent on a base greater than 1. For example, the decay function $f(x) = \left(\dfrac{1}{2}\right)^x$ can be expressed as $f(x) = 2^{-x}$ since $2^{-x} = \dfrac{1}{2^x} = \left(\dfrac{1}{2}\right)^x$.

8. Exponential functions obey all the rules of function translation. That is, if $f(x)$ is an exponential function and c is a positive constant, then $f(x) + c$ represents an upward vertical translation of $f(x)$, $f(x) - c$ represents a downward vertical translation of $f(x)$, $f(x + c)$ represents a leftward horizontal translation of $f(x)$, and $f(x - c)$ represents a rightward horizontal translation of $f(x)$.

9. The graph of the function $g(x) = A_0 b^{-x}$ is an expansion or contraction of the graph of the function $f(x) = b^{-x}$ by the factor A_0.

ILLUMINATOR SET A

The graphs of $f(x) = 2^x$ and $g(x) = f(x + 3) = 2^{x+3}$ are illustrated, respectively, in Window 5.2. Notice that the graph of $g(x) = f(x + 3)$ is precisely the graph of $f(x)$ shifted to the left 3 units.

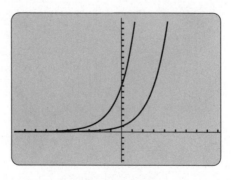

Window 5.2

It is worthwhile noting that since

$$f(x+3) = 2^{x+3}$$

$$= 2^x \cdot 2^3$$

$$= 8 \cdot 2^x$$

$$= 8 \cdot f(x)$$

the graph of $f(x+3)$ can be obtained from the graph of $f(x)$ by multiplying each output value of $f(x)$ by 8. Thus, for the function $f(x) = 2^x$, a horizontal translation to the left by 3 is also an expansion by 2^3.

Applications of Exponential Functions

When observing phenomena, people most often notice change; people observe growth and decay, and exponential functions can serve as good instruments for modeling such behavior.

As a first example of an application of an exponential function, we consider an application from finance.

ILLUMINATOR SET B

At 7.2% interest compounded annually, an investment of $1 doubles in value about every 10 years. We can tabulate the values for this investment for several 10-year periods.

Period	Value
0	$1 = $1
1	$2 \cdot \$1 = \$2$
2	$2 \cdot 2 \cdot \$1 = \$4$
3	$2 \cdot 2 \cdot 2 \cdot \$1 = \$8$
4	$2 \cdot 2 \cdot 2 \cdot 2 \cdot \$1 = \$16$
5	$2 \cdot 2 \cdot 2 \cdot 2 \cdot 2 \cdot \$1 = \$32$

The initial value is $1. After five ten-year periods (50 years), the value is $2 \cdot 2 \cdot 2 \cdot 2 \cdot 2 \cdot 1 = 32$ times as large, or $32.

Table 5.1

P	Value
0	$2^0 = 1$
1	$2^1 = 2$
2	$2^2 = 4$
3	$2^3 = 8$
4	$2^4 = 16$
5	$2^5 = 32$

The table does not give us the value of an investment after fifteen years, or 1.5 time periods. In order to determine this value, we shall create a function for which the values at 0, 1, 2, 3, 4, and 5 are those recorded in the table. We can then evaluate this function at 1.5. To help us create this function, we carefully observe the table, looking for patterns. We notice that, in each case, the number of 2's is precisely the same as the time period. Table 5.1 gives the same information using standard exponential notation (where P represents Period).

The pattern apparent in the table indicates that the value at any time period can be obtained by raising 2 to that time-period number. Thus, the function defined by

$$f(x) = 2^x \quad x \ge 0$$

seems to be the required function.

The value of the investment after 15 years, or 1.5 ten year periods is then obtained by computing $f(1.5)$.

$$f(1.5) = 2^{1.5}$$

$$\approx 2.83$$

Of course, there are many other exponential functions. For example, you may wish to consider the time it takes for an investment to triple rather than double. In this case, the function you create would be

$$f(x) = 3^x, \quad x \ge 0$$

EXAMPLE SET A

If the interest rate allows an investment to triple every 10 years, (1) find the value of a $100 investment after 15 years, and (2) with an error of at most 0.01 years, determine when the investment will be worth $4800.

Solution:

1. The underlying exponential function is $f(x) = 3^x$, but this is the function for a $1 investment. The function for a $100 investment is defined by

$$f(x) = 100 \cdot 3^x, \quad x \ge 0$$

where x represents the number of ten-year time periods. (Can you convince yourself that this is the appropriate function? Construct a table similar to the last one we constructed using $100 as the initial investment rather than $1, and you will see that it is.) Using this function, the value of the investment after 15 years can be obtained by computing $f(1.5)$.

$$f(1.5) = 100 \cdot 3^{1.5}$$

$$= 519.62$$

2. Using the calculator to construct the graphs of both $f(x) = 100 \cdot 3^x$ and $g(x) = 4800$, constructing boxes around the intersection of the two curves, and applying the approximation techniques of Section 1.4, we find that, with an error of at most 0.01 time periods, the investment will be worth $4800 after 3.52 ten-year time periods, which translates to about 35.2 years.

The Natural Exponential Function

When we examined variation in Section 2.6, we found that two quantities vary directly if their quotient is always constant, that is, if y and x vary directly, then $\dfrac{y}{x} = k$, where k is a constant (they vary inversely if their product is a constant). When this equation is solved for y, we get $y = kx$. When constructing mathematical models to describe the behavior of many types of applied phenomena, what is observed is continuous change. Very often the continuous change in the output quantity y, per unit change in the input quantity x, varies directly with the amount of the output quantity currently present. That is,

$$\frac{\text{change in } y}{\text{change in } x} = k \cdot y$$

This equation expresses the relationship between the continuous change in the input quantity and the continuous change in the output quantity, and not the input and output quantities themselves. The problem of constructing a model that relates the input quantity x and the output quantity y is solved using the methods of calculus. The solution to this equation is

$$y = A_0 e^{kx}$$

Continuous Growth Constant
Decay Constant

where A_0 represents the initial amount of the output quantity present, and both k and e are constants. The constant of variation k is called the **continuous growth constant**, and it is always positive. To model decay, or negative growth, $-k$ is used and is then called the **decay constant**. The number e is an irrational number that is approximately equal to 2.718281828459045. The constant e usually appears in the solution of all continuous growth and decay problems. The importance of this number was first recognized by the Swiss mathematician Leonard Euler (pronounced "Oiler") (1707–1783), and in his honor, it is denoted by e. The number e is important because as the input value x increases, the value of $y = e^x$ increases at a rate precisely equal to e^x.

The Natural Exponential Function
The exponential function having base e, $f(x) = e^x$, is called the **natural exponential function**.

The Natural Exponential Function

Continuous Change Model

The exponential function $y = A_0 e^x$, or more appropriately, $f(x) = A_0 e^x$, is called the **continuous change model**.

Since $e > 1$, $f(x) = e^x$ represents a growth function, and $f(x) = e^{-x}$ represents a decay function. The figure on the left of Window 5.3 shows the graph of the growth function $f(x) = e^x$, and the figure on the right, the graph of the decay function $f(x) = e^{-x}$.

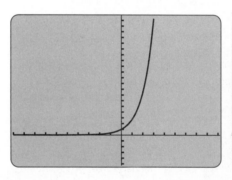 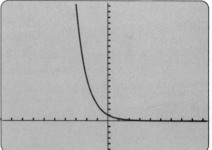

Window 5.3

EXAMPLE SET B

A radioactive compound decays, approximately, according to the function

$$A(t) = A_0 e^{-0.2t}$$

where A_0 is the amount of the compound initially present, and t is the time in seconds from the initial observation. If there are initially 4 grams of the compound present, how much will be present at the end of 8 seconds? (Round to the nearest hundredth.)

Solution:
We need to find $A(8)$. Substitute 4 for A_0 and 8 for t and compute.

$$A(t) = A_0 e^{-0.2t}$$

$$A(8) = 4e^{-0.2(8)}$$

$$= 0.807586072\ldots$$

Rounding to the nearest hundredth, we see that the amount of the radioactive compound that remains after 8 seconds is 0.81 grams.

Graphs of the Natural Exponential Function

You can get a feel for the behavior of $f(t) = A_0 e^{kt}$ by examining graphs for a specific value of A_0 and for various values of k. For convenience, choose $A_0 = 1$ (any other number would do). The calclulator screen on the left of Window 5.4 shows the graphs of $f_1(t) = e^{0.1t}$, $f_2(t) = e^{0.3t}$, $f_3(t) = e^{0.5t}$, and $f_4(t) = e^{1t}$ when plotted on the same window. The calculator screen on the right of Window 5.4 shows the graphs of $f_4(t) = e^{1t}$, $f_5(t) = e^{1.5t}$, and $f_6(t) = e^{2t}$, when plotted on the same window. (It is worthwhile to watch your calculator when graphing these functions several at a time on the same coordinate system. You will be able to experience the effect of the growth constant k.)

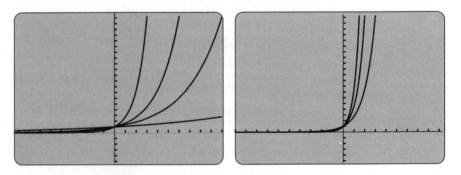

Window 5.4

You can see that each graph displays the recognizable upward trend of the exponential growth function. You should also notice that as the value of the growth constant k increases, the curve bends more sharply upward. This indicates that for small values of k, growth is increasing, but relatively slowly, whereas for large values of k, growth is increasing relatively quickly.

Constructing Modeling Functions Using the Calculator

You can use your calculator to construct exponential functions to model phenomena that exhibit growth or decay behavior.

Each function that you construct will be only as accurate as the data you collect. Also, the more data you collect and use, the more effective your function will be at modeling the behavior of the phenomena you are describing. If, on the graph, the data are scattered all about the coordinate system, there is probably no trend, and therefore no modeling function. If the data appear to lie along a straight line, they indicate a linear relationship. If they appear to lie along a curve, they may indicate an exponential or a polynomial relationship.

1. Collect some data (ordered input-output pairs) that will help you see the behavior of the relationship. On the calculator screen or on a piece of graph paper, plot the input-output pairs (ordered pairs) and look to see what type of trend they exhibit, if any. This graph is called a *scatter diagram*. The process for doing so is as follows: your calculator can help you decide which type of relationship your data exhibit and, therefore, help you decide which type of modeling function to construct.

2. Use the STAT (statistics) feature of your calculator to input the ordered pairs (being careful to clear out any previously entered data). (See Section 2.7.)

3. Continue using the STAT feature to construct the function by (a) selecting the linear model if the data appear to lie along a straight line. The linear model has the form $y = a + bx$; (b) selecting the exponential model if the data exhibit the growth or decay trend of the exponential function. The exponential model has the form $y = ab^x$. Or (c) selecting the power/polynomial model if the data appear to lie along a curve. The power model has the form $y = ax^b$. Some calculators may have the capability of constructing 2nd-, 3rd-, and 4th-degree polynomial models. Such models have the respective forms, $y = a_2x^2 + a_1x + a_0$, $y = a_3x^3 + a_2x^2 + a_1x + a_0$, and $y = a_4x^4 + a_3x^3 + a_2x^2 + a_1x + a_0$.

   To help you decide which is the appropriate model, the calculator will return, along with the values of a and b, a number r that is a measure of the strength of the relationship. It will always be the case that $-1 \le r \le 1$. The closer r is to -1 or 1, the stronger the relationship. An r-value of 0 indicates no relationship. An r-value above 0.9 or below -0.9 indicates a significantly strong relationship. To determine which model best describes the relationship, simply choose each, and select the one with the r-value that is closest to -1 or 1. (With information about the value of r, you

Table 5.2

t	$P(t)$
0	610
1	1,098
2	1,976
3	3,557
4	6,403
5	11,526
6	20,747
7	37,345
8	67,221
9	120,000
10	217,798

can, in fact, omit the step of graphing the ordered pairs to "see" which model, if any, is most appropriate. The graph, after all, presents only an eyeball measure of the strength of the relationship. The r-value is a quantitative measure of the strength of the relationship and is, therefore, more accurate.)

EXAMPLE SET C

For a 10-hour period, a biologist's observations and weighings of a growing population of bacteria show the relationship given in Table 5.2 between the time t, in hours since the initial observation, and the population P of bacteria.

   Construct a function that models this growth behavior.

Solution:

Begin by entering the 10 ordered pairs using the STAT feature of your calculator. Use x as the representation for the input value (the value to predict from), and y as the representation for the output value (the value to predict). Once the data are entered, select each regression model, noting the value of r to determine which model produces the strongest relationship.

   For this set of data, the linear model (LinReg) produces $r \approx 0.811$, the exponential model (ExpReg) produces $r \approx 0.999$, the logarithmic model (LnReg) and the power/polynomial (ab^n) models produce error messages indicating that the calculator cannot construct such a model. (We discuss the logarithm function and why an error message is produced in the next section). The clear choice is the exponential model, with $r \approx 0.999$, $a \approx 610.4$, and $b \approx 1.8$. We can use P to represent the output, the population size in this case, and t to represent the time in hours from the initial observation. Recalling that the exponential model has the form $y = ab^x$, we can very nicely model these data with the function

$$P(t) = 610 \cdot (1.8)^t$$

It is natural to wonder why a continuous growth function does not use the number e. We show later in this chapter how to convert a constant such as 1.8 to a power of e. In this case, $1.8 \approx e^{0.588}$, so that $(1.8)^t = (e^{0.588})^t = e^{0.588t}$. Thus, the modeling function can be described using the base e as

$$P(t) = 610e^{0.588t}$$

To check that this function does indeed provide a good model, choose an input value somewhere within the observed values and compute the corresponding output value. That computed output value should be within the appropriate range of observed output values. For example, if we choose $t = 4.5$, we expect that $P(4) < P(4.5) < P(5)$. Computing, we get $P(4.5) \approx 8599$, which is between $P(4)$ and $P(5)$.

EXERCISES

1. Investigate the relationship between the behavior of the power function $f(x) = x^2$ and the exponential function $g(x) = 2^x$ by completing the table and specifying on which intervals the power function or the exponential function is smaller or larger, and at which x-value, if any, they are equal.

x	x^2	2^x	$x^2 > 2^x$ **or** $2^x > x^2$ **?**
0			
1			
2			
3			
4			
5			
6			
10			

2. Investigate the relationship between the behavior of the power function $f(x) = x^3$ and the exponential function $g(x) = 3^x$ by completing the table and specifying on which intervals the power function or the exponential function is smaller or larger, and at which x-value, if any, they are equal.

x	x^3	3^x	$x^3 > 3^x$ **or** $3^x > x^3$ **?**
0			
1			
2			
3			
4			
5			
6			
10			

For exercises 3–7, express each function so that only positive exponents appear. Specify each function as a growth or a decay function.

3. $f(x) = 10^{-x}$.

4. $f(x) = 4^{-x}$.

5. $f(x) = \left(\frac{1}{6}\right)^{-x}$.

6. $f(x) = \left(\frac{1}{3}\right)^{-x}$.

7. $f(x) = \left(\frac{1}{e}\right)^{-x}$.

8. The function $P(x) = \frac{1}{\sqrt{2\pi}}e^{-x^2/2}$ is called the *standard normal probability density function* and is used often in computing probabilities of events that are normally distributed. Use the window $-6 \le x \le 6$ and $-2 \le y \le 1$ to complete each of the following parts. (a) Specify the intervals upon which $P(x)$ is increasing and the intervals upon which $P(x)$ is decreasing, (b) with an error of at most 0.01 units, specify any relative maximum or minimum points. (c) Estimate the domain of $P(x)$. (You have graphed $P(x)$ on $[-6, 6]$, but this is not the complete graph.) (d) Estimate the range of $P(x)$.

On the same coordinate system, use your calculator to construct the graphs of each pair of functions in exercises 9–16. Specify, in terms of symmetry, any relationship between them.

9. $f(x) = 3^x$ and $f(x) = \left(\frac{1}{3}\right)^x$.

10. $f(x) = 2^x$ and $f(x) = \left(\frac{1}{2}\right)^x$.

11. $f(x) = 3^x$ and $f(x) = 3^{-x}$.

12. $f(x) = 2^x$ and $f(x) = 2^{-x}$.

13. $f(x) = 3^{-x}$ and $f(x) = \left(\frac{1}{3}\right)^x$.

14. $f(x) = 2^{-x}$ and $f(x) = \left(\frac{1}{2}\right)^x$.

15. $f(x) = e^x$ and $f(x) = \left(\dfrac{1}{e}\right)^x$.

16. $f(x) = e^{-x}$ and $f(x) = \left(\dfrac{1}{e}\right)^x$.

17. In calculus, the value of e is computed by considering the long-term behavior of the function $f(x) = \left(1 + \dfrac{1}{x}\right)^x$. Show, by completing the table, that as the values of x get bigger and bigger positively, the values of $f(x) = \left(1 + \dfrac{1}{x}\right)^x$ get closer and closer to the number denoted by e. (This exercise is related to exercise 19.)

x	$f(x) = (1 + 1/x)^x$
1	
10	
100	
1000	
10,000	
100,000	
1,000,000	
1,000,000,000	

18. In calculus, the value of e is computed by considering the long-term behavior of the function considered in exercise 17, $f(x) = \left(1 + \dfrac{1}{x}\right)^x$. It can also be computed by considering the short-term behavior of the function $f(x) = (1 + x)^{1/x}$. Show, by completing the table, that as the values of x get closer and closer to zero, the values of $f(x) = (1 + x)^{1/x}$ get closer and closer to the number denoted by e. (This exercise is related to exercise 20.)

x	$f(x) = (1 + x)^{1/x}$
1/1	
1/10	
1/100	
1/1000	
1/10,000	
1/100,000	
1/1,000,000	
1/1,000,000,000	

19. In exercise 17, you demonstrated that as the values of x got bigger and bigger, the values of $f(x) = \left(1 + \dfrac{1}{x}\right)^x$ got closer and closer to the number denoted by e. Since $f(x)$ is a function, you should be able to demonstrate this fact graphically. Noting from the table you constructed in exercise 17 that $f(x)$ is never negative and that it approaches e rather quickly, use your calculator to construct the graph of $f(x)$ to demonstrate that as x gets bigger and bigger, $f(x)$ approaches e.

20. In exercise 18, you demonstrated that as the values of x got closer and closer to zero positively, the values of $f(x) = (1 + x)^{1/x}$ got closer and closer to the number denoted by e. Since $f(x)$ is a function, you should be able to demonstrate this fact graphically. Noting from the table you constructed in exercise 18 that $f(x)$ is never negative and that it approaches e rather quickly, use your calculator to construct the graph of $f(x)$ to demonstrate that as x gets smaller and smaller, $f(x)$ approaches e.

On the same coordinate system, construct the graph of each pair of functions by graphing the first using your calculator, and the second using the first and what you know about translations of functions in exercises 21–32.

21. $f(x) = e^x$ and $g(x) = e^{x+4}$.

22. $f(x) = 2^x$ and $g(x) = 2^{x-2}$.

23. $f(x) = 3^x$ and $g(x) = 3^x + 2$.

24. $f(x) = e^x$ and $g(x) = e^x - 5$.

25. $f(x) = e^x$ and $g(x) = e^x + 2$.

26. $f(x) = e^x$ and $g(x) = e^{x+1} - 4$.

27. $f(x) = 32 \cdot 2^x$ and $g(x) = 2^{x+5}$.

28. $f(x) = 9 \cdot 3^x$ and $g(x) = 3^{x+2}$.

29. $f(x) = e^x$ and $g(x) = -e^x$.

30. $f(x) = 2^x$ and $g(x) = -2^x$.

31. $f(x) = 3^x$ and $g(x) = -3^{x+4} + 5$.

32. $f(x) = e^x$ and $g(x) = -e^{x-3} - 4$.

In exercises 33–38, use your calculator to find, with an error of at most 0.01 units, the zeros of each function.

33. $f(x) = e^x - 4$.

34. $f(x) = e^x - 25$.

35. $f(x) = .06e^x - 105$.

36. $f(x) = 3 \cdot 2^x - 4$.

37. $f(x) = 3 \cdot 2^{x-3} - 31$.

38. $f(x) = 5^{-2x+10} - 20$.

39. Let $f(x) = e^x$ and $g(x) = x - 5$. Find expressions for the functions $(f \circ g)(x)$ and $(g \circ f)(x)$, and graph each one. Are they the same? How do they differ from $f(x) = e^x$?

40. A research project is carried out to determine if, in a certain region of the country, there is an exponential relationship between the value V, in thousands of dollars, of real estate and the time t since the purchase of the real estate. The data are $(3, 246)$, $(25, 298)$, $(6, 328)$, $(7, 360)$, $(8, 396)$, $(9, 436)$, $(10, 480)$. The input variable is the time since purchase, and the output variable is the value of the real estate. Determine if an exponential relationship exists, and if it does, construct the function using the STAT feature, and find the expected value of the real estate 15 years from the time of purchase.

t	$V(t)$
3	246
5	298
6	328
7	360
8	396
9	436
10	480

41. To determine if there is an exponential relationship between the amount of undissolved salt in a water solution and the time t, in minutes after the salt is placed into the water, a precalculus student placed 10 grams of salt into a beaker of water, measured the amount of undissolved salt at various times, and collected the following data: $(1, 8)$, $(2, 6.4)$, $(3, 5.12)$, $(4, 4)$, $(5, 3.27)$, $(10, 1.07)$, $(15, 0.35)$. The input variable is the time since the salt was placed into the water, and the output variable is the quantity of undissolved salt that remains in the solution. Determine if an exponential relationship exists, and if it does, construct the function using the STAT feature, and find the expected quantity of salt that remains in a water solution 8 minutes after 10 grams of it was placed into the solution.

t	$Q(t)$
1	8
2	6.4
3	5.12
4	4
5	3.27
10	1.07
15	0.35

42. The following data represent the amount, $Q(t)$, in grams, of a radioactive substance that is present t hours from the beginning of the initial observation. Initially, there are 500 grams present. The input variable is

t	$Q(t)$
1	475
2	450
3	430
4	409
5	390
10	303
20	184

the time from the initial observation, and the output variable is the amount of material that remains at time t. Determine if an exponential relationship exists between t and $Q(t)$, and if one does, construct the function using the STAT feature, and find the expected quantity of the material that remains 30 hours after the initial observation.

43. Research carried out by the state to determine if the population of a certain city in the state is growing exponentially produces the following data. The input variable t is the number of years from 1985. The output variable $P(t)$ (in thousands) is the population of the city at time t.

t	$P(t)$
1	6.12
2	6.24
3	6.37
4	6.50
5	6.63
6	6.76
7	6.90

Determine if an exponential relationship exists between t and $P(t)$, and if one does, construct the function using the STAT feature, and find the expected population of the city in the year 2000.

44. Data collected on a group of college students applying to medical school indicate a relationship between the student's grade-point average (GPA) and the number of applications for admission the student submits. The data indicate that the higher the GPA, the fewer applications he or she must submit. Construct a function that predicts the number of applications for admission a student must submit from the student's GPA. The student's GPA is represented by x and is restricted to values between 2.0 and 4.0. The number of applications submitted is represented by $N(x)$.

x	$N(x)$
2.0	40
2.2	31
2.4	24
2.6	18
3.0	11
3.1	10
3.3	7
4.0	3

45. It is known that when a drug is injected into the bloodstream, its concentration diminishes as time goes by. To determine if the concentration $C(t)$ of a particular drug decreases exponentially over time t, a researcher collects the following set of data. In the data set, the input variable is the time in hours since injection of the drug, and the output variable is the concentration $C(t)$ of the drug in the bloodstream.

t	$C(t)$
1	2.5
3	1.7
5	1.2
7	0.85
10	0.50
12	0.35
15	0.20
24	0.04

Construct a function that predicts the concentration of the drug in the bloodstream from the time since the injection of the drug.

46. If the value of a \$300 investment doubles every five years, determine the value of the investment after 20 years.

47. If the value of a \$200 investment doubles every seven years, determine the value of the investment after 25 years.

48. An initial investment is held for 15 years at an interest rate that triples the investment every 9 years. In terms of the original investment, what is the value of the investment after 15 years?

49. An initial deposit in a savings account is held at an interest rate that triples the deposit every 4 years. In terms of the original investment, what is the value of the investment after 10 years?

50. The number of bacteria in a colony is halved every 5 hours if the temperateure is held constant at 15° C. At this temperature, how long will it take for the colony to be 1/32 its original size?

51. The number of bacteria in a colony is halved every 4 hours if the acidity of the medium in which the bacteria live has a pH value of 7.8. At this level of acidity, how long will it take for the colony to be 1/16 its original size?

52. Suppose the number N of bacteria in a culture at a particular time t is approximated by the function

$$N(t) = N_0 5^{0.03t}$$

(a) If there are initially $8,000$ bacteria present, how many will be present after 20 hours?
(b) With an error of at most 1, how many hours after the initial observation will there be $15,000$?

53. The atmospheric pressure, P, in inches of mercury, is given by

$$P(h) = 30 \cdot 10^{-0.09h}$$

where h is the height, in miles, above sea level. Find the atmospheric pressure (a) 5 miles above sea level, (b) 1/2 mile above sea level, and (c) 1/4 mile above sea level. (d) With an error of at most 0.1 mile, find the height above sea level for which the pressure is 25 inches of mercury.

54. A radioactive compound decays according to the function

$$A(t) = A_0 e^{-0.3t}$$

where A_0 is the initial amount of the compound present, and t is the number of seconds that have passed since the initial observation. (a) If there are initially 20 grams of such a compound present, how much will be present at the end of 12 seconds? (b) With an error of at most 0.1 seconds, how much time must pass since the initial observation until there are 10 grams left?

55. A loaf of bread is removed from an oven when its temperature is 300°F. After sitting on the countertop in a room with a constant temperature of 70° for 3 minutes, its temperature is 200°F. The temperature T of the loaf of bread at any time t is given by

$$T(t) = 70 + 230e^{-0.19108t}$$

(This function is a version of Newton's Law of Cooling.) (a) What will be the temperature of the bread after 28 minutes? (b) How long after the bread is removed from the oven and placed on the countertop will its temperature be 80°?

56. Psychologists have developed a formula that approximately relates the number N of nonsense symbols a person can memorize to the time t, in minutes, that the person studies the symbols. The formula is

$$N(t) = 16(1 - e^{-t})$$

Find the approximate number of nonsense symbols a person can memorize if she studies them for 5 minutes.

57. A cable such as a telephone line hung between two poles forms a curve called a *catenary*. The equation of the catenary is

$$h(x) = a(e^{x/200} + e^{-x/200})$$

where h represents the height of the cable, x is the position from the midpoint of the poles horizontally out, and a is a constant determined by the weight of the cable. How high above the ground is the cable 50 feet from either pole if the poles are 300 feet apart? (Assume $a = 13$ feet.)

58. In their research about the participation of members of a discussion group, the sociologists Stephan and Mischler ranked the members according to the number of times each participated in the discussion. Using the data they had collected, they established a relationship between the number N_1 of times the first-ranked person participated and the number of times the kth-ranked person participated. The relationship is given by the function

$$N(N_1, k) = N_1 e^{-0.11(k-1)}, \quad 1 \le k \le 10$$

(a) If the first-ranked person participated 80 times, how many times did the 4th-ranked person participate? (b) What rank was assigned to the person who participated 41 times?

59. A bacteria colony in a dish is doubling in size every day until bacteria fill the dish on the 21st of June. When is the dish half full? When is it one quarter full? If the colony is started with four bacteria, what is the expression for the number of bacteria, t days from the start?

60. Two sisters, Lisa and Misa, inherit $10,000 each. Lisa invests her money at 5% compounded quarterly while Misa invests her money at 7.5% compounded annually. If both keep their money in these accounts for three years, who comes out ahead? How

much money does each of them have at the end of these three years?

61. (Refer to the previous exercise) When did the other sister catch up? Solve by graphing.

62. A sapling is planted in the fall. On the first day of the following spring the tree has 14 leaves. A gardener observes that the tree is doubling its number of leaves every week. How many leaves does the tree have 26 weeks from that first day of spring?

63. The population of a small island in the Bay of Bengal is increasing at the rate of 1.5%. The island had 2000 people in 1980. What is the expected population for the year 2000? Write an expression for that population in terms of the number n in years.

64. Unfortunately, the island mentioned in the previous exercise is losing its arable soil at an exponential rate. The loss of soil is given by the formula $A(t) = A_0 e^{-0.002t}$ where t is the time in years, A_0 is the amount of arable soil in acres at the beginning, and $A(t)$ is the amount that is left now. Find the amount of arable soil that will be left in the year 2000 if $A_0 = 5000$ in 1980.

65. (Refer to the previous two exercises) If an acre of land is needed to support a family of four (augmented with seafood), will there be sufficient acreage in the year 2000 to support the burgeoning population?

66. (Refer to the previous three exercises) Graph the two functions: population growth and soil erosion. Check their point of intersection. When do the two curves intersect? (Since one acre supports one person, this point of intersection has a valid meaning.) What does that point indicate?

67. Ovi's parents set aside $10,000 when he was 8 years old and put it into an account that paid 7% compounded continuously. They were hoping that by the time Ovi was college age (i.e., 18 years old), the money would have doubled. Were they right? (Use $A(t) = A_0 e^{rt}$.) How much money is there in ten year's time?

68. (Refer to the previous exercise) How long would the parents have to keep the money in the bank until the total reaches $25,000?

5.2   The Logarithm Function

Introduction

Constructing the Logarithm Function

Expressing the Inverse of the Exponential Function

The Graph of the Logarithm Function

Logarithms Are Exponents

Notational Implications of the Inverse Notation

Common and Natural Logarithms

Introduction

The exponential function $f(x) = b^x$, $b > 0$, $b \neq 1$ is one-to-one as is seen from its graph, and, therefore, its inverse is a function. In this section we construct the inverse of the exponential function, give it a name, the *logarithm function*, see that logarithms are exponents, and investigate how one function can be converted to the other.

Constructing the Logarithm Function

In our study of the exponential functions, $f(x) = b^x$, $b > 0$, $b \neq 1$, in Section 5.1, we noted that they are strictly increasing when the base is greater than 1 (growth functions), and strictly decreasing when the base is between 0 and 1 (decay functions). Because of this behavior, exponential functions are one-to-one, and, therefore, their inverses are functions. From our study of inverse functions in Chapter 4, we know that the inverse of a particular function can be obtained by interchanging the order of the components that make up the ordered pairs that define the function. We can construct the inverse of the exponential function using this method.

Strategy to Construct the Inverse   We construct the inverse of the exponential function by considering a particular case, the growth function $f(x) = 2^x$. This one example should provide us with a feel for the general behavior of the inverse of the exponential function. We use the graphing calculator to construct the graph of $f(x) = 2^x$ by entering the function parametrically. This allows us to control the order of the components of each ordered pair $(x, 2^x)$. We then construct the graph of the inverse function, again parametrically, by reversing the order of the components to $(2^x, x)$.

The Construction   To construct the graph of $f(x) = 2^x$, we input the function using the parametric mode of the calculator. Place your calculator into parametric mode, and enter T for X1T and 2^T for Y1T. Select RANGE or WINDOW and enter 0 for Tmin, 95 for Tmax, 1 for Tstep, -7 for Xmin, 25 for Xmax, -7 for Ymin, and finally 25 for Ymax. Entering the function parametrically like this helps us see the ordered pairs that the calculator graphs. The assignments Tmin = 0, Tmax = 95, and Tstep = 1 direct the calculator to assign the integers 0, 1, 2, $\cdots$, 95 to the parameter T. Then as t steps through the 96 values (adjustments covered previously will need to be made for the TI-82 and 85) from 0 to 95, the first component in the ordered pair, T, takes on each value, and the second component in the ordered pair, Y, takes on the value 2^T. In essence, these assignments direct the calculator to plot the ordered pairs $(0, 2^0)$, $(1, 2^1)$, $(2, 2^2)$, $\cdots$, $(95, 2^{95})$. The range assignments provide a suitable window in which we can view the graph of the function. Since T does not assume negative values, the graph will not be a complete graph. With these assignments made, press GRAPH to construct the graph of $f(x) = 2^x$. Your graph should look something like Window 5.5. Press TRACE to activate the trace cursor and use the left-arrow key to move it to $x = 0$.

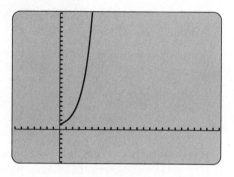

Window 5.5

x	$f(x)$
0	1
1	2
2	4
3	8
4	16
5	32
6	64
7	128

The screen should display the values of the parameter T and the components of the ordered pairs, (x, y), that lie on the curve. We can construct a table of input/output values by pressing the right-arrow key once and writing down the x-value and the corresponding y-value. Continue doing this until you have several ordered pairs that describe the function. The table in the margin displays eight ordered pairs.

We can use the calculator to construct the graph of the inverse function by interchanging the order of the components of the ordered pairs of the exponential function. To construct the graph of $f(x) = 2^x$, we entered T for X1T and 2^T for Y1T. This assignment made x the first component and 2^x the second component. To construct the graph of $f^{-1}(x)$, enter 2^T for X2T and T for Y2T. This assignment makes 2^x the first component and x the second component, just the reverse of $f(x) = 2^x$. The graphs of both $f(x) = 2^x$ and its inverse, $f^{-1}(x)$, appear in Window 5.6. The window has been *squared* by selecting the Square command from the ZOOM menu. (The viewing window is constructed so that it is 96 pixels across, but only 64 pixels high. The Square command adjusts the viewing window so that pictures drawn in it do not appear skewed.)

You can get a feel for the inverse relationship of these two functions by using the TRACE key. Press the TRACE key. Although the trace cursor will not be visible (it is out of the range of the viewing window), it is located on the curve $f(x) = 2^x$, as this is the first function to be plotted. Use the left-arrow key to move the cursor to $x = 0$. The cursor is now located at the point $(0, 1)$

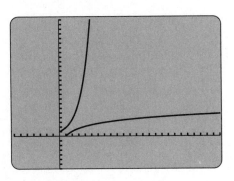

Window 5.6

x	$f^{-1}(x)$
1	0
2	1
4	2
8	3
16	4
32	5
64	6
128	7

on the exponential function $f(x) = 2^x$. Recalling that the up- and down-arrow keys move the trace cursor from one curve to the other, move the trace cursor to the inverse function. The cursor should be located at the point $(1, 0)$. As you made a table of ordered pairs for $f(x) = 2^x$, you can also make a table for $f^{-1}(x)$. The table in the margin displays some ordered pairs of the inverse function.

Comparing the components of these two ordered pairs, we see that they have been interchanged. Now, move the cursor back to the function $f(x) = 2^x$ and repeat the process several times. You should notice two things.

1. The order of the components of corresponding points on the two functions is reversed.

2. As the cursor moves from one curve to the other, it appears to move symmetrically about the 45° line, $y = x$, that separates the two curves.

You might also notice that the exponential function increases very quickly for x-values greater than 1, but its inverse increases very slowly there. (You can see this very clearly by placing the trace cursor on the inverse function and moving it sequentially one unit to the right, each time noticing that it takes more and more input units to generate a one-unit increase in the output.)

Expressing the Inverse of the Exponential Function

We will obtain an expression for the inverse of the exponential function in the usual algebraic way.

1. Replace $f(x)$ with y: $y = 2^x$.

2. Interchange x and y: $x = 2^y$.

3. Solve for y. Now we have a problem: we cannot solve for y using any of the algebraic operations we have studied so far. We notice that

   y is the power to which 2 must be raised to produce x

The word "is" in this statement indicates an equation.

$$y = \underbrace{\text{the power to which 2 must be raised to produce } x}_{\text{We need a notation for this expression}}$$

The notation "$\log_2 x$" has been created for this purpose. We then have

$$y = \underbrace{\text{the power to which 2 must be raised to produce } x}_{\log_2 x}$$

This means that $x = 2^y$ can be written with y explicitly in terms of x producing the function

$$y = \log_2 x$$

$y = \log_2 x$ defines a function and is the inverse of the function $y = 2^x$.

We now define the general logarithm function.

The Logarithm Function

The Logarithm Function

The inverse of the exponential function with base b, $f(x) = b^x$, is the **logarithm function** with base b, $g(x) = \log_b x$. The allowable values of the base b are $b > 0$ and $b \neq 1$. The domain of $g(x) = \log_b x$ is the range of the exponential function, namely, $x > 0$.

The definition indicates that if $f(x) = b^x$ and $g(x) = \log_b x$, then $g(x) = f^{-1}(x)$. It is also true that the inverse of the logarithm function is the exponential function, that is, $f(x) = g^{-1}(x)$, provided $b > 0$.

EXAMPLE SET A

For the bases 3, 5, 1/2, and m, express the exponential function and its inverse.

Solution:
1. For the base 3, $f(x) = 3^x$ and $f^{-1}(x) = \log_3 x$.
2. For the base 5, $f(x) = 5^x$ and $f^{-1}(x) = \log_5 x$.
3. For the base $\frac{1}{2}$, $f(x) = \left(\frac{1}{2}\right)^x$ and $f^{-1}(x) = \log_{1/2}(x)$.
4. For the base m, $f(x) = m^x$ and $f^{-1}(x) = \log_m x$.

EXAMPLE SET B

Find the inverse of the function $f(x) = 6^{4x-1} + 8$.

Solution:

1. Replace $f(x)$ with y.
   $$y = 6^{4x-1} + 8$$

2. Interchange x and y.
   $$x = 6^{4y-1} + 8$$

3. Solve for y.

$$6^{4y-1} = x - 8 \quad \text{Convert to logarithmic form}$$

$$\log_6(x - 8) = 4y - 1 \quad \text{Solve for y}$$

$$4y = \log_6(x - 8) + 1$$

$$y = \frac{\log_6(x - 8) + 1}{4}$$

4. Replace y with $f^{-1}(x)$.

$$f^{-1}(x) = \frac{\log_6(x - 8) + 1}{4}$$

The Graph of the Logarithm Function

Since the exponential and logarithm functions are inverses of each other, we can obtain much information about the graph of the logarithm function from the graph of the exponential function.

1. The domain of the logarithm function is the set of all positive real numbers. This follows from the fact that the range of the exponential function is the set of all positive real numbers. The logarithm function is, therefore, not defined for 0 or negative numbers. ($f(0) = \log_b(0)$ and $f(-5) = \log_b(-5)$ do not produce real numbers or nonreal complex numbers.)

2. The range of the logarithm function is the set of all real numbers. This follows from the fact that the domain of the exponential function is the set of all real numbers. When $0 < x < 1$, $\log_b x$ is negative. When $x > 1$, $\log_b x$ is positive. When $x = 1$, $\log_b x = 0$.

3. The graph of the logarithm function is a smooth continuous curve in the sense that it has no holes, breaks, gaps, or corners. This follows from the fact that the graph of the exponential function exhibits the same behavior.

4. A logarithm function with base greater than 1 is increasing at a decreasing rate, and a logarithm function with base between 0 and 1 is decreasing at a decreasing rate. This behavior follows from the corresponding behavior of the exponential function.

5. The logarithm function passes through the point $(1,0)$. This follows from the fact that the exponential function passes through the point $(0,1)$.

The graph on the left of Figure 5.2 shows the graph of the $f(x) = \log_b x$, where $b > 1$, and the graph on the right of Figure 5.2 shows the graph of $f(x) = \log_b x$, where $0 < b < 1$.

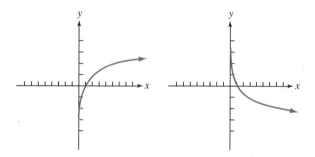

Figure 5.2

The logarithm function obeys all the rules of function translation. For example, the graph of $g(x) = \log_{10}(x - 4) + 3$ is obtained from the graph of $f(x) = \log_{10} x$ by vertically translating the graph of $f(x) = \log_{10} x$ upward 3 units and to the right 4 units. Figure 5.3 shows the graphs of each of these functions on the same coordinate system. (You should try to graph these functions using your calculator and the LOG key. Something appears odd about the graph of $f(x) = \log_{10}(x - 4) + 3$. Can you explain what has happened?)

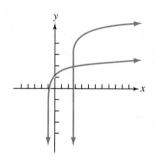

Figure 5.3

Logarithms Are Exponents

In our development of the logarithm function, we encountered $x = b^y$, and wrote $y = \log_b x$ to represent the number to which b must be raised to produce x. Notice that the logarithm of x is y, and that y is the exponent of the base b. Thus, we can loosely say that *a logarithm is an exponent*, and we can write the following equivalency.

| | **Logarithms Are Exponents** |

Logarithms Are Exponents

If $x > 0$, $b > 0$, and $b \neq 1$, then the function $y = \log_b x$ is called the **logarithm function with base** b, and is equivalent to the function represented by $x = b^y$.

EXAMPLE SET C

Convert each exponential form to logarithmic form.

1. $8 = 2^3$.

    Solution:
    Since 3 is the exponent of 2 that yields 8, $\log_2(8) = 3$, that is,

    $$8 = 2^3 \text{ is equivalent to } 3 = \log_2 8$$

2. $1296 = 6^4$.

    Solution:
    Since 4 is the exponent of 6 that yields 1296, $\log_6(1296) = 4$, that is,

    $$1296 = 6^4 \text{ is equivalent to } 4 = \log_6 1296$$

3. $\dfrac{16}{25} = \left(\dfrac{4}{5}\right)^2$.

    Solution:
    Since 2 is the exponent of (4/5) that yields (16/25), $\log_{4/5}(16/25) = 2$, that is,

    $$\frac{16}{25} = \left(\frac{4}{5}\right)^2 \text{ is equivalent to } 2 = \log_{4/5}(16/25)$$

4. $\dfrac{1}{9} = 9^{-1}$.

    Solution:
    Since -1 is the exponent of 9 that yields 1/9, $\log_9(1/9) = -1$, that is,

    $$\frac{1}{9} = 9^{-1} \text{ is equivalent to } -1 = \log_9(1/9)$$

5. $1 = b^0, b \neq 0$.

    Solution:
    Since 0 is the exponent of b that yields 1, $\log_b(1) = 0$, that is,

    $$1 = b^0, b \neq 0 \text{ is equivalent to } 0 = \log_b 1$$

EXAMPLE SET D

Convert each logarithmic form to exponential form.

1. $2 = \log_{13} 169$.

   Solution:
   The base is 13. Since $\log_{13} 169$ is the exponent of 13 that yields 169, and this exponent is 2,

   $$2 = \log_{13} 169 \text{ is equivalent to } 169 = 13^2$$

2. $\log_{10} 1000 = 3$.

   Solution:
   The base is 10. Since $\log_{10} 1000$ is the exponent of 10 that yields 1000, and this exponent is 3,

   $$\log_{10} 1000 = 3 \text{ is equivalent to } 10^3 = 1000$$

3. $\dfrac{1}{2} = \log_{16} 4$.

   Solution:
   The base is 16. Since $\log_{16} 4$ is the exponent of 16 that yields 4, and this exponent is 1/2,

   $$\frac{1}{2} = \log_{16} 4 \text{ is equivalent to } 16^{1/2} = 4$$

4. $\log_b 1 = 0$.

   Solution:
   The base is b. Since $\log_b 1$ is the exponent of b that yields 1, and this exponent is 0,

   $$\log_b 1 = 0 \text{ is equivalent to } b^0 = 1$$

5. $-1 = \log_7\left(\dfrac{1}{7}\right)$.

   Solution:
   The base is 7. Since $\log_7\left(\dfrac{1}{7}\right)$ is the exponent of 7 that yields 1/7, and this exponent is -1,

   $$-1 = \log_7\left(\frac{1}{7}\right) \text{ is equivalent to } 7^{-1} = \frac{1}{7}$$

EXAMPLE SET E

Determine each value.

1. $\log_2 8$.

   Solution:

   It is helpful to represent the value of $\log_2 8$ with y, that is, we'll let $y = \log_2 8$. We can solve for y by converting to exponential form. $y = \log_2 8$ is equivalent to $8 = 2^y$. From our experience with exponents we know that $y = 3$. Thus,

   $$\log_2 8 = 3$$

2. $\log_3\left(\dfrac{1}{3}\right)$.

   Solution:

   Represent the value of $\log_3\left(\dfrac{1}{3}\right)$ with y by letting $y = \log_3\left(\dfrac{1}{3}\right)$. Solve for y by converting to exponential form.

   $y = \log_3\left(\dfrac{1}{3}\right)$ is equivalent to $3^y = \dfrac{1}{3}$. From our experience with exponents, we know that $y = -1$. Thus,

   $$\log_3\left(\dfrac{1}{3}\right) = -1$$

3. $\log_3(-3)$.

   Solution:

   We recognize that -3 is not in the domain of the logarithm function. Therefore, $\log_3(-3)$ is not defined. This is further reinforced by noting that if $y = \log_3(-3)$, then $3^y = -3$ and there is no value of y that will make this statement true.

Notational Implications of the Inverse Notation

Recall from your study of inverse functions that

$$f[f^{-1}(x)] = x \quad \text{and} \quad f^{-1}[f(x)] = x$$

Because $f(x) = b^x$ and $f^{-1}(x) = \log_b x$ are inverses of each other, they satisfy these conditions. Evaluating these composite functions we have:

$$f[f^{-1}(x)] = x$$

$$f[\log_b x] = x$$

$$b^{\log_b x} = x$$

$$f^{-1}[f(x)] = x$$

$$f^{-1}[b^x] = x$$

$$\log_b(b^x) = x$$

We summarize these results in the following table.

Exponential Form	Logarithm Form
$b^{\log_b x} = x$	$\log_b(b^x) = x$

Notice that in each case, when the base of the logarithm form matches the base of the exponential form, the argument x results. This is an important feature of the inverse notation as it can allow quick and easy evaluation of some logarithmic and exponential functions. You can think of one function *undoing* the effect of the other function.

EXAMPLE SET F

Explain why each of the following expressions is a true statement.

1. $\log_6(6^8) = 8$.

   Solution:
   Notice that the bases of the logarithmic and exponential expressions match. The logarithm function *undoes* the effect of the exponential function.

2. $\left(\dfrac{1}{14}\right)^{\log_{1/14}(9)} = 9$.

   Solution:
   Notice that the bases of the logarithmic and exponential expressions match. The exponential function *undoes* the effect of the logarithm function.

Common and Natural Logarithms

Common Logarithms

Logarithms with base 10 are called **common logarithms**. To simplify notation, the base is omitted. Thus,

$$\log_{10} x \quad \text{is equivalent to} \quad \log x$$

We know, from our study of logarithms, that $\log_{10} 10^m = m$, the exponent of 10. This means that common logarithms of powers of 10 are easy to determine. The logarithm of a power of 10 is simply the exponent of 10. Table 5.3 illustrates several such cases.

Table 5.3		
$\log x$	$\log x^m$	**Value**
$\log 1$	$\log 10^0$	0
$\log 10$	$\log 10^1$	1
$\log 100$	$\log 10^2$	2
$\log 1000$	$\log 10^3$	3
$\log 10,000$	$\log 10^4$	4
$\log(.1)$	$\log 10^{-1}$	-1
$\log(.01)$	$\log 10^{-2}$	-2
$\log(.001)$	$\log 10^{-3}$	-3

You can use your calculator to approximate $\log_{10} x$ for values of x other than powers of 10 and then convert from logarithmic form to exponential form. Then, using the fact that $y = \log_{10} x$ and $y = 10^x$ are inverses of each other, you can write the corresponding exponential form of $y = \log_{10} x$.

ILLUMINATOR SET A

Use the LOG key on your calculator to find each logarithm to four decimal places. In each case, write the corresponding exponential form.

1. $\log(14) \approx 1.1461$. This means that $10^{1.1461} \approx 14$.

2. $\log(514) \approx 2.7110$. This means that $10^{2.7110} \approx 514$.

3. $\log(6014) \approx 3.7792$. This means that $10^{3.7792} \approx 6014$.

4. $\log(500,014) \approx 5.6990$. This means that $10^{5.6990} \approx 500,014$.

Just as $y = \log_{10} x$ and $y = 10^x$ are inverses of each other, so are $y = \log_e x$ and $y = e^x$, where e is the irrational number we examined in Section 5.1. Logarithms with base e are called **natural logarithms** and are denoted by $\ln x$.

Natural logarithms

ILLUMINATOR SET B

Use the LN key on your calculator to find each logarithm to four decimal places. In each case, write the corresponding exponential form.

1. $\ln(14) \approx 2.6391$. This means that $e^{2.6391} \approx 14$.
2. $\ln(514) \approx 6.2422$. This means that $e^{6.2422} \approx 514$.
3. $\ln(6014) \approx 8.7018$. This means that $e^{8.7018} \approx 6014$.
4. $\ln(500,014) \approx 13.1224$. This means that $e^{13.1224} \approx 500,014$.

EXERCISES

For exercises 1–6, write the inverse of each function.

1. $f(x) = 7^{2x}$
2. $f(x) = 3^{-8x}$
3. $f(x) = 2^{x+4}$
4. $f(x) = 10^{x-1}$
5. $f(x) = 5^{6x-3} + 2$
6. $f(x) = e^{3x+8} - 12$

For exercises 7–21, write each exponential expression in logarithmic form.

7. $6^2 = 36$
8. $4^5 = 1024$
9. $7^3 = 343$
10. $3^8 = 6561$
11. $3^{-4} = \dfrac{1}{81}$
12. $4^{-3} = \dfrac{1}{64}$
13. $(0.02)^3 = 0.000008$
14. $(0.6)^4 = 0.1296$
15. $27^{2/3} = 9$

16. $16^{3/4} = 8$
17. $b^c = a$
18. $(2m)^{3n} = k$
19. $1 = 10^0$
20. $100 = 10^2$
21. $e^x = y$

For exercises 22–30, write each logarithmic expression in exponential form.

22. $\log_3 27 = 3$
23. $\log_2 32 = 5$
24. $\log_5 625 = 4$
25. $\log_{1/2} \left(\dfrac{1}{8} \right) = 3$
26. $\log_{2/3} \left(\dfrac{8}{27} \right) = 3$
27. $\log_4 \left(\dfrac{1}{16} \right) = -2$
28. $\log_2 \left(\dfrac{1}{64} \right) = -6$
29. $\log_e e = 1$
30. $\log_6 6 = 1$

For exercises 31–38, find x, y, or b.

31. $\log_3 81 = y$

32. $\log_4 16 = y$

33. $\log_b 216 = 6$

34. $\log_b 81 = 4$

35. $\log_6 \left(\dfrac{1}{36} \right) = y$

36. $\log_{5/2} \left(\dfrac{4}{25} \right) = y$

37. $\log_8 x = 1$

38. $\log_{10} x = -3$

For exercises 39–49, simplify each expression.

39. $\log_5 5^7$

40. $\log_4 4^3 = 0$

41. $\log_e e^e$

42. $\log_{10} 10^5$

43. $\log_6 (\log_6 6)$

44. $\log_2 (\log_2 2)$

45. $\log_3 (\log_3 27)$

46. $\log_4 (\log_4 81)$

47. $\log_4 [\log_2 (\log_3 81)]$

48. $\log_4 [\log_3 (\log_2 8)]$

49. $\log_b (\log_a a^b)$

For exercises 50–53, use your calculator to display the graph of $f(x) = \log x$ with $[0, 10]$ for x, and $[-1, 5]$ for y. Using the TRACE key, approximate, to four decimal places, each logarithm.

50. $\log(4.210)$

51. $\log(8)$

52. $\log(0.7368)$

53. $\log(0.4211)$

For exercises 54–59, enter the exponential function $f(x) = 3^x$ and the logarithm function $g(x) = \log_3 x$ parametrically by placing your calculator in parametric mode and entering T

for X1T and 3^T for Y1T, then 3^T for X2T and T for Y2T. For the Range, enter 0 for Tmin, 25 for Tmax, 0.1 for Tstep, -7 for Xmin, 25 for Xmax, -7 for Ymin, and finally 25 for Ymax. The graph of the exponential function will be displayed first, and the graph of the logarithm function second. Activate the TRACE cursor.

Using the exponential function, approximate each logarithm.

54. $\log_3 7.2247$

55. $\log_3 2.6879$

56. $\log_3 302.7126$

Using the logarithm function, approximate the exponent of each exponential function.

57. $3^x = 5.1961$

58. $3^x = 1.7321$

59. $3^x = 15.5885$

For exercises 60–68, ~~without the use of your calculator~~, construct each graph based on your knowledge of the graph of $f(x) = \log x$.

60. $g(x) = \log(x - 3)$

61. $g(x) = \log(x + 2)$

62. $g(x) = \log(x + 5) + 3$

63. $g(x) = \log(x - 3) - 4$

64. $g(x) = -\log(x)$

65. $g(x) = -\log(x - 2)$

66. $g(x) = \log(x - 5) + 5$

67. $g(x) = -\log(x + 3) + 2$

68. $g(x) = \log |x|$

69. The doubling time t, in years, for an investment of P dollars made at interest rate r, compounded semiannually is given by the function

$$t(r) = \frac{\ln 2}{2 \ln(1 + \frac{1}{2}r)}$$

Find the doubling time of an investment if the interest rate is 7%.

70. A manufacturer can sell D units at the price x per unit (dollars) given by the demand function

$$D(x) = \frac{65}{\ln(x + 3.4)}$$

Approximate the number of units that will be sold when the price per unit is $17.80.

71. The daily cost C, in dollars, to a manufacturer to produce x electronic video components is given by the function

$$C(x) = 1000 \ln\left(\sqrt[3]{3x^2 - 700}\right)$$

Find the daily cost of producing 60 components.

72. The population P of a city and the average speed S, in feet per second, at which people walk in that city is given by the function

$$S(P) = 1.98 \log P + 0.05$$

Find the average walking speed in a city of 305,000 people.

73. Graph the two functions $f(x) = \ln(x) + \ln(2x + 1)$ and $g(x) = \ln(2x^2 + x)$. Describe what you see and give the reasons why you see the graphs in this way.

74. An architect in a moment of creativity decides to use logarithmic curves in his designs. He is using an arch in front of a building and he curves it using the logarithmic function (see figure).

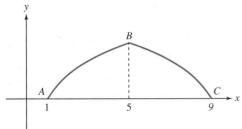

The graph of AB is the logarithm function from $x = 1$ to $x = 5$. This part is then reversed at BC. Write down the equation of ABC in two parts. (Hint: Remember reflection and translation.) What is the greatest height of the arch (Point B) above the horizontal?

75. The architect finds the height of the arch in the previous exercise to be too low, but he likes the shape of it. How can he increase the height without changing the logarithmic nature of the curve?

76. (Refer to the previous two exercises) Encouraged by the results of his experimentation, the architect wants to try his hand at an exponential curve. He designs some wrought-iron support for the balconies as shown.

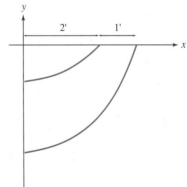

Both the curves are exponential. Find the equation of the inner curve as well as that of the outer curve, if these curves are based on $y = e^x$.

77.

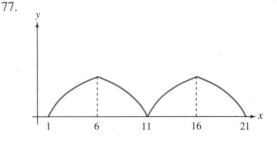

In the accompanying figure, the shape of the four parts of the arches are the same portion of the graph of $y = ln(x^3)$ intercepted between $x = 1$ and $x = 6$. Write down the equation of the curve with the proper domain (try graphing them).

78. A ball drops from a 200m-high building. When it hits the ground, it bounces up but only up to half the previous height. Every time the ball bounces it repeats the same pattern. What is the maximum height the ball achieves after the nth bounce?

5.3 Properties of Logarithms

Introduction

Properties of Logarithms

The Change-of-Base Formula

Introduction

We are familiar with the rules for simplifying exponential expressions. It is natural to ask if a set of rules exist for simplifying logarithmic expressions. In this section, we examine five properties that help us simplify logarithmic expressions. One of the properties is the change-of-base formula, a formula that provides a method of converting from one logarithmic base to another.

Properties of Logarithms

We begin our investigation of the properties of logarithms by considering the logarithmic statements

$$\log_3(9) = 2, \quad \log_3(27) = 3, \quad \text{and} \quad \log_3(243) = 5$$

Let's investigate the relationship that exists, if any, between the numbers 9, 27, and 243, and 2, 3, and 5, and hence, $\log_3(9)$, $\log_3(27)$, and $\log_3(243)$.

$$\log_3(9 \cdot 27) = \log_3(243)$$

$$= 5$$

$$= 2 + 3$$

$$= \log_3(9) + \log_3(27)$$

Thus, for the numbers 9 and 27,

$$\log_3(9 \cdot 27) = \log_3(9) + \log_3(27)$$

which might lead us to believe that the logarithm of a product is the sum of the logarithms of the factors.

Is this a coincidence or is it a specific example of a more general case? Let's see if we can generalize this result. Suppose that for $m, n > 0$

$$\log_b(m) = s$$

$$\log_b(n) = t$$

We can translate each of these logarithmic equations into exponential equations.

$$\log_b(m) = s \quad \longrightarrow \quad b^s = m \quad \text{for some } s$$

$$\log_b(n) = t \quad \longrightarrow \quad b^t = n \quad \text{for some } t$$

We wish to determine an expression for $\log_b(m \cdot n)$.

$$\log_b(m \cdot n) = \log_b(b^s \cdot b^t) = \log_b(b^{s+t}) = s + t$$

since the bases of the logarithm form and the exponential form match

$$= \log_b(m) + \log_b(n)$$

Although this analysis demonstrates that the logarithm of a product of two factors is equal to the sum of the logarithms of the individual factors, it is also true that the logarithm of a product of two *or more* factors is equal to the sum of the logarithms of the individual factors, that is

$$\log_b(x_1 \cdot x_2 \cdot \ldots \cdot x_n) = \log_b(x_1) + \log_b(x_2) + \ldots + \log_b(x_n)$$

It is worthwhile to look at this result pictorially in terms of the graph of $f(x) = \log_b(x)$. Figure 5.4 illustrates the relationship graphically. Let's carefully consider the meaning of the statement

$$\log_b(mn) = \log_b(m) + \log_b(n)$$

In the figure,

1. The numbers m, n, and mn are input values and represent distances (greater than 1) to the right of the origin, and are located on the horizontal, input axis. Thus, the values m and n are located m and n units, respectively, to the right of the origin, and the value $m \cdot n$ is located $m \cdot n$ units to the right of the origin.

2. The logarithms of these numbers, $\log_b(m)$, $\log_b(n)$, and $\log_b(mn)$, are output values and represent distances above or below the origin, and are located on the vertical, output axis. The values $\log_b(m)$ and $\log_b(n)$ are located $\log_b(m)$ and $\log_b(n)$ units, respectively, above the origin, and the value $\log_b(mn)$ is located $\log_b(m) + \log_b(n)$ units above the origin.

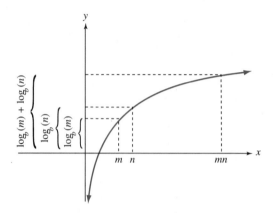

Figure 5.4

Constructing arguments similar to the arguments that motivated the logarithm property $\log_b(mn) = \log_b(m) + \log_b(n)$, we can prove that

$$\log_b\left(\frac{m}{n}\right) = \log_b(m) - \log_b(n)$$

Another important property of logarithms states that the logarithm of a power is the exponent of the power times the logarithm of the base of the power. Symbolically, if r represents any rational number, then

$$\log_b(m^r) = r\log_b(m)$$

We can prove this assertion in the following way. Assume that m is a positive real number, and that r is any rational number. Since m is a positive real number, $\log_b(m)$ represents some real number, say x. That is

$$\log_b(m) = x$$

Converting to exponential form we get

$$b^x = m$$

But since we want to investigate m^r, we'll raise both sides to the r power. Then we have

$$(b^x)^r = m^r$$

which, when simplified is

$$b^{rx} = m^r$$

Converting back to logarithmic form, we get

$$\log_b(b^{rx}) = \log_b(m^r)$$

But since $\log_b(b^{rx}) = rx$ (the bases of the logarithm and the exponential match), we have

$$rx = \log_b(m^r)$$

Then, since $\log_b(m) = x$, we get the result we are looking for:

$$r\log_b(m) = \log_b(m^r)$$

It is important not to mistake $\log_b(mn^r)$ to be the same as $r\log_b(mn)$, which is actually $\log_b(m) + r\log_b(n)$.

Another property of logarithms states that if two logarithms with the same base are equal, then their arguments must be equal. That is, if m and n are positive real numbers and if $\log_b(m) = \log_b(n)$, then $m = n$, since the $f(x) = \log_b(x)$ function is a one-to-one function.

We now summarize these four properties of logarithms.

Properties of Logarithms

Properties of Logarithms
Let m and n be positive real numbers.
1. $\log_b(m \cdot n) = \log_b(m) + \log_b(n)$
2. $\log_b\left(\dfrac{m}{n}\right) = \log_b(m) - \log_b(n)$
3. $\log_b(m^r) = r\log_b(m)$
4. $\log_b(m) = \log_b(n)$, then $m = n$.

EXAMPLE SET A

Write each expression as a single logarithm with coefficient 1.

1. $3\log_2 5 + \log_2 9 - \log_2 15$.

Solution:

$$3\log_2 5 + \log_2 9 - \log_2 15 = (3\log_2 5 + \log_2 9) - \log_2 15$$

$$= (\log_2 5^3 + \log_2 9) - \log_2 15 \quad \text{(by Property 3)}$$

$$= \log_2(5^3 \cdot 9) - \log_2 15 \quad \text{(by Property 1)}$$

$$= \log_2 \left(\frac{5^3 \cdot 9}{15} \right) \quad \text{(by Property 2)}$$

$$= \log_2 75$$

2. $\dfrac{3}{5} \log_8 a - \dfrac{5}{9} \log_8 b - \dfrac{1}{4} \log_8 c.$

Solution:

$$\frac{3}{5} \log_8 a - \frac{5}{9} \log_8 b - \frac{1}{4} \log_8 c = \frac{3}{5} \log_8 a - \left\{ \frac{5}{9} \log_8 b + \frac{1}{4} \log_8 c \right\}$$

$$= \log_8 a^{3/5} - \{ \log_8 b^{5/9} + \log_8 c^{1/4} \}$$

$$= \log_8 \sqrt[5]{a^3} - \{ \log_8 \sqrt[9]{b^5} + \log_8 \sqrt[4]{c} \}$$

$$= \log_8 \frac{\sqrt[5]{a^3}}{\sqrt[9]{b^5} \cdot \sqrt[4]{c}}$$

EXAMPLE SET B

Expand each logarithmic expression so that no logarithm of a product, quotient, or power appears. Assume a, b, and x represent positive real numbers only.

1. $\log_2 \dfrac{a^2 \cdot b^3}{x^{-2}}.$

Solution:

$$\log_2 \frac{a^2 \cdot b^3}{x^{-2}} = \log_2 [a^2 \cdot b^3] - \log_2 x^{-2}$$

$$= \log_2 a^2 + \log_2 b^3 - \log_2 x^{-2}$$

$$= 2 \log_2 a + 3 \log_2 b - (-2) \log_2 x$$

$$= 2 \log_2 a + 3 \log_2 b + 2 \log_2 x$$

2. $\log_{10} \sqrt[5]{\dfrac{a^3}{bx^4}}.$

Solution:

$$\log_{10} \sqrt[5]{\frac{a^3}{bx^4}} = \log_{10}\left[\frac{a^3}{bx^4}\right]^{1/5}$$

$$= \frac{1}{5}\log_{10}\frac{a^3}{bx^4}$$

$$= \frac{1}{5}[\log_{10}a^3 - \log_{10}bx^4]$$

$$= \frac{1}{5}[\log_{10}a^3 - (\log_{10}b + \log_{10}x^4)]$$

$$= \frac{1}{5}\log_{10}a^3 - \frac{1}{5}\log_{10}b - \frac{1}{5}\log_{10}x^4$$

$$= \frac{3}{5}\log_{10}a - \frac{1}{5}\log_{10}b - \frac{4}{5}\log_{10}x$$

The Change-of-Base Formula

The last property of logarithms that we will examine is the *change-of-base* formula. The change-of-base formula is a property of logarithms that provides us with a method for converting from one logarithm base to another.

Change-of-Base Formula

Change-of-Base Formula
If x is a positive real number and a and b are permissible logarithm bases, then $$\log_a x = \frac{\log_b x}{\log_b a}$$

We prove this assertion in the following way.

Consider the expressions $\log_b x$ and $\log_a x$. The expression $\log_a x$ is some real number that we will represent by y, that is, $y = \log_a x$. This is equivalent to $a^y = x$.

Take the logarithm, base b of each side of the equation $a^y = x$.

$$\log_b(a^y) = \log_b x$$

Then, by logarithm property 3,

$$y \log_b a = \log_b x$$

Solve for y by dividing each side by $\log_b a$

$$y = \frac{\log_b x}{\log_b a}$$

But $y = \log_a x$. Therefore,

$$\log_a x = \frac{\log_b x}{\log_b a}$$

and the assertion is established.

EXAMPLE SET C

1. Express $\log_2 9$ in terms of $\log_{10}$.

   Solution:

   $$\log_2 9 = \frac{\log_{10} 9}{\log_{10} 2}$$

2. Express $\log_3 2$ in terms of $\log_2$.

   Solution:

   $$\log_3 2 = \frac{\log_2 2}{\log_2 3}$$

   $$= \frac{1}{\log_2 3}$$

3. Express $\ln x$ in terms of $\log_{10}$.

   Solution:

   $$\ln x = \log_e x$$

   $$= \frac{\log_{10} x}{\log_{10} e}$$

   $$\approx \frac{\log x}{0.4343}$$

   $$\approx 2.3026 \log x$$

4. Use your calculator to approximate, to four decimal places, $\log_2 20$.

Solution:

$$\log_2 20 = \frac{\log_{10} 20}{\log_{10} 2}$$

$$\approx 4.3219 \quad \text{to four decimal places}$$

5. Using the viewing window $0 \leq x \leq 10$ and $-3 \leq y \leq 3$, (a) display the graphs of $f(x) = \log x$ and $f(x) = \log_2 x$, (b) use the graph to approximate, to four decimal places, $\log 6$ and $\log_2 6$, and (c) express the results of part (b) in exponential form.

Solution:
(a) Window 5.7 shows the graphs of both $f(x) = \log x$ and $f(x) = \log_2 x$.

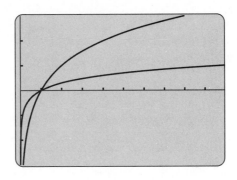

Window 5.7

(b) Using the TRACE key and running the trace cursor to $x = 6$, we see that $\log 6 \approx 0.7782$ and $\log_2 6 \approx 2.5850$.
(c) In exponential form, the results of part (b) translate respectively to

$$10^{0.7782} \approx 6 \quad \text{and} \quad 2^{2.5850} \approx 6$$

EXERCISES

For exercises 1–13, write each logarithmic expression as a single logarithm with a coefficient of 1.

1. $\log_5 4 + \log_5 7 - \log_5 2$

2. $\log_9 4 + \log_9 8 - \log_9 16$

3. $\log_3 x^4 - \log_3 y^2 - \log_3 z^3$

4. $\log_2 a^{1/3} + \log_2 a^{2/3} - \log_2 b^{3/5} - \log_2 b^{2/5}$

5. $3 \log_e x + \frac{1}{4} \log_e y + 2/9 \log_e z - \frac{5}{6} \log_e w$

6. $\frac{1}{3} \log_{10} 8 + \frac{2}{3} \log_{10} 27 + \frac{1}{2} \log_{10} 4 - 2 \log_{10} 6^2$

7. $-3 \ln x - 2 \ln y$

8. $-\ln 35 - \ln 4x + \ln 70$

9. $2 \log_4 (x - 3) - \frac{1}{2} [\log_4 (x - 4) - \log_4 (x + 2)]$

10. $\ln(\ln x + \ln y) + \ln(\ln x - \ln y)$

11. $2\ln(\ln x^2) - 2\ln x^2$

12. $\dfrac{1}{2}\left[\dfrac{1}{2}(\ln a + \ln b) - \ln c\right]$

13. $\log_7 7^{4\log x} - 2^{\log_2 x^4} + 14$

For exercises 14–22, expand each logarithmic expression so that no products, quotients, or powers appear.

14. $\log_5 \dfrac{xyz}{u}$

15. $\log_6 \dfrac{a^2 b^3}{c^2 d^4}$

16. $\ln \dfrac{(x-1)^2 (x-2)^3}{(x+3)^4}$

17. $\ln \dfrac{15x^2}{[(3x-5)^2(x-5)^3]^3}$

18. $\log_2 \sqrt[5]{\dfrac{x^2 b}{5z^3}}$

19. $\log \sqrt[8]{\dfrac{3a^2 b^4 c^6}{4x^2 y^{1/4}}}$

20. $\ln \left(\dfrac{6x^4 y^{2/5}}{5a^{1/2} b^{3/5}}\right)^{3/4}$

21. $\log \dfrac{\log x^2 y^3}{\log a^3 b^4}$

22. $\log_6 \dfrac{\log_6 xy^3}{6^{xy^3}}$

23. Using the viewing window $0 \le x \le 10$ and $-3 \le y \le 3$, (a) display the graphs of $f(x) = \log x$ and $f(x) = \log_3 x$, (b) use the graphs to approximate, to four decimal places, $\log 6$ and $\log_3 6$, and (c) express the results of part (b) in exponential form.

24. Using the viewing window $0 \le x \le 10$ and $-3 \le y \le 3$, (a) display the graphs of $f(x) = \log x$ and $f(x) = \log_4 x$, (b) use the graphs to approximate, to four decimal places, $\log 6$ and $\log_4 6$, and (c) express the results of part (b) in exponential form.

25. Using the viewing window $0 \le x \le 10$ and $-3 \le y \le 3$, (a) display the graphs of $f(x) = \log x$ and $f(x) = \log_5 x$, (b) use the graphs to approximate, to four decimal places, $\log 6$ and $\log_5 6$, and (c) express the results of part (b) in exponential form.

26. Using the viewing window $0 \le x \le 10$ and $-3 \le y \le 3$, (a) display the graphs of $f(x) = \log x$ and $f(x) = \log_6 x$, (b) use the graphs to approximate, to four decimal places, $\log 6$ and $\log_6 6$, and (c) express the results of part (b) in exponential form.

27. In exercises 23–26, you displayed the graphs of $\log_3 x$, $\log_4 x$, $\log_5 x$ and $\log_6 x$, and compared them to the graph of $\log_{10} x$. Write what you notice about the graphs of $\log_n x$ as n approaches 10.

28. On the same coordinate system, display the graphs of $f(x) = \log_{10} x$ and $f(x) = \log_{1/10} x$. Use the TRACE feature to compute both $\log_{10}(4)$ and $\log_{1/10}(4)$. Accepting some calculator roundoff, what do you notice about $\log_{10}(4)$ and $\log_{1/10}(4)$? (Write your observation in terms of $\log_{10}(4)$ and $\log_{1/10}(4)$.)

29. On the same coordinate system, display the graphs of $f(x) = \log_{10} x$ and $f(x) = \log_{1/10} x$. Use the TRACE feature to compute both $\log_{10}(6)$ and $\log_{1/10}(6)$. Accepting some calculator roundoff, what do you notice about $\log_{10}(6)$ and $\log_{1/10}(6)$? (Write your observation in terms of $\log_{10}(6)$ and $\log_{1/10}(6)$.)

30. In the previous two exercises, you made observations about a relationship between $\log_{10}(4)$ and $\log_{1/10}(4)$ or $\log_{10}(6)$ and $\log_{1/10}(6)$. Generalize the results you observed in terms of $\log_b x$ and $\log_{1/b} x$, where $b > 0$, $b \ne 1$. (You may find it helpful to let $\log_b x = a$ and $\log_{1/b} x = c$.)

31. Prove logarithm property 2. That is, prove that the logarithm of a quotient is the difference of the logarithms.

32. Prove that $\log_b(xyz) = \log_b x + \log_b y + \log_b z$.

33. In the text we used the change-of-base formula to express $\ln x$ in terms of $\log_{10} x$. Use the same procedure to express $\log_{10} x$ in terms of $\ln x$.

34. Find $\log_6 15$.

35. Find $\log_2 25$.

36. Find $\log_8 2$.

37. Find $\log_3 \sqrt[5]{7}$.

38. Find $\log_{2/3} \dfrac{3}{2}$.

39. Select values of x, y, and b to demonstrate that $\log_b(x + y) \neq \log_b x + \log_b y$.

40. Select values of x, y, and b to demonstrate that $\log_b\left(\dfrac{x}{y}\right) \neq \dfrac{\log_b x}{\log_b y}$.

41. Select values of x and y to demonstrate that $\log_b(x \cdot y) \neq (\log_b x)(\log_b y)$.

42. The expression $\log_b x$, where $b > 1$, expresses the output value for the input value of x. $\log_b x$ can be either positive or negative. What values of x will produce a positive output, and what values of x will produce a negative output?

43. Explain why it is not possible to compute $\log(\log(0.5))$.

44. Explain why it is not possible to compute $\ln(\ln(1))$.

45. The logarithm property $\log_b\left(\dfrac{m}{n}\right) = \log_b n - \log_b n$ can be viewed pictorially in a way similar to the way we looked at the property $\log_b(mn) = \log_b n + \log_b n$ earlier in the section. Using the figure that was constructed in the section as a guide, construct a figure that illustrates the quotient relationship.

5.4 Exponential and Logarithmic Equations

Introduction

Exponential and Logarithmic Equations

Applications of Exponential and Logarithmic Equations

Introduction

Exponential and logarithmic functions model a great variety of physical phenomena, including compound interest, population growth and radioactive decay, activation energy in chemical processes, skin-healing processes, intensity of sound in the human ear, and earthquake magnitude. In this section we investigate such models, and in so doing, examine methods for determining the value of an input variable that produces a specified value of the output variable, that is, we examine methods for solving exponential and logarithmic equations.

Exponential and Logarithmic Equations

We begin by defining exponential and logarithmic equations and specifying the most common methods of solving them.

Exponential Equations

Exponential Equations
An **exponential equation** is an equation that involves one or more exponential expressions that contain a variable(s) in an exponent(s).

Exponential equations are commonly solved by taking logarithms of both sides of the equation.

Logarithmic Equations

Logarithmic Equations
A **logarithmic equation** is an equation that involves one or more logarithmic functions whose argument contains a variable(s).

Logarithmic equations are commonly solved by expressing the logarithmic expressions as a single logarithmic expression with coefficient 1, then converting this logarithmic form to its equivalent exponential form.

Solving exponential equations by taking logarithms of both sides works well because the logarithm function is one-to-one (logarithm property 4 of the previous section), that is,

$$x = y \longrightarrow \log_b x = \log_b y \quad \text{and} \quad \log_b x = \log_b y \longrightarrow x = y$$

It is this property of logarithm functions that ensures that no extraneous solutions are introduced when logarithms are taken.

EXAMPLE SET A

1. Solve $5^x = 12$. Round the result to four decimal places.

Solution:

We solve this equation by taking the common logarithm of each side, then using logarithm property 3 to obtain an expression free of variable exponents.

$$5^x = 12$$

$$\log(5^x) = \log(12)$$

$$x\log(5) = \log(12)$$

$$x = \frac{\log(12)}{\log(5)}$$

$$x \approx \frac{1.0792}{0.6990}$$

$$x \approx 1.5440$$

(handwritten notes: "notes for alternate way")

(handwritten: $x = \log_5 12$)

(handwritten: $x = \dfrac{\log 12}{\log 5}$)

Thus, $5^{1.5440} \approx 12$. (You can check this result by making the computation, or by graphing the constant function $f(x) = 5^{1.5440}$. The result should be a horizontal line that intercepts the output axis close to 12.)

2. Solve $3^{x+1} = 9 \cdot 10^x$.

Solution:

We solve this exponential equation by taking the common logarithm of each side and using the first and third properties of logarithms.

$$3^{x+1} = 9 \cdot 10^x$$

$$\log 3^{x+1} = \log(9 \cdot 10^x)$$

$$(x + 1)\log 3 = \log 9 + \log 10^x$$

$$x\log 3 + \log 3 = \log 9 + x\log 10$$

$$x\log 3 - x\log 10 = \log 9 - \log 3$$

$$x(\log 3 - \log 10) = \log 9 - \log 3$$

$$x(\log 3 - 1) = \log 9 - \log 3$$

$$x = \frac{\log 9 - \log 3}{\log 3 - 1}$$

$$x \approx \frac{.9542 - .4771}{.4771 - 1}$$

$$x \approx \frac{.4771}{-.5229}$$

$$x \approx -0.9124$$

Thus, $3^{-0.9124+1} \approx 9 \cdot 10^{-0.9124}$.

3. Solve $8e^{4x-5} = 120$.

Solution:
Since this exponential equation involves the base e, it might be helpful to use the natural logarithm rather than the common logarithm (since $\ln(e) = 1$).

$$8e^{4x-5} = 120$$

$$\ln(8e^{4x-5}) = \ln(120)$$

$$\ln 8 + \ln(e^{4x-5}) = \ln(120)$$

$$\ln 8 + (4x - 5)\ln e = \ln(120)$$

$$\ln 8 + (4x - 5) \cdot 1 = \ln(120)$$

$$\ln 8 + 4x - 5 = \ln(120)$$

$$4x = \ln(120) - \ln 8 + 5$$

$$x \approx \frac{4.7875 - 2.0794 + 5}{4}$$

$$x \approx 1.9270$$

Thus, $8e^{4(1.9270)-5} \approx 120$.

4. Solve for x: $(1.8)^t = e^x$.

Solution:
In Example Set C of Section 5.1, we used the calculator to produce, from a set of data, the continuous exponential growth function $P(t) = 610 \cdot (1.8)^t$. At that time we wondered why such a continuous growth function did not involve the number e, and noted that we would later show how to make such a conversion to e.

We wish to find a number x such that $(1.8)^t$ is equal to e^x. This is easily done by solving for x in terms of t. Since the number e is involved, we'll solve for x by taking the natural logarithm of each side.

$$(1.8)^t = e^x$$

$$\ln(1.8)^t = \ln e^x$$

$$t\ln(1.8) = x\ln e$$

$$t\ln(1.8) = x \cdot 1$$

$$t\ln(1.8) = x$$

$$x \approx 0.5878t$$

Thus, $x \approx 0.5878t$ and $P(t) = 610 \cdot (1.8)^t$ can be converted to $P(t) = 610e^{0.5878t}$.

5. Solve $0.6(e^{3x} - e^{-x}) = -1.61$.

Solution:

Since this exponential equation involves the base e, we again use the natural logarithm. We begin by dividing each side by 0.6.

$$0.6(e^{3x} - e^{-x}) = -1.61$$

$$e^{3x} - e^{-x} = -2.68333$$

At this point, we would probably want to take the natural log of each side so as to simplify e^{3x} to $3x$, and e^{-x} to $-x$. But taking the natural log of each side now would result in $\ln(e^{3x} - e^{-x})$, which is *not* equal to $\ln(e^{3x}) + \ln(e^{-x})$. In other words, introduction of the logarithm function does not help simplify sums and differences—only products, quotients, and powers. However, since $e^x \cdot e^{-x} = e^{x-x} = e^0 = 1$, we might try multiplying each side by e^x.

$$e^x(e^{3x} - e^{-x}) = e^x(-2.68333)$$

$$e^{4x} - 1 = -2.68333e^x$$

$$e^{4x} + 2.68333e^x - 1 = 0$$

This equation does not seem to be any easier to solve than the original. We could make this a polynomial-type equation with the substitution $u = e^x$, but this would produce the equation $u^4 - 2.6333u - 1 = 0$, and we currently do not have any techniques for determining the exact solution to this equation. Our best effort will be to approximate the solution with a maximum error of say, 0.01. Graphing either $f(x) = e^{4x} + 2.68333e^x - 1$, or $f(x) = 0.6(e^{3x} - e^{-x}) + 1.61$ (a version of the original function), and locating the zero gives $x = -1.0052$, with a maximum error of 0.01.

This last example illustrates that not all equations have known techniques for finding solutions. In such cases, we have to settle for approximations. Notice that we could have approximated the solutions of each of the equations in the previous example set. Approximation techniques are very important when it comes to solving equations that arise from functions constructed from data. Before looking at some applications, we'll solve some logarithmic equations.

EXAMPLE SET B

1. Solve the logarithmic equation $\log_5 x - \log_5(x - 4) = 1$.

   Solution:

   We first note that because the domain of the logarithmic function is the set of positive real numbers, we must have that $x > 0$, from the first term of the left member, and $x - 4 > 0$, or $x > 4$, from the second term of the left member. Thus, $x > 0$ and $x > 4$, which implies that $x > 4$. Then

   $$\log_5 x + \log_5(x - 4) = 1$$

   $$\log_5[x(x - 4)] = 1$$

   $$x(x - 4) = 5^1$$

   $$x^2 - 4x = 5$$

   $$x^2 - 4x - 5 = 0$$

   $$(x - 5)(x + 1) = 0$$

   $$x = -1, 5$$

   Since the domain is the set of all x-values greater than 4, the number -1 can be eliminated from consideration. We can check the value 5 by substituting $x = 5$ into the original equation and computing to see if a true statement results.

   $$\log_5 5 - \log_5(5 - 4) = 1 - \log_5 1 = 1 - 0 = 1$$

   so that the solution checks, and we conclude that $x = 5$.

2. Solve the logarithmic equation $\ln(x - 4) - \ln(x + 5) = 2$.

Solution:

We first note that because the domain of the logarithmic function is the set of positive real numbers, we must have that $x - 4 > 0$, or $x > 4$, from the first term of the left member, and $x + 5 > 0$, or $x > -5$, from the second term of the left member. Thus, $x > -5$ and $x > 4$, which implies that $x > 4$. Then

$$\ln(x - 4) - \ln(x + 5) = 2$$

$$\ln\left(\frac{x - 4}{x + 5}\right) = 2$$

Converting to exponential form,

$$\frac{x - 4}{x + 5} = e^2$$

$$x - 4 = e^2(x + 5)$$

$$x - 4 = e^2 x + 5e^2$$

$$x - e^2 x = 5e^2 + 4$$

$$x(1 - e^2) = 5e^2 + 4$$

$$x = \frac{5e^2 + 4}{1 - e^2}$$

$$x \approx -6.4087$$

Since this value of x is outside the domain, we exclude it. Then, since there is no other value of x to choose from, we conclude that this equation has no solution.

Applications of Exponential and Logarithmic Equations

We can investigate applied problems using methods that involve solving exponential and logarithmic equations. The next several examples are designed to illustrate these methods.

ILLUMINATOR SET A

A colony of bacteria grows exponentially with time. Under certain conditions, the colony doubles in size in some period of time. The equation describing this growth is

$$A = A_0 \cdot 2^n$$

where A_0 is the initial number of bacteria present, n is the number of time periods (the number of times the bacteria double) being considered, and A is the number of bacteria present after n time periods have passed.

If at 22°C, the colony doubles in size every five days, determine (1) the size of the colony after 10 days, and (2) the length of time necessary for the colony to reach five times its original size.

1. We need to determine A when $n = \dfrac{10}{5} = 2$ doubling time periods have passed.

$$A = A_0 \cdot 2^2$$

$$= 4A_0$$

We interpret this result as: After 10 days (or two time periods) the size of the population of bacteria will be four times the original size of the population.

2. We wish to determine the value of n so that the number of bacteria, A, will be 5 times the original number of bacteria, A_0. Thus, we wish to find n when $A = 5A_0$.

$$A = A_0 \cdot 2^n$$

$$5A_0 = A_0 \cdot 2^n$$

$$5 = 2^n$$

$$\log 5 = \log 2^n$$

$$\log 5 = n \log 2$$

$$n = \frac{\log 5}{\log 2}$$

$$n \approx 2.3219$$

Since n is the number of time periods, we need to multiply 2.3219 by 5 to find the actual number of days required for the colony to reach five times its original size. $(2.3219)(5) \approx 11.61$. Thus, it will take this particular colony of bacteria about 11.61 days to reach five times its original size.

The next example shows how logarithmic equations that arise in chemistry can be solved.

ILLUMINATOR SET B

The pH of a solution is a measure of the acidity of the solution. Solutions with a high concentration of hydrogen ions are more acidic than are solutions with a lower concentration of hydrogen ions. The pH of a solution is defined as

$$pH = -\log\left[H^+\right]$$

where $\left[H^+\right]$ is a numerical value that represents the concentration of hydrogen ions. The units of $\left[H^+\right]$ are *gram-ions per liter* of solution.

1. The hydrogen ion concentration of pure water is very close to 10^{-7} gram-ions per liter. Find the pH of water.
   Using the formula $pH = -\log\left[H^+\right]$, we get

$$pH = -\log(10^{-7})$$

$$= 7$$

   Thus, the pH of pure water is approximately 7.

2. Find the hydrogen ion concentration of a solution with pH = 8.46.
   We wish to find $\left[H^+\right]$ when pH = 8.46. Using the formula $pH = -\log\left[H^+\right]$, we get

$$8.46 = -\log\left[H^+\right]$$

$$-8.46 = \log\left[H^+\right]$$

Converting to exponential form,

$$10^{-8.46} = \left[H^+\right]$$

$$3.4674 \times 10^{-9} \approx \left[H^+\right]$$

   Thus, the hydrogen ion concentration of a solution having pH = 8.46 is approximately 3.4674×10^{-9} gram-ions per liter.

The next example demonstrates how logarithms are used to compare the intensity of one sound to a standard intensity. The scale used for comparison is the decibel (db) scale.

ILLUMINATOR SET C

The **decibel** is the unit of measure for the intensity i of sound and is defined by the logarithmic function

$$D(i) = 10 \log \left(\frac{i}{i_0} \right)$$

where i_0 is the standard intensity and is the least intensity that can be heard by the human ear.

1. What is the approximate decibel level of a sound that is 7500 times the least intensity?

    We wish to find D for $i = 7500 i_0$. Substitute $7500 i_0$ into the function $D(i) = 10 \log \left(\dfrac{i}{i_0} \right)$, and compute $D(7500 i_0)$.

$$D(7500 i_0) = 10 \log \left(\frac{7500 i_0}{i_0} \right)$$

$$= 10 \log(7500)$$

$$\approx 38.7506$$

    Thus, the decibel level of sound that is 7500 times as intense as the least intense sound that can be heard by the human ear is about 38.75.

2. How many times the least intensity i_0 does a decibel level of 120 db represent?

    If we let x represent the number of times the least intensity, then, with $D = 120$, we get

$$120 = 10 \log \left(\frac{x i_0}{i_0} \right)$$

$$120 = 10 \log(x)$$

$$12 = \log(x)$$

$$10^{12} = x$$

Thus, $x = 10^{12} = 1,000,000,000,000$, which is one trillion times the least intensity. It is interesting to note that some music is played at about 120 db.

3. How much louder, in terms of sound intensity, is music played at 110 db than two people talking at 65 db?

Suppose we let I_m represent the sound intensity of the music, and I_c represent the sound intensity of the conversation. The difference in intensity between the two sounds is the difference in their decibel levels. The decibel level of the music is

$$\text{db}_\text{m} = 10 \log \left(\frac{I_m}{I_0} \right)$$

and the decibel level of the conversation is

$$\text{db}_\text{c} = 10 \log \left(\frac{I_c}{I_0} \right)$$

Thus, the difference in the decibel levels is

$$\text{db}_\text{m} - \text{db}_\text{c} = 10 \log \left(\frac{I_m}{I_0} \right) - 10 \log \left(\frac{I_c}{I_0} \right)$$

$$110 - 65 = 10 \log \left(\frac{I_m}{I_0} \right) - 10 \log \left(\frac{I_c}{I_0} \right)$$

$$45 = 10 \left[\log \left(\frac{I_m}{I_0} \right) - \log \left(\frac{I_c}{I_0} \right) \right]$$

$$4.5 = \log \left(\frac{I_m}{I_0} \right) - \log \left(\frac{I_c}{I_0} \right)$$

$$4.5 = \log \left(\frac{\frac{I_m}{I_0}}{\frac{I_c}{I_0}} \right)$$

$$4.5 = \log \left(\frac{I_m}{I_c} \right)$$

$$10^{4.5} = \frac{I_m}{I_c}$$

$$10^{4.5} I_c = I_m$$

Thus, the intensity of the music is $10^{4.5} \approx 31,600$ times the intensity of the conversation, and we interpret this to mean that the music is about 31,600 times as loud as the conversation.

EXERCISES

For exercises 1–24, solve each equation. If the equation is not solvable by algebraic means, approximate the solution with an error of at most 0.01.

1. $6^x = 15$

2. $5^x = 105$

3. $3^{x+4} = 20$

4. $e^{x-5} = 46$

5. $e^{7x-3} = 0.0412$

6. $15^{2x+5} = -0.067$

7. $3^x = 5(4)^x$

8. $1.7^x = (4.2)(1.6)^x$

9. $6^{x-2} = 3^{2x-1}$

10. $5^{4x+1} = 2^{-3x+1}$

11. $e^{-2x+1} = 15e^x$

12. $4e^{-x-7} = -2e^x$

13. $0.3(e^x + e^{-x}) = -2$

14. $0.8(e^{5x} + e^{-5x}) = 10$

15. $0.2(e^{3x} + e^{-2x}) = 14$

16. $6(e^{1.5x} + e^{-0.5x}) = 2.6$

17. $\log x + \log(x - 15) = 2$

18. $\log(x - 4) + \log(x - 1) = 1$

19. $\log_2(3x - 2) + \log_2(x + 1) = 1$

20. $\ln(x - 4) = \ln(x + 1) + 1.2$

21. $\log(x - 6) = 15 - \log(x - 4)$

22. $\log x + \log(x + 7) = \log(x + 12) + 100$

23. $\log(x^2) = (\log x)^2$

24. $\log(\log x) = 1$

25. Solve $y = \dfrac{e^x + e^{-x}}{2}$ for x.

26. Solve $y = \dfrac{e^{2x} - e^{-2x}}{2}$ for x.

27. Solve $y = \dfrac{e^{3x} + e^{-3x}}{e^{3x} - e^{-3x}}$ for x.

28. Solve $y = \dfrac{e^{1.5x} - e^{-1.5x}}{e^{1.5x} + e^{-1.5x}}$ for x.

29. Convert the exponential expression $16(2.4)^{0.712}$ to an exponential expression in which the base of the exponential factor is e.

30. Convert the exponential expression $4(5.2)^{0.03}$ to an exponential expression in which the base of the exponential factor is e.

31. In exercise 41 of Section 5.1, you constructed an exponential function that modeled the relationship between the amount, Q, of undissolved salt in a water solution t minutes after the salt was placed into the solution. Convert that exponential function to an exponential function in which the base of the exponential factor is e.

32. In exercise 42 of Section 5.1, you constructed an exponential function that modeled the relationship between the amount Q of radioactive substance present and the time t since the beginning of the observation. Convert that exponential function to an exponential function in which the base of the exponential factor is e.

33. In exercise 43 of Section 5.1, you constructed an exponential function that modeled the relationship between the population P of a city and the time t in years from 1985. Convert that exponential function to an

exponential function in which the base of the exponential factor is e.

34. In exercise 44 of Section 5.1, you constructed an exponential function that modeled the relationship between the student's grade-point average and the number N of applications sent to medical schools. Convert that exponential function to an exponential function in which the base of the exponential factor is e.

35. The exponential function $P(t) = \dfrac{C}{1 + C_0 e^{-kt}}$ is called the logistic growth function and models the growth of a population that is in some way constrained. Solve this equation for t.

36. Show that if $\ln i = \ln 60 + 2 \ln T + \dfrac{B}{T}$, then $i = 60 T^2 e^{B/T}$.

37. Show that if $\ln Q = \ln 15 = \dfrac{1}{2} \ln P + \dfrac{3}{T}$, then $Q = 15 e^{3/T} \sqrt{P}$.

Versions of the following group of exercises were presented (and hopefully solved) in Section 5.1. At that time, solutions were approximated using the graphing feature of a calculator. We now have algebraic techniques for obtaining exact solutions or very good approximations without the use of the graphing feature of a calculator.

38. The atmospheric pressure P in inches of mercury is given by

$$P(h) = 30 \cdot 10^{-0.09h}$$

where h is the height, in miles, above sea level. Find the height above sea level for which the pressure is 25 inches of mercury. Compare your solution to the solution you obtained in exercise 53 in Section 5.1.

39. Suppose the number N of bacteria in a culture at a particular time t is approximated by the function

$$N(t) = N_0 5^{0.03t}$$

How many hours after the initial observation will there be 15,000 bacteria if there are 8000 initially? Compare your solution to the solution you obtained in exercise 52 in Section 5.1.

40. A radioactive compound decays according to the function

$$A(t) = A_0 e^{-0.3t}$$

where A_0 is the initial amount of the compound present, and t is the number of seconds that have passed since the initial observation. How much time elapses for 10 grams to remain if you start with 40 grams? Compare your solution to the solution you obtained in exercise 54 in Section 5.1.

41. A loaf of bread is removed from an oven when its temperature is 300°F. After sitting on the countertop in a room with a constant temperature of 70° for 3 minutes, its temperature is 200°F. The temperature T of the loaf of bread at any time t is given by

$$T(t) = 70 + 230 e^{-0.19108t}$$

(This function is a version of Newton's Law of Cooling.) How long after the bread is removed from the oven and placed on the countertop will its temperature be 80°? Compare your solution to the solution you obtained in exercise 55 in Section 5.1.

Exercises 42–45 deal with chemical pH, which was considered in Illuminator Set B.

42. Find the pH of a solution with hydrogen ion concentration 2.6×10^{-7}.

43. Find the pH of a solution with hydrogen ion concentration 1.8×10^{-5}.

44. Find the hydrogen ion concentration of a solution with pH of 5.18.

45. Find the hydrogen ion concentration of a solution with pH of 7.36.

Exercises 46–49 deal with the decibel (db) level of sound considered in Illuminator Set C.

46. What is the decibel level of a sound that is 525,000 times the least intensity?

47. What is the decibel level of a sound that is 1,500,000 times the least intensity?

48. How many times the least intensity is a sound with decibel level of 55?

49. How many times the least intensity is a sound with decibel level of 100?

50. The **Weber-Fechner Law** in physiology relates the apparent loudness L of a sound and the natural logarithm of the sound intensity i by the function

$$L(i) = k \ln(i)$$

where k is a constant. Find how the intensity of the sound must be changed in order to cut the apparent loudness of a sound by one-half.

51. How much louder is a sound, in terms of intensity, of 95 db than a sound of 45 db?

52. How much louder is a sound, in terms of intensity, of 130 db than a sound of 70 db?

Exercises 53–56 deal with the Richter scale, and can be solved using much the same approach as used with decibel level exercises. The magnitude R (on the Richter scale) of an earthquake is given by the logarithm function

$$R(i) = \log\left(\frac{i}{i_0}\right)$$

where i represents the intensity of the earthquake, and i_0, the minimum intensity that can be measured and is used as a standard for a comparison.

53. Find the magnitude of an earthquake that measures 5.2 on the Richter scale.

54. Find the ~~magnitude~~ intensity of an earthquake that measures 7.2 on the Richter scale.

55. if two earthquakes have magnitudes that differ by 2, how much more intense is one than the other?

56. If two earthquakes have magnitudes that differ by 3, how much more intense is one than the other?

57. With each day of use of a computer typing tutorial program, the number N of words per minute an average student can type increases according to the function

$$N(t) = 65(1 - e^{-0.35t})$$

where t is the time in days the program has been used. How long will it take an average student to be typing 60 words per minute?

58. The electric current I in amperes that flows in a series circuit with inductance L henrys, resistance R ohms, and electromotive force E volts, is given by the natural exponential function

$$I = \frac{E}{R}(1 - e^{-Rt/L})$$

where t is the time in seconds after the start of the flow of current. How long will it take the current to reach 1.3 amperes in a circuit in which $E = 7.5$ volts, $L = 0.03$ henry, and $R = 3$ ohms?

59. At 21° C, a colony of bacteria doubles in size every five hours. How long will it take this colony to triple in size?

60. The interest rate on a bank deposit is such that it allows the deposit to double every 7 years. How long will it take the deposit to quadruple in size?

61. A radioactive substance decays according to the function

$$Q(t) = A_0 2^{-0.000175t}$$

where Q represents the amount of the substance present t years from now. How long will it take a quantity of the substance to decay so that only one-third of its original amount is present?

Summary and Projects

Summary

The Exponential Function The exponential function $f(x) = b^x$, where $b > 0$ and $b \neq 1$, is one of the most commonly used functions for modeling or simulating certain types of behavior. When the base b is greater than 1, we obtain a graph that increases more and more rapidly as you move from left to right. When the base is less than one (but positive), the graph decreases less and less rapidly. Typical graphs are shown in Figure 5.5.

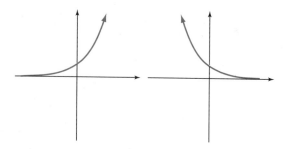

Figure 5.5

You can see from the graphs that the domain of every standard exponential function is the set of all real numbers, $\{x | x \in R\}$, and the range is the set of all positive real numbers, $\{y | y > 0\}$. The graphs also show that all standard exponential functions are one-to-one functions. Exponential growth behavior is represented by an equation of the type

$$A(t) = A_0 e^{kt}$$

where $k > 0$ with the most frequently occurring base $e \approx 2.71828$. Exponential decay behavior is represented by an equation of the type

$$A(t) = A_0 e^{-kt}, \quad k > 0$$

A table of data will exhibit an exponential behavior if the differences in consecutive domain values are constant and the ratios of consecutive range values are constant.

The Logarithm Function   Since all exponential functions ($f(x) = b^x$) are one-to-one functions, their inverses are also functions. The inverse of the exponential function is called the logarithm function. From our knowledge that the domains and ranges of a function and its inverse are interchanged, we can easily state the domain and range of all standard logarithm functions. Namely, the domain is the set of all positive real numbers, $\{x | x > 0\}$ and the range is the set of all real numbers, $\{y | y \in R\}$.

You can obtain the logarithm function by interchanging the variables in the exponential function $y = b^x$ to obtain $x = b^y$. Since we cannot solve for y using the rules of algebra, we invent a notation to express y explicitly. The notation is $y = \log_b(x)$. So that

$$y = \log_b(x) \quad \text{if and only if} \quad x = b^y$$

You can understand the standard logarithm function best by first looking at the more familiar exponential function and interchanging variables and domains and ranges. This connection, along with the fact that functions and their inverses are reflections of each other about the line $y = x$, will help you gain insight into the logarithm function. In addition, your graphing calculator can display both the exponential and logarithm function very quickly and should be relied upon whenever appropriate. The logarithm function that is the inverse of the standard exponential growth function shown above is displayed in Figure 5.6.

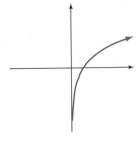

Figure 5.6

Properties of Logarithms   It would be reasonable to expect that the properties of logarithms are closely connected to the properties of exponents, as indeed they are. The three major properties of logarithms are ($m, n > 0$):

$$\log_b(m \cdot n) = \log_b(m) + \log_b(n)$$

$$\log_b\left(\frac{m}{n}\right) = \log_b(m) - \log_b(n)$$

$$\log_b(m^r) = r \log_b(m) \quad r \quad \text{a rational number}$$

In addition, $\log_b(1) = 0$ and $\log_b(b) = 1$.

Exponential and Logarithmic Equations Equations with an exponent containing a variable are exponential equations. The normal approach to solving such an equation (if it is solvable) is to change over to logarithms. For example, you can approach the solution to the equation

$$7^x = 2$$

by taking the natural logarithm of both sides and proceeding as follows:

$$\ln(7^x) = \ln(2)$$

$$x\ln(7) = \ln(2)$$

$$x = \frac{\ln(2)}{\ln(7)}$$

Notice that taking the natural logarithm of both sides removes the variable x from the exponent, which is the basic goal of this approach.

Equations with the argument of a logarithm function containing a variable are logarithmic equations. The normal approach to solving such an equation (when solvable) is to change to exponential form. For example, consider the logarithmic equation

$$\log(x) + \log(x + 3) = 1$$

The approach here is to first use the properties of logarithms to obtain a single logarithm with a coefficient of 1. This can be accomplished as follows:

$$\log(x) + \log(x + 3) = 1$$

$$\log(x(x + 3)) = 1$$

$$\log(x^2 + 3x) = 1$$

At this point, you can change to exponential form to obtain

$$x^2 + 3x = 10^1$$

$$x^2 + 3x - 10 = 0$$

$$(x + 5)(x - 2) = 0$$

$$x = 2 \quad \text{and} \quad x = -5$$

Since the argument of a logarithm function cannot be negative, we discard -5 and so the solution is $x = 2$.

Projects

1. In this chapter it was stated that the number e is important because as the input value x increases, the values of $f(x) = e^x$ increase at a rate precisely equal to the function value itself, e^x. For example, if x is currently equal to 3 and is increased by 1 unit, from 3 to 4, then e^x will increase by a factor of e.

   Prove symbolically that this is true by considering $f(x) = e^x$ and comparing the output values at x and $x + 1$. Recall that two numbers can be compared by division or by subtraction. Comparison by division shows how many times more one number is than another, and comparison by subtraction shows how much more one number is than another. You will want to use division. Specifically, you will want to look at

$$\frac{f(x+1)}{f(x)}$$

   Generalize this result to any function that has this property, namely, that the ratio of two arbitrary function values whose input values are one unit apart is constant.

2. For $b > 0$, and on the interval $(0, \infty)$, the graphs of the power function, $f(x) = x^b$, and the exponential function, $f(x) = b^x$, appear to exhibit the same behavior: they both increase at an increasing rate. This might lead one to believe that they are the same type of function. Although they both increase at an increasing rate, they are entirely different types of functions as they exhibit different types of behavior. Complete Table 5.4 and Table 5.5.

   In the table columns,
   A represents the phrase *As x increases 1 unit from*
   B, the phrase *f(x) increases from*
   C, the phrase *An increase of,*
   D, the phrase *Which is (how many) times the previous increase* and,
   NA, the phrase *Not Applicable*

   These tables can be used to examine the behavior of the power function $f(x) = x^2$ and the exponential function $f(x) = 2^x$. Use columns 3 and 4 of the tables to complete the statements (a) and (b) below. (Hint: You will want to use both subtraction and division to make the required comparisons).

   (a) For each one-unit increase in the value of the input variable, the exponential function, $f(x) = b^x$, increases in value by a factor equal to (write the appropriate word here), and the power function, $f(x) = x^2$, increases in value by a factor equal to (write the appropriate word here).

   (b) For the exponential function, each increase in the output is (how many) times the previous increase. In fact, this increase is (write the appropriate word here). For the power function, each increase in the output is (write the appropriate word here) times the previous increase. In fact, this increase appears to approach (what number).

Table 5.4			
Behavior of $f(x) = 2^x$			
A	B	C	D
0 to 1	1 to 2	1 unit	NA
1 to 2	_ to _	_ units	_ times
2 to 3	_ to _	_ units	_ times
3 to 4	_ to _	_ units	_ times
4 to 5	_ to _	_ units	_ times
5 to 6	_ to _	_ units	_ times

Table 5.5			
Behavior of $f(x) = x^2$			
A	B	C	D
0 to 1	0 to 1	1 unit	NA
1 to 2	_ to _	_ units	_ times
2 to 3	_ to _	_ units	_ times
3 to 4	_ to _	_ units	_ times
4 to 5	_ to _	_ units	_ times
5 to 6	_ to _	_ units	_ times

CHAPTER 6

Trigonometric Preliminaries

6.1 Angles and the Central Angle—Arc Length Connection

Introduction

Description of an Angle

The Measure of an Angle

Radian Measure of an Angle

Introduction

The study of trigonometry requires a background in both algebra and geometry. We review the key concepts of geometry in this section since you already know the required algebra from your study of the earlier chapters of this book. By its nature, a review is generally brief, so you may wish to borrow a geometry book if you need to review any particular topic in greater depth than is presented here.

Description of an Angle

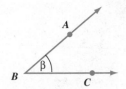

Figure 6.1

We need to recall the two aspects of an angle—its geometric definition and its measure. Geometrically, an angle is the union of two rays with a common endpoint. The common endpoint is called the vertex of the angle and the rays are the sides of the angle. Naming conventions for angles are best illustrated with a picture as in Figure 6.1. Here, the angle can be identified as $\angle B$ or $\angle ABC$ or $\angle \beta$.

When more than one angle occurs at a single vertex as in Figure 6.2, then identifying the angle using $\angle B$ is not sufficient and one of the other two descriptions would be required.

In this case,

$$\angle \beta = \angle ABC \text{ and } \angle \alpha = \angle CBD$$

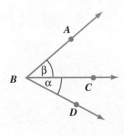

Figure 6.2

The Measure of an Angle

The measure of an angle can be thought of as the "size" of the angle or amount of sweep needed to rotate one of the sides of the angle onto the other side. The most familiar unit of measure is the degree, where $90°$ is the measure of a right angle.

A degree can be further subdivided into minutes and seconds, where 60 minutes equals one degree and 60 seconds equals one minute. We will often find it useful to work with degree measure in decimal form rather than in minutes and seconds. The conversion is obtained through the definition so that an angle whose measure is 78° 48′ 15″ (in degree, minutes, and seconds) would be equivalent to 78.825° in decimal form. This conversion is accomplished as follows:

$$\frac{15''}{60} = 0.25'$$

$$\text{Hence,} \quad 48'15'' = 48.25'$$

$$\frac{48.25'}{60} \approx 0.804°$$

$$\text{so that} \quad 78°48'15" \approx 78.804°$$

Your calculator can be very helpful in these conversions. You can perform this computation on your graphing calculator by entering the following expression:

$$(15/60+48)/60 + 78$$

Radian Measure of an Angle

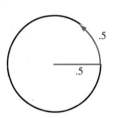

Figure 6.3

Other units can be used to measure the "size" of angles. For example, an angle can be measured by the fraction of a complete revolution in the plane that it represents. Thus, an angle whose measure is 90° is an angle that measures $\frac{1}{4}$ of a revolution. Another unit of measure that you will use extensively in this book is the radian unit. To get a feel for an angle whose measure is one radian, draw a circle with a radius of about 0.5 inches. Now draw an arc with a length of about 0.5 inches on the circle as in Figure 6.3. Starting at the center of the circle, draw two rays that intercept this arc at its endpoints.

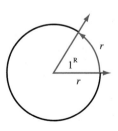

Figure 6.4

The acute angle you have constructed at the center of the circle has a measure of one radian. Angles with their vertex at the center of a circle are called central angles. Your diagram should look something like Figure 6.4. This angle is said to have a measure of one radian, written 1^R. It is interesting to ask, "About how big is this angle in degrees?" It is clearly less than a right angle but seems to be more than 45°. It would appear that an angle of 1^R is about two-thirds of a right angle, and is actually about 57.296°. In general, an angle of measure x radians will be symbolically represented as x^R.

To delve more deeply into the meaning of radian measure, let us consider two concentric circles with arbitrary radii, say r_1 and r_2, and an acute central angle as in Figure 6.5. The arc lengths *subtended* (cut off) by the rays of the angle are used to determine the radian measure of the angle. We cannot, however, use the arc lengths themselves as the measure since they are clearly different. Since these two circles are similar, the ratios of the subtended arc lengths to the lengths of the radii are the same. Thus, the arc length is to the radius in the small circle as the arc length is to the radius in the larger circle. This relationship can be expressed as a proportion:

$$\frac{\text{small arc length}}{\text{small radius}} = \frac{\text{large arc length}}{\text{large radius}}$$

Or, using letters,

$$\frac{s_1}{r_1} = \frac{s_2}{r_2}$$

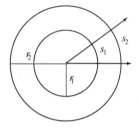

Figure 6.5

Call the number represented by these ratios θ. The number θ is used for the radian measure of the angle. We can also consider this angle measure as the number of radii contained in the arc length since the arc length divided by the radius yields the number of radii in the arc length. (For example, if the radius is 2 and the arc length is 3, then there are 3/2 radii contained in the arc length and the angle measure is $1\frac{1}{2}$ radians.) This notion is the reason for the name radian.

Radian Measure of an Angle

Radian Measure of an Angle

Given the centralangle θ subtending an arc of length s on a circle of radius r then the measure of angle θ in radians is

$$m\angle\theta^R = \frac{s}{r}$$

where the superscript R on θ indicates that the angle measure is in radians.

We note that if the radius of the circle is 1 then the radian measure of the central angle θ is the length s of the arc. We will, therefore, find it very useful to use circles with radius one.

EXAMPLE SET A

1. Determine the radian measure of each of the following central angles.

a.

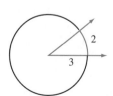

b.

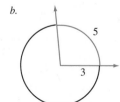

c.   *d.*

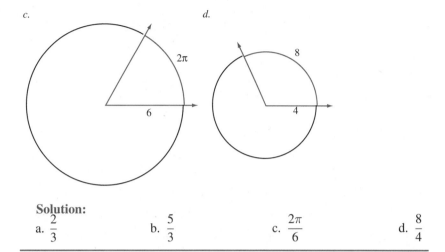

Solution:

a. $\dfrac{2}{3}$ b. $\dfrac{5}{3}$ c. $\dfrac{2\pi}{6}$ d. $\dfrac{8}{4}$

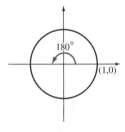

Figure 6.6

You will need to know how to convert from degrees to radians and from radians to degrees. You know that one-half a revolution is 180 degrees. Consider a central angle of 180 degrees in a circle of radius 1. See Figure 6.6. What is the measure of this angle in radians? The radian measure of this angle is its associated arc length divided by the radius. The circumference of a circle with radius 1 unit is 2π, so the associated arc length of half this circle is π. Thus, an angle swept out by one-half a revolution whose degree measure is 180 has radian measure $\frac{\pi}{1}$ or π.

You can convert an angle of 240 degrees to radians as follows: Dividing both sides of $180° = \pi^R$ by 180 gives

$$\frac{180°}{180} = \frac{\pi^R}{180}$$

Multiplying both sides by 240 yields

$$240\left(\frac{180}{180}\right)^{\!\circ} = 240\left(\frac{\pi}{180}\right)^{\!R}$$

Finally, reducing this fraction to lowest terms produces $\dfrac{4\pi}{3}$ radians. Hence, 240 degrees equals $\dfrac{4\pi}{3}$ radians.

To convert an angle of $\dfrac{3\pi}{4}$ radians to degrees, you follow a similar procedure. You know that $180° = \pi^R$ so that $\dfrac{180°}{\pi} = \dfrac{\pi^R}{\pi}$. Then multiplying both sides by $\dfrac{3\pi}{4}$ gives $\dfrac{3\pi}{4}\left(\dfrac{180}{\pi}\right)^{\!\circ} = \dfrac{3\pi}{4}\left(\dfrac{\pi}{\pi}\right)^{\!R}$. Finally, reducing $\dfrac{3\pi}{4}\left(\dfrac{180}{\pi}\right)$ to lowest terms results in 135. Hence, $\dfrac{3\pi}{4}$ radians equals 135 degrees.

EXAMPLE SET B

1. Determine the radian measure of the following central angles:

   a. 60°   b. 150°   c. 240°

   Solution:

   a. $60°(\frac{180}{180}) = 60°(\frac{\pi}{180}) = \frac{\pi}{3}$. The associated arc length of this angle is a little more than one radius long.

   b. $150°(\frac{180}{180}) = 150°(\frac{\pi}{180}) = \frac{5\pi}{6}$. The associated arc length of this angle is about $\frac{5(3.14)}{6} \approx 2.6$ radii long.

   c. $240°(\frac{180}{180}) = 240°(\frac{\pi}{180}) = \frac{4\pi}{3}$. The associated arc length of this angle is about 4.2 radii long.

2. Determine the degree measure of the following central angles:

   a. $\frac{\pi}{4}$ b. $\frac{2\pi}{3}$ c. $\frac{\pi}{6}$

   Solution:

   a. Since $\frac{\pi^R}{\pi} = \frac{180}{\pi}, \frac{\pi}{4}(\frac{\pi}{\pi}) = \frac{\pi}{4}(\frac{180}{\pi}) = 45°$. Since $\frac{180}{\pi}$ is greater than 1, the degree measure is much larger than the corresponding radian measure.

   b. Since $\frac{\pi}{\pi} = \frac{180}{\pi}, \frac{2\pi}{3}(\frac{\pi}{\pi}) = \frac{2\pi}{3}(\frac{180}{\pi}) = 120°$.

   c. Since $\frac{\pi}{\pi} = \frac{180}{\pi}, \frac{\pi}{6}(\frac{\pi}{\pi}) = \frac{\pi}{6}(\frac{180}{\pi}) = 30°$.

There are several terms you need to be familiar with. An angle is con-

Standard Position sidered to be in **standard position** in a Cartesian coordinate system when its vertex is placed at the origin and one of the sides of the angle lies along **Initial Side** the positive x–axis. The side that lies along the positive x–axis is called **Terminal Side** the **initial side** while the other side is called the **terminal side**. Figure 6.7 shows angle $\angle AOB$ in standard position with ray OB as initial side and ray OA as terminal side. Extending our knowledge of geometry, we consider angles whose measures are not restricted between 0 and 180 degrees. To start with, consider a 40° angle in standard position formed by rotating initial side OB onto terminal side OA in a counterclockwise direction as shown in the diagram on the left in Figure 6.8.

Now, in Figure 6.7 rotate initial side OB clockwise onto terminal side OC so that the geometric measure of $\angle BOC$ is equal to the geometric measure of $\angle AOB$. We consider angles generated by rotating in a clockwise direction to be negative. Thus, the angle indicated in the diagram on the right in Figure 6.8 has measure −40°.

Now if you consider rotating terminal side OA counterclockwise onto initial side OB as in the left diagram of Figure 6.9 , you sweep out an angle

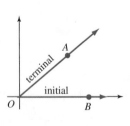

Figure 6.7

of measure $360 - 40 = 320°$. The measure of this angle is positive since counterclockwise rotation will be considered positive while clockwise rotation will be considered negative.

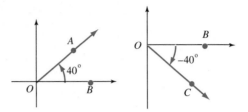

Figure 6.8

Finally, if you rotate side *OB* in Figure 6.7 clockwise onto side *OA* as in the right diagram of Figure 6.9, the measure of the angle is $-320°$. While the measures of all three of these angles whose initial and terminal sides pass through *B* and *A*, respectively, are different, all three have the same sides. Angles with the same sides are called *coterminal*.

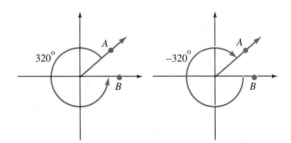

Figure 6.9

EXAMPLE SET C

1. Determine a positive angle measure for each of the given angles.

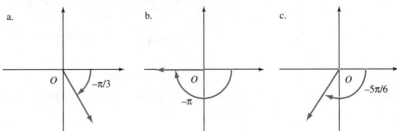

Solution:

a.

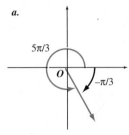

b.

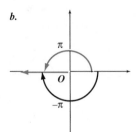

c.

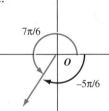

$$2\pi + \left(\frac{-\pi}{3}\right) = \frac{5\pi}{3}$$ $$2\pi + (-\pi) = \pi$$ $$2\pi + \left(\frac{-5\pi}{6}\right) = \frac{7\pi}{6}$$

2. Determine a positive angle measure for each of the given angles.

a.

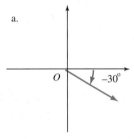

b.

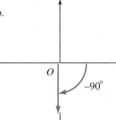

c.

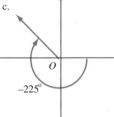

Solution:
a. $360° + (-30°) = 330°$
b. $360° + (-90°) = 270°$
c. $360° + (-225°) = 135°$

3. To the angle whose measure is q as illustrated in the diagram on the left of Figure 6.10, give a positive angle whose terminal side is in the same position.

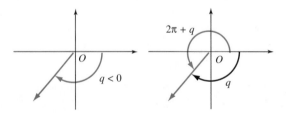

Figure 6.10

Solution:
$2\pi + q$ is the radian measure as shown in the diagram on the right of Figure 6.10.

EXERCISES

In exercises 1–10, determine the exact radian measure for each of the central angles. (Note: The length of one unit (scale) changes from problem to problem.)

1.

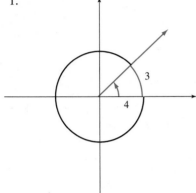

2.

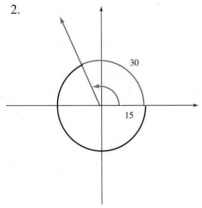

3.

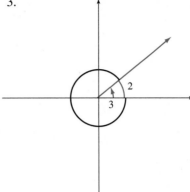

4.

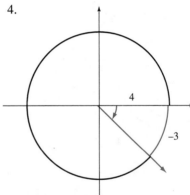

5.

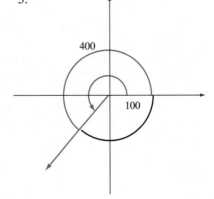

6.

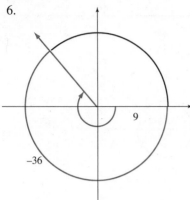

7.

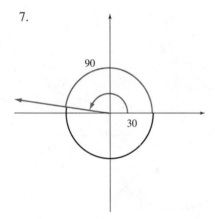

8.

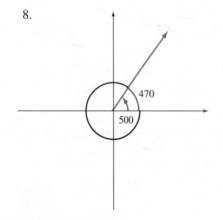

9.

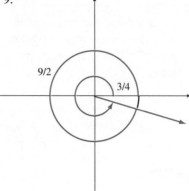

10.

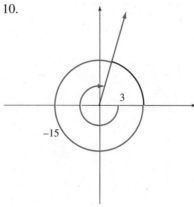

In exercises 11–16, convert each of the angle measures from degrees to radians.

11. 60°   12. 135°

13. 50°   14. −210°

15. −75°   16. 315°

In exercises 17–22, convert each of the angle measures from radians to degrees.

17. $\frac{\pi}{3}$ 18. $\frac{-7\pi}{6}$

19. 5   20. 3

21. −6   22. $\frac{5\pi}{4}$

In exercises 23–26, for each given angle, indicate a coterminal angle with positive radian measure and give its measure.

23. $\frac{2\pi}{3}$

24. $\frac{-5\pi}{4}$

25. -4

26. 2

In exercises 27–30, for each given angle, indicate a coterminal angle with positive degree measure and give its measure.

27. $160°$

28. $-330°$

29. $-15°$

30. $-240°$

In exercises 31–46, use your calculator to determine in which quadrant each of the following radian measure angles lie. On paper, duplicate the unit circle pictured in the accompanying diagram, and place the approximate measure of each angle in the appropriate location.

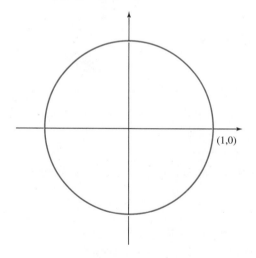

31. 0.5

32. 1

33. 1.5

34. 1.6

35. 1.7

36. 3

37. 3.1

38. 3.2

39. 4

40. 4.6

41. 4.7

42. 5

43. 5.4

44. 6

45. 6.2

46. 6.3

In exercises 47–54, place the numbers that follow (radian angle measures) on the number line.

47. $\frac{\pi}{3}$

48. $\frac{2\pi}{3}$

49. π

50. $\frac{5\pi}{4}$

51. $\frac{5\pi}{3}$

52. $\frac{-7\pi}{6}$

53. $\frac{-3\pi}{4}$

54. $\frac{-11\pi}{6}$

In exercises 55–58, locate the quadrantal angle measures on the number line given. A quadrantal angle is an angle whose terminal side coincides with the positive or negative x- or y-axes.

55. $\frac{\pi}{2}$

56. $\frac{3\pi}{2}$

57. $-\pi$

58. 2π

59. Two mechanical cars are put on a semicircular track of radius 10 feet, where the start point is taken as the origin of a rectangular coordinate system and the end point is located at (20, 0). One car is fitted with a meter that gives its horizontal distance x from the start point, and the other car has a meter that gives the perpendicular distance y from the start point. If the readings from both cars are 5, have both cars traveled the same distance? What are the angles subtended at the center C by the paths of these two cars? (Assume that both angles are less than 90 degrees.) Which

car is faster if these readings are observed after the same interval of time?

60. (Refer to the previous problem) Is it possible for the two cars to have the same reading at the same time, assuming they start together? If so, what is the reading and what are the angles subtended at the center? Why is there a requirement that the angles be less than 90 degrees? What would happen if this were not a requirement?

61. Mike has an assignment to build a toy racing car for his class science project. The car that goes the farthest wins a prize. Mike builds a

car that consists of a wood-block body and four wheels. The power mechanism he devises provides enough energy to turn the rear axle 15 times. How far will the car go if the radius of each wheel is 4 cm?

62. (Refer to the previous problem) Mike believes that bigger wheels will make his car go farther. To this end, he puts in wheels of radius 5 cm. However, he fails to consider that because bigger wheels are heavier, his power pack can only provide for 12 rotations. Which set of wheels should he use?

6.2 Formulas and Functions

Introduction

The Pythagorean Theorem and the Distance Formula

The Unit Circle Equation

The Associated Right Triangle of an Angle

Introduction

In this section, we study two formulas from algebra that are useful in trigonometry. We then use these formulas in working with central angles of a circle of unit radius.

The Pythagorean Theorem and the Distance Formula

The result of the Pythagorean Theorem can be stated algebraically as

$$a^2 + b^2 = c^2$$

for a right triangle with hypotenuse of length c and legs of lengths a and b.

EXAMPLE SET A

What is the length of the hypotenuse c of the right triangle in the accompanying diagram if $a = 3$ and $b = 4$?

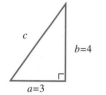

Solution:

$$3^2 + 4^2 = c^2$$

$$25 = c^2$$

$$c = \pm 5$$

$$c = 5 \quad \text{(since } c \text{ is positive as a length)}$$

Thus, the length of the hypotenuse is 5 units.

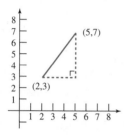

Figure 6.11

If we wish to find the length of a line segment between the points (2,3) and (5,7), we can draw a line parallel to the x-axis through (2,3) and a line parallel to the y-axis through (5,7) as indicated in the diagram (Figure 6.11). The triangle created in this way is a right triangle with legs of length 3 and 4. Then, using the Pythagorean Theorem and the results of the previous example, we note that the line segment between (2,3) and (5,7) is 5 units long. We can generalize this notion. The distance between two points $P_1(x_1, y_1)$ and $P_2(x_2, y_2)$ (Figure 6.12) is a function of the coordinates based on the Pythagorean Theorem:

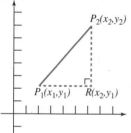

Figure 6.12

$$d(x_1, y_1, x_2, y_2) = \sqrt{(x_2 - x_1)^2 + (y_2 - y_1)^2}$$

where $x_2 - x_1$ is the distance between x_1 and x_2 on the x-axis and $y_2 - y_1$ is the distance between y_1 and y_2 on the y-axis. This is the Distance Formula and it is frequently written

$$d(P_1, P_2) = \sqrt{(x_2 - x_1)^2 + (y_2 - y_1)^2}$$

The Unit Circle Equation

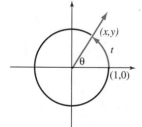

Figure 6.13

Recall that the general form of any circle with center at the origin is also based on the Pythagorean Theorem and is $x^2 + y^2 = r^2$ where r is the radius of the circle. The unit circle is the name given to the circle with its center at the origin and a radius of one unit. Its equation is $x^2 + y^2 = 1$.

The radian measure of the acute angle θ in standard position from the definition is $\dfrac{t}{1} = t$, the length of the arc subtended by the angle on the unit circle (Figure 6.13) so that $m\angle\theta = t$.

We can develop a number of interesting functions based on the unit circle.

The Associated Right Triangle of an Angle

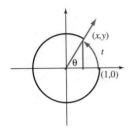

Figure 6.14

In Figure 6.14, the endpoint of the arc having length t is (x, y). An interesting function that we will name $\dfrac{opposite}{adjacent}$ is defined as

$$\frac{opposite}{adjacent}(t) = \frac{y}{x}$$

This unusual function name can be explained by drawing a perpendicular line segment from the point (x, y) to the x-axis. The right triangle so formed will be called the associated right triangle (see Figure 6.14.) Note that it has one side opposite angle θ of length y, the second coordinate of (x, y). The other side of the triangle is called the side adjacent to angle θ and lies along the horizontal axis. It has length x, the first coordinate of (x, y). Consider the arc of length t given in Figure 6.15 which terminates at the point $(0.6, 0.8)$. In this case, we have that $\dfrac{opposite}{adjacent}(t) = \dfrac{0.8}{0.6}$.

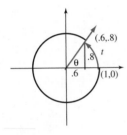

Figure 6.15

EXAMPLE SET B

Give the value of the function $\dfrac{opposite}{adjacent}$ for the measure t of each acute central angle θ.

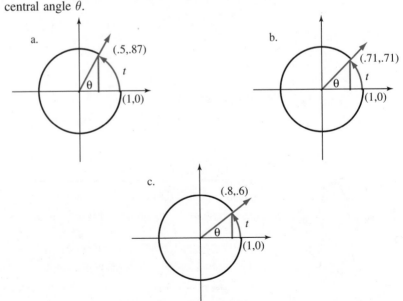

Solution:

a. $\dfrac{opposite}{adjacent}(t) = \dfrac{0.87}{0.5} = 1.74$

b. $\dfrac{opposite}{adjacent}(t) = \dfrac{0.71}{0.71} = 1$

c. $\dfrac{opposite}{adjacent}(t) = \dfrac{0.6}{0.8} = \dfrac{3}{4}$

If the angle is not in the first quadrant, we can still draw a perpendicular line from the point (x, y) to the x-axis. This triangle is the associated right triangle. In the triangle so formed there will also be an $\dfrac{opposite}{adjacent}$ function. Although θ is not inside the associated right triangle in such cases, the $\dfrac{opposite}{adjacent}$ function will still be defined as $\dfrac{y}{x}$ where, again, (x, y) is the point of intersection of the terminal side of angle θ and the unit circle. Thus, we extend the definition of $\dfrac{opposite}{adjacent}$ to angles that are not in the first quadrant.

EXAMPLE SET C

1. Draw in the perpendicular from the given point to the x-axis. Indicate the directed lengths of the legs of the right triangle formed. Then estimate the value of the $\dfrac{opposite}{adjacent}$ function for each angle measure t.

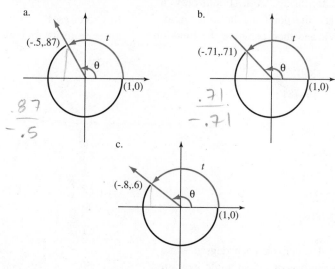

Solution:

a. $\dfrac{opposite}{adjacent}(t) = \dfrac{0.87}{-0.5} = -1.74$

b. $\dfrac{opposite}{adjacent}(t) = \dfrac{0.71}{-0.71} = -1$

c. $\dfrac{opposite}{adjacent}(t) = \dfrac{0.6}{-0.8} = -\dfrac{3}{4}$

2. Show that each point in the previous example is on the unit circle.

Solution:
The point in (a) is $(-0.5, 0.87)$. If this point is on the unit circle, it must satisfy the equation $x^2 + y^2 = 1$. Using your graphing calculator, you can calculate $(-0.5)^2 + (0.87)^2$. The value displayed (1.0069) is not exactly 1 because the given coordinates are estimates of the actual values. In the same way, for the point in (b) $(0.71)^2 + (0.71)^2$ is approximately 1 and for the point in (c) $(0.8)^2 + (0.6)^2$ is exactly 1.

EXERCISES

In exercises 1–8, $\angle\alpha$ in standard position subtends an arc of length t as measured from the point $(1,0)$. Find the value of the $\dfrac{opposite}{adjacent}$ function for the arc length t where the terminal side of $\angle\alpha$ intersects the unit circle at the point P with the indicated coordinates. You should draw a unit circle, place the point on the unit circle (approximately), and then compute the function value.

1. $P(-.9422, .3350)$

2. $P(-.7259, -.6878)$

3. $P(.4685, -.8835)$

4. $P(-.2108, -9775)$

5. $P(.1700, .9854)$

6. $P(.8347, -.5507)$

7. $P(.9422, .3350)$

8. $P(-.5885, .8085)$

In exercises 9–16, convert each degree angle measure to its equivalent radian angle measure.

9. $150°$

10. $240°$

11. $45°$

12. $-132.6°$

13. $57.31°$

14. $-315°$

15. $2.35°$

16. $249°$

In exercises 17–29, convert each radian angle measure to its equivalent degree angle measure.

17. $\dfrac{\pi}{3}$

18. $\dfrac{-7\pi}{6}$

19. $\dfrac{5\pi}{3}$

20. $\dfrac{3\pi}{2}$

21. $-\dfrac{\pi}{8}$

22. $\dfrac{7\pi}{15}$

23. $\dfrac{3\pi}{20}$

24. $\dfrac{17\pi}{9}$

25. 1.57

26. 2.8

27. -5

28. 7

29. 4.7331

32. 50°

33. −200○

In exercises 30–37, draw the angle in standard position with the given angle measure and then give the measure of three coterminal angles with at least one of them being negative.

34. 2.5^R

35. -5.4^R

36. 0.4^R

37. 6.1^R

30. 30°

31. 130°

$$360 + 130 = 490°$$

6.3   Arcs and Their Associated Angles and Triangles

Introduction

Measure of Angles on Different Circles

Programming the Unit Circle

Introduction

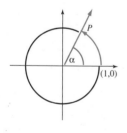

Figure 6.16

In studying angles or rotational motion, we can use circles and right triangles. We start with the unit circle. Figure 6.16 illustrates an arbitrary angle, $\angle\alpha$, in standard position.

Notice that the terminal side of $\angle\alpha$ intersects the unit circle at the point P. The length of the arc from the point $(1,0)$ to P on the unit circle is the radian measure of the angle. If we draw a perpendicular line segment from P to the x-axis, we form the right triangle $\triangle OQP$ illustrated in Figure 6.17. The point P has coordinates (x,y). Thus, we have established an association between $\angle\alpha$, the intercepted arc s, and the right triangle $\triangle OQP$. The point

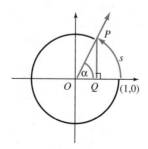

Figure 6.17

of intersection of the terminal side of the angle and the unit circle is the key point in the construction of the right triangle. This point exists for each central angle. If the angle's terminal side is *not* on an axis, then a right triangle can be associated with it by drawing a perpendicular line segment from the key intersection point to the *x*-axis.

Measure of Angles on Different Circles

Consider the measure of $\angle\alpha$ in the diagram on the left of Figure 6.18 to be s. Let's consider a circle with radius $r_1 > 1$ along with its associated right triangle. The arc length associated with $\angle\alpha$ is s_1 as shown in the diagram on the right of Figure 6.18.

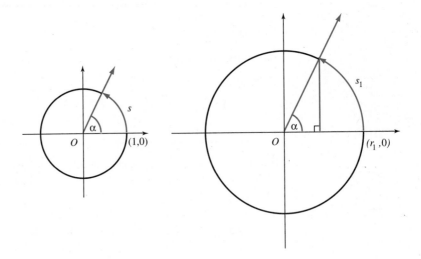

Figure 6.18

How is this new arc length associated with the measure of $\angle\alpha$? To answer this question, recall that all circles are similar. Thus, the arc length is to the radius in the small circle as the arc length is to the radius in the larger circle. This relationship can be expressed as a proportion:

$$\frac{\text{small arc length}}{\text{small radius}} = \frac{\text{large arc length}}{\text{large radius}}$$

$$\text{or} \quad \frac{s}{1} = \frac{s_1}{r_1}$$

$$\text{or} \quad s = \frac{s_1}{r_1}$$

The measure of the angle α is s or $\dfrac{s_1}{r_1}$. So, the measure of $\angle \alpha$ is the arc length s_1 divided by the radius r_1, (as we saw in Section 6.1). Thus, the radian measure of an angle can also be thought of as the number of radii contained in the arc length.

EXAMPLE SET A

Angle α has the same measure in both diagrams of Figure 6.19. The radius of the smaller circle is 2 units and the radius of the larger is 4 units. Show that the radian measure of $\angle \alpha$ is the same for both circles.

Solution:
Angle α in Figure 6.19 can be measured using either of the two circles.

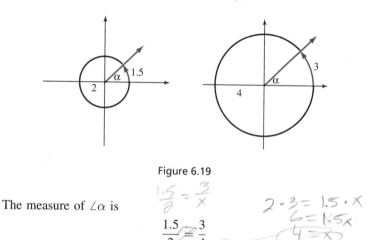

Figure 6.19

The measure of $\angle \alpha$ is

$$\frac{1.5}{2} \overset{?}{=} \frac{3}{4}$$

$$\frac{1.5}{2} \overset{3}{=} \frac{3}{x}$$

$$2 \cdot 3 = 1.5 \cdot x$$
$$6 = 1.5x$$
$$\boxed{4 = x}$$

In Figure 6.20, the terminal side of $\angle \alpha$ intersects the unit circle at P (we'll designate this point as $P(x, y)$). Constructing the associated triangle, we have a right triangle with sides of lengths x and y and hypotenuse of length 1.

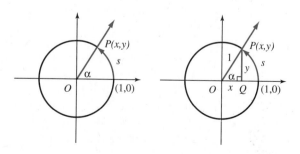

Figure 6.20

Using the Pythagorean Theorem, we again obtain the familiar equation of the unit circle: $x^2 + y^2 = 1$. This equation can be used to determine the value of x if the value of y is known or vice versa. We can also use the equation to determine the values of x and y for some particular arcs.

EXAMPLE SET B

What are the coordinates (x, y) of the point of intersection of the terminal side of an angle of radian measure $\dfrac{\pi}{4}$ with the unit circle?

Solution:

If $\angle \alpha$ has radian measure $\dfrac{\pi}{4}$ $\left(\dfrac{\pi}{4} = 45°\right)$, then the x and y values are the same, that is, $y = x$. (Remember that the 45° line is the line $y = x$.) Thus,

$$x^2 + y^2 = x^2 + x^2 = 1, \text{ or}$$

$$2x^2 = 1$$

Thus, $x^2 = \frac{1}{2}$ or $x = \pm\frac{1}{\sqrt{2}}$. Since the point P is in the first quadrant, both x and y are positive. Thus, $x = y = \frac{1}{\sqrt{2}}$ and the coordinates of the point with radian measure $\frac{\pi}{4}$ is $\left(\frac{1}{\sqrt{2}}, \frac{1}{\sqrt{2}}\right)$.

Programming the Unit Circle

You can draw a unit circle on the display of your TI graphing calculator. To do this you need to become familiar with the PRGM key. Also, you will use the RNG or Window submenu of the VARS key.

To begin, press the PRGM key. Press the right-arrow key to highlight EDIT on the TI-81 or NEW on the TI-82. On the TI-81, select 1 if the space to the right of Prgm1 is blank. If not, press the first number where it is blank. On the TI-82, press ENTER after you have highlighted NEW. Type UNITCIR and press ENTER to name this program. The screen should look something like Window 6.1.

There should be a blinking rectangle next to a colon (:) on the second line. The second line will be

    −1.8 →Xmin

To do this, you will press many keys. Of course, you should look at the screen each time you press a key. First, press the gray negative sign (−) and 1.8. Then press the black STO key to produce the assignment arrow. You cannot enter Xmin by typing letters on the TI-81. The Xmin symbol is obtained

Window 6.1

by pressing the VARS key and selecting RNG or Window, and then pressing the number next to Xmin. Now press ENTER to go to the next line.

The following lines should be entered in the same way.

1.8 →Xmax
-1.2 →Ymin
1.2 →Ymax
.1 →Tstep

Tstep can be found by looking around in the menus that access Xmin. You should press ENTER at the end of each line.

Be sure that your calculator is in both Rad(ian) and Param(etric) modes. To do this, the last two lines of the program are

Rad
Param

on the TI-81 and

Radian
Param

on the TI-82.

You can access these symbols with the MODE key.

The entire program should now be displayed on your screen, which should look something like Window 6.2.

We can use the INS and DEL keys to make any changes that are necessary.

To see how this program works, press 2nd CLEAR to quit the programming mode. Now press PRGM and then the number of the program you just entered. When we do this, Prgm and a number or name appears in the computation window. Now press ENTER to run this program. We know that the program has run when Done is displayed on the screen. We can check the results of this program by pressing the RANGE key on the TI-81 or the WINDOW key on the TI-82 and observing the Tstep, Xmin, Xmax, Ymin, and Ymax values. The Tmin

```
PROGRAM:UNITCIR
:-1.8->Xmin
:1.8->Xmax
:-1.2->Ymin
:1.2->Ymax
:0.1->Tstep
:Radian
:Param
```

Window 6.2

and Tmax values should be 0 and 6.283185307, respectively, which are their standard values provided they were not changed previously.

To finish the preliminary work for drawing a unit circle, we need to enter expressions in the Y= menu. Press Y=. Enter COS(T) for X1T and SIN(T) for Y1T. If there are other expressions below these two, clear them out. Now press GRAPH.

We have displayed the unit circle. Now let's use it.

EXAMPLE SET C

Using the TRACE and arrow keys, write down the coordinates of four points on the unit circle you just drew, one in each of the four quadrants. Show that these points are on the unit circle.

Solution:
The point in the fourth quadrant might be approximately $(0.6347, -0.7728)$ using four-decimal-place accuracy. This point should satisfy the equation $x^2 + y^2 = 1$. We calculate $0.6347^2 + (-0.7728)^2$ to be 1.00006393, which is approximately 1. If we use greater accuracy for the coordinates of the point, we will obtain values closer to 1. When you use the TRACE feature, the X and Y values shown at the bottom of the screen are stored as the x and y values in your graphing calculator. You can achieve the most accurate results with your TI graphing calculator by entering (using the ALPHA key) X^2 + Y^2 in the computation window. You should be sure to complete the computations for the three other points. You will be using the UNITCIR program again in this chapter.

EXERCISES

In exercises 1–8, draw in the corresponding
right triangles and indicate the corresponding arc
lengths for each central angle. Estimate the radian
measure in each case to the nearest integer.

1.

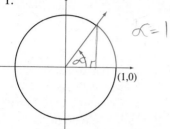

2.

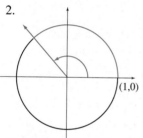

3.

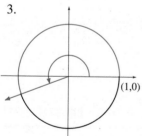

4.

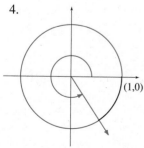

5.

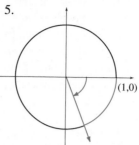

6.

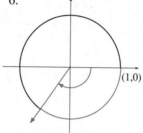

7.

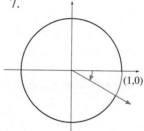

8.

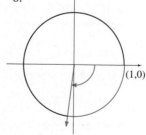

In exercises 9–16, run the UNITCIR program and then, in the RANGE or WINDOW menu, change Tmax to 6.3 and Tstep to 0.1 in drawing a unit circle. On the unit circle, use TRACE to move to the indicated points given below and write down the measure of the associated angle. Convert the radian measure of the associated angle to degree measure.

9. (.9801, .1987) $t = .2$ $(11.5°)$

10. (−.3233, .9463)

11. (−.8569, .5155) $t = 2.6$ $(149°)$

12. (−.9668, −.2555)

13. (−.7259, −.6878) $t = 3.9$ $(223.5°)$

14. (−.4008, −.9162)

15. (.4685, −.8835) $t = 5.2$ $(298°)$

16. (.7087, −.7055)

In exercises 17–24, run the UNITCIR program and then, in the RANGE or WINDOW menu, change Tmax to 360 and Tstep to 10. If Tmin is not 0, change it. In the MODE menu, change radians to degrees so that computations are done in degree measure. These changes cause the unit circle to be drawn with degree measure for T (rather than radian measure). Draw the unit circle and use

TRACE to find the measure of the associated angle and the quadrant for each of the following points.

17. (.766, .6428)   18. (.1736, .9848)
   $t = 40°$ I

19. (−.6428, .766)   20. (−.866, .5)
   $t = 130°$ II

21. (−.9397, −342)   22. (−.342, −.9397)
   $t = 200°$ III

23. (.5, −.866)   24. (0, −1)
   $t = 300°$ IV

25. In the old days, some people had to get their drinking water from wells dug into the ground. A bucket hangs from a rope down into a deep well, and the water is brought to the surface in the bucket by winding the rope onto a roller. If a roller has a radius of 6 inches and the water is at a depth of 24 feet, how many times will the crank have to be turned in order to raise the water?

26. Modify the UNITCIR program to change the RANGE or WINDOW menu values for Tmin to -2π and Tmax to 2π. Draw a unit circle with these values and discuss why the trace cursor starts at (1,0). You may need to consult your graphing calculator manual to help you do this.

6.4   A Moving Point with Its Associated Triangles

Introduction

A Moving Point on the Unit Circle

Programming the Associated Right Triangle

Symmetric Points

Programming Symmetric Points

Introduction

A point on a unit circle has been associated with its central angle and associated right triangle. As a point moves around the unit circle, it brings its associated right triangle with it. Now we investigate how the associated right triangle changes as a point moves around the unit circle.

A Moving Point on the Unit Circle

One of the applications of trigonometry is in the investigation of rotational motion. In this section, we consider a point that moves along the circumference of the unit circle. The coordinates of this point will change as the point moves along the circle. These changing coordinates will always satisfy the Pythagorean Theorem for the right triangle (when it exists) associated with the point. In either case, if the coordinates of the moving point are (x, y), then $x^2 + y^2 = 1$. In Figure 6.21, you can see a "snapshot" of the moving point P with its associated right triangle.

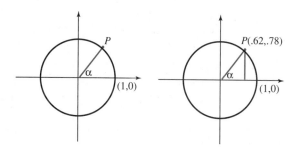

Figure 6.21

The coordinates are given as decimal approximations to two places. The arc length is *always* measured from the point $(1, 0)$ to P. The measure of the arc is considered positive if the point is moving in a counterclockwise direction and negative if the movement is clockwise. The ray from the origin through the point P is the terminal side of the angle whose initial side is the positive x-axis. The terminal side OP of this angle rotates about the origin as the point P moves along the circle. The central angle at the origin increases as the arc length from $(1, 0)$ to P increases.

Programming the Associated Right Triangle

You can draw an associated triangle on a unit circle with your graphing calculator. To do this, first run the UNITCIR program and graph a unit circle. The following program will draw the associated triangle for any non-axis point you select with the TRACE feature. Enter this program and name it TRIANGLE in the same way as the UNITCIR program. Refer back to those instructions, if necessary. You can access the desired commands through the DRAW menu.

```
Line(X,Y,0,0)
Line(X,Y,X,0)
```

The first statement of this program joins your selected point to the origin. The second statement draws a perpendicular line from your selected point to the *x*-axis.

EXAMPLE SET A

Draw the associated right triangle for the arc lengths 0.9 and 4.5.

Solution:
Clear the graphing window with the ClrDraw selection (1) from the DRAW menu (press 2nd and then PRGM). Run the UNITCIR program and press GRAPH to graph the unit circle. Now use TRACE to select the point whose T value is about 0.9. Press PRGM and run the TRIANGLE program. You can use the TRACE feature to select the point whose T value is about 4.5 and repeat the process. In this way, you can draw an associated triangle for every (non-axis) point on the unit circle in the first revolution. You might wish to estimate the arc length that gives an associated right triangle that is isosceles.

Let's look at the angle whose terminal side intersects the unit circle at the point $P_1(x, y)$ where x and y are equal. Figure 6.22 shows both the angle by itself and the angle with its associated right triangle. Recall that the measure of this angle is $\frac{\pi}{4}$.

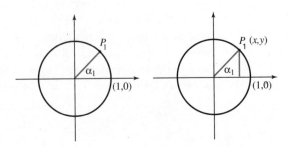

Figure 6.22

As we saw in the last section, the coordinates of the point P_1 are $(\frac{1}{\sqrt{2}}, \frac{1}{\sqrt{2}})$. As the point continues moving counterclockwise along the circle, it will go up and then come down again to a point P_2 that has the same height as P_1. The points P_1 and P_2 are symmetric with respect to the y-axis. (See Figure 6.23).

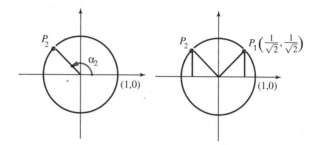

Figure 6.23

EXAMPLE SET B

What are the coordinates of P_2 in Figure 6.24?

Solution:
Since the points are symmetric about the y-axis, the second coordinate of P_2, $\frac{1}{\sqrt{2}}$, will be the same as the second coordinate of P_1. Looking at Figure 6.24, the right triangle associated with P_2 is in the second quadrant and is congruent to the triangle associated with P_1 in the first quadrant. The measure of the central angle in standard position whose terminal side passes through P_2 is $\pi - \frac{\pi}{4} = \frac{3\pi}{4}$. Since the points are symmetric about the y-axis, point P_2 must be the same distance from the y-axis as point P_1. This means that the first coordinate for P_2 must be $-\frac{1}{\sqrt{2}}$. The coordinates of P_2 are $(-\frac{1}{\sqrt{2}}, \frac{1}{\sqrt{2}})$.

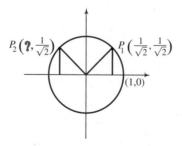

Figure 6.24

Symmetric Points

Now move along the unit circle to the points P_3 and P_4 that are symmetric to P_1 and P_2 about the x-axis, as shown in Figure 6.25. The coordinates of P_3 and P_4 are $(-\frac{1}{\sqrt{2}}, -\frac{1}{\sqrt{2}})$ and $(\frac{1}{\sqrt{2}}, -\frac{1}{\sqrt{2}})$, respectively. The points P_3 and P_4 are associated with central angles of $\frac{5\pi}{4}$ and $\frac{7\pi}{4}$, respectively. Can you tell how the measure of these central angles is obtained? The congruence of the four triangles indicates that we should add $\frac{\pi}{4}$ to π and subtract $\frac{\pi}{4}$ from 2π to obtain $\frac{5\pi}{4}$ and $\frac{7\pi}{4}$, respectively. Note again that all four points P_1, P_2, P_3, and P_4 are symmetric points on the unit circle.

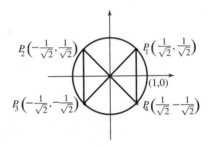

Figure 6.25

EXAMPLE SET C

What are the coordinates of the points symmetric to $P_1(\frac{1}{2}, \frac{\sqrt{3}}{2})$ in Figure 6.26?

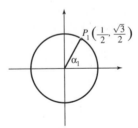

Figure 6.26

Solution:
Let's consider the point symmetric to P_1 in the third quadrant.

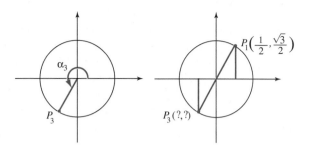

Figure 6.27

From the previous example, we see that the coordinates for all the symmetric points are the same as P_1 except for the sign. By noting what quadrant the symmetric point is in, we can easily determine the signs of its coordinates. For example, P_3 is in the third quadrant, so both its coordinates must be negative, as we can see from Figure 6.27. Thus, the coordinates of P_3 are $(-\frac{1}{2}, -\frac{\sqrt{3}}{2})$.

Using a similar procedure, all the symmetry points are illustrated in Figure 6.28.

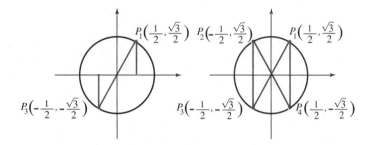

Figure 6.28

Negative Arc Lengths

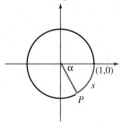

Figure 6.29

Both the measure of the angle and the arc length generated by a point rotated in the clockwise direction are negative. For example, the point P in Figure 6.29, has been rotated one-sixth of a revolution in the clockwise direction, and, therefore, both the measure of α and the associated arc length s are negative.

Let's estimate the coordinates of P. The arc length s on the unit circle is $-\frac{\pi}{3}$ ($\frac{\pi}{3} = \frac{1}{6} \times 2\pi$, where the circumference of the unit circle is 2π. Recall that on a unit circle the measure of a central angle in radians is equal to the arc length subtended by the angle. Thus, $m\angle\alpha = -\frac{\pi}{3}$). We can reasonably assure ourselves that P is in the same position as P_4 in Figure 6.28. We accomplish this by first running the UNITCIR program. Then use the RANGE or WINDOW menu to change the values of Tmin, Tmax, and Tstep to -6.2832, 6.2832, and $.10472$, respectively. Press GRAPH and use the TRACE feature to move to a T value of approximately -1.047 (the approximate value of $-\frac{\pi}{3}$). We note that the coordinates associated with this T value are about $(.5, -.866)$, the approximate values of P_4. Therefore, the coordinates of P are $(\frac{1}{2}, -\frac{\sqrt{3}}{2})$. P clearly has a right triangle associated with it.

Programming Symmetric Points

Any point you choose on the unit circle has three other symmetric points associated with it. You can use the SYMPTS program in Window 6.3, along with the UNITCIR program, to locate these symmetric points.

```
PROGRAM:SYMPTS
:Line(0,0,-X,Y)
:Line(0,0,X,-Y)
:Line(0,0,-X,-Y)
:Pt-Off(X,Y)
```

Window 6.3

The Line(and Pt-Off(symbols are entered using the DRAW menu. The negative sign $(-)$ is on a gray key as you know.

Once you have entered this program, you should modify the UNITCIR program by replacing the statement

.1 →Tstep

with

2π/60 →Tstep

This change makes the arc lengths that you select evenly spaced around exactly one revolution. Clear the graphics window using ClrDraw on the DRAW menu. Now run UNITCIR and then press GRAPH to draw the unit circle. Use TRACE and the arrow keys to move to a point on the unit circle. Now we will run the SYMPTS program. To do this, press PRGM and select the number or name of the program. You are now in the computation window with Prgm followed by the number or name of the SYMPTS program. Press ENTER. You should see three lines drawn (assuming your chosen point is not on an axis) to the symmetric points of your chosen point. If you look carefully at the screen, you will see the point you initially chose on the unit circle represented by a missing point.

EXAMPLE SET D

Use your graphing calculator to check the coordinates of the symmetric points associated with the arc length 0.41887902 (be sure to use $2\pi/60 \rightarrow$ Tstep).

Solution:
Clear the graphing window. Draw the unit circle, use TRACE and the arrow keys to move the cursor to a point whose T value is 0.41887902. Record the coordinates of this point. Now run the SYMPTS program. You will see three lines drawn in the graphing window. Use TRACE and the arrow keys to move to each of the symmetric points and record their coordinates and associated arc lengths. The coordinates should be the same (in absolute value) as the coordinates of the original point. How are the arc lengths of the symmetric points related to the original arc length? You know or can establish that the distance along the unit circle from the x-axis to any of the four points in question is the same. The values are $\pi - 0.41887902$, $\pi + 0.41887902$, and $2\pi - 0.41887902$.

(.91354546, .40673664)

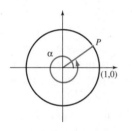

Figure 6.30

Both the angle and the arc generated by a point rotated more than one revolution in the counterclockwise direction are larger than 2π. Figure 6.30 is an example of such a rotation of the point P.

Notice that there is still an associated right triangle. It is difficult to indicate that the point P has moved more than one revolution.

EXERCISES

In problems 1–6, draw a unit circle on a piece of paper and then place the endpoints of the indicated arc length (starting from $(1,0)$) on the unit circle.

1. $t = 2$
2. $t = 4.9$
3. $t = 5$
4. $t = 0.8$
5. $t = -2.4$
6. $t = 3.3$

7. On the unit circles in exercises 1–6, draw in the associated right triangles.

In exercises 8–13, the terminal point of an arc length t is located on the unit circle at the indicated point. Draw a unit circle on a piece of paper and on the unit circle, sketch in the arc t (from $(1,0)$ to the indicated intersection point) and the associated right triangle.

8. $\left(\frac{1}{3}, \frac{\sqrt{8}}{3}\right)$
9. $\left(\frac{1}{4}, \frac{\sqrt{15}}{4}\right)$
10. $\left(-\frac{2}{3}, \frac{\sqrt{5}}{3}\right)$
11. $(.6, .8)$
12. (a, b)
13. $(-c, d)$

14. On the unit circles in exercises 8–13, plot the symmetric points and indicate their coordinates.

In exercises 15–20, the first or second coordinate of a point on the unit circle and its quadrant are given. Compute the other coordinate and plot the point on the unit circle.

15. (0.23,),
    Quadrant I

16. (−0.63,),
    Quadrant III

17. (, 0.81),
    Quadrant II

18. (, −0.38),
    Quadrant IV

19. $(a,)$, Quadrant I   20. $(, d)$, Quadrant I

For problems 21–23, use the UNITCIR and TRIANGLE programs.

21. Make sure that 0.1 is the Tstep value. Draw a unit circle on your graphing calculator. Move the trace cursor to $(1, 0)$. Hold down the right-arrow key and write down a description of the behavior of the arc length T alone. Next, write down a description of the behavior of T and the associated X values as you move the trace cursor from $(1, 0)$ through positive values of T. Finally, describe the behavior of T and the associated Y values as you move the trace cursor from $(1, 0)$ through positive values of T.

22. Change the Tmax value to 12.56 and make sure that 0.1 is the Tstep value. Write down a description of the behavior of the arc length T alone as you move the trace cursor starting at $(1, 0)$ and moving twice around the circle. Next, write down a description of the behavior of T and the associated X values for the same movement of the trace cursor. Finally, describe the behavior of T and the associated Y values as you move the trace cursor from $(1, 0)$.

23. Change the Tmin value to −6.3, the Tmax value to 6.3, and make sure that 0.1 is the Tstep value. Use the unit circle to complete Table 6.1. Describe a pattern for X and Y

Table 6.1		
T	X	Y
0.1	.99500417	.09983342
−0.1	.99500417	−.09983334
0.7	.76484219	.64421769
−0.7	.76484219	−.6442177
1.3	.26749883	.96355819
−1.3	.26749883	−.96355582
1.9	−.3232896	.94630009
−1.9	−.3232896	−.9463001

from this table. If you extended this table, would your pattern still hold?

For problems 24–26, run the UNITCIR program. Then change the Tmin value to 0, the Tmax value to 12.6, and the Tstep value to .10472. Press GRAPH to draw the unit circle and move the trace cursor to the indicated arc length value. Record the X and Y coordinates for the indicated arc length T. Move the trace cursor to $T + 2\pi$. (You may wish to approximate this value with your calculator.) How do the X and Y values of this new arc length compare with these values for T?

24. $T = 0.4189$

25. $T = 2.094$

26. $T = 3.351$

For the next four problems, consider the following situation: The ultramodern Symmetry City was built in the shape of a circle of radius 12 miles, with Town Hall in the center and 12 main boulevards radiating out from this central hub. The six cross streets are equally spaced concentric circles, the last one encircling the city.

27. Mike, a cyclist, starts to ride around the outside edge of the city, traveling at about 10 mph. He stops to rest after 4 hours. How far has he traveled? At this rate, how much longer will it take for him to complete the round?

28. Tyke, being less ambitious, picks one of the inner circles. He knows he can only ride for 3 hours, at an average speed of 13 mph. Which is the longest street he can choose and still complete the whole circle? What is his angular speed (angle described per unit time)?

29. Suppose Moe is traveling at 15 mph on the 5th street from the center, and Joe is traveling at 10 mph on the 3rd street. If they start from the same boulevard (same radius of the circle) at the same time, who has the greater angular speed? What is the difference in their times for completing their respective circles?

30. If Moe and Joe (from the previous exercise) start on the same boulevard and travel in opposite directions on the 4th street (radius 8 miles) where and when will they meet? What angles will each of them have described?

Summary and Projects

Summary

Measure of a Central Angle $m\angle\theta^R = \dfrac{s}{r}$ where θ is a central angle in radians on a circle of radius r intercepting an arc of length s. See Figure 6.31.

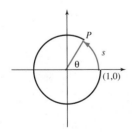

Figure 6.31

Conversion of Angle Measures To convert the measure of an angle from degrees to radians and vice versa, simply recall the measure of a straight angle in degrees and radians. That will spur the thought that $180° = \pi^R$. Appropriate application of this equation allows us to convert either way.

Unit Circle Equation The unit circle is the circle with center at the origin and radius 1. Its equation is $x^2 + y^2 = 1$.

Angle/Arc Connection Every arc on the unit circle has an associated central angle with the same measure (see Figure 6.32). We can see this by recalling that $m\angle\theta^R = \dfrac{s}{r}$ and replacing r with 1.

Figure 6.32

Associated Right Triangle Every point on the unit circle that is not on an axis has a right triangle associated with it. The right triangle is formed by drawing a perpendicular from the point on the circle to the x-axis (see Figure 6.33). The x and y coordinates of each point are the lengths of the adjacent and opposite sides, respectively, of the associated central angle or its supplement.

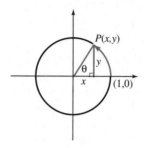

Figure 6.33

Symmetric Points on the Unit Circle   Every point on the unit circle that is not on an axis has three other points on the circle symmetric to it. The coordinates of all four points are numerically the same but their signs are determined by the quadrant they are in.

Projects

1. Use your calculator to draw the top half of the unit circle. Estimate the measure of the angle that the graph of each of the following linear functions makes with the positive x-axis (the initial side). Estimate the tangent of each subtended arc length.

$$f(x) = (4/3)x$$

$$f(x) = (1/2)x$$

$$f(x) = (7/4)x$$

$$f(x) = (11/3)x$$

2. The Game Company is developing a new game. The game involves a circular boundary wall with a point on it from which balls are shot. These balls bounce off the wall at the same angle with which they hit it. The goal of the game is to get the ball back to its start point after between 5 and 10 bounces.

(a) At what angle to the initial radius should the ball be aimed so that it returns to its start point after exactly 5 bounces? After exactly 10 bounces?

(b) Would any angle between these two values you found be suitable for winning the game? Give reasons for your answer.

CHAPTER 7

Circular Functions

7.1   The Sine and Cosine Functions

Introduction

The Unit Circle Approach

The Right Triangle Approach

A Moving Point on the Unit Circle

Special Arc Lengths

Introduction

Periodic phenomena like the movement of the earth around the sun, the action of a piston inside an internal combustion engine, and the movement of a needle in a sewing machine can be modeled by what are called circular or trigonometric functions. We can use models constructed with such functions to make predictions about the future behavior of periodic phenomena and to design mechanisms that have a periodic behavior.

We define the circular functions sine and cosine using points on the unit circle and their associated right triangles. We see that the sine and cosine values are simply new names for the coordinates of the point P on the unit circle with corresponding arc length s.

The Unit Circle Approach

We begin with a unit circle and a central angle, α, in standard position whose measure is s radians. We will define the cosine and sine functions of the arc length s, and denote them by $\cos(s)$ and $\sin(s)$. In Figure 7.1, $\angle\alpha$ is in standard position and has measure s. Its terminal side intersects the unit circle at $P(x, y)$, where x and y are the lengths of the horizontal and vertical sides, respectively, of the associated right triangle.

It is very common to abbreviate the names of the sine and cosine functions. The sine function is abbreviated **sin** while the cosine function is abbreviated **cos**. Thus, sine(s) will normally be written $\sin(s)$ and cosine(s) will normally be written $\cos(s)$. We will use $\cos(\alpha)$ and $\cos(s)$ interchangeably even though, technically, $\cos(m\angle\alpha)$ is correct.

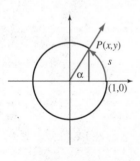

Figure 7.1

The Cosine and Sine Functions

The Cosine and Sine
Functions

Consider $\angle\alpha$ in standard position with measure s. Let $P(x, y)$ be the point on the unit circle $x^2 + y^2 = 1$ through which the terminal side of $\angle\alpha$ passes. Let the right triangle associated with point $P(x, y)$, angle α, and arc length s be the triangle formed by drawing a perpendicular from P to the x-axis. We define the **cosine** of s as:

$$\text{cosine}(s) = x \qquad\qquad \text{cosine}(\alpha) = \frac{\text{side adjacent to } \angle\alpha}{\text{hypotenuse}} = \frac{x}{1}$$

and the **sine** of s as:

$$\text{sine}(s) = y \qquad\qquad \text{sine}(\alpha) = \frac{\text{side opposite } \angle\alpha}{\text{hypotenuse}} = \frac{y}{1}$$

Notice that on the unit circle (radius 1), the $\frac{\text{side adjacent}}{\text{hypotenuse}}$ is $\frac{x}{1}$ and $\frac{\text{side opposite}}{\text{hypotenuse}}$ is $\frac{y}{1}$. Thus, we can look at these functions either in terms of the coordinates of the point P on the unit circle or in terms of the ratios of the appropriate sides of the associated right triangles. In addition, we know that the radian measure of an angle in standard position is the same as the length of the arc it intercepts on the unit circle. The connection here is the relation between a central angle, the arc it intercepts, and the radius of the circle $\left(m\angle\alpha = \dfrac{s}{r}\right)$.

For every arc length s there is an $\angle\alpha$ with s as its radian measure. Thus, the cosine and sine functions are defined for every real number s.

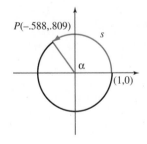

$P(-.588, .809)$

Figure 7.2

EXAMPLE SET A

1. Determine the sine and cosine function values for the angle whose terminal side passes through the point P on the unit circle (Figure 7.2).

   Solution:

   The x-coordinate of a point P on the unit circle and the cosine of the angle with measure s in standard position and whose terminal side passes through P are the same. Therefore, $\cos(s) \approx -0.588$. Likewise, the y-coordinate of the point and the sine of the arc are the same so that $\sin(s) \approx 0.809$.

The Right Triangle Approach

In geometry, the sine and cosine functions are sometimes defined for angles in terms of ratios of sides of right triangles. Similarity of triangles can be used to modify these geometric definitions to correspond to the definitions given above. These functions, when defined in terms of right triangles, are used to measure the sides and angles of triangles, thus the name trigonometric (triangle measurement) functions.

In Figure 7.3, recall that $\cos(\alpha)$ is 4/5 (the side adjacent over the hypotenuse) and $\sin(\alpha)$ is 3/5 (the side opposite over the hypotenuse).

In Figure 7.4, we superimpose the unit circle so that angle α is in standard position. By similarity, the associated right triangle has sides of length 4/5 and 3/5 since

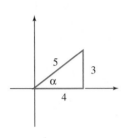

Figure 7.3

$$\frac{4}{5} = \frac{x}{1} \quad \text{and} \quad \frac{3}{5} = \frac{y}{1}$$

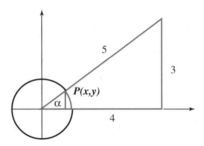

Figure 7.4

Thus, the coordinates of P are $\left(\frac{4}{5}, \frac{3}{5}\right)$ and since the cosine function is defined as the first coordinate of P, the cosine value is $\cos(\alpha) = \frac{4}{5}$. Similarly, $\sin(\alpha) = \frac{3}{5}$, the second coordinate of P. We can also convince ourselves that the point $\left(\frac{4}{5}, \frac{3}{5}\right)$ lies on the unit circle by noting that

$$\left(\frac{4}{5}\right)^2 + \left(\frac{3}{5}\right)^2 = 1$$

EXAMPLE SET B

1. Determine the sine and cosine function values for arc lengths s that begin at $(1,0)$ and end at the indicated points on the unit circle.

   a. $(0.866, 0.500)$

   b. $(-0.407, 0.914)$

   c. $(0.978, -0.208)$

Solution:

Since the first coordinate of a point on the unit circle represents the function value of the cosine and the second coordinate the function value of the sine, we write the appropriate coordinate by observation.

a. $\cos(s) = 0.866$, $\sin(s) = 0.5$

b. $\cos(s) = -0.407$, $\sin(s) = 0.914$

c. $\cos(s) = 0.978$, $\sin(s) = -0.208$

2. Determine the sine and cosine function values for arc lengths s from $(1, 0)$ to P with associated right triangles as indicated in Figure 7.5.

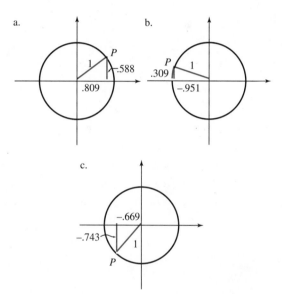

Figure 7.5

Solution:

Since we know the sides of the associated triangles, we can write the coordinates of the point P in each case and then determine the sine and cosine function values.

a. The coordinates of P are $(0.809, 0.588)$ so that $\cos(s) = 0.809$ and $\sin(s) = 0.588$.

b. In this case, P has coordinates $(-0.951, 0.309)$ so that $\cos(s) = -0.951$ and $\sin(s) = 0.309$.

c. The coordinates of P are $(-0.669, -0.743)$ so that $\cos(s) = -0.669$ and $\sin(s) = -0.743$.

3. Find the sine and cosine of angle α in each part of Figure 7.6. Also show that $\sin^2(\alpha) + \cos^2(\alpha) = 1$.

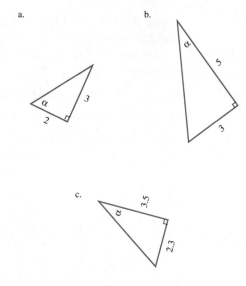

Figure 7.6

Solution:

First we label the opposite and adjacent sides of the right triangle with respect to angle α as in Figure 7.7.

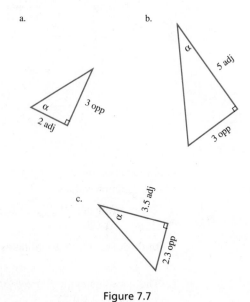

Figure 7.7

Next, we determine the hypotenuse in each case and use the definition.

a. The hypotenuse is $\sqrt{2^2 + 3^2} = \sqrt{13}$. Then,

$$\sin(\alpha) = \frac{\text{opposite}}{\text{hypotenuse}} = \frac{3}{\sqrt{13}}$$

Also,

$$\cos(\alpha) = \frac{\text{adjacent}}{\text{hypotenuse}} = \frac{2}{\sqrt{13}}$$

In addition,

$$\sin^2(\alpha) + \cos^2(\alpha) = \left(\frac{3}{\sqrt{13}}\right)^2 + \left(\frac{2}{\sqrt{13}}\right)^2 = \frac{9}{13} + \frac{4}{13} = 1$$

b. The hypotenuse is $\sqrt{3^2 + 5^2} = \sqrt{34}$. Then,

$$\sin(\alpha) = \frac{\text{opposite}}{\text{hypotenuse}} = \frac{3}{\sqrt{34}}$$

Also,

$$\cos(\alpha) = \frac{\text{adjacent}}{\text{hypotenuse}} = \frac{5}{\sqrt{34}}$$

In addition,

$$\sin^2(\alpha) + \cos^2(\alpha) = \left(\frac{3}{\sqrt{34}}\right)^2 + \left(\frac{5}{\sqrt{34}}\right)^2 = \frac{9}{34} + \frac{25}{34} = 1$$

c. The hypotenuse is $\sqrt{(2.3)^2 + (3.5)^2} = \sqrt{17.54} \approx 4.2$. Then,

$$\sin(\alpha) = \frac{\text{opposite}}{\text{hypotenuse}} = \frac{2.3}{4.2} \approx 0.55$$

Also,

$$\cos(\alpha) = \frac{\text{adjacent}}{\text{hypotenuse}} = \frac{3.5}{4.2} \approx 0.83$$

In addition,

$$\sin^2(\alpha) + \cos^2(\alpha) = (0.55)^2 + (0.83)^2 = 0.3025 + 0.6889 \approx 0.99$$

which is approximately 1.

4. Given that $\sin(s) = \dfrac{2}{3}$ for a first quadrant arc length s, find $\cos(s)$.

Solution:

We know that $\sin(s)$ is the y-coordinate of a point on the unit circle and that $\cos(s)$ is the x-coordinate. Thus, we are considering the point $\left(?, \dfrac{2}{3}\right)$

on the unit circle. Recalling that $x^2 + y^2 = 1$ for points on the unit circle, we can substitute $\frac{2}{3}$ for x and solve for y.

$$\left(\frac{2}{3}\right)^2 + y^2 = 1$$

$$y^2 = 1 - \frac{4}{9}$$

$$y^2 = \frac{5}{9}$$

$$y = \frac{\pm\sqrt{5}}{3}$$

We choose the positive value since s is in the first quadrant. You should check these values using the UNITCIR program.

A Moving Point on the Unit Circle

Arc lengths on the unit circle are determined by moving the point P in either the positive or negative direction (counterclockwise or clockwise, respectively). Since the point P can revolve indefinitely along the unit circle, infinitely many arc lengths can be determined. These arc lengths wind around the circumference of the unit circle and their lengths represent domain values of the sine and cosine functions. The maximum for the function values of sine and cosine is 1 since the coordinates of any point on the unit circle can be at most 1. Likewise, the minimum value for the sine and cosine is -1. In Figure 7.8, the moving point P can go up and down no more than 1 unit and left and right no more than 1 unit. Therefore, the range of the sine and cosine functions is the interval $[-1, 1]$.

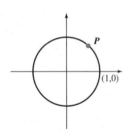

Figure 7.8

The Domain and Range of the Sine and Cosine Functions

The Domain and Range of the Sine and Cosine Functions

The domain of the sine function is the set of all real numbers.
The range of the sine function $= [-1, 1]$, so that $-1 \leq \sin(s) \leq 1$.
The domain of the cosine function is the set of all real numbers.
The range of the cosine function $= [-1, 1]$, so that $-1 \leq \cos(s) \leq 1$.

Special Arc Lengths

There are three special arc lengths (angles) that play an important part in the study of trigonometry. They are $\frac{\pi}{6}$, $\frac{\pi}{4}$, and $\frac{\pi}{3}$. It is useful to have the function values of sine and cosine for these angles as part of your resource bank.

EXAMPLE SET C

1. Determine the function values $\sin(\frac{\pi}{4})$ and $\cos(\frac{\pi}{4})$.

   Solution:

   The solution to this problem is straightforward, provided we can make a connection to similar information in the previous chapter. Recall that we determined the coordinates on the unit circle to be $(\frac{1}{\sqrt{2}}, \frac{1}{\sqrt{2}})$ when the arc length ending at this point and starting at $(1,0)$ is $\frac{\pi}{4}$. As soon as we make this connection, we can write down the solution, which is $\sin(\frac{\pi}{4}) = \frac{1}{\sqrt{2}}$ and $\cos(\frac{\pi}{4}) = \frac{1}{\sqrt{2}}$.

2. Determine the function values $\sin(\frac{\pi}{6})$, $\cos(\frac{\pi}{6})$, $\sin(\frac{\pi}{3})$, and $\cos(\frac{\pi}{3})$.

   Solution:

   Though there is not a similar connection for this problem with the previous chapter, we can use our unit circle resources. Place an angle with measure $\frac{\pi}{6}$ in standard position. Now we know that $\frac{\pi}{6} = 30°$ so that the associated triangle is a 30-60-90 right triangle as shown in Figure 7.9.

   We again need to tap our resource bank from geometry to recall that the length of the side opposite the 30° angle in a 30-60-90 right triangle is half the hypotenuse. Therefore, the side opposite the 30° angle (the y-coordinate of P on the unit circle) is $\frac{1}{2}$ since the hypotenuse is 1. We can either recall similar information about the side opposite the 60° angle in a 30-60-90 right triangle or use the Pythagorean Theorem to obtain $\frac{\sqrt{3}}{2}$ for the side opposite the 60° angle (the x-coordinate of P on the unit circle). The coordinates of P are, therefore, $(\frac{\sqrt{3}}{2}, \frac{1}{2})$. We can now easily write down half of the solutions provided our resources contain a clear recall of the sine and cosine functions.

   $$\sin\left(\frac{\pi}{6}\right) = \frac{1}{2} \quad \text{and} \quad \cos\left(\frac{\pi}{6}\right) = \frac{\sqrt{3}}{2}$$

   To obtain the remaining solutions, we can use the same figure and apply our knowledge regarding right triangles. We obtain $\sin\left(\frac{\pi}{3}\right) = \frac{\sqrt{3}}{2}$ using the fact that $\frac{\sqrt{3}}{2}$ is the length of the side opposite angle $\frac{\pi}{3}$ (60°) and the hypotenuse is 1. Using similar reasoning, $\cos\left(\frac{\pi}{3}\right) = \frac{1}{2}$.

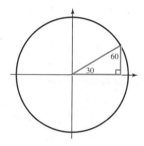

Figure 7.9

3. Find the following function values.

   a. $\sin\left(\dfrac{2\pi}{3}\right)$

   b. $\cos\left(\dfrac{5\pi}{4}\right)$

   c. $\cos\left(\dfrac{11\pi}{6}\right)$

   d. $\sin\left(-\dfrac{\pi}{4}\right)$

Solution:

a. Recall the discussion of symmetric points on the unit circle in Section 6.4. The point with arc length $\dfrac{2\pi}{3}$ is symmetric to the point with arc length $\dfrac{\pi}{3}$ whose coordinates are $\left(\dfrac{1}{2},\dfrac{\sqrt{3}}{2}\right)$. Thus, the coordinates associated with the arc $\dfrac{2\pi}{3}$ are $\left(\dfrac{-1}{2},\dfrac{\sqrt{3}}{2}\right)$. Therefore, $\sin\left(\dfrac{2\pi}{3}\right)=\dfrac{\sqrt{3}}{2}$.

b. The point with arc length $\dfrac{5\pi}{4}$ is symmetric to the point with arc length $\dfrac{\pi}{4}$ whose coordinates are $\left(\dfrac{1}{\sqrt{2}},\dfrac{1}{\sqrt{2}}\right)$. Thus, the coordinates associated with the arc $\dfrac{5\pi}{4}$ are $\left(-\dfrac{1}{\sqrt{2}},-\dfrac{1}{\sqrt{2}}\right)$. Therefore, $\sin\left(\dfrac{5\pi}{4}\right)=-\dfrac{1}{\sqrt{2}}$.

c. The point with arc length $\dfrac{11\pi}{6}$ is symmetric to the point with arc length $\dfrac{\pi}{6}$ whose coordinates are $\left(\dfrac{\sqrt{3}}{2},\dfrac{1}{2}\right)$. Thus, the coordinates associated with the arc $\dfrac{11\pi}{6}$ are $\left(\dfrac{\sqrt{3}}{2},-\dfrac{1}{2}\right)$. Therefore, $\sin\left(\dfrac{11\pi}{6}\right)=-\dfrac{1}{2}$.

d. The point with arc length $-\dfrac{\pi}{4}$ is symmetric to the point with arc length $\dfrac{\pi}{4}$ whose coordinates are $\left(\dfrac{1}{\sqrt{2}},\dfrac{1}{\sqrt{2}}\right)$. Thus, the coordinates associated with the arc $-\dfrac{\pi}{4}$ are $\left(\dfrac{1}{\sqrt{2}},-\dfrac{1}{\sqrt{2}}\right)$. Therefore, $\sin\left(-\dfrac{\pi}{4}\right)=-\dfrac{1}{\sqrt{2}}$.

4. Find an arc length whose sin is 2.

Solution:
From our knowledge of the unit circle, the largest sine value is 1. Thus, there is no real number whose sine is 2. This should remind us that the range of the sine (and cosine) function is $[-1,1]$.

EXERCISES

In exercises 1–10, determine the sine and cosine function values for the angle whose terminal side passes through the point P on the unit circle.

1. $P(\frac{1}{2}, \frac{\sqrt{3}}{2})$

2. $P(-\frac{1}{2}, -\frac{\sqrt{3}}{2})$

3. $P(\frac{1}{\sqrt{2}}, -\frac{1}{\sqrt{2}})$

4. $P(-\frac{\sqrt{2}}{2}, \frac{\sqrt{2}}{2})$

5. $P(0, 1)$

6. $P(-1, 0)$

7. $P(0.452, 0.892)$

8. $P(-0.763, 0.646)$

9. $P(0.115, -0.993)$

10. $P(-0.582, -0.813)$

In exercises 11–18, determine the sine and cosine function values for the given arc length s, which begins at $(1, 0)$ and ends at the indicated points on the unit circle. Be sure to draw a unit circle and place the given information on it.

11. $s = 0.98$, $P(0.557, 0.830)$

12. $s = 3.60$, $P(-0.898, -0.441)$

13. $s = 5.90$, $P(0.928, -0.372)$

14. $s = -4.47$, $P(-0.244, 0.970)$

15. $s = -5.62$, $P(0.786, 0.618)$

16. $s = 11.3$, $P(0.344, -0.939)$

17. $s = 15.7$, $P(-0.999, -0.037)$

18. $s = 0.56$, $P(0.847, 0.532)$

In exercises 19–22, determine the length of the hypotenuse, specify the length of the adjacent and opposite sides of θ, and then find the sine and cosine of angle θ in each of the following diagrams. Also show that $\sin^2(\theta) + \cos^2(\theta) = 1$.

19.

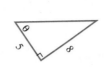

20.

21.

22.

23. Specify the domain and range of the sine function.

24. Specify the domain and range of the cosine function.

In exercises 25–28, we are given the function value of either the sine or cosine function and information that can determine what quadrant the angle (arc length) is in. Determine the function value of the other function and check your answers using UNITCIR.

25. $\sin(s) = \dfrac{2}{5}$ and s ends in the second quadrant.

26. $\cos(s) = \dfrac{3}{4}$ and $\sin(s) < 0$.

27. $\cos(\alpha) = -\dfrac{1}{3}$ and α is in the third quadrant.

28. $\sin(\alpha) = -\dfrac{2}{7}$ and $\cos(\alpha) > 0$.

29. For what angle θ (arc length) is $\sin(\theta) = \cos(\theta)$?

30. Using the fact that in a right triangle if one of the acute angles is θ, then the other acute angle is $\frac{\pi}{2} - \theta$, show that $\sin(\theta) = \cos\left(\frac{\pi}{2} - \theta\right)$.

31. In exercise 30, you showed that in a right triangle for angle θ, $\sin(\theta) = \cos\left(\frac{\pi}{2} - \theta\right)$. Use this fact to compute (a) $\sin^2(38°) + \sin^2(52°)$, (b) $\sin^2(17°) + \sin^2(73°)$, (c) $\sin^2(10°) + \sin^2(20°) + \sin^2(30°) + \sin^2(40°) + \sin^2(50°) + \sin^2(60°) + \sin^2(70°) + \sin^2(80°)$.

32. A regular polygon on n sides that is inscribed inside a circle of radius r has a perimeter P that is given by the trigonometric function

$$P(n, r) = 2nr \sin\left(\frac{180°}{n}\right)$$

Find the perimeter of a regular polygon inscribed in a circle of radius 5 if the polygon has (a) 3 sides, (b) 5 sides, (c) 10 sides, (d) 100 sides, (e) 180 sides, and (f) 360 sides.

33. Prove that for all integers n,
$$\sin\left(\frac{(2n + 1)\pi}{2}\right) = (-1)^n.$$

34. Prove that for all integers n,
$$\cos\left(\frac{\pi}{3}(1 + 3n)\right) = \frac{(-1)^n}{2}.$$

35. Use the unit circle to show that (a) $\sin(\pi - \theta) = \sin(\theta)$, (b) $\sin(\pi + \theta) = -\sin(\theta)$, (c) $\cos(\pi - \theta) = -\cos(\theta)$, and (d) $\cos(\pi + \theta) = -\cos(\theta)$.

36. If θ is in radians, then $\sin(\theta)$ can be approximated by the polynomial function $s(t) = \frac{1}{120}t^5 - \frac{1}{6}t^3 + t$, where the maximum error of the approximation is $\frac{1}{5040}t^7$. Use your calculator to approximate $\sin\left(\frac{\pi}{3}\right)$, and determine the maximum error of the approximation.

37. If θ is in radians, then $\cos(\theta)$ can be approximated by the polynomial function $c(t) = 1 - \frac{1}{2}t^2 + \frac{1}{24}t^4$, where the maximum

error of the approximation is $\frac{1}{720}t^6$. Use your calculator to approximate $\cos\left(\frac{\pi}{3}\right)$, and determine the maximum error of the approximation.

38. The work W (in foot-pounds) required to move an object weighing w pounds up a d-foot-long ramp that is inclined $\theta°$ from the horizontal is given by the trignometric function

$$W(w, d, \theta) = wd \sin(\theta)$$

How much work is required to move a 75-pound object up a 12-foot-long ramp that is inclined 7° from the horizontal?

39. When an object is projected from the ground at an angle of $\theta°$ with an initial velocity of v feet per second, then, if the resistance due to the air is negligible, the maximum height h attained by the object is given by

$$h(v, \theta) = \frac{v^2 \sin^2(\theta)}{64} \text{ feet}$$

and the distance from the point of origin attained by the object is given by

$$d(v, \theta) = \frac{v^2 \sin(2\theta)}{32} \text{ feet}$$

Suppose that a soccer ball is kicked off the ground at a 35° angle with an initial velocity of 60 feet per second. Find the maximum height and distance attained by the ball.

40. Chutima wants to find the angle of elevation of the sun at 12:00 noon on the day of summer solstice. She has a 9-foot-long vertical flagpole in her back yard. At exactly 12:00 noon she marks the end of the flagpole's shadow. Then she measures the distance from the mark to the base of the pole. How will she use this information to find the angle of the elevation of the sun? See the accompanying figure.

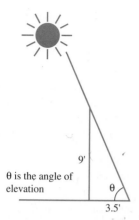

θ is the angle of elevation

way that his eye level (which is 5'8" high) is exactly in the middle. If he is standing five feet behind the window, is he able to do his job effectively? If not, where should he stand?

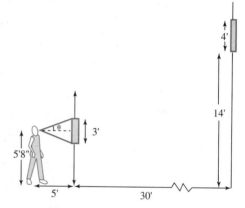

41. A detective on a stakeout is keeping his eyes on the window of the house exactly 30 feet across the street from the apartment building he is standing in. The top of the house's second story window is at a height of 18 feet above the street level and the bottom is 14 feet above the street level. The detective is inside the apartment building, standing in front of a three-foot-high window placed in such a

(Hint: Find the viewing angle. Check whether this viewing angle spreads to cover the area to be observed. Picture is not to scale).

7.2   Graphs of the Standard Sine and Cosine Functions

Introduction

The Standard Sine Function Graph

Using the TI Graphing Calculator

Behavior of the Sine Function

The Standard Cosine Function Graph

Behavior of the Cosine Function

Introduction

In this section, we study the behavior of the standard sine and cosine functions by constructing their graphs in the rectangular coordinate system. We can accomplish this by connecting our knowledge of their behavior on the unit circle to the standard rectangular coordinate system.

The Standard Sine Function Graph

Recall from the previous section that the domain of the sine and cosine functions is the set of all real numbers. This domain could be pictured by visualizing a point P traveling around the unit circle and noting the associated arc length. We realize that the point can travel in either direction and can revolve indefinitely. As the point P moves around the circle, it generates associated arcs from $(1, 0)$ to wherever the point P is on the unit circle. The associated arc lengths are the domain values for both the cosine and sine functions.

Using the TI Graphing Calculator

The graphing calculator will give an excellent first look at the graphs of the sine and cosine functions. Be sure your calculator is in Function mode. Then access the Y= menu and enter SIN X using the SIN key. Press the ZOOM key and then 7 to use the Trig graphing option. The sine graph displayed looks like Window 7.1.   Now press the RANGE or WINDOW key and look at the

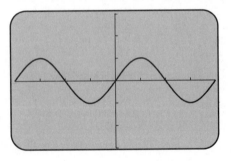

Window 7.1

graphing window values the Trig option provides. Provided the calculator is in radian mode, the x values are from -2π to 2π (approximately) and the y values are from -3 to 3.

We can also graph the sine function in the traditional way. We first develop a table of points, plot the points, and then draw a smooth curve through those points. Table 7.1 shows the values of the sine function for multiples of $\frac{\pi}{6}$ between 0 and 2π.

Table 7.1	
x	$\sin(x)$
0	0
$\dfrac{1\pi}{6}$	$\dfrac{1}{2}$
$\dfrac{2\pi}{6}$	$\dfrac{\sqrt{3}}{2}$
$\dfrac{3\pi}{6}$	1
$\dfrac{4\pi}{6}$	$\dfrac{\sqrt{3}}{2}$
$\dfrac{5\pi}{6}$	$\dfrac{1}{2}$
$\dfrac{6\pi}{6}$	0
$\dfrac{7\pi}{6}$	$-\dfrac{1}{2}$
$\dfrac{8\pi}{6}$	$-\dfrac{\sqrt{3}}{2}$
$\dfrac{9\pi}{6}$	-1
$\dfrac{10\pi}{6}$	$-\dfrac{\sqrt{3}}{2}$
$\dfrac{11\pi}{6}$	$-\dfrac{1}{2}$
$\dfrac{12\pi}{6}$	0

As we can see from Table 7.1, the range values have a maximum of 1 and a minimum of -1. The range values for $x > \frac{3\pi}{6}$ can be obtained using symmetric points on the unit circle. In Figure 7.10, the numbers around the circle are the arc lengths s from the point $(1,0)$ and the lengths of the vertical line segments are the corresponding sine values.

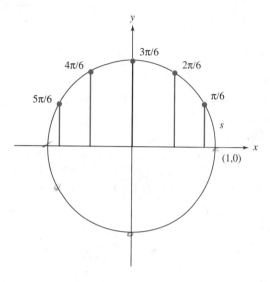

Figure 7.10

In Figure 7.11, we have plotted $(s, \sin(s))$ by moving horizontally s units and vertically $\sin(s)$ units for values of s between 0 and 2π. The points are the ordered pairs in Table 7.1. In Figure 7.10, s is the distance on the unit circle starting at $(1,0)$ and ending at a point P and $\sin(s)$ is the distance from the x-axis to P. These two quite different representations are very important because they are each useful in understanding the behavior of the sine function. The graph of the sine function is conventionally given as the graph in Figure 7.11.

From a dynamic point of view, the point moving around the unit circle gives a visual image of the sine function increasing and decreasing between and including the values 1 and -1 in a cyclical (periodic) manner. The point moving along the standard graph of the sine function in Figure 7.11 gives this same information in a conventional manner represented on the Cartesian coordinate plane where s is moving linearly along the horizontal axis. You should make the connection between these two representations of the sine function by running the DYNSINE program shown in Window 7.2 on your TI graphing calculator. This program shows the relationship between the unit circle and the conventional sine function graph. In watching the simultaneous graphing of the unit circle and the sine function, notice that the y-coordinate

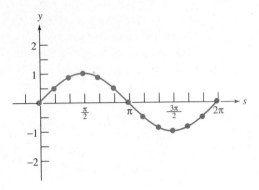

Figure 7.11

of the moving point on the unit circle is the same as the y-coordinate of the sine function. Use the TRACE feature to convince yourself that the arc length (T value) is the same as the x-coordinate of the conventional sine function graph.

```
PROGRAM:DYNSINE
:Param
:Simul
:"cos T-1"->X₁T
:"sin T"->Y₁T
:"T"->X₂T
:"sin T"->Y₂T
:2π/60->Tstep
```

Window 7.2

The remainder of the program lines are given below

$-2 \rightarrow$ Xmin
$2\pi \rightarrow$ Xmax
$-2.6 \rightarrow$ Ymin
$2.6 \rightarrow$ Ymax

You should press Graph after you execute the DYNSINE program. As a point P moves beyond one revolution on the unit circle, the values of the sine function start to repeat since P arrives at the same points it traversed in the first revolution. Therefore, the values of the sine function repeat every 2π

units since the circumference of a unit circle is 2π. The period of a cyclical function is the smallest positive number for which the function values begin to repeat. The period of the sine function, therefore, is 2π.

$$\sin(s \pm 2\pi) = \sin(s)$$

EXAMPLE SET A

1. Find the values of

   a. $\sin(\frac{13\pi}{6})$ and

   b. $\sin(\frac{-2\pi}{3})$

   This is showing values remaining after 2π has been extracted

   Solution:

   a. $\sin(\frac{13\pi}{6}) = \sin(\frac{\pi}{6} + \frac{12\pi}{6}) = \sin(\frac{\pi}{6} + 2\pi) = \sin(\frac{\pi}{6}) = \frac{1}{2}$

   b. $\sin(\frac{-2\pi}{3}) = \sin(\frac{4\pi}{3} - \frac{6\pi}{3}) = \sin(\frac{4\pi}{3} - 2\pi) = \sin(\frac{4\pi}{3}) = -\frac{\sqrt{3}}{2}$

Recall that the value of the sine function for $\frac{4\pi}{3}$ can be obtained from its value at $\frac{\pi}{3}$ using symmetric points.

We have used only multiples of $\frac{\pi}{6}$ in generating the graph of the sine function. We also know the exact values for multiples of $\frac{\pi}{4}$. We can obtain additional approximate values for the sine function using our calculator or a table of sine values. You should be sure in either case that you are using radian measure. The graph of the sine function using multiples of $\frac{\pi}{4}$ should look something like the graph in Figure 7.12.

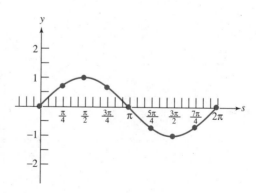

Figure 7.12

We should be able to determine the domain and range of a function from its graph. The standard graph of the sine function for two cycles (0 to 4π)

is given in Figure 7.13. This graph, along with the periodicity of the sine function, tells us that the graph continues on indefinitely in both directions. This reaffirms the fact that this function's domain is all real numbers as we saw in Section 7.1. The range values of the sine function can also be read from this graph and are seen to vary between -1 and 1, inclusive. This agrees with the range values given in Section 7.1.

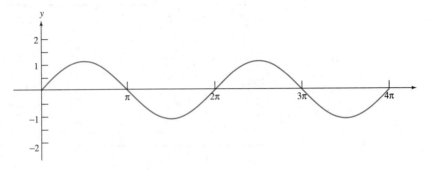

Figure 7.13

Behavior of the Sine Function

The unit circle is divided into four equal parts (corresponding to the four quadrants of the rectangular coordinate axes) by the points whose distances along the unit circle from $(1, 0)$ are 0, $\frac{\pi}{2}$, π, and $\frac{3\pi}{2}$, respectively. These values on the number line also divide the interval from 0 to 2π into four equal parts as well. Figure 7.14 shows the relationship between these arc

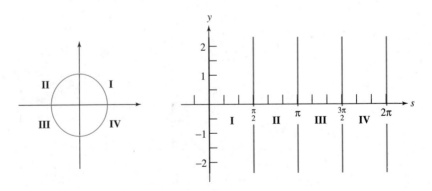

Figure 7.14

lengths and points on the number line. When the sine function is graphed on this interval, it is divided into four equal parts at these points, as shown in Figure 7.15.

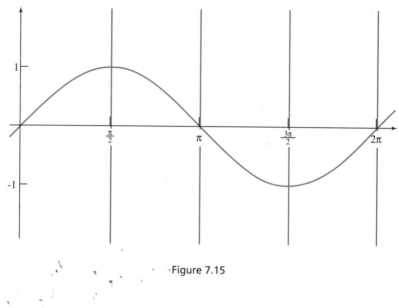

·Figure 7.15

The behavior of the sine function can be described in terms of these four subintervals of $[0, 2\pi]$. The sine function is positive and increasing in the first subinterval (quadrant), positive and decreasing in the second, negative and decreasing in the third, and negative and increasing in the fourth, going from left to right on the number line (counterclockwise on the unit circle). It is important to note that the sine function is 0 at the endpoints and the midpoint of the interval $[0, 2\pi]$ and 1 or -1 at $\frac{\pi}{2}$ and $\frac{3\pi}{2}$, respectively.

The Standard Cosine Function Graph

The cosine function can be graphed in a similar way. It is still very helpful to get a quick, accurate graph on the graphing calculator. Following the same idea as for the sine function above, we obtain the following graph for the cosine function. If you compare this graph with that of the standard sine function, you should be able to see that if you shift the cosine graph to the right by $\pi/2$ units, it becomes the sine graph.

Table 7.2

x	$\cos(x)$
0	1
$\dfrac{1\pi}{6}$	$\dfrac{\sqrt{3}}{2}$
$\dfrac{2\pi}{6}$	$\dfrac{1}{2}$
$\dfrac{3\pi}{6}$	0
$\dfrac{4\pi}{6}$	$-\dfrac{1}{2}$
$\dfrac{5\pi}{6}$	$-\dfrac{\sqrt{3}}{2}$
$\dfrac{6\pi}{6}$	-1
$\dfrac{7\pi}{6}$	$-\dfrac{\sqrt{3}}{2}$
$\dfrac{8\pi}{6}$	$-\dfrac{1}{2}$
$\dfrac{9\pi}{6}$	0
$\dfrac{10\pi}{6}$	$\dfrac{1}{2}$
$\dfrac{11\pi}{6}$	$\dfrac{\sqrt{3}}{2}$
$\dfrac{12\pi}{6}$	1

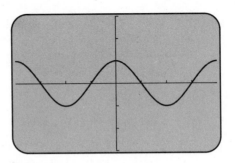

Window 7.3

Table 7.2 shows the values of the cosine function.

Behavior of the Cosine Function

We know that the arc lengths 0, $\frac{\pi}{2}$, π, $\frac{3\pi}{2}$, and 2π divide the circle into four equal parts and, as with the sine function graph, produce four equal parts on the number line as well. These dividing point values are indicated in Figure 7.16. The behavior of the cosine function can be described in terms of these four

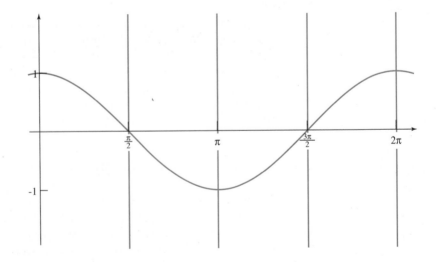

Figure 7.16

subintervals of $[0, 2\pi]$. The cosine function is positive and decreasing in the first subinterval (quadrant), negative and decreasing in the second, negative and increasing in the third, and positive and increasing in the fourth, going

Table 7.3

x	$\cos(x)$
0	1
$\dfrac{1\pi}{4}$	$\dfrac{1}{\sqrt{2}}$
$\dfrac{2\pi}{4}$	0
$\dfrac{3\pi}{4}$	$-\dfrac{1}{\sqrt{2}}$
$\dfrac{4\pi}{4}$	-1
$\dfrac{5\pi}{4}$	$-\dfrac{1}{\sqrt{2}}$
$\dfrac{6\pi}{4}$	0
$\dfrac{7\pi}{4}$	$\dfrac{1}{\sqrt{2}}$
$\dfrac{8\pi}{4}$	1

from left to right on the number line (counterclockwise on the unit circle). It is important to note that the cosine function is 1 at the endpoints and -1 at the midpoint of the interval $[0, 2\pi]$ and 0 at $\frac{\pi}{2}$ and $\frac{3\pi}{2}$, the other two dividing points.

EXAMPLE SET B

Make a table of values for the cosine function for multiples of $\frac{\pi}{4}$ on the interval $[0, 2\pi]$. On graph paper, graph the cosine function on this interval from these values.

Solution:
From our work on the unit circle, we know the values of the cosine function and its symmetric points at $\frac{\pi}{4}$. We begin with $\cos\left(\dfrac{\pi}{4}\right) = \dfrac{1}{\sqrt{2}}$. The rest of the table is easy to complete using the notion of symmetric points. See Table 7.3.

Notice that there are fewer points in Table 7.3 than in Table 7.2, but the general shape of the graph is the same. We can get a more accurate graph of the function by including both multiples of $\frac{\pi}{6}$ and multiples of $\frac{\pi}{4}$. Additional accuracy can be obtained by graphing even more points with the help of our calculator.

EXERCISES

In exercises 1–4, graph the function on $[0, 2\pi]$.

1. $f(x) = -\sin(x)$
2. $f(x) = -\cos(x)$

3. $f(x) = \sin(x) + 1$
4. $f(x) = \cos(x) + 2$

In exercises 5–8, use graph paper to graph the following functions on $[-2\pi, 4\pi]$. Be sure to label tick marks on the horizontal axis in terms of π.

5. $f(x) = \sin(x)$
6. $f(x) = \cos(x)$

7. $f(x) = -1 + \sin(x)$
8. $f(x) = 2 + \cos(x)$

9. Describe the behavior of the function $\sin(x)$ on $\left[\frac{\pi}{2}, \frac{3\pi}{2}\right]$.

10. Describe the behavior of the function $\cos(x)$ on $\left[-\frac{\pi}{2}, \frac{\pi}{2}\right]$.

In exercises 11–18, find the *exact* function value.

11. $\sin\left(\frac{15\pi}{6}\right)$
12. $\sin\left(\frac{15\pi}{4}\right)$

13. $\sin\left(-\frac{8\pi}{3}\right)$
14. $\cos\left(-\frac{10\pi}{3}\right)$

15. $\cos\left(\frac{17\pi}{6}\right)$
16. $\cos\left(\frac{11\pi}{4}\right)$

17. $\sin\left(\frac{38\pi}{3}\right)$
18. $\sin\left(\frac{45\pi}{6}\right)$

In exercises 19–26, determine the quadrant of the angle (given in degrees) and then find a three-decimal approximation of the value of the function.

19. $\sin(472°)$

20. $\sin(1420°)$

21. $\sin(-195°)$

22. $\cos(561°)$

23. $\cos(-82°)$

24. $\cos(2290°)$

25. $\cos(580°)$

26. $\sin(-1645°)$

In exercises 27–34, determine the quadrant of the angle (given in radians) and then find a three-decimal approximation of the value of the function.

27. $\sin(8.87)$

28. $\sin(16.4)$

29. $\sin(-10.2)$

30. $\cos(-8.64)$

31. $\cos(-13.5)$

32. $\cos(42.3)$

33. $\sin(71.6)$

34. $\sin(-19.8)$

7.3   The Tangent Function

Introduction

The Tangent Function

The Standard Tangent Function Graph

Introduction

The tangent function is another important function. It is related to the sine and cosine functions. In this section, we will explore that relationship as well as construct the graph of the tangent function.

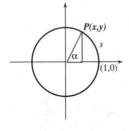

Figure 7.17

The Tangent Function

Consider $\angle \alpha$ in standard position (see Figure 7.17). Its terminal side intersects the unit circle at $P(x, y)$, where x and y are the lengths of the horizontal and vertical sides, respectively, of the associated right triangle.

The tangent function is normally abbreviated **tan**. We will use $\tan(\alpha)$ and $\tan(s)$ interchangeably even though, technically, $\tan(m\angle\alpha)$ is correct.

**The Tangent
Function**

The Tangent Function

Consider $\angle \alpha$ in standard position with measure s. Let $P(x, y)$ be the point on the unit circle $x^2 + y^2 = 1$ through which the terminal side of $\angle \alpha$ passes. Let the right triangle associated with point $P(x, y)$, angle α, and arc length s be the triangle formed by drawing a perpendicular from P to the x-axis. We define the **tangent** of s as:

$$\text{tangent}(s) = \frac{y}{x}, \qquad\qquad \text{tangent}(\alpha) = \frac{\text{side opposite}}{\text{side adjacent}}$$

We should note that, as a ratio, the tangent function will be defined only when the denominator, x, is not zero.

Notice that this definition is the same as the $\dfrac{opposite}{adjacent}$ function we studied in Section 6.2. It is defined in terms of the coordinates of the point P on the unit circle, namely, the ratio of the y value over the x value. It is important to observe that the tangent function is the ratio of the sine function over the cosine function. That is, since $\sin(s) = y$ and $\cos(s) = x$, then

$$\frac{\sin(s)}{\cos(s)} = \frac{y}{x} = \tan(s)$$

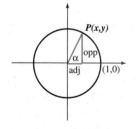

Figure 7.18

This can also be verified for the alternate definition using ratios of sides of the associated right triangle as seen in Figure 7.18. You should perform this verification yourself. You might very well ask why we introduce another function that is made up of two functions already defined. While we could use the ratio of sine over cosine, a new name is useful since this function occurs frequently.

The Standard Tangent Function Graph

We can obtain the graph of the tangent function using the graphing calculator. This tool gives us a quick, accurate picture of the tangent function. Use the Y= menu and the TAN key along with the Trig graphing option under ZOOM to graph the tangent function. In the graphing window, we see that the tangent function seems to have vertical asymptotes every so often (see Window 7.4).

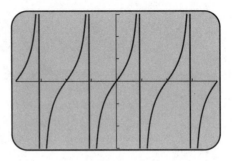

Window 7.4

We should recall, however, that the lines that appear vertical are really points connected by our graphing calculator. If we use the Trig option on the ZOOM menu when graphing, we can determine that the asymptotes occur at $\pm\frac{\pi}{2}$ and $\pm\frac{3\pi}{2}$, the points at which $x = \cos(s) = 0$.

The graph gives a great deal of additional information as well. Looking carefully at the graph, we note that the tangent function repeats itself every π units and so its period is π. Interestingly, the tangent function repeats in half the distance that the cosine (or sine) function repeats. The fact that the period of the tangent function is π allows us to determine where all of its asymptotes occur. A compact way of writing the locations of the asymptotes is:

$$s = \frac{\pi}{2} + k\pi$$

where k is an integer. You should verify that the asymptotes restricted to the window above correspond to the values $k = 0, 1, -1$, and -2.

The graph shows us the total behavior of the tangent function. We note that, moving left to right, the function is increasing on every interval that does not contain an asymptote. The graph also shows us that the function values of the tangent are positive in the first quadrant, negative in the second, positive in the third, and negative in the fourth quadrant. As in the previous section, we need to keep in mind that the four quadrants we are familiar with in the unit circle representation become the four strips on our graph described by the intervals 0 to $\frac{\pi}{2}$, $\frac{\pi}{2}$ to π, π to $\frac{3\pi}{2}$, and $\frac{3\pi}{2}$ to 2π.

Observe that the tangent function is undefined at the same angles (arc lengths) at which the asymptotes occur.

To repeat, the tangent function is periodic with period π. This can be compactly written as:

$$\tan(s + k\pi) = \tan(s)$$

where k is any integer. It is very useful for us to get a mental image of the tangent function. When we bring up a picture of its graph in our mind's eye, we can easily describe its behavior. We can reach this state by using our graphing calculator frequently to display the tangent function graph and reviewing all aspects of its behavior each time we do.

Producing a table of tangent function values would not provide any great advantage for graphing purposes in this case. However, it would provide a chance to see what the tangent values are for the same special arc lengths (angles) we used for sine and cosine.

EXAMPLE SET A

1. Make a table of tangent function values for multiples of $\frac{\pi}{6}$ and $\frac{\pi}{4}$ for one period.

   Solution:
   We begin by noting that since $\sin(0) = 0$ and $\cos(0) = 1$, then

   $$\tan(0) = \frac{\sin(0)}{\cos(0)} = \frac{0}{1} = 0$$

   Next, we recall that $\sin(\frac{\pi}{6}) = \frac{1}{2}$ and $\cos(\frac{\pi}{6}) = \frac{\sqrt{3}}{2}$. This gives:

   $$\tan\left(\frac{\pi}{6}\right) = \frac{\sin(\frac{\pi}{6})}{\cos(\frac{\pi}{6})} = \frac{1/2}{\sqrt{3}/2} = \frac{1}{\sqrt{3}}$$

   We can continue in this manner to obtain $\tan(\frac{\pi}{4}) = 1$ and $\tan(\frac{\pi}{3}) = \sqrt{3}$. We can fill out the rest of the table to π (the period of the tangent function) by taking the appropriate ratios of sine and cosine already computed before. You should do this now and write down a table for the tangent function similar to the ones for sine and cosine.

2. Determine the function values of tangent for the following arc lengths (angles):

   a. $\frac{3\pi}{4}$

   b. $\frac{5\pi}{6}$

   c. $\frac{4\pi}{3}$

   d. $\frac{3\pi}{2}$

   Solution:

   a. $\tan(\frac{3\pi}{4}) = \dfrac{\sin(\frac{3\pi}{4})}{\cos(\frac{3\pi}{4})} = \dfrac{\sin(\frac{\pi}{4})}{-\cos(\frac{\pi}{4})} = \dfrac{1/\sqrt{2}}{-1/\sqrt{2}} = -1.$

   b. $\tan(\frac{5\pi}{6}) = \dfrac{\sin(\frac{5\pi}{6})}{\cos(\frac{5\pi}{6})} = \dfrac{\sin(\frac{\pi}{6})}{-\cos(\frac{\pi}{6})} = \dfrac{1/2}{-\sqrt{3}/2} = -1/\sqrt{3}.$

   c. $\tan(\frac{4\pi}{3}) = \dfrac{\sin(\frac{4\pi}{3})}{\cos(\frac{4\pi}{3})} = \dfrac{-\sin(\frac{\pi}{3})}{-\cos(\frac{\pi}{3})} = \dfrac{-1\sqrt{3}/2}{-1/2} = \sqrt{3}.$

d. $\tan(\frac{3\pi}{2}) = \tan(\frac{\pi}{2} + \pi) = \tan(\frac{\pi}{2}) = \dfrac{\sin(\frac{\pi}{2})}{\cos(\frac{\pi}{2})} = \dfrac{1}{0}$, which is undefined.

Notice that $\tan(\frac{4\pi}{3}) = \tan(\frac{\pi}{3} + \pi) = \tan(\frac{\pi}{3})$, giving further verification that the period of the tangent function is π.

EXAMPLE SET B

Determine the angles where the tangent function is undefined.

Solution:
The tangent function is defined by

$$\tan(s) = \frac{\sin(s)}{\cos(s)}$$

We know that the tangent function is not defined when $\cos(s)$ is zero. From our knowledge of the cosine function, we know that the cosine is zero when s is an odd multiple of $\dfrac{\pi}{2}$ or $s = k\dfrac{\pi}{2}$ where k is an odd integer. Thus, the solution is the set $\{s | s \neq k\dfrac{\pi}{2}\}$ where k is an odd integer, or the set $\{\ldots, -\dfrac{3\pi}{2}, -\dfrac{\pi}{2}, \dfrac{\pi}{2}, \dfrac{3\pi}{2}, \ldots\}$.

EXERCISES

In exercises 1–4, use your graphing calculator to plot the following functions. Describe the difference you observe between each of them and the standard tangent function $y = \tan(s)$. You may need to adjust the graphing window to distinguish the behavior of the entire function.

1. $f(s) = \tan(2s)$

2. $g(s) = \tan(\frac{s}{3})$

3. $h(s) = \tan(s + \frac{\pi}{2})$

4. $k(s) = \frac{1}{3}\tan(s)$

In exercises 5–12, determine *exact* function values for the tangent of the given arc lengths (angles).

5. $\frac{2\pi}{3}$

6. $\frac{5\pi}{3}$

7. $\frac{11\pi}{4}$

8. $\frac{13\pi}{4}$

9. $\frac{11\pi}{6}$

10. $\frac{7\pi}{6}$

11. 3π

12. $\frac{5\pi}{2}$

In exercises 13–22, approximate, to three decimals, the tangent of the given angles measured in radians.

13. 1.52

14. 1.56

15. 1.57

16. 4.71

17. 2.38

18. 3.04

19. 10.2

20. 12.5

21. 52.7

22. 53.5

In exercises 23–32, approximate, to three decimals, the tangent of the given angles measured in degrees. *calculator*

23. 51.3

24. 78.2

25. 90.1

26. 132

27. 211

28. 264

29. 281

30. 345

31. 1250

32. 975

In exercises 33–38, determine whether the tangent is positive or negative for the given radian measured angles.

33. 1.58

34. 4.76 → *Quadrant IV tangent is neg.*

35. 12.9

36. 13.1

37. −3.52

38. 27.4

In exercises 39–44, determine whether the tangent is positive or negative for the given degree measured angles.

39. 460

40. 545

41. 272

42. −930

43. 2000

44. −112

45. On learning that a picture should be hung at eye level, Agatha, an art lover, hangs her picture so that its center is five feet from the floor (which also happens to be her eye level height). Also, she knows that for the best viewing one should stand at a distance from the wall that will make the viewing angle between 40° and 45°. What is her viewing angle if she is standing (a) 6 feet from the wall, (b) 7 feet from the wall and (c) 5 feet from the wall? Which position(s), supposedly, will give her the best viewing angle?

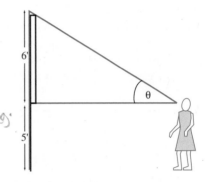

46. A satellite is orbiting the earth at a height of 300 km. Find the viewing angle θ. Assume that the earth is a sphere of radius 6400 km. How much of the surface of the earth can be seen by the satellite at each instant? Answer this question by finding ϕ. The diagram below is not to scale.

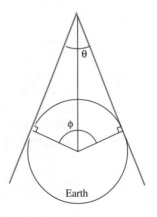

Earth

The equation for the path of a projectile thrown at an angle θ with a velocity v is given by $y = x \cdot \tan \theta - \dfrac{16}{v^2 \cos^2 \theta} \cdot x^2$ where x is the horizontal distance in feet from the point of projection and y is the height from the same point. Use this equation for exercises 47-49.

47. If the initial velocity is 300 ft/sec and the angle of projection is 40°, what is the range of the function? How far does the projectile travel horizontally? (Hint: Set the RANGE or WINDOW variables Xmax, Ymax to large values in order to get an idea of the graph. Then use TRACE.)

48. If the angle of projection is made 45° with the same initial velocity, what is the maximum vertical height and the horizontal distance?

49. Change the angle to 50°. What are the numbers this time? Do you see any patterns? What conclusion can you make about the horizontal distance? Check your conclusions with more input angles.

7.4 The Secant, Cosecant, and Cotangent Functions

Introduction

The Secant, Cosecant, and Cotangent Functions

The Graph and Behavior of the Secant Function

The Graphs and Behavior of the Cosecant and Cotangent Functions

Introduction

We have defined three functions so far using a point on the unit circle and its associated right triangle. But there are also other ratios of the sides of the triangle. In this section, we define and name these other ratios, see how they are related to the sine and cosine functions, use our calculators to construct their graphs, and use the graphs to describe their behavior.

The Secant, Cosecant, and Cotangent Functions

We defined the tangent function to be the ratio of the side opposite over the side adjacent an angle in a right triangle. What about the ratio of the side adjacent over the side opposite? So far, we have considered the ratios $\dfrac{\text{side opposite}}{\text{hypotenuse}}$,

$\dfrac{\text{side adjacent}}{\text{hypotenuse}}$, and $\dfrac{\text{side opposite}}{\text{side adjacent}}$, which we have named the sine, cosine, and tangent functions, respectively. Now we will consider the reciprocals of these ratios, $\dfrac{\text{hypotenuse}}{\text{side adjacent}}$, $\dfrac{\text{hypotenuse}}{\text{side opposite}}$, and $\dfrac{\text{side adjacent}}{\text{side opposite}}$. We will name these ratios secant, cosecant, and cotangent, respectively, and define them next.

The Secant, Cosecant, and Cotangent Functions

The Secant, Cosecant, and Cotangent Functions

Consider $\angle\alpha$ in standard position with measure s. Let $P(x, y)$ be the point on the unit circle $x^2 + y^2 = 1$ through which the terminal side of $\angle\alpha$ passes. Let the right triangle associated with point $P(x, y)$, angle α, and arc length s be the triangle formed by drawing a perpendicular from P to the x-axis. We define the secant of s as:

$$\textbf{secant}(s) = \frac{1}{x}, \quad x \neq 0 \qquad \textbf{secant}(\alpha) = \frac{\text{hypotenuse}}{\text{side adjacent}}$$

the cosecant of s as:

$$\textbf{cosecant}(s) = \frac{1}{y}, \quad y \neq 0 \qquad \textbf{cosecant}(\alpha) = \frac{\text{hypotenuse}}{\text{side opposite}}$$

and the cotangent of s as:

$$\textbf{cotangent}(s) = \frac{x}{y}, \quad y \neq 0 \qquad \textbf{cotangent}(\alpha) = \frac{\text{side adjacent}}{\text{side opposite}}$$

Shortened names are used for these new functions as well. The secant, cosecant, and cotagent functions are expressed **sec**, **csc**, and **cot**, respectively.

We again note that these three definitions are just the reciprocals of the three previously defined functions. They are abbreviated and related in the following way.

$$\sec(s) = \frac{1}{\cos(s)}, \quad s \neq \frac{\pi}{2} + \pi k, k \in I$$

$$\csc(s) = \frac{1}{\sin(s)}, \quad s \neq \pi k, k \in I$$

$$\cot(s) = \frac{1}{\tan(s)}, \quad s \neq \pi k, k \in I$$

Note also the connection we can make for the cotangent function because of its relation to the tangent function, namely:

$$\cot(s) = \frac{1}{\tan(s)} = \frac{1}{\dfrac{\sin(s)}{\cos(s)}} = \frac{\cos(s)}{\sin(s)}$$

These three new functions are defined in terms of the three previous functions and this connection can be very helpful in recalling their behavior.

The Graph and Behavior of the Secant Function

A picture of these new functions is again an excellent start to understanding their behavior. There are no keys on our graphing calculator corresponding to these new functions so we will need to use their relationships to the other trigonometric functions to graph them. For example, we can display the graph of $\sec(s)$ by entering $\dfrac{1}{\cos(s)}$ in the Y= menu and then graphing it using the Trig option under the ZOOM menu. Do that now and spend a few moments studying the behavior of the graph. The graph should look something like this:

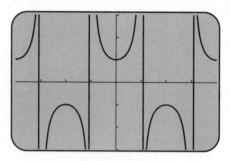

Window 7.5

Again, the apparent vertical lines look like the actual asymptotes but we need to remember they are not, and closer observation reveals that they do not cross the x-axis at the actual asymptote locations. For example, the first asymptote to the right of 0 should occur at $\frac{\pi}{2}$ (where $\cos(s) = 0$), but you can see that the apparent vertical line is just to the right of $\frac{\pi}{2}$. The location of the asymptotes of the secant function occur at the same places as the tangent function. This is clear when we recall that both of these functions, when written in terms of sines and cosines, have $\cos(s)$ in the denominator and, therefore, become undefined at the same values. Other aspects of the behavior of the secant function can be seen by studying its graph. The domain of the secant function is the set of all real numbers excluding those that cause the $\cos(s)$ to become zero. That means the replacement values for s cannot be $\frac{\pi}{2} + \pi k, k \in I$. Further observation indicates that the range of the secant function excludes values between -1 and 1. That is, the range of the secant function is $\sec(s) \leq -1$ or $\sec(s) \geq 1$.

The Graphs and Behavior of the Cosecant and Cotangent Functions

We should also study the graphs of the cosecant and cotangent functions in order to get a picture of their behavior. Their graphs can be displayed by using their definitions in terms of sines and cosines. We can study the general shape of these graphs and the location of their asymptotes. Their domains and ranges should also be observed and placed into our resource bank. Do not side step this useful means of learning the details of these functions; rather, consider using the graphing calculator whenever you need to recall one or more details. The figure on the left of Window 7.6 shows the graph of the cosecant functions, and the figure on the right of Window 7.6 the graph of the cotangent functions.

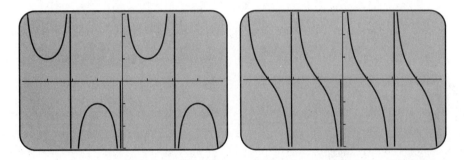

Window 7.6

The domain of both the cosecant and cotangent functions are the same since their definitions have the sine function in the denominator. Hence, they both become undefined when $\sin(s)$ is zero, which happens when $s = \pi k$ where $k \in I$. These values describe the location of the asymptotes and are excluded in the description of the domain. The range of the cotangent function is all real numbers and that of the cosecant function is the same as the secant function, namely, $\csc(s) \leq -1$ or $\csc(s) \geq 1$. This information gains much more significance by looking at their corresponding graphs. Notice also that the graph of the cosecant function would become the secant graph if you shift it $\pi/2$ units to the left.

EXAMPLE SET A

Fill in Table 7.4.

Table 7.4			
x	$\sec(x)$	$\csc(x)$	$\cot(x)$
0			
$\dfrac{1\pi}{6}$			
$\dfrac{2\pi}{6}$			
$\dfrac{3\pi}{6}$			
$\dfrac{4\pi}{6}$			
$\dfrac{5\pi}{6}$			
$\dfrac{6\pi}{6}$			
$\dfrac{7\pi}{6}$			
$\dfrac{8\pi}{6}$			
$\dfrac{9\pi}{6}$			
$\dfrac{10\pi}{6}$			
$\dfrac{11\pi}{6}$			
$\dfrac{12\pi}{6}$			

Solution:

We can fill in this table using the definitions of these functions. For example, we know that $\sin\left(\frac{\pi}{6}\right) = \frac{1}{2}$, so that $\csc\left(\frac{\pi}{6}\right) = \dfrac{1}{\sin\left(\frac{\pi}{6}\right)} = \dfrac{1}{\frac{1}{2}} = 2$. Likewise, we know that $\cos(\frac{\pi}{4}) = \frac{1}{\sqrt{2}}$, so that $\sec(\frac{\pi}{4}) = \frac{1}{\cos\left(\frac{\pi}{4}\right)} = \frac{1}{\frac{1}{\sqrt{2}}} = \sqrt{2}$. To obtain

the value of $\cot(\frac{\pi}{3})$, we could use the fact that $\cot(\frac{\pi}{3}) = \frac{\cos(\frac{\pi}{3})}{\sin(\frac{\pi}{3})} = \frac{\frac{1}{2}}{\frac{\sqrt{3}}{2}} = \frac{1}{\sqrt{3}}$.

We can determine the remaining values by recalling the values of the sine and cosine functions and using the definition. You should fill in the rest of the table now.

EXERCISES

In exercises 1–12, determine the *exact* value.

1. $\sec(\frac{\pi}{3})$

2. $\sec(\frac{-\pi}{4})$

3. $\csc(\frac{\pi}{3})$

4. $\csc(\frac{5\pi}{3})$

5. $\cot(\frac{\pi}{6})$

6. $\cot(\frac{-3\pi}{4})$

7. $\sec(\frac{23\pi}{6})$

8. $\sec(\frac{7\pi}{6})$

9. $\csc(\frac{-5\pi}{3})$

10. $\csc(\frac{11\pi}{4})$

11. $\cot(\frac{7\pi}{2})$

12. $\cot(5\pi)$

In exercises 13–18, determine approximate values to three decimal places using your calculator. The angles are in radians.

13. $\sec(22.35)$

14. $\sec(-6.15)$

15. $\csc(1002)$

16. $\csc(18.5)$

17. $\cot(8.62)$

18. $\cot(-4.38)$

In exercises 19–24, determine approximate values to three decimal places using your calculator.

19. $\sec(22.35°)$

20. $\sec(605°)$

21. $\csc(-37.2°)$

22. $\csc(114°)$

23. $\cot(822°)$

24. $\cot(-132°)$

In exercises 25–28, give the domain and range of each of the following functions.

25. $f(x) = 2\csc(x)$

26. $g(x) = \sec(2x)$

27. $h(x) = \cot(2x)$

28. $k(x) = 5\sec(3x)$

7.5   **Characteristics of Sinusoidal Functions**

Introduction

The Amplitude of Sinusoidal Functions

The Period of Sinusoidal Functions

The Phase Shift of Sinusoidal Functions

Introduction

Functions of the form $f(t) = A \sin(Bt + C)$ or $f(t) = A \cos(Bt + C)$ occur regularly in science and engineering applications of mathematics. Graphs of these functions are often called sinusoidal or harmonic curves and are closely related to the standard sine and cosine function graphs. We consider several of their uses in the next section; but, in the meantime, we study their behavior to prepare us for these coming applications.

The Amplitude of Sinusoidal Functions

We can look at how various values of A, B, and C transform the standard sine and cosine graphs by considering the effect of each of these constants individually. Let's consider the effect of A first since it is the easiest to work with. We can get a fast look at A's effect by graphing $y = \sin(x)$, $y = 3\sin(x)$, and $y = (1/2)\sin(x)$ with the ZOOM-Trig option in the same graphics window. We observe that the shape is the same for all three graphs, but the highest value is 3 for the function that has a 3 multiplier and $\frac{1}{2}$ for the function that has a $\frac{1}{2}$ multiplier (see Window 7.7). Use the TRACE feature at several values of x to compare the values of the three functions using the up- and down-arrow keys. In general, the absolute value of the multiplier of the standard sine or cosine function is the highest value reached by the function and is called the amplitude of the function. You should be able to connect the effect of A on the sine and cosine functions to the expansion and contraction effects you studied in Section 2.4.

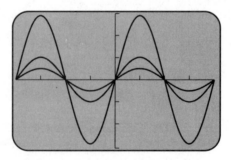

Window 7.7

Next we define the amplitude more formally.

Amplitude of the Sinusoidal Functions

Amplitude of the
Sinusoidal Functions

The absolute value of A for the function $y = A\sin(Bt + C)$ or $y = A\cos(Bt + C)$ is the **amplitude** of these functions and represents the maximum height of the function.

The transformation occurs because each function value of $\sin(t)$ is multiplied by A. The amplitude concept is enhanced by studying Table 7.5, which contrasts the function values of $\sin(t)$ and $3\sin(t)$ at several domain values.

Table 7.5

x	$\sin(x)$	$3\sin(x)$
$\dfrac{\pi}{6}$	$\dfrac{1}{2}$	$\dfrac{3}{2}$
$\dfrac{\pi}{3}$	$\dfrac{\sqrt{3}}{2}$	$\dfrac{3\sqrt{3}}{2}$
$\dfrac{\pi}{2}$	1	3

EXAMPLE SET A

1. Determine the amplitude and sketch the graph of $y = 4\cos(t)$.

   Solution:
   The amplitude is 4, the absolute value of A. We can sketch the graph using our calculators. We can also sketch the graph by plotting the points $(0, 4)$, $(2\pi, 4)$, and $(\pi, -4)$, recalling that the maximum for the cosine function occurs at the endpoints of one cycle and its minimum at the midpoint. We then plot the points $\left(\frac{\pi}{2}, 0\right)$ and $\left(\frac{3\pi}{2}, 0\right)$, recalling that the cosine function is zero between the maximum points and the minimum point. Using these points as guides, and knowing the shape of the cosine function, we can sketch a fairly accurate graph of $y = 4\cos(t)$. This (mental) graphing method is very effective in strengthening our understanding of sinusoidal functions. This stengthening, in turn, builds our resource bank of mathematical concepts.

2. Determine the amplitude and sketch the graph of of $y = -3\sin(t)$.

Solution:
The amplitude is 3, the absolute value of A. We need to think about the effect of the negative multiplier. We should recall that it reflects the curve $y = 3\sin(t)$ about the x-axis. If this does not come to mind, you can plot $y = 3\sin(t)$ and $y = -3\sin(t)$ in the same window on your graphing calculator to observe this effect. You can now use the same approach as in the previous example in sketching the graph by hand. We plot the points $(0,0)$, $(2\pi, 0)$, and $(\pi, 0)$ by recalling that the sine function has zeros at its endpoints and midpoint, respectively, for one cycle. We then plot the points $\left(\frac{\pi}{2}, 3\right)$ and $\left(\frac{3\pi}{2}, -3\right)$ recalling that the sine function attains its maximum and minimum values between its zeros. Using these points as guides and knowing to reflect the shape of the sine function about the x-axis, we can sketch a fairly accurate graph of $y = -3\sin(t)$.

The Period of Sinusoidal Functions

Next, we consider the effect that B has on the standard sine function. We use a similar initial approach as we did for the amplitude. Graph the following functions in the same window: $y = \sin(t)$, $y = \sin(2t)$, $y = \sin(\frac{1}{2}t)$ and watch as they are being constructed. It may not be as easy to tell this time, but we should eventually be able to see that all three functions have the same shape. The difference is in how frequently the pattern is repeated.

Observing the graphs as they are drawn indicates that $\sin(2t)$ repeats more frequently than the standard sine graph. Further study of the graphs shows that $\sin(2t)$ repeats every π units, which is half the period of the standard sine function. The pattern of $y = \sin(\frac{1}{2}t)$ repeats every 4π units, which is twice the period of the standard sine function. Although the pattern has not started to repeat, we can observe that a full sine wave has been drawn and is about to be repeated. We can use the TRACE feature and the right-arrow key to scroll the graph through additional repetitions of the pattern for $\sin(\frac{1}{2}t)$.

Can we draw any conclusions from our observations of these graphs? The fact that a 2 multiplier of t causes the standard sine period to be halved and a $\frac{1}{2}$ multiplier doubles the period of the standard sine function could lead us to conclude that we can obtain the period of $y = \sin(Bt)$ if we divide the period of the standard sine function 2π by the multiplier of t. That is, the period of $y = \sin(Bt)$ is $\frac{2\pi}{B}$. Through this development of the information about B, we are almost ready to give the formal definition of the period of the sinusoidal functions. We need only decide how we will incorporate a negative multiplier. The absolute value again will cover this case.

Period of the Sinusoidal Curves

Period of the
Sinusoidal Curves

Period of the Sinusoidal Curves

The **period** (or repeat distance) of $y = A\sin(Bt + C)$ or $y = A\cos(Bt + C)$ is defined to be $\dfrac{2\pi}{|B|}$.

EXAMPLE SET B

1. Determine the amplitude and period of $y = 6\sin(3t)$.

   Solution:

   The amplitude is 6, the absolute value of A. The period is the period of the standard sine function divided by 3 the absolute value of B, and, thus, is $\frac{2\pi}{3}$.

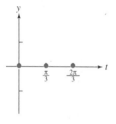

Figure 7.19

We might recall that the standard sine function is zero at the endpoints and midpoint of one cycle and reaches its maximum and minimum halfway between these values. The general sine and cosine (sinusoidal) curves have exactly the same features. We can incorporate these facts in graphing the above example using pencil and paper. We first plot the points where t is 0, $\frac{2\pi}{3}$, and $\frac{\pi}{3}$, the endpoints and midpoint of one cycle, respectively, as shown in Figure 7.19.

Then we graph the points at $\frac{\pi}{6}$, and $\frac{3\pi}{6}$, the maximum and minimum values halfway between the points above. Finally, we would draw one cycle of the sine function through these points using our mental image of the shape of the standard sine function as a guide, as shown in Figure 7.20.

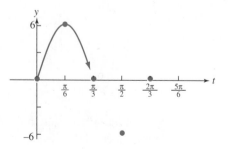

Figure 7.20

We would use the same method for the general cosine function except we would need to use the attributes of the standard cosine function instead. Recall that the standard cosine function attains its maximum value at the endpoints of its period and its minimum value at the midpoint. It passes through 0 halfway between these points. These guides are extremely useful when plotting sinusoidal curves using pencil and paper.

EXAMPLE SET C

1. Graph $y = 6\cos(3t)$ using pencil and paper and check your results with your calculator.

   Solution:

   We note that the amplitude is 6 and the period is $\dfrac{2\pi}{3}$. We can plot the first set of guide points $(0, 6)$, $\left(\frac{2\pi}{3}, 6\right)$ and $\left(\frac{\pi}{3}, -6\right)$, which are the endpoints and midpoint of one period, respectively. We plot the last guide points $\left(\frac{\pi}{6}, 0\right)$ and $\left(\frac{3\pi}{6}, 0\right)$, the zeros between the above points. Finally, we use the shape of the standard cosine function to sketch in the graph of $y = 6\cos(3t)$ as shown in Figure 7.21.

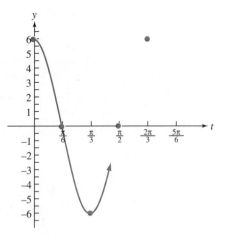

Figure 7.21

We can repeat this pattern in either direction if we desire a more extensive graph. In checking this function with our graphing calculator, we may need to change the range to accommodate the amplitude. Perhaps a choice of $[-10, 10]$ for the range might be satisfactory. In addition, $[-\pi, \pi]$ for the domain could improve the display even further. As always, the dimensions we choose for our graphing window can significantly influence which details we can easily observe.

The Phase Shift of Sinusoidal Functions

We now consider the effect that C has on the standard sine and cosine graphs for the sinusoidal functions. We will use our graphing calculators to get a quick and accurate look at this effect. To this end, we graph $y = \sin(t)$ and $y = \sin(t + \frac{\pi}{6})$ on the same window. Here we consider only one transformed sinusoidal graph against the standard sine graph to avoid screen clutter. We again select Trig from the ZOOM menu and observe the graphs as they are being constructed. We can see that $y = \sin(t + \frac{\pi}{6})$ is the same as $y = \sin(t)$ except that it is shifted to the left, which we expect from our knowledge of horizontal translations of functions. The graphs are displayed in Window 7.8.

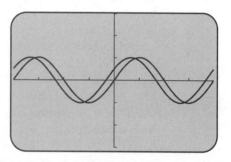

Window 7.8

We might ask how far has it shifted? You can reinforce your answer by using the TRACE feature to move to the curve $y = \sin(t + \frac{\pi}{6})$ and then use the arrow keys to move to the zero of the function nearest the origin. The approximate x value is -0.463. This is not a familar number. Let's look at the decimal value of C to see if it has some connection with this unfamiliar number. So we calculate the decimal value of $\frac{\pi}{6}$, which is approximately 0.524. This doesn't seem very close to 0.463. However, if we had taken note of what the y value was when $x = -0.463$, we would have seen that it was not zero but approximately 0.06. This should encourage us to use the ZOOM feature to get a more accurate value of the zero of this function. We do obtain a value -0.524 if we zoom enough times. Our graphical approach leads us to believe that the shift is indeed $\frac{\pi}{6}$ to the *left*. However, before drawing a conclusion, let's contrast the graphs of $y = \sin(t)$ and $y = \sin(2t + \frac{\pi}{6})$. We will need to change the shifted function and return to the Trig option of the

ZOOM menu. The difference in periodicity should be no surprise since $B = 2$ in the shifted curve. Our display looks like Window 7.9.

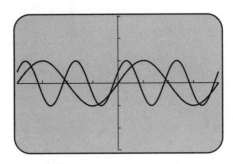

Window 7.9

We again use the TRACE feature to find approximately how far the shift is by moving to the zero nearest the origin of the function $y = \sin(2t + \frac{\pi}{6})$. The approximations seem very rough, so that zooming in is necessary to obtain a reasonable approximation. It becomes apparent very quickly that the zero occurs at about -0.26, which is not $\frac{\pi}{6}$. However, by study and observation, we might note that this value is about half of -0.524, so that the 2 multiplier of t must be involved somehow. We need to pay attention to the form of the sinusoidal curve, therefore.

Rather than focusing our attention on the difference in the two forms $y = A\sin(Bt + C)$ and $y = A\sin[B(t + C)]$, let us approach this difficulty with a

Finding the Period and Phase Shift

method that may be easier to recall. We know that the period of the standard sine function $y = \sin(t)$ is 2π. In other words, the argument t must change by 2π units before the function starts to repeat. Applying this logic to either $y = A\sin(Bt + C)$ or $y = A\sin[B(t + C)]$, the argument $Bt + C$, or $B(t + C)$ must move a distance of 2π before the function repeats. We can handle this nicely by solving a compound inequality as follows for $Bt + C$:

$$0 \leq Bt + C \leq 2\pi$$

$$-C \leq Bt \leq 2\pi - C$$

$$\frac{-C}{B} \leq t \leq \frac{2\pi - C}{B} \quad \text{where} \quad B > 0$$

The form $B(t + C)$ is handled in the same way. The advantage of this method is that it is relatively easy to remember and, more importantly, the shift and period are available as soon as we simplify the compound inequality. **The period is the difference between the outside values of the compound inequality.**

The left side of the inequality specifies the number of units the curve is shifted horizontally. It is common to call $-\dfrac{C}{B}$ the phase shift.

In those cases where $B < 0$, we can use the notion that $\sin(-t) = -\sin(t)$ and $\cos(-t) = \cos(t)$ to obtain a positive value B, the coefficient of t.

EXAMPLE SET D

1. Determine the phase shift and period of $y = \sin(2t + \frac{\pi}{6})$.

   Solution:

   We begin by solving the compound inequality.

   $$0 \le 2t + \frac{\pi}{6} \le 2\pi$$

   $$-\frac{\pi}{6} \le 2t \le 2\pi - \frac{\pi}{6}$$

   $$\frac{-\frac{\pi}{6}}{2} \le t \le \frac{\frac{11\pi}{6}}{2}$$

   $$-\frac{\pi}{12} \le t \le \frac{11\pi}{12}$$

   We can now calculate the period by the difference $\frac{11\pi}{12} - (-\frac{\pi}{12}) = \pi$. The phase shift is $\frac{-\pi}{12}$. We can now graph this function using paper and pencil by plotting points at $(\frac{-\pi}{12},0),(\frac{11\pi}{12},0)$, and $(\frac{5\pi}{12},0)$ for the endpoints and midpoint of one cycle, respectively. Since the amplitude is 1, we plot the maximum and minimum values at $(\frac{\pi}{6},1)$ and $(\frac{2\pi}{3},1)$. With these guide points, we can sketch one cycle of $y = \sin(2t + \frac{\pi}{6})$ and then repeat the pattern if necessary. See Figure 7.22.

   Although we can use our calculator to quickly graph these complicated functions, we plot them by hand to enhance our understanding of the effect of the constants A, B, and C. We need this understanding to effectively use sinusoidal curves as tools in working with real-world phenomena.

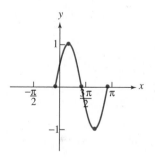

Figure 7.22

2. Determine the amplitude, phase shift, and period of $y = 2\cos(3t - \frac{\pi}{2})$ and sketch its graph.

   Solution:

   It should be clear that the amplitude is 2, since it is the multiplier of $\cos(3t - \frac{\pi}{2})$. Solving the double inequality will give us the period and phase shift.

   $$0 \le 3t - \frac{\pi}{2} \le 2\pi$$

$$\frac{\pi}{2} \le 3t \le 2\pi + \frac{\pi}{2}$$

$$\frac{\frac{\pi}{2}}{3} \le t \le \frac{\frac{5\pi}{2}}{3}$$

$$\frac{\pi}{6} \le t \le \frac{5\pi}{6}$$

The period is the difference $\frac{5\pi}{6} - (\frac{\pi}{6}) = \frac{2\pi}{3}$. The phase shift is $\frac{\pi}{6}$. Graphing this function by hand requires only five guide points. We plot $(\frac{\pi}{6}, 2),(\frac{5\pi}{6}, 2)$, and $(\frac{\pi}{2}, -2)$ for the endpoints and midpoint of one cycle, respectively. Then we plot zeros at $(\frac{\pi}{3}, 0)$ and $(\frac{2\pi}{3}, 0)$. With these guide points, we can sketch one cycle of $y = 2\cos(3t - \frac{\pi}{2})$ and then repeat the pattern, if necessary.

You should plot $y = \cos(t)$ and $y = 2\cos(3t - \frac{\pi}{2})$ on your graphing calculator now to get a feel for the change in behavior of the standard cosine function when these constants are introduced.

It is useful to realize that the effects of these constants on the sinusoidal curves are the same for the other four trigonometric functions as well. For example, in graphing $y = 3\tan(2x)$, the 3 multiplier would expand the standard tangent curve by 3 units so that $y = 3\tan(2x)$ would pass through the point $(\frac{\pi}{4}, 3)$ rather than $(\frac{\pi}{4}, 1)$. This is the same effect as the one that the multiplier has for the sinusoidal curves except that the multiplier is not called the amplitude in this case, since the tangent function has no maximum value. Also, the 2 multiplier of x changes the frequency with which the function repeats. The period for $y = 3\tan(2x)$ is $\frac{\pi}{2}$, which is half the period of the standard tangent function, π. The graph of this function looks like Window 7.10

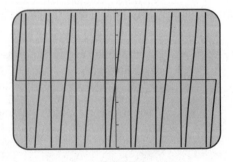

Window 7.10

In general, the period of the function $y = \tan(Bx)$ is $\frac{\pi}{|B|}$, since the standard tangent function is periodic with period π. As before, the period can be determined by solving the double inequality $0 \le Bx \le \pi$, since it takes a distance of π for the tangent function to repeat. The shift is also handled in the same way as for the sinusoidal curves. You need only remember the fact that the period of the sine, cosine, secant, and cosecant functions is 2π, while the period of the tangent and cotangent is π.

EXERCISES

In exercises 1–17, determine the amplitude, period, and phase shift of each function and make a careful sketch by hand of the function over two cycles. Use your calculator to check your results only after you have completed the exercise.

1. $y = 4\sin(3x)$

2. $y = -2\sin(\frac{3}{2}x)$

3. $y = 5\cos(2x)$

4. $y = -4\cos(4x)$

5. $y = \frac{3}{2}\sin(4x)$

6. $y = -\frac{5}{3}\sin(2x)$

7. $y = -\frac{5}{2}\cos(\frac{1}{3}x)$

8. $y = \frac{4}{3}\cos(\frac{5}{3}x)$

9. $y = -3\sin(2x + \frac{\pi}{3})$
10. $y = \frac{2}{3}\sin(3x - \frac{\pi}{6})$
11. $y = 2\cos 4(x + \frac{\pi}{2})$
12. $y = -5\cos 3(x + \frac{\pi}{6})$
13. $y = -\frac{2}{5}\sin \frac{1}{2}(x - \frac{\pi}{8})$
14. $y = -\frac{3}{4}\sin(\frac{5}{2}x + \pi)$
15. $y = \frac{3}{5}\sin 3(x - \frac{\pi}{4})$
16. $y = -7\cos(5x + \frac{3\pi}{4})$
17. $y = 4\cos(\frac{3}{2}x - \frac{\pi}{8})$

In exercises 18–21, make a careful sketch by hand of the function over two cycles. Use your calculator to check your results only after you have completed the exercise.

18. $y = 2\tan(3x)$

19. $y = 3\sec(2x)$

20. $y = 5\cot(2x + \frac{\pi}{6})$

21. $y = \frac{1}{2}\csc 3(x - \frac{\pi}{3})$

For exercises 22–27, use your graphing calculator to graph each function in the following way. Use Y1 to graph one of the trigonometric terms, Y2 to graph the other, and Y3 to graph the original function. The angle *theta* is in radians.

22. $f(\theta) = \sin(\theta) + \cos(\theta)$

23. $f(\theta) = \frac{1}{2}\sin(\theta) - 2\cos(\theta)$

24. $f(\theta) = 3\sin(\theta) + \cos(\theta)$

25. $f(\theta) = \sin\left(\frac{1}{2}\theta\right) - \cos(2\theta)$

26. $f(\theta) = \sin(\pi\theta) - \ln(\theta)$

27. $f(\theta) = \cos(2\theta) - |\frac{\theta}{2}|$

J.W. is sitting in his backyard when he sees some strange, pulsating lights emanating from a distant hill. Fascinated, he decides to record the intensity of these lights. He sets up some equipment, records his data, and obtains the following graph. Exercises 28–33 deal with J.W. trying to fit a

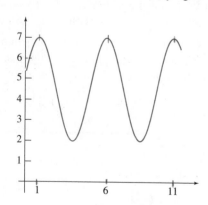

known function to the observed graph.

28. Since the graph looks sinusoidal, he draws a graph of $y = \sin(x)$ for two cycles. Do you agree with his observation? Is there any other function that you would use?

29. He checks the period of his observed graph. He measures the horizontal distance between the two peaks (in seconds). Is this amount the period of the graph? Since the period is not 2π the equation of $y = \sin(x)$ has to be modified accordingly. Write down the new function whose period is the same as that of the graph's.

30. The vertical distance between the peak and valley is measured. Why? How would you incorporate this distance into modifying your equation? (Hint: Think about the amplitude.)

31. Of course, he has to take into account the vertical shift of the graph. What is the vertical shift and how does it change the equation?

32. Lastly, the phase shift has to be included. What is the phase shift? J.W. is very happy that he is able to find an equation that exactly describes the graph. Can you find one, too?

33. If instead of $y = \sin(x)$ you used the equation $y = \cos(x)$, in which of the last five steps would you get a different answer? What would the final equation be?

The next challenge for J.W. is to find a function that will fit the picture in Figure 7.23. Exercises 34–38 walk through the process he needs to use.

34. Is there a standard trigonometric function whose graph is similar to this? Is there more than one such function? If so, what are they?

Then, as before, find the period and modify the graph.

35. What is the next modification? How do you find it?

36. Is there any vertical displacement? If there is, include it in your equation.

37. What is the phase shift? How does that change your equation? What is the final function?

38. The number of daylight hours $D(t)$ at a place at any particular time of the year is approximated by the following formula $D(t) = \dfrac{k}{2} \sin \dfrac{2\pi}{365}(t - 79) + 12$, where t is the number of days since January 1 (when $t = 0$). Knowing that at the North Pole $D(t) = 0$ on January 1, find the value of k for the North Pole (use radian measure). Graph the function with the value of k you found. Use TRACE to find the day that has the maximum daylight.

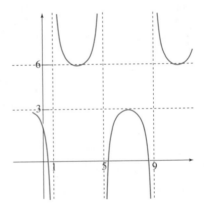

Figure 7.23

7.6   **Inverse Trigonometric Functions**

Introduction

The Inverse Sine Function

The Inverse Cosine Function

The Inverse Tangent Function

Introduction

It may be a good idea to review inverse functions from Chapter 4 now so that the central ideas regarding inverse functions are fresh in your mind. Recall that the inverse of a function is obtained by interchanging the components of the ordered pairs of the function. Thus, the domains and ranges of inverse functions are interchanged. Also, the graphs of inverse functions are reflections of each other about the line $y = x$. We can find the equation of the inverse of a function given by $y = f(x)$ by interchanging the variables x and y and then solving for y, if possible. Finally, we recall that only one-to-one functions have inverses that are also functions.

The Inverse Sine Function

We now consider the inverse of the sine function $y = \sin(x)$. We can obtain a quick accurate picture of the graph of this function by using 2nd and SIN to graph $\sin^{-1}(x)$ on your graphing calculator. The graph looks like Window 7.11.

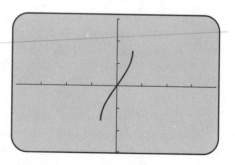

Window 7.11

Study the graph with regard to the behavior of this function. Notice that it is heavily restricted in contrast to the standard sine function. If we use the Trig option of the ZOOM menu and the TRACE feature, we can predict that the domain is the interval $[-1, 1]$. The range, however, will not be as clear. When we try to obtain the coordinates of the "last" point with the TRACE feature, we find that the y coordinate becomes blank. We are off the graph of the function when that happens. The "last" y coordinate is not a familiar number, but by zooming in on one of the points where the graph ends, we can eventually conclude that the range of $y = \sin^{-1}(x)$ is $[-\frac{\pi}{2}, \frac{\pi}{2}]$. We can always get a quick refresher of its behavior on the graphing calculator but an in-depth understanding requires that we also study this function from the theoretical basis.

We already know that the sine function is not one-to-one. If we take the inverse of the standard sine function, the inverse will not be a function. We can restrict the domain of the standard sine function to the interval $[-\frac{\pi}{2}, \frac{\pi}{2}]$ (see Figure 7.24).

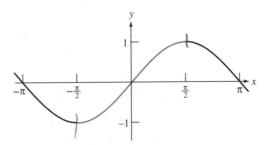

Figure 7.24

This restriction yields a function that is one-to-one and still retains the original range of $[-1, 1]$. The inverse of this function is obtained by interchanging the variables x and y as follows:

$$x = \sin(y)$$

The domain is $-1 \leq x \leq 1$ and the range is $-\frac{\pi}{2} \leq y \leq \frac{\pi}{2}$, which are the interchange of the domain and range of the restricted sine function. The rule $x = \sin(y)$ indeed represents the inverse of the restricted sine function given above. We can graph this function by hand by first building a table of xy-pairs and then plotting the points. We accomplish this by giving allowable values to y and calculating the corresponding values of x. The graph would be the same as the one we obtained using the calculator. While the form $x = \sin(y)$, with its domain $[-1, 1]$ and range $[\frac{-\pi}{2}, \frac{\pi}{2}]$, needs to be a part of our resource bank, it is not always the most convenient form. One reason that comes to mind immediately is that we cannot use this form with Y1= on the calculator. It can be useful, at times, to have y in terms of x, and there are two commonly used forms to represent the inverse of the restricted sine function: they are $y = \sin^{-1}(x)$ and $y = \arcsin(x)$. The first is the form we find on our calculator and is a more recent notation that is becoming dominant. The second is an older notation that is still used frequently. The formal definition of the inverse of the restricted sine function is given here.

The Inverse Sine Function

The Inverse Sine Function

The **inverse sine function** is defined by

$$y = \sin^{-1}(x) \text{ if and only if } x = \sin(y),$$

where $-1 \leq x \leq 1$ and $-\frac{\pi}{2} \leq y \leq \frac{\pi}{2}$.

It is important for us to carry the restrictions along with our image of the inverse sine function. We often make more use of the range restriction $-\frac{\pi}{2} \leq y \leq \frac{\pi}{2}$ than of the domain restriction but both are clearly important. We can recall these facts with the graphing calculator or by visualizing the standard sine graph and asking ourselves what restrictions we would place on the domain to cause it to be a one-to-one function. The interval resulting from the domain restriction becomes the range of its inverse. Understanding the inverse sine function starting from the original sine function stocks our resource bank with more complete and readily accessible information (and, therefore, increases our analyzing ability) than does memorizing the information or relying on the graphing calculator.

EXAMPLE SET A

1. Determine the value of each of the following using pencil and paper and check your answers with your calculator.

   a. $\sin^{-1}\left(\frac{\sqrt{3}}{2}\right)$

   b. $\sin^{-1}\left(-\frac{1}{2}\right)$

   c. $\arcsin\left(\frac{\sqrt{2}}{2}\right)$

   Solution:

   a. One approach to the solution is to write $y = \sin^{-1}\left(\frac{\sqrt{3}}{2}\right)$ and then rewrite it, using the definition, as $\frac{\sqrt{3}}{2} = \sin(y)$ recalling the restriction $-\frac{\pi}{2} \leq y \leq \frac{\pi}{2}$. We might recall now that replacing y by $\frac{\pi}{3}$ will make the statement true. Unlike our experience with the standard sine function which would give us an infinite number of solutions, the required restrictions allow but one solution.

b. We use the approach above and rewrite $y = \sin^{-1}(-\frac{1}{2})$ as $-\frac{1}{2} = \sin(y)$ with $-\frac{\pi}{2} \leq y \leq \frac{\pi}{2}$. This problem strongly tests our depth of understanding of the inverse sine function. We may be tempted to write the answer as $y = \frac{7\pi}{6}$ and perhaps even list two answers. But as we focus our attention on the restriction $-\frac{\pi}{2} \leq y \leq \frac{\pi}{2}$ required of the inverse sine function, we realize that the answers we are initially tempted to write do not lie in this interval. Since $f(t) = \sin^{-1}(t)$ is a function, $\sin^{-1}(-\frac{1}{2})$ can have only one value. Further thought leads us to the only value possible, which is $-\frac{\pi}{6}$; that is, $\sin^{-1}(-\frac{1}{2}) = -\frac{\pi}{6}$.

c. We now have a solid approach to such problems and we continue to rewrite $\arcsin(\frac{\sqrt{2}}{2})$ as $\frac{\sqrt{2}}{2} = \sin(y)$. The value $\frac{\sqrt{2}}{2}$ is also a familiar number and brings to mind the corresponding number $\frac{\pi}{4}$. And, indeed, $y = \frac{\pi}{4}$ is the unique value of $\arcsin(\frac{\sqrt{2}}{2})$.

We can use SIN^{-1} on our calculators to check our answers.

2. Approximate, to three decimal places, the value of each of the following.

   a. $\sin^{-1}(0.75)$

   b. $\sin^{-1}(-0.53)$

   c. $\arcsin(-0.91)$

   d. $\arcsin(1.1)$

Solution:
We first make sure that our calculator is set in radian mode. Then we make use of the SIN^{-1} feature to solve each problem.

   a. 0.848

   b. -0.559

   c. -1.143

   d. No solution. The argument is outside of the domain of the inverse sine function.

The Inverse Cosine Function

We use the COS^{-1} feature on our calculators to view the graph of the inverse cosine function. The domain is relatively easy to predict as the interval $[-1, 1]$ since this is the range of the cosine function. Predicting the range is more difficult. It will require some careful zooming to arrive at the true range, which is the interval $[0, \pi]$. It will be easier to arrive at the range, in this case, as we study the theoretical approach to the inverse cosine function. The graph, however, gives us a visual look at the behavior of this function and we should study it frequently enough to recall this behavior when needed. We can see that the inverse cosine function is never negative (Window 7.12).

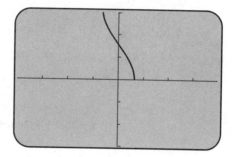

Window 7.12

We begin our theoretical approach by observing that the cosine function, like all of the trigonometric functions, is not one-to-one. We look at the standard cosine function on the interval $[-\pi, 2\pi]$ to see where we might restrict its domain so that it will be one-to-one. Looking at Window 7.13, it would seem that a reasonable choice would appear to be the interval $[0, \pi]$. This choice, fortunately, includes the first quadrant angles we use as reference angles (and that appear in trigonometric tables) and retains the range of the standard cosine function. Once again, it is useful to visualize the standard cosine function and the restrictions needed to make it a one-to-one function when

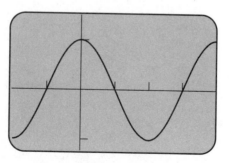

Window 7.13

we wish to recall the range of the inverse cosine function. The connection is the interchange of the domain and range of a function and its inverse. This connection facilitates our recall of one of the forms of the inverse cosine function, $x = \cos(y)$, which is simply the interchange of the variables in the standard cosine function. This preliminary investigation of the inverse cosine function helps us to formulate the definition of the inverse cosine function.

The Inverse Cosine Function

The Inverse Cosine Function

The **inverse cosine function** is defined by

$$y = \cos^{-1}(x) \text{ if and only if } x = \cos(y),$$

where $0 \leq x \leq \pi$ and $-1 \leq y \leq 1$.

Once again, graphing the inverse cosine function using pencil and paper can reinforce and enhance our understanding. We can do so relatively easily by building a table of values using the form $x = \cos(y)$, choosing convenient values for y and calculating the corresponding x values. We need to restrict our choices for y to $0 \leq y \leq \pi$. It would also be very helpful to choose equal units on the x- and y-axes. Thus, the tick mark for 1 on the y-axis is about two-thirds of the distance from the origin that the tick mark for $\frac{\pi}{2}$ is on the x-axis. We can then plot the restricted standard cosine function, the inverse cosine function, and the line $y = x$ to obtain a useful image of all three and their connection. Following through with this important step, we obtain a solid relationship between the standard restricted cosine function and its inverse as in Figure 7.25. The inverse cosine function is drawn as a solid colored line and the standard cosine function and line of symmetry $y = x$ as dashed lines.

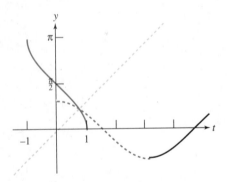

Figure 7.25

As we approach problems involving inverse trigonometric functions, it pays to remember that there is exactly one function value associated with an allowable domain value. This can avoid giving more than one answer, which can be a temptation from our previous experience with the unrestricted trigonometric functions. Also, more care is generally required when considering negative argument values for inverse trigonometric functions than for positive ones.

EXAMPLE SET B

Determine the *exact* value of each of the following using pencil and paper and check your answers with your calculator.

1. $\cos^{-1}(\frac{\sqrt{3}}{2})$
2. $\cos^{-1}(-\frac{1}{2})$
3. $\arccos(-\frac{\sqrt{2}}{2})$

Solution:

We can take the same approach as we used in the example involving the inverse sine function.

1. First write the problem $y = \cos^{-1}(\frac{\sqrt{3}}{2})$ as $\frac{\sqrt{3}}{2} = \cos(y)$ and remember the restriction $0 \leq y \leq \pi$ for the inverse cosine function. Think about what arc length has a cosine value of $\frac{\sqrt{3}}{2}$. Tapping your resource bank should bring to mind the value $\frac{\pi}{6}$. Thus, $\cos^{-1}(\frac{\sqrt{3}}{2}) = \frac{\pi}{6}$.

2. The function $y = \cos^{-1}(-\frac{1}{2})$ can be written equivalently as $-\frac{1}{2} = \cos(y)$ with $0 \leq y \leq \pi$. We can give only one value and the only value that falls in the interval $0 \leq y \leq \pi$ is $\frac{2\pi}{3}$.

3. Setting $y = \arccos(-\frac{\sqrt{2}}{2})$, rewriting it equivalently as $-\frac{\sqrt{2}}{2} = \cos(y)$, and recalling that $0 \leq y \leq \pi$ should allow us to see that the solution is $\frac{3\pi}{4}$.

The Inverse Tangent Function

We graph the inverse tangent function on our calculators to get an initial impression of its shape and behavior. The graph looks like Window 7.14. The domain would appear to be all real numbers while the range values appear to level off as the x values both decrease toward $-\infty$ and increase toward $+\infty$. It turns out that the inverse tangent function has horizontal asymptotes at $\frac{\pi}{2}$ and $\frac{-\pi}{2}$.

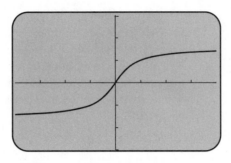

Window 7.14

We recall the shape of the standard tangent function in order to decide how to restrict its domain so that it will be a one-to-one function. The connection between the standard trigonometric functions and their inverses remains a very useful tool. The choice would appear to be a restriction to the interval $(-\frac{\pi}{2}, \frac{\pi}{2})$ excluding the endpoints since the tangent function is undefined there. This domain restriction will become the range of the inverse tangent function, which is the same as for the inverse sine function *except* for the exclusion of the endpoints. The restriction continues to include the first quadrant angles and also retains the range of the function.

The Inverse Tangent Function

The Inverse Tangent Function

The **inverse tangent function** is defined by

$$y = \tan^{-1}(x) \text{ if and only if } x = \tan(y),$$

where $-\frac{\pi}{2} < x < \frac{\pi}{2}$ and $-\infty \leq y \leq \infty$.

The value of graphing the inverse functions using pencil and paper cannot be over-emphasized in developing understanding and in making connections. We again approach this activity from the standpoint of creating a table using the form $x = \tan(y)$ and choosing convenient values of y in the interval $(-\frac{\pi}{2}, \frac{\pi}{2})$. Choosing equal units on each axis and using the table and our knowledge of the tangent function leads to the graph in Figure 7.26. The inverse tangent function is drawn as a solid line and the standard tangent function and the line of symmetry $y = x$ are drawn as dashed lines.

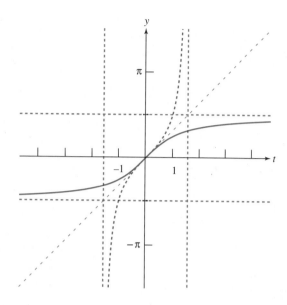

Figure 7.26

EXAMPLE SET C

Determine the *exact* value of each of the following using pencil and paper and check your answers with your calculator.

1. $\tan^{-1}(\sqrt{3})$
2. $\tan^{-1}(-\frac{1}{\sqrt{3}})$
3. $\arctan(-1)$

Solution:

We take the same approach as in the previous examples.

1. Rewrite $\tan^{-1}(\sqrt{3})$ as $\sqrt{3} = \tan(y)$. Now we ask ourselves: what one value of y in the interval $(-\frac{\pi}{2}, \frac{\pi}{2})$ will make this statement true? We may need to review the tangent function values associated with the special angles $\frac{\pi}{6}$, $\frac{\pi}{4}$, and $\frac{\pi}{3}$. We should be able to convince ourselves that $\frac{\pi}{3}$ is the answer.

2. Using the same strategy as in part 1 on $\tan^{-1}(-\frac{1}{\sqrt{3}})$ leads to the solution $-\frac{\pi}{6}$. (Don't forget the negative argument and the restrictions on y.)

3. We know the answer is associated with $\frac{\pi}{4}$. Taking into account the range of the inverse tangent function, we determine the answer to be $-\frac{\pi}{4}$.

EXERCISES

In exercises 1–10, find the *exact* value.

1. $\sin^{-1}(1/2)$

2. $\sin^{-1}(-\sqrt{3}/2)$

3. $\arcsin(-1/\sqrt{2})$

4. $\arccos(-1/\sqrt{2})$

5. $\arccos(-\sqrt{3}/2)$

6. $\cos^{-1}(1/2)$

7. $\tan^{-1}\left(\dfrac{1}{\sqrt{3}}\right)$

8. $\tan^{-1}(-\sqrt{3})$

9. $\arctan(1)$

10. $\arcsin(-\dfrac{1}{2})$

In exercises 11–20, approximate each value to three decimal places.

11. $\sin^{-1}(0.25)$

12. $\sin^{-1}(-0.25)$

13. $\arcsin(-0.625)$

14. $\arccos(-0.625)$

15. $\arccos(-2.25)$

16. $\cos^{-1}(0.91)$

17. $\tan^{-1}(5.25)$

18. $\tan^{-1}(-1.45)$

19. $\arctan(0.45)$

20. $\arcsin(-0.36)$

21. A function is defined by $f(x) = 3 + 0.1\tan^{-1}x$. What is the domain of this function? What is the range? Graph the function.

For exercises 22–25, verify the truth of each equality.

22. $\sin^{-1}(-x) = -\sin^{-1}(x)$. (Hint: Let $x = \sin^{-1}(-x)$ and show that $\sin^{-1}(x) = -x$.)

23. $\sin^{-1}(x) = \tan^{-1}\left(\dfrac{x}{\sqrt{1-x^2}}\right)$, for $x > 0$

24. $\cos^{-1}(x) = \tan^{-1}\left(\dfrac{\sqrt{1-x^2}}{x}\right)$, for $x > 0$

25. Simplify $\sin^{-1}(x) + \cos^{-1}(x)$. (Hint: Let $\theta = \sin^{-1}(x)$ so that $x = \sin(\theta) = \cos\left(\dfrac{\pi}{2} - \theta\right)$, where $-\dfrac{\pi}{2} \le \theta \le \dfrac{\pi}{2}$.) Use your calculator to check your result.

For exercises 26–35, use your calculator to construct each graph. Use the graph to summarize the behavior of the function.

26. $f(x) = \dfrac{1}{2}\sin^{-1}(x)$

27. $f(x) = 2\cos^{-1}(x)$

28. $f(x) = 2\tan^{-1}(x)$

29. $f(x) = 2\cot^{-1}(x)$ (Hint: $\cot^{-1}(x) = \tan^{-1}(1/x)$)

30. $f(x) = 2\sin^{-1}(x) + \cos^{-1}(x)$

31. $f(x) = \tan^{-1}(x) + \cos^{-1}(x)$

32. $f(x) = \sec^{-1}(x) + \csc^{-1}(x)$ (Hint: $\sec^{-1}(x) = \cos^{-1}(1/x)$)

33. $f(x) = \sin(x) + \sin^{-1}(x)$

34. $f(x) = |\sin^{-1}(x)|$

35. $f(x) = |\tan^{-1}(x)|$

7.7   **Applications of Circular Functions**

Introduction

Alternating Current

Sinusoidal Linear Motion

The Motion of a Spring

Sound Vibrations

Introduction

Table 7.6	
secs.	**volts**
0.0	0.15
0.1	0.74
0.2	1.21
0.3	1.46
0.4	1.47
0.5	1.21
0.6	0.76
0.7	0.16
0.8	−0.46
0.9	−1.00
1.0	−1.37
1.1	−1.49
1.2	−1.37
1.3	−1.01
1.4	−0.47
1.5	0.15
1.6	0.74
1.7	1.21
1.8	1.45
1.9	1.47
2.0	1.22

period

We can use circular functions in describing many natural periodic or cyclical situations. We usually begin by taking measurements of some physical aspect of the situation to discover a function that might model it. Then we analyze the data to determine whether it is truly cyclical. If the collected data look cyclical, we can begin to match the situation with a specific circular function based on our resource bank of information about circular functions. Thus, we would try to determine the period and amplitude of the data. The phase shift and vertical displacement can also be factors in the final function we use to approximate or model the situation. We apply this process to four applied cases.

Alternating Current

Many uses of current in electrical devices are periodic. Here we study the generation of alternating current.

EXAMPLE SET A

A small electrical generator is used to produce alternating current. The energy for the generator is provided by rotating a crank at 100 revolutions per minute. The electrical output for 2 seconds is measured using a voltmeter and recorded on a TI graphing calculator as indicated in Table 7.6. Find a trigonometric function that models the behavior of the current from this generator.

Solution:

Use the Stat Edit menu to enter the data. Use the Scatter feature to plot the data on the appropriate graphing window. From the data, you might determine the graphing window to be Xmin=0, Xmax=2, Ymin= −2, and Ymax=2. Your graph should look something like Window 7.15.

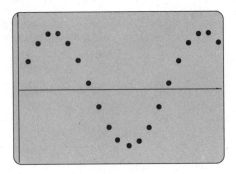

Window 7.15

The data seem to be periodic with a period of about 3/2 seconds. Looking at the table, the largest output, in absolute value, is 1.49. This will serve as our first estimate of the amplitude. The graph appears to be sinusoidal. Thus, our first approximation would be

$$f(t) = 1.49 \sin\left(\frac{2\pi}{3/2}t\right)$$

We find that this function provides a fair approximation to the data when we plot it against the scatter diagram. Zooming in on the least positive zero of our function, we find that the zeros of our function are about 0.02 away from the apparent zeros of the data. Thus, we adjust our function as follows:

$$f(t) = 1.49 \sin\left(\frac{2\pi}{3/2}(t + 0.02)\right) = 1.49 \sin\left(\frac{4\pi}{3}(t + 0.02)\right) \qquad (7.1)$$

This function matches very closely when we plot it against our scatter diagram.

Sinusoidal Functions

The alternating current flowing into a motor will also have a sinusoidal character as long as the motor is running under a fixed load.

You may recall that functions like the one in equation 7.1 are called **sinusoidal functions** and they have the general form

$$f(t) = A \sin [B(t + C)] + D$$

Sinusoidal Linear Motion

There are many situations in which it is useful to turn linear motion into circular or rotational motion. The internal combustion, piston-driven engine in automobiles is an example of such a case.

Table 7.7	
secs.	**position**
0.000	3.24
0.005	5.97
0.010	4.73
0.015	1.21
0.020	0.28
0.025	3.17
0.030	5.96
0.035	3.72
0.040	1.18
0.045	0.09
0.050	3.20
0.055	5.94
0.060	3.76
0.065	1.17
0.070	0.27
0.075	3.21
0.080	5.97
0.085	3.12
0.090	1.22
0.095	0.28
0.100	3.22

EXAMPLE SET B

The linear motion of a piston in an automobile engine is converted into the rotational motion of the drive shaft. Automobile engineering shops have devices to measure the movement of pistons. The data recorded in Table 7.7 give the distance that the top of the piston is from the top of the cylinder in an engine that is running at highway speeds in high gear. Develop a sinusoidal function that will model the motion of the piston.

Solution:

Enter the data as in the previous example and obtain a scatter diagram of the data with an appropriate graphing window. The plot of the data points should look something like Window 7.16. From this scatter diagram, we can estimate

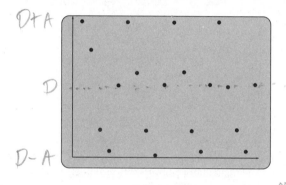

Window 7.16

that the motion of the piston is sinusoidal with period of approximately 0.025 and amplitude $(5.97 - 0.09)/2 = 2.94$. The center of the data is the average of the maximum and minimum values of the position coordinates and will be the vertical shift. This average is $(5.97 + 0.09)/2 = 3.03$. Thus, our first estimate of the function model of the motion of the piston is

$$f(t) = 2.94 \sin\left(\frac{2\pi}{0.025}t\right) + 3.03$$

The plot of this curve closely approximates the shape of the data points, but it seems to be translated to the right a small amount. We might wish to make the horizontal correction by adding 0.01 to the argument. Thus, our new estimate of the function model is

$$f(t) = 2.85 \sin\left(\frac{2\pi}{0.025}(t + 0.01)\right) + 3.12$$

This new function model is one of many that give a very close match with the original data. There are several questions that are of interest to the engineer designing engines. When is the piston moving fastest? Does the piston ever stop? These questions are addressed in calculus courses. If this piston is connected to a drive shaft with a camshaft, how many revolutions per minute is the drive shaft making? Since the drive shaft revolves once each time the piston goes through a cycle, the drive shaft is rotating at approximately 2400 revolutions per minute.

The Motion of a Spring

Table 7.8	
secs.	**position**
0.00	6.96
0.05	7.19
0.10	7.40
0.15	8.28
0.20	8.84
0.25	8.45
0.30	8.20
0.35	7.85
0.40	7.23
0.45	7.03
0.50	7.25
0.55	7.85

Springs play an important part in our lives. There are spring-loaded doors, our cars have springs in them to smooth out the ride, and some expensive watches still have springs in them. The motion involving a spring can sometimes be modeled by trigonometric functions.

EXAMPLE SET C

A spring is hung from the ceiling of an industrial laboratory with a heavy object or weight attached to it. The purpose is to study the behavior of the spring under certain loading to determine if it reacts according to the engineers' design. The weight on the spring moves up and down when the weight is pulled downward and released. The data collection begins when the weight is released. Data are collected and recorded with a TI graphing calculator and a motion detector placed on the floor beneath the spring and weight. The data collected are displayed in Table 7.8. Create a function that models the motion of the weight on the spring.

Solution:

Enter and plot the data points. After you have adjusted the graphing window, your scatter diagram should look something like Window 7.17.

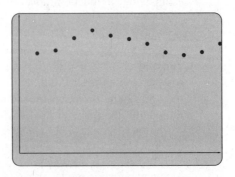

Window 7.17

The period appears to be about 0.43 seconds and the amplitude about 0.83 ft. The vertical offset can be seen to be about 8 ft. Thus, our modeling function should be

$$f(t) = 0.83 \sin\left(\frac{2\pi}{0.43}t\right) + 8$$

The plot of this function does not match the data points, because the phase shift has not been taken into account. The maximum point on our function graph should be moved to the right about 0.12 to match the maximum point of the data. Our adjusted function is

$$f(t) = 0.83 \sin\left(\frac{2\pi}{0.43}(t - 0.12)\right) + 8$$

The plot of this function closely matches the data points.

Sound Vibrations

The sound of pianos and stringed instruments is created from the vibration of strings. This vibration can be modeled by trigonometric functions. We study simple vibration here although the more complex musical sounds of pianos and violins can be modeled as well.

EXAMPLE SET D

Table 7.9	
msec.	**pressure**
0	2.66
1/3	−3.53
2/3	3.96
1	−3.91
4/3	3.39
5/3	−2.46
2	1.24
7/3	0.14
8/3	−1.50
3	2.68

A tuning fork is struck and data are collected that measure changes in the magnitude of the force of air pressure. The data are displayed in Table 7.9 where msec indicates thousandths of a second. Determine a sinusoidal model of the sound vibrations created by the tuning fork as displayed in the table.

Solution:
You can determine that the period of the data is about 0.6 thousandths of a second and the amplitude is about 4 units once you have entered the data and created a scatter diagram. This gives a first estimate of the sinusoidal function as

$$f(t) = 4 \sin\left(\frac{2\pi}{0.6}t\right)$$

When we have graphed this function against the data points, we find that our estimates for the period and the amplitude are correct. Once again, we need to adjust the position of the function graph to match the data points. To match the zeros of the data and our function, we must move the graph about 0.07 thousandths of a second to the left. This can be seen by joining two adjacent data points on either side of the horizontal axis to approximate a zero of the data. This phase shift gives us the following function

$$f(t) = 4\sin\left(\frac{2\pi}{0.6}(t + 0.07)\right)$$

The graph of this function closely matches the graph of the air pressures. Can you determine how many vibrations the tuning fork makes each second? If you are a musician, what note might this be?

EXERCISES

1. Give five examples from business or economics of phenomena that are cyclical.

2. Give five examples of biological phenomena that are periodic.

In exercises 3–10, graph the given sinusoidal function and indicate the amplitude, period, frequency, and phase shift.

3. $f(t) = 0.05\sin(2\pi \cdot t)$

4. $g(t) = 2.93\sin(2\pi \cdot 3t)$

5. $f(t) = 5\sin(2\pi \cdot 1.397t)$

6. $g(t) = 1570\sin(2\pi \cdot 0.005t)$

7. $f(t) = 103\sin(341t)$

8. $g(t) = 421.5\sin(0.095(t - 3.74))$

9. $d(t) = 32\sin(0.0003(t - 0.0001))$

10. $P(t) = 4.3\sin(2004(t - 0.0043))$

In exercises 11–15, determine if the given table represents a sinusoidal function.

11.

time	height
0	4
1/3	2
2/3	4
1	5
4/3	3
5/3	2
2	4
7/3	6
8/3	4
3	2

12.

time	pressure
0	4
1/3	2
2/3	5
1	2
4/3	4
5/3	2
2	3
7/3	2
8/3	2
3	2

13.

time	temperature
0	0.2
1	0.3
2	0.4
3	0.5
4	0.4
5	0.3
6	0.2
7	0.1
8	0.2
9	0.3

14. Time and number.

t	N
0	0
1	4
2	0
3	−4
4	0
5	4
6	0
7	−4
8	0
9	4

15. Time and volume.

t	V
0	2
1	2
2	2
3	2
4	2
5	2
6	2
7	2
8	2
9	2

Exercises 16–19 involve data tables you are to use to generate a sinusoidal function.

16. Ecologists investigate environments in which there are predators and creatures these predator prey upon. In such situations, the prey population rises and falls based on the number of predators. The table below gives the population of prey in Big Creek Valley from April 1985 to March 1990 every quarter year. Create a sinusoidal function that models these data.

quarter	number
0	502
1	626
2	501
3	372
4	505
5	625
6	492
7	380
8	515
9	618

17. The ecologists involved in the Big Creek Valley study investigation in the previous exercise are awarded a grant to study a predator/prey environment in Little Creek Valley. The accompanying table gives the population of prey in Little Creek Valley from April 1990 to March 1993 every quarter. Are these data periodic? If you think the data are not quite periodic, make up an explanation for why the data might vary from the sinusoidal function graph.

quarter	number
0	402
1	526
2	401
3	272
4	405
5	525
6	392
7	220
8	395
9	490
10	388
11	283
12	420

minutes	temperature
0	71
20	68
40	67
60	69
80	72
100	73
120	72
140	67
160	67
180	70

18. The temperatures inside a house with a thermostatically controlled furnace on a cold winter day are given in the accompanying table every 20 minutes for 3 hours. Are these data periodic? If so, generate a sinusoidal function that closely matches the table. How often does the furnace come on? Why would the temperature in such a house be periodic?

19. In the year 2020, a modern house will assume that its occupants are conversant with trigonometry. The furnace will be set by entering the amplitude and period of a sinusoidal function you wish the temperature to match. Create a function for your home. Indicate the maximum and minimum temperatures you are willing to accept and how often you wish the furnace to come on.

Summary and Projects

Summary

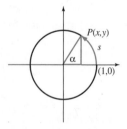

Figure 7.27

The Cosine and Sine Functions Consider a point $P(x, y)$ on the unit circle. Let s be the arc length on the circle from the point $(1, 0)$ to P and α the associated central angle (see Figure 7.27). Then we simply rename the x coordinate $\cos(s)$ or $\cos(\alpha)$ and rename the y coordinate $\sin(s)$ or $\sin(\alpha)$. That is, $\cos(s) = x$ and $\sin(s) = y$.

The Right Triangle Association   Use the information above. Draw a perpendicular from P to the x-axis forming a right triangle (see Figure 7.28). Then x and y become the adjacent and opposite sides, respectively, of the central angle or the supplement of the central angle. We define

$$\cos(\alpha) = \frac{adjacent}{hypotenuse} = \frac{x}{1} \quad \text{and} \quad \sin(\alpha) = \frac{opposite}{hypotenuse} = \frac{y}{1}$$

The First Basic Trigonometric Identity   Since $\cos(\alpha)$ and $\sin(\alpha)$ are the coordinates of a point on the unit circle, we have

$$\cos^2(\alpha) + \sin^2(\alpha) = 1$$

for all angles α.

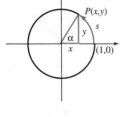

Figure 7.28

The Domain and Range of the Sine and Cosine Functions   We need only remember that $\cos(\alpha)$ and $\sin(\alpha)$ are coordinates of a point on the unit circle associated with an arc s starting at $(1,0)$ and its associated central angle α (See Figure 7.27). Thinking about the restricted motion of a point on the unit circle but with unrestricted arc lengths, we can recall that $s \in R$ is the domain for both functions. This consideration also gives $-1 \le \cos(s) \le 1$ and $-1 \le \sin(s) \le 1$ for their ranges.

Sine and Cosine Function Values for Special Arc Lengths (Angles)

$$\cos\left(\frac{\pi}{6}\right) = \cos(30°) = \frac{\sqrt{3}}{2}$$

$$\sin\left(\frac{\pi}{6}\right) = \sin(30°) = \frac{1}{2}$$

$$\cos\left(\frac{\pi}{4}\right) = \sin\left(\frac{\pi}{4}\right) = \sin(45°) = \frac{1}{\sqrt{2}}$$

$$\cos\left(\frac{\pi}{3}\right) = \cos(60°) = \frac{1}{2}$$

$$\sin\left(\frac{\pi}{3}\right) = \sin(60°) = \frac{\sqrt{3}}{2}$$

Symmetric Points on the Unit Circle   The points on the unit circle associated with the special arc lengths (angles) above have three symmetric points. This information allows us to recall the sine and cosine function values for the special arc lengths (angles) associated with these symmetric points.

Graphs of the Trigonometric Functions   Whenever we cannot bring up a mental image of a trigonometric function, we should use our graphing calculators to do so. However, we should also strive to understand how these graphs come about. Making the connection with the theory involved can be accomplished by recalling the definitions of the trigonometric functions along with the function values at the special arc lengths. This allows us to create a table of values we can then graph.

Periodicity of the Sine and Cosine Functions   We can recall that the period of both the sine and cosine functions is 2π by picturing a point moving around the unit circle. We observe that the point arrives at a point it passed before, after each revolution. We can, therefore, write $\sin(\alpha + k \cdot 2\pi) = \sin(\alpha)$ and $\cos(\alpha + k \cdot 2\pi) = \cos(\alpha)$, where k is any integer.

Behavior of the Sine and Cosine Functions   It is most helpful to study the behavior of these functions by studying their graphs. It is useful to observe that the standard sine function is 0 at the endpoints and midpoint of one cycle and reaches its maximum and minimum between these values. The standard cosine function attains its maximum value at the endpoints of one cycle and its minimum at the midpoint. It is 0 between these values.

The Tangent Function   If we can recall that $\tan(s) = \dfrac{\sin(s)}{\cos(s)}$, then we are in a position to derive essentially any facts we may need concerning the tangent function. As always, a mental picture of its graph produces a great deal of information about its behavior. Such resources allow us to recall that the domain of the tangent function is $s \neq \dfrac{\pi}{2} + k \cdot \pi$, where k is any integer. These excluded domain values are precisely where the asymptotes occur. The graph of the standard tangent function also reminds us that its period is π.

The Secant, Cosecant, and Cotangent Functions   Recollection of the behavior of these functions is most rapidly achieved through their definitions and graphs. We need to remember that $\sec(s) = \dfrac{1}{\cos(s)}$, $\csc(s) = \dfrac{1}{\sin(s)}$, and $\cot(s) = \dfrac{1}{\tan(s)}$. If we think about the meaning of these definitions along with either a recalled mental picture of their graphs or displaying their graphs on our calculators, we can bring forward any facts about these functions we may need.

The Sinusoidal Functions   The graphs of $f(x) = A\sin(Bx + C)$ and $g(x) = A\cos(Bx + C)$ are certainly easy to display on our graphing calclulators. However, it remains important to make the connection with the theory involved. For example, if we contrast the graph of $f(x) = \sin(x)$ with the graph of

$g(x) = 2\sin(x)$ on our graphing calculators, we can convince ourselves that A affects the amplitude (high point reached) of the function. Similar exploration will show the effect B and C have on the standard sine (or cosine) function regarding how often the function repeats and how far it is shifted. Recall that placing the entire argument, $Bx + C$, of the sine or cosine function between 0 and 2π as in $0 \le Bx + C \le 2\pi$ and solving the inequality for x will determine the period and phase shift of the function.

The Inverse Trigonometric Functions   We can quickly and easily display the graphs of the inverses of the sine, cosine, and tangent functions on our graphing calculators. Since it is so easy to do this, we should display their graphs a sufficient number of times to get a solid mental image of them. Such mental images give us basically all the information we might need in any given situation. It continues to be important to connect the theory with our graphical information.

One way to make this connection is to consider the original function, say $f(x) = \sin(x)$, at the start. We then proceed to determine what restrictions on the domain of the original function will cause it to be a one-to-one function and retain all of its range values. This restricted domain becomes the range of the inverse function while the range of the original function becomes the domain of the inverse function. The equation of the inverse function is obtained by interchanging the x and y variables in the original equation. In addition, the graphs of the original function and its inverse are reflections of each other about the line $y = x$. If we have this information in our resource bank, we can derive or summon up any details we need for these inverse functions. More importantly, such information in our resource bank will allow us to more fully understand the mathematics as well as become better problem solvers.

Projects

1. This project develops a geometric notion of the secant and tangent functions as line segment lengths. First draw a unit circle and construct a vertical line tangent to the unit circle at the point $(1, 0)$. Next draw an arc length, s, of less than 1, on the unit circle starting at $(1, 0)$. Finally, draw a ray from the center of the unit circle through the endpoint of the arc. The intersection of this ray and the tangent line is a point whose distance from the point $(1, 0)$ is the tangent of s and whose distance from the origin is the secant of s. Using similar triangles, show that these definitions of the tangent and secant of a first quadrant arc length, s, are the same as the definitions given in the text. Extend these definitions to arc lengths that are in the second, third, and fourth quadrants. Be careful to consider whether the secant and tangent values are positive or negative.

2. This project develops a geometric notion of the cosecant and cotangent functions as line segment lengths. First draw a unit circle and construct a horizontal line tangent to the unit circle at the point $(0, 1)$. Next draw an arc length, s, of less than 1, on the unit circle starting at $(1, 0)$. Finally, draw a ray from the center of the unit circle through the endpoint of the arc. The intersection of this ray and the tangent line is a point whose distance from the point $(0, 1)$ is the cotangent of s and whose distance from the origin is the cosecant of s. Using similar triangles, show that these definitions of the cotangent and cosecant of a first quadrant arc length, s, are the same as the definitions given in the text. Extend these definitions to arc lengths that are in the second, third, and fourth quadrants. Be careful to consider whether the cosecant and cotangent values are positive or negative.

3. The identity $\sec^2(\alpha) - \tan^2(\alpha) = 1$ is similar to the basic identity $\cos^2(\alpha + \sin^2(\alpha) = 1$. When we graph the unit circle in parametric form, we use $x(t) = \cos(t)$ and $y(t) = \sin(t)$. Using Dot mode, what kind of geometric figure do we get when we graph parametrically, $x(t) = \sec(t)$ and $y(t) = \tan(t)$? Give the rectangular form of this parametric representation. Explain the difference between the equation for the unit circle in rectangular form and this new geometric figure in rectangular form.

4. Prove that

$$\tan(\alpha + \beta) = \frac{\tan(\alpha) + \tan(\beta)}{1 - \tan(\alpha)\tan(\beta)}$$

Hint:

$$\tan(\alpha + \beta) = \frac{\sin(\alpha + \beta)}{\cos(\alpha + \beta)}$$

5. Prove that

$$\tan(2\alpha) = \frac{2\tan(\alpha)}{1 - \tan^2(\alpha)}$$

6. Make a careful sketch by hand of the following functions, being careful to label the axes and the tick marks. Be sure to graph them on your graphing calculator as well and note, describe, and analyze any differences between your hand graphs and the calculator graphs.

   a. $f(t) = \sin(\pi t)$

   b. $g(t) = \sin(\pi(t + 1))$

   c. $h(t) = \cos(63t)$

   d. $k(t) = \sin(628t)$

CHAPTER 8

Trigonometric Laws, Identities, and Equations

8.1 Solving Right Triangles

Introduction

Solving Right Triangles

Introduction

We need to measure only some of the sides and angles of a triangle to determine the sizes of the remaining sides and angles. This makes triangles useful in numerous situations including determining boundaries in real estate developments and finding the size of objects that cannot be measured directly. *Solving a triangle* means determining the measures of all three angles and the lengths of all three sides.

Solving Right Triangles

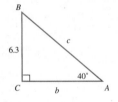

Figure 8.1

Our resource bank currently contains sufficient information for us to formulate methods to solve *right* triangles. We can solve any triangle if we know three of the six parts of the triangle including at least one side. For example, suppose we are given a right triangle with an acute angle of 40° and the side opposite of length 6.3 inches and we wish to find the remaining parts. Figure 8.1 shows this right triangle.

Let's analyze the problem. Can you think of a way to obtain the measure of the acute angle *B* knowing that one acute angle *A* is 40° and the right angle *C* is 90°? You can if you recall from geometry that the sum of the angles of any triangle is 180°. With this information, we determine that the measure of angle *B* is 50°. We move forward with our analysis by asking ourselves how we might determine the length of one of the other sides, say *b*. The idea here is to involve two known pieces of information with one that is not known in an equation that contains a trigonometric function:

$$\text{trig}(\textit{angle measure}) = \frac{\textit{one side}}{\textit{another side}}$$

We select one of the six trigonometric functions (and its defining ratio) that we studied in Chapter 7.

In this situation, we know angle *A* and side *a* and we are trying to determine side *b*. These three parts of the right triangle involve the side opposite and the side adjacent to angle *A*. Connecting these three parts of the triangle, we select the tangent function from the six trigonometric functions studied in Chapter 7. Thus we write

$$\tan(A) = \frac{a}{b}$$

or

$$\tan(40°) = \frac{6.3}{b}$$

The cotangent function would work just as well since it involves the same three parts of the right triangle.

We now see that this equation contains but one unknown b and we solve for it algebraically to obtain

$$b = \frac{6.3}{\tan(40°)}$$

Using the calculator, we obtain $b \approx 7.5$.

How might we determine the length of c? We have two options: use the Pythagorean Theorem or involve c with two of the known values in one of the six trigonometric ratios. If we choose the ratio method, we might ask ourselves what trigonometric ratio involves the hypotenuse c, angle $A = 40°$, and side $a = 6.3$? We again study the diagram and note that these three pieces of information involve the hypotenuse and side opposite angle A. This should call to mind the sine function, in which case we write

$$\sin(40°) = \frac{6.3}{c}$$

Solving for c gives

$$c = \frac{6.3}{\sin(40°)} \quad \text{or} \quad c \approx 9.8$$

This method of solving right triangles is very effective. Again, it is a matter of drawing a right triangle and labeling all the known and unknown quantities. Then we need to think about what three pieces of information, two known and one unknown, we can involve in one of the six trigonometric ratios. The remainder of the solution involves a repetition of this process and the algebra to isolate the unknown. We can then use our calculators, if necessary, to find or approximate the value of each of the unknown parts of the right triangle.

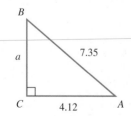

Figure 8.2

EXAMPLE SET A

Solve the right triangle given in Figure 8.2.

Solution:

Consider unknown angle A along with known adjacent side b and hypotenuse c. This combination leads us to consider the cosine function, which involves the side adjacent over the hypotenuse. We write

$$\cos(A) = \frac{4.12}{7.35} \approx 0.5605$$

We can now use $\cos^{-1}$ on the calculator to approximate the measure of angle A as $55.9°$ or 0.976 radians (if the calculator is in radian mode). We can now determine angle B by summing angle A and right angle C and subtracting the total from $180°$. This gives us $34.1°$ for the measure of angle B. We can determine the length of side a by either involving it in an appropriate trigonometric ratio or by using the Pythagorean Theorem. Using the Pythagorean Theorem gives

$$a \approx \sqrt{7.35^2 - 4.12^2} \approx 6.09$$

Thus, $A \approx 55.9°, B \approx 34.1°, C = 90°, a \approx 6.09, b = 7.35,$ and $c = 4.12$.

Solving right triangles is useful in many applications. The following example illustrates one use.

EXAMPLE SET B

The angle of elevation from the ground to a hot-air balloon is determined to be $33.6°$ using a transit. We also know that the distance from a point on the ground directly below the balloon to the transit is 1430 feet (see Figure 8.3). How high is the balloon?

Solution:

We see that a right triangle is formed. We further observe that the two known quantities and the unknown quantity h are involved with the side opposite and the side adjacent to the angle $33.6°$. This combination leads us to the tangent function and we then write

$$\tan(33.6°) = \frac{h}{1430} \quad \text{or} \quad h = 1430\tan(33.6°) \approx 950$$

Therefore, the balloon is about 950 feet above the ground.

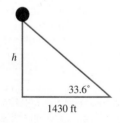

h

$33.6°$

1430 ft

Figure 8.3

EXERCISES

In exercises 1–14, find the remaining three parts of right triangle ABC where angle C is the right angle and two other parts of the triangle are given.

1. Side $a = 12$ and $m\angle A = 64°$.

2. Side $a = 5.24$ and $m\angle B = 27.6°$.

3. Side $b = 3.17$ and $m\angle A = 14.5°$.

4. Side $b = 773$ and $m\angle B = 35.2°$.

5. Hypotenuse $c = 4.28$ and $m\angle A = 41°$.

6. Hypotenuse $c = 15.9$ and $m\angle B = 11.6°$.

7. Hypotenuse $c = 8.86$ and $m\angle A = 71.3°$.

8. Hypotenuse $c = 1.25$ and $m\angle B = 23.2°$.

9. Side $a = 24.7$ and $m\angle A = 43.5°$.

10. Side $b = 17.7$ and $m\angle A = 38.4°$.

11. Sides $a = 1.46$ and $b = 3.52$.

12. Sides $a = 431$ and $c = 768$.

13. Sides $b = 38.5$ and $c = 52.3$.

14. Sides $a = 17.9$ and $b = 12.6$.

15. Mark measures the shadow of a flagpole to be 34 feet and estimates the angle of elevation of the sun to be approximately 45°. Approximate the height of the flagpole.

16. Manoj decides to get an approximate value for the height reached by a spotlight advertising the opening of a new store. He walks down the street from the spotlight and estimates the distance from the spotlight and himself to be 100 yards. He then estimates the angle of elevation of what appears to be the end of the vertical beam of light to be 60°. Based on this information, what value should Manoj come up with for the height of the light?

17. Ermanno is standing outside his store one day when he notices an inflated blimp in the air. He knows that a new store two miles down the road is due to open and figures that this is advertising for the grand opening. He estimates the angle of elevation of the blimp to be 30°. What is the approximate height of the blimp?

18. Kristin is looking out the window of a hotel next to the ocean. She sees Rich on a surfboard riding a wave. She looks straight out and then down to his location and guesses the angle to be 30°. A little while later, she guesses the angle down to Rich to be 45°. She is on the 35th floor and estimates that each floor is about 12 feet above the one below it. With the knowledge that her eyes are about five feet six inches above the floor, what value would she calculate for the distance Rich travelled?

19. Mina is flying a kite and has all of the 300 yards of string out. Her eyes are about five feet above the ground and she estimates the angle from the horizontal to the kite to be 60°. Her hands are about four feet above the ground. She decides that the distance from her hands to the kite is 20 yards less due to the sag in the string. What estimate should she calculate for the height of the kite?

20. Describe how you could use right triangle trigonometry to estimate the height of a tree if all you have is a 25-foot measuring tape and you do not wish to climb the tree.

21. Last night a sky watcher observed a point of light cross her line of sight (vertically upwards). The point of light moved steadily across the sky until it disappeared below the horizon in exactly six minutes.

    (a) Assuming the radius of the earth is 6400 km, write an equation that expresses the angle (see diagram) as a function of h. Assume the point is describing a circular orbit around the earth.

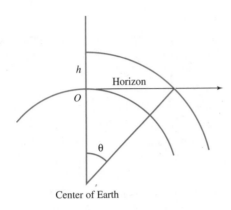

Center of Earth

(b) The fact that it took six minutes to describe the angle θ is insufficient information to solve for h. What other data are needed?

(c) The sky watcher kept watching the sky and saw the object again crossing the vertical at about 65 minutes from the first observation. What is the height of the object?

22. A huge douglas fir tree is struck by lightning, causing a break. The part broken is still attached to the main trunk, but the top has fallen to the ground. Ranger Roger measured the distance x between the trunk and the top of the tree and the angle made by the top with the ground. What is the original height of the tree if $\theta = 33°$ and $x = 133'$?

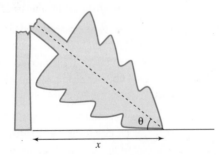

23. Two boats will leave the dock at the same time, one traveling 40 mph in the NE direction and the other going 20 mph in the SE direction. How far apart will they be in one hour? From this calculation can you estimate how far apart they will be in two hours? Three hours?

24. Mary, the lighthouse keeper, wants to measure the height of the lighthouse. She walks 20' away from the base of the lighthouse and measures the angle of elevation of the top of the lighthouse; this measurement is 43°. If her eyes are at the height of 5'8", what is the height of the lighthouse?

25. (Refer to the previous exercise) Wanting to be sure that her calculations are correct, Mary devises a different method. She measures the length of the shadow of the lighthouse at noon on a day when the sun's highest altitude is 31° (the sun's altitude = the angle of elevation of the sun) and finds it to be 40.4 feet. Using the accuracy of her measurements, do the results of these two exercises agree?

26. Joe climbs to the top of the tallest building in his city. While enjoying the view, he spots his car parked on the street quite a distance away. His telescope allows him to measure the angle of depression of the cars (which is 59°). The height of the building is well-known—320 feet. How far away is the car parked from the building?

27. (Refer to the previous exercise) If Joe's car is parked on top of a parking garage that is 22 feet high, what is the distance of the garage from the building if the angle of depression observed by Joe is the same, namely 59°?

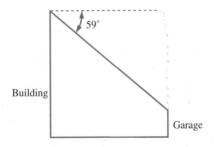

28. The building code of Flat City, USA, requires that the pitch of a roof be at most one in ten. (Pitch = 1/10 means that as you go 10 units along a slope you rise by 1 unit, which is different from the slope of a line.) The WeBuild Company is planning a new development. One style of houses is designed to be rectangular in shape ($20' \times 35'$). The side view is shown in the graph below. The roof is positioned symmetrically with respect to the sides. What is the lowest legal height of the top edge of the roof above the ground? (Hint: $\sin \theta = \dfrac{1}{10}$.)

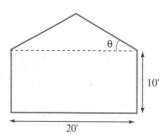

29. (Refer to the previous exercise) The WeBuild Company is also planning to build a different kind of house whose side view is shown below. The front wall is ten feet high. How high should the back wall be if they still maintain the least pitch for the roof?

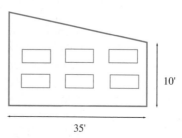

30. A rain gutter is made by folding a 9″ wide metal sheet (any length) into three equal parts and then bending the two sides at an angle with the vertical (as shown).

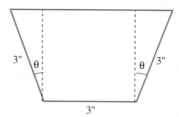

Since the gutter is needed to obtain water from the roof, the maximum value of the cross-sectional area is desired (this will make the volume of the trough a maximum). Find the area of the trapezoid as a function of θ. Then draw the graph of the area function to find the best possible value for θ. Is it possible for you to solve this problem analytically (like in the previous exercise)?

8.2   Law of Sines

Introduction

Law of Sines

A General View of the Ambiguous Case

The Derivation of the Law of Sines

Introduction

There are numerous other applications involving triangles but not necessarily right triangles. Solving non-right triangles still requires that three parts of the triangle be known, at least one of which must be the length of a side. We need to develop two methods to solve non-right triangles because neither method alone can handle all the possible cases that can arise. In this section, we state, examine, and derive the law of sines. In the next section, we study the other method, the law of cosines.

Law of Sines

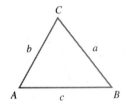

Law of Sines

The law of sines is so named because the formula involves the sines of the angles of the triangle. We state the law first, use it in some examples, and derive it last.

Law of Sines

Given triangle ABC with sides a, b, and c, the law of sines states that

$$\frac{a}{\sin(A)} = \frac{b}{\sin(B)} = \frac{c}{\sin(C)}$$

You might remember this formula by saying it to yourself in words as: a is to $\sin(A)$ as b is to $\sin(B)$ as c is to $\sin(C)$. This shorthand representation is called a double equality and states that any two of the three possible pairs you choose are equal. It is also useful to realize that we can employ the law of sines only when the information given includes a side and the angle opposite that side, that is, one of the ratios. Let us apply this formula in an example.

EXAMPLE SET A

Find the remaining three parts of triangle ABC given $m\angle A = 37°$, $m\angle B = 48°$, and side $a = 11.2$ inches. See Figure 8.4.

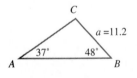

Figure 8.4

Solution:
We start by drawing the triangle roughly to scale and labeling the sides and angles because it can give us insight into the expected solution. We should develop the habit of sketching the triangle roughly to scale.

We know one of the ratios in the law of sines since we know side a and angle A. This is a sure sign for us to use this method. Since we also know angle B, we begin with:

$$\frac{a}{\sin(A)} = \frac{b}{\sin(B)}$$

Next we substitute the given values to obtain:

$$\frac{11.2}{\sin(37°)} = \frac{b}{\sin(48°)}$$

Then we solve for b:

$$b = \frac{11.2\sin(48°)}{\sin(37°)}$$

We can now use the calculator (making sure it is in degree mode) to obtain:

$$b \approx \frac{11.2(0.7431)}{0.6018} \approx 13.8$$

We can determine angle C since we know the other two angles, so $m\angle C = 180 - (37 + 48) = 95$. Now we use

$$\frac{a}{\sin(A)} = \frac{c}{\sin(C)}$$

and substitute to obtain:

$$\frac{11.2}{\sin(37°)} = \frac{c}{\sin(95°)}$$

Solving for c gives us

$$c = \frac{11.2\sin(95°)}{\sin(37°)}$$

Using our calculator gives us

$$c \approx \frac{11.2(0.9962)}{0.6018} \approx 18.5$$

The law of sines has allowed us to determine the three unknown parts of this triangle because we knew one of the ratios in the formula. Thus, $A = 37°$, $B = 48°$, $C = 95°$, $a = 11.2$, $b \approx 13.8$, and $c \approx 18.5$, where the sides are in inches.

It is important to note that whenever we are given two angles of a triangle and the length of any side, we can use the law of sines even if the given side is not opposite one of the given angles. This is possible because we can always find the measure of the third angle using the fact that the sum of the three angles of a triangle is 180°. Once the third angle is known, we then know one of the ratios.

EXAMPLE SET B

Find the remaining three parts of triangle ABC given $m\angle B = 78.2°$, $m\angle C = 34.8°$, and side $a = 4.85$ meters. See Figure 8.5.

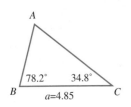

Figure 8.5

Solution:
We start by drawing the triangle roughly to scale and labeling the sides and angles.

We think about the information given and look toward the possibility of using the law of sines. We realize we need to know one of the ratios in the formula and, at first glance, that does not seem to be the case. However, there is a fair chance we will connect the fact that we know two of the angles of the triangle thereby allowing us to obtain the third. This discovered, we know the law of sines will apply. First we find the measure of angle A using $m\angle A = 180 - (78.2 + 34.8) = 67$. Next we apply the law of sines. We use:

$$\frac{a}{\sin(A)} = \frac{b}{\sin(B)}$$

Substituting the given values into the law of sines yields:

$$\frac{4.85}{\sin(67°)} = \frac{b}{\sin(78.2°)}$$

We solve for b to obtain:

$$b = \frac{4.85 \sin(78.2°)}{\sin(67°)}$$

We can use our calculators to evaluate this expression.

$$b \approx \frac{4.85(0.9789)}{0.9205} \approx 5.16$$

We continue to use the law of sines to find c.

$$\frac{a}{\sin(A)} = \frac{c}{\sin(C)}$$

Substitution yields:

$$\frac{4.85}{\sin(67°)} = \frac{c}{\sin(34.8°)}$$

Algebraic manipulations are needed once again to solve for c:

$$c = \frac{4.85 \sin(34.8°)}{\sin(67°)} \approx \frac{4.85(0.5705)}{0.9205} \approx 3.01$$

Thus, $A = 67°$, $B = 78.2°$, $C = 34.8°$, $a = 4.85$, $b \approx 5.16$, and $c \approx 3.01$, where the sides are in meters.

Let's look at a practical application of the law of sines.

EXAMPLE SET C

Two people are stationed on a road that a hot-air balloon race will pass over. A special award is given for the balloon that is the highest when it crosses the road. Each person has a sextant that can measure the angle of elevation of an object. The two people are stationed a distance of 1250 feet apart. The first balloon that crosses the road makes angles of 32.8° and 43.4° as indicated in Figure 8.6. Find the height of this balloon.

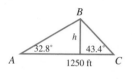

Figure 8.6

Solution:

We look at each of the two right triangles containing h and realize that we know only angles but no sides of the right triangles. After some thought, we see that we can calculate h using right-triangle trigonometry if we find the distance of either observer to the foot of the balloon. We can compute the distance AB of triangle ABC using the law of sines because we have two angles and a side.

First we find angle ABC as follows: $B = 180° - (32.8° + 43.4°) = 103.8°$. Then we apply the law of sines to obtain:

$$\frac{c}{\sin(C)} = \frac{b}{\sin(B)}$$

$$\frac{c}{\sin(43.4°)} = \frac{1250}{\sin(103.8°)}$$

or

$$c = \frac{1250\sin(43.4°)}{\sin(103.8°)} \approx 884.4$$

Using right-triangle trigonometry,

$$\sin(32.8°) = \frac{h}{884.4}$$

and

$$h = 884.4\sin(32.8°) \approx 479$$

Thus, the height of the balloon, to the nearest foot, is 479 feet.

We need to discuss one more case to complete our study of the law of sines. This is the case in which the given information of a triangle can lead to two solutions. This case is often called the ambiguous case of the law of sines. Two possible solutions for a triangle can arise when the information given is two sides of a triangle and the angle opposite one of them. Since one of the ratios is known, it clearly involves the law of sines. We can get a fairly good notion of when this situation occurs by studying the following example.

EXAMPLE SET D

Find the remaining parts of triangle ABC given side $a = 4$, side $b = 6$, and angle $A = 40°$.

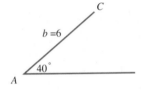

Figure 8.7

Solution:
As before, we sketch the triangle roughly to scale. The following procedure can help illuminate the ambiguous case. Draw a horizontal line segment of any length. At the left endpoint of the segment, draw a ray that roughly represents the given angle $A = 40°$. Along this ray, mark off a length that you decide represents the length of the non-opposite side, $b = 6$. We now have the sketch in Figure 8.7.

We know side $a = 4$ and that it lies opposite angle A. Now try to imagine a length that is two-thirds of the length you marked off for side b and spread your thumb and first finger to approximate that length of 4 for a. Now with your thumb and finger spread to this length, place your finger on the endpoint of side b away from angle A and swing your hand like a compass with the first finger fixed at the endpoint of b. Can you see that your thumb will cross the horizontal line in two places without going past the triangle vertex at A? Your imagery should look something like Figure 8.8.

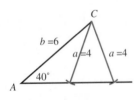

Figure 8.8

Hence, we see that we obtain two triangles that satisfy the original information given; these are triangle ABC and triangle $AB'C'$ as shown in Figure 8.9.

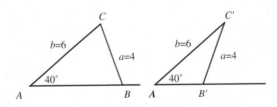

Figure 8.9

We use the law of sines in triangle ABC to obtain

$$\frac{4}{\sin(40°)} = \frac{6}{\sin(B)}$$

$$\sin(B) = \frac{3}{2}\sin(40°)$$

$$B \approx 74.6°$$

Using this value of B, we can obtain angle $C = 65.4°$ and side $c = 5.66$. Now how do we go about solving triangle $AB'C'$? Referring to Figure 8.8, we might be able to see that obtuse angle B' and angle B are supplementary angles (angles whose sum is $180°$) since the base angles of the isosceles triangle formed are equal. Hence, $m\angle B' = 180° - m\angle B$ or $B' = 180 - 74.6 = 105.4°$. We obtain angle C' using $C' = 180 - (A + B') = 180 - (40 + 105.4) = 34.6°$. Finally, we can find side c' using the law of sines as

$$\frac{4}{\sin(40°)} = \frac{c'}{\sin(34.6°)}$$

This yields $c' \approx 3.53$. Thus, $A = 40°$, $B \approx 74.6°$, $C \approx 65.4°$, $a = 4$, $b = 6$, and $c \approx 5.66$, or $A = 40°$, $B' \approx 105.4°$, $C' \approx 34.6°$, $a = 4$, $b = 6$, and $c' \approx 3.53$.

Had side a in the previous example been 7 instead of 4, your finger compass would have crossed the horizontal line once on the right of vertex A and again on the left of vertex A thus giving only one triangle that satisfies the original conditions. You should try it to convince yourself.

A General View of the Ambiguous Case

The previous example gives a fairly good visual picture of what is going on in the ambiguous case. However, it is not a proof and rough sketches can be inconclusive depending upon the particular information given. Therefore, we need to make a theoretical connection to handle these situations. We consider the same case in general where sides a and b are given as well as angle A. Figure 8.10 shows the horizontal line with angle A constructed and side b marked off appropriately.

If we draw a perpendicular line segment from the endpoint of side b to the horizontal line, we can see that side a must be at least this long to even form a triangle. This length can be calculated and is $h = b\sin(A)$. (Check this with your right triangle information.) Thus if $a = b\sin(A)$, there is only one triangle and it is a right triangle. Looking at Figure 8.10 also shows that if a

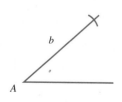

Figure 8.10

is greater than $b\sin(A)$ but less than b, two triangles will result. Finally, if a is greater than $b\sin(A)$ and bigger than or equal to b, then only one triangle will result.

The ambiguous case can arise when two sides and an angle opposite one of them is given. We need to be able to deal with the situation regardless of the particular sides and angle given.

The Derivation of the Law of Sines

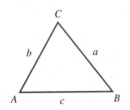

Figure 8.11

In Figure 8.11, a general triangle is shown and labeled. We derive the law of sines for this acute-angle triangle and leave the derivation for the general triangle with one obtuse angle as a project.

The approach we will use is to construct two different altitudes of triangle ABC. Each altitude will form a pair of right triangles in triangle ABC. First draw the altitude from vertex C to side AB as in Figure 8.12. The altitude h can be computed using right triangle-trigonometry in two ways.

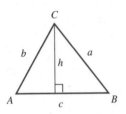

Figure 8.12

$$\sin(A) = \frac{h}{b} \qquad \sin(B) = \frac{h}{a}$$

or   or

$$h = b\sin(A) \qquad h = a\sin(B)$$

Since h is the same, we obtain

$$a\sin(B) = b\sin(A)$$

Then, dividing both sides by $\sin(A)\sin(B)$, we obtain

$$\frac{a}{\sin(A)} = \frac{b}{\sin(B)} \tag{8.1}$$

Next we compute another altitude of the triangle using a as the base and letting k represent the altitude (see Figure 8.12).

From right-triangle trigonometry, we have

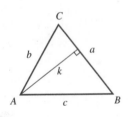

Figure 8.13

$$\sin(C) = \frac{k}{b} \qquad \sin(B) = \frac{k}{c}$$

or   or

$$k = b\sin(C) \qquad k = c\sin(B)$$

Since k is the same, we obtain

$$b\sin(C) = c\sin(B)$$

Then, dividing both sides by $\sin(B)\sin(C)$, we obtain

$$\frac{b}{\sin(B)} = \frac{c}{\sin(C)} \tag{8.2}$$

Using the results from 8.1 and 8.2, we obtain the law of sines:

$$\frac{a}{\sin(A)} = \frac{b}{\sin(B)} = \frac{c}{\sin(C)}$$

We indicated at the beginning of this section that we will need two methods to solve non-right triangles. The law of sines can be used to solve triangles in which two sides and an angle opposite one of them is known or two angles and any side is known. This is true because we can determine one of the ratios in the formula for the law of sines when such information is given. The first case in which two sides and an angle opposite one of them is given is easily seen. The second case with two angles and any side given is also easy, provided we recall that we can determine the third angle of a triangle whenever we know two of them. We can always use the law of sines when we know four parts of the triangle.

Let us think about why we might need another method to solve some triangles. Consider the case in which we are given the lengths of all three sides. After some thought, we might realize that we need to know an angle in order to know any one of the ratios in the law of sines. Remember

$$\frac{a}{\sin(A)} = \frac{b}{\sin(B)} = \frac{c}{\sin(C)}$$

To obtain an equation that allows us to solve for a single unknown, we need at least one of these ratios. Thus, the law of sines cannot help us solve a triangle when three sides are given.

The only remaining situation is when we know two sides and the included angle of a triangle. Why doesn't the law of sines work in this case? Again, we need to convince ourselves that we cannot determine any one of the ratios in the law of sines. We can only determine a fourth part of a triangle directly when two of the given parts are angles and we compute the third angle by subtracting the sum of the two given angles from 180°. Perhaps we can now more clearly see that the law of sines cannot handle the cases when either (1) three sides or (2) two sides and the included angle are given. We will develop the law of cosines to solve these cases in the next section.

EXERCISES

In exercises 1–12, find the remaining parts of the given triangle *ABC*.

1. $\angle A = 58.0°$, $\angle C = 38.7°$, and $c = 1.91$.
2. $\angle A = 66.9°$, $\angle B = 61.8°$, and $a = 3.16$.
3. $\angle B = 54.7°$, $\angle C = 42.5°$, and $a = 3.68$.
4. $\angle A = 24.5°$, $\angle C = 36.1°$, and $a = 3.03$.
5. $\angle B = 52.6°$, $b = 30.3$, and $c = 36.8$.
6. $\angle A = 18.2°$, $a = 261$, and $c = 175$.
7. $\angle C = 31.0°$, $b = 22.2$, and $c = 12.0$.
8. $\angle B = 40.3°$, $b = 1.65$, and $c = 2.58$.
9. $\angle A = 56.3°$, $a = 1.90$, and $c = 1.55$.
10. $\angle A = 77.5°$, $a = 31.5$, and $b = 19.8$.
11. $\angle C = 32.1°$, $b = 3.45$, and $c = 1.93$.
12. $\angle A = 135°$, $a = 3.71$, and $c = 1.53$.

13. Two naval ships that are 175 miles apart get a distress call from a freighter. The naval ships are able to determine the angles from the line joining themselves to the ship in distress as in the accompanying figure. Determine how far the closest naval ship is to the freighter.

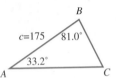

14. Janice and Susie are 680 yards apart on opposite sides of the air strip at Edward's Air Force Base in the desert. They sight the Columbia space shuttle coming in for a landing between them. Just as the shuttle passes between them, they determine the angles of elevation to be $4.9°$ and $7.8°$, respectively. Determine whether Janice or Susie is furthest from the shuttle and by how much. Also determine the distance the shuttle was off the ground when it passed between them.

15. Jane is on the tenth floor of a hotel. Her friend Gail is directly below on the sixth floor. Jane figures that with the standard room height of eight feet, she is 80 feet and Gail is 48 feet above the street. Jane looks out her window to the building across the street and guesses that the angle from the horizontal to the top of that building is about $30°$. She calls Gail and asks her to estimate the angle from the horizontal to the top of the building from her window. Gail estimates that the angle is about $50°$. Jane then determines the approximate height of the building across the street. What value do you think she got?

16. John decides to get some idea as to the height of a vertical tree in front of his house. His house is on a street that is not level. Standing next to the tree, he looks up and then down the street and guesses that the angle between the tree and the street is about $110°$. He then walks down the street about 100 feet and looks up the street and then up to the top of the tree estimating that angle to be about $30°$. What value should John get as the approximate height of the tree?

17. Lisa decides to use her math to estimate the distance between her house, which is next to a river, and her friend Kim's house on the other side of the river. She estimates the angle from her house to Kim's to be about $40°$ by looking down a line parallel to the river's bank and then looking at Kim's house. She then estimates the angle between her house and a house that is closer but still on the other side of the river to be about $60°$. Lisa then decides she needs the distance between the two houses on the other side of the river so she crosses the river on a bridge and paces off 96 yards between the houses. She also decides that her eyeball estimates of the angles could be off by as much as $5°$ high or low for each angle. What are the approximate maximum and minimum distances between her house and Kim's using this information? What would be a faster way for Lisa to estimate this distance?

18. The dispatcher at the main office of a trucking company knows that a driver has just left the company's warehouse in Tracy headed for their warehouse in Sweetly as shown in the accompanying figure. The dispatcher, whose radio has a range of 50 miles, wants the driver to return to the Tracy warehouse for a special delivery item. The driver is travelling at 55 mph. How soon will she be able to contact the driver?

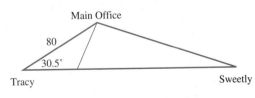

19. Standing directly in front of his family's house, Mark estimates that the angle of elevation to the top of a tree in front of the house is 50°. When he left town 25 years ago, the angle of elevation was estimated to be about 30°. He paces off the distance from the house to the tree and finds that it is about 102 feet. How much has the tree grown?

20. Ann is on a cliff that is 44 feet high. Using a clinometer (a pocket-size instrument that can measure angles) she sights her brother swimming in to shore. She determines the angles from the horizontal down to her brother and to the shore's edge to be 34.9° and 53.4°, respectively. How far does her brother have to swim to reach the shore's edge?

21. A real estate developer bought a triangular piece of land in the shape given in the diagram.

(a) Find the length of the other two boundary lines correct to the nearest foot. (Hint: Use the sine formula.)

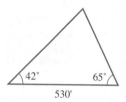

(b) What is the area of this land?

22. A surveyor sent one of his assistants to chart the diagram of a piece of land. His assistant brought back the following information about the triangular piece of land: If he marked the vertices A, B, C, then $a = 310$ feet, $C = 23°$, $c = 44$ feet. The boss immediately got angry with him. Give two reasons for his anger.

8.3   Law of Cosines

Introduction

Law of Cosines

The Derivation of the Law of Cosines

Introduction

We saw in the last section that we can use the law of sines to solve a non-right triangle provided we know three pieces of information, including at least one side and one of the ratios. This situation occurs when we are given two sides and an angle opposite one of them or we are given two angles and any side. We cannot employ the law of sines when the information involves three sides or two sides and the included angle. These are the remaining possibilities, and we need to develop yet another approach to solve such triangles. As in the previous section, we state the law (of cosines), use it to solve several examples, and then derive the formula.

Law of Cosines

The law of cosines is so named because the formulas involve the cosines of the angle of a triangle.

<table>
<tr><td>**Law of Cosines**</td></tr>
<tr><td>Given triangle ABC with sides a, b, and c, the law of cosines states that
$$c^2 = a^2 + b^2 - 2ab\cos(C)$$</td></tr>
</table>

Law of Cosines

Notice that the cosine is taken of the angle between the two sides appearing on the right of the equal sign and opposite the side appearing on the left. Thus, the formula could be written in either of the following ways.

$$a^2 = b^2 + c^2 - 2bc\cos(A)$$

or

$$b^2 = a^2 + c^2 - 2ac\cos(B)$$

Enough of notational conventions. Let's work an example.

EXAMPLE SET A

Solve triangle ABC given $a = 26.3$, $b = 30.7$, and $\angle C = 41.3°$ in Figure 8.14.

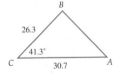

Figure 8.14

Solution:
As before, it is useful to sketch the triangle roughly to scale.
    Since angle C is given, we would use the following form of the law of cosines:

$$c^2 = a^2 + b^2 - 2ab\cos(C)$$

and upon substituting the given values we have

$$c^2 = 26.3^2 + 30.7^2 - 2(26.3)(30.7)\cos(41.3°)$$

Computing the square root of the expression on the right gives $c \approx 20.5$. Now that we have the side opposite $\angle C$, we can use the law of sines to find the remaining angles using the full computational accuracy of our calculators. Applying the law of sines, we obtain $A \approx 57.8°$ and $B \approx 80.9°$. Thus, $A \approx 57.8°$, $B \approx 80.9°$, $C = 41.3°$, $a = 26.3$, $b = 30.7$, and $c \approx 20.5$. If we use the rounded value of 20.5 for c in the calculation of A and B, those values can differ considerably from $57.8°$ and $80.9°$.

EXAMPLE SET B

Find the angle opposite side b in a triangle with sides $a = 5.2$, $b = 3.1$, and $c = 6.8$.

Solution:
We use the formula

$$b^2 = a^2 + c^2 - 2ac\cos(B)$$

because we know all three sides and we wish to find $\angle B$. Substituting the values, we obtain

$$3.1^2 = 5.2^2 + 6.8^2 - 2(5.2)(6.8)\cos(B), \quad B \text{ is measured in degrees.}$$

Solving for $\cos(B)$ we obtain

$$\cos(B) = \frac{5.2^2 + 6.8^2 - 3.1^2}{2(5.2)(6.8)} \approx 0.9003$$

Using the inverse cosine function, we have that $B \approx 25.8°$. (The other angles can be determined by repeated use of the law of cosines or the law of sines.)

In those cases where we compute an additional side from a given angle and the two sides adjacent to it, the law of sines can be used to compute the remaining parts of the triangle. We must be careful in doing this as the next example illustrates.

EXAMPLE SET C

Solve triangle ABC given $a = 2.15$, $b = 5.38$, and $\angle C = 38.4°$.

Solution:
As before, it is useful to sketch the triangle roughly to scale as in Figure 8.15. We start with

$$c^2 = a^2 + b^2 - 2ab\cos(C)$$

Upon substituting the given values we have

$$c^2 = 2.15^2 + 5.38^2 - 2(2.15)(5.38)\cos(38.4°)$$

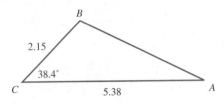

Figure 8.15

Using our calculator, we find $c \approx 3.93$. Now that we have the side opposite $\angle C$, we can use the law of sines to find the remaining parts. Suppose we decide to find $\angle B$ next. The law of sines gives us

$$\frac{b}{\sin(B)} = \frac{c}{\sin(C)}$$

and substituting

$$\frac{5.38}{\sin(B)} = \frac{3.93}{\sin(38.4°)}$$

$$\sin(B) = \frac{5.38\sin(38.4)}{3.93} \approx 0.850$$

This yields $B \approx 58.2°$ using the inverse sine function. Is this answer reasonable from our rough sketch above? It seems clear that $\angle B$ is obtuse. What went wrong?

There are two possible angles between $0°$ and $180°$ (triangle angles) that are solutions to $\sin(B) = 0.850$. One is acute and has measure $58.2°$; the other is obtuse and has an approximate measure of $121.8°$. The correct value for B is $121.8°$. The value of A is approximately $19.8°$ by subtracting the sum of B and C from $180°$.

This example shows that we need to be careful in using the law of sines, especially after using the law of cosines, to find the parts of a triangle. Sketching a triangle with the given information roughly to scale will often help in determining which solution to an equation of the form $\sin(x) = k$ is appropriate for the given problem. If we are unable to determine which value is correct from our diagram, we use the law of cosines to find the other angles even though this requires more computation.

The law of cosines can be applied in a number of practical situations where two sides and the included angle are known.

EXAMPLE SET D

A radar station detects an SOS signal from a ship in distress and determines the location of the distressed ship and the location of another ship that is in the area. This information is illustrated in the diagram below. If the help ship travels at 15 knots, how long will it take to reach the ship in distress?

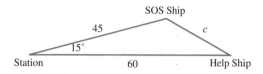

Figure 8.16

Solution:
We label the distance between the ships as c in Figure 8.16. We begin with the law of cosines because we know two sides and the included angle.

$$c^2 = a^2 + b^2 - 2ab\cos(C) \tag{8.1}$$

Substituting the values from the diagram, we have

$$c^2 = 45^2 + 60^2 - 2(45)(60)\cos(15°) \tag{8.2}$$

This gives a value of $c \approx 20.2$ nautical miles. Now dividing 20.2 nautical miles by 15 nautical miles per hour gives approximately 1.35, which is about 1 hour and 21 minutes. Based on this information, the radar station radios the distressed ship to inform them that help is on the way and will arrive in about an hour and 20 or 30 minutes.

The Derivation of the Law of Cosines

In Figure 8.17, a general triangle has been placed with one side along the positive x-axis and one vertex at the origin.

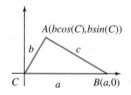

Figure 8.17

We need to tap our resource bank of unit circle concepts to be able to label the coordinates of the vertex A. Recall that a point on the unit circle on the terminal side of angle C (in standard position) has coordinates $(\cos(C), \sin(C))$. Thus, from the similarity of circles, the coordinates of A at a distance b from the origin are $(b\cos(C), b\sin(C))$. We now compute the square of the distance between A and B using the distance formula and set it equal to c^2 (the length of the side opposite angle C).

$$c^2 = [b\cos(C) - a]^2 + [b\sin(C) - 0]^2$$

$$c^2 = b^2 \cos^2(C) - 2ab \cos(C) + a^2 + b^2 \sin^2(C)$$

$$c^2 = b^2 [\cos^2(C) + \sin^2(C)] - 2ab \cos(C) + a^2$$

$$c^2 = a^2 + b^2 - 2ab \cos(C)$$

The same derivation holds for an obtuse triangle as well, since the coordinates of vertex A will still be the same.

EXERCISES

In exercises 1–14, find the remaining parts of the given triangle ABC.

1. $a = 2.54$, $b = 3.85$, and $\angle C = 25.8°$.

2. $a = 18.7$, $b = 14.2$, and $\angle C = 41.6°$.

3. $a = 548$, $c = 490$, and $\angle B = 97.4°$.

4. $a = 0.56$, $c = 0.47$, and $\angle B = 14.8°$.

5. $b = 7.83$, $c = 11.2$, and $\angle A = 25.8°$.

6. $b = 48.5$, $c = 37.9$, and $\angle A = 137°$.

7. $a = 1290$, $c = 2100$, and $\angle B = 55.8°$.

8. $b = 22.5$, $c = 15.8$, and $\angle A = 115°$.

9. $a = 2.96$, $b = 3.78$, and $c = 4.54$.

10. $a = 541$, $b = 453$, and $c = 388$.

11. $a = 23.8$, $b = 36.2$, and $c = 23.7$.

12. $a = 0.35$, $b = 0.42$, and $c = 0.29$.

13. $a = 7.83$, $b = 5.49$, and $c = 6.22$.

14. $a = 10.0$, $b = 14.0$, and $c = 17.5$.

15. Two helicopter stations that are 137 miles apart receive a distress call from a ship. By the location given by the ship, the two helicopter stations determine that they are 283 and 314 miles, respectively, from the ship. Determine the direction the closer helicopter should head by finding the angle joining the line between the stations and the line joining the ship and the closer helicopter.

16. The sides of a triangular plot of land are measured to be 340 feet, 285 feet, and 221 feet, respectively. Find the measure of the angle opposite the longest side.

17. Two ships leave the same port at the same time along paths that make an angle of 115° with each other. If one ship travels at 22 nautical miles per hour and the other at 18 nautical miles per hour, how many nautical miles apart are the ships after two hours?

18. Two cars leave from the same place at the same time along straight roads that make an angle of 75° with respect to each other. If one car travels at 70 miles per hour and the other at 55 miles per hour, what is the distance between the cars after one hour?

19. An ant started walking at one corner of a rectangular box, traveled along a diagonal to the other corner of the box, and then went up along the other diagonal as shown. What is the angle between these two diagonals?

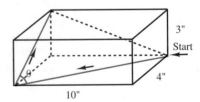

20. A missile tracking station just spotted one
of their planes coming back after a sortie.
Right behind that plane they see an enemy
plane. The instruments at the station give the
readings for the distances of the two planes
and the angle between the two directions.
What is the distance between the planes at
that instant?

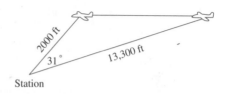

Station

8.4 **Trigonometric Identities**

Introduction

Basic Trigonometric Identities

Generating Identities

Introduction

Trigonometric identities are useful in many areas of mathematics and engineering such as calculus, differential equations, and circuit analysis to name
just a few. Identities are equations that are true for all allowable replacement
values of the variables. An algebraic identity such as:

$$\frac{1}{x^2 - 4} = \frac{1}{(x-2)(x+2)}$$

is true for all real numbers except $\{-2, 2\}$. A conditional equation like

$$\frac{1}{x^2 - 4} = 1$$

has the same allowable set of replacement values but is true for only two values,
namely, $-\sqrt{5}$ and $\sqrt{5}$. A mathematical equation is said to be conditional if it
is true for only some of the allowable replacements of the variables. We seek
solutions of conditional equations whereas we need to prove that a suspected
identity is indeed an identity whose solution set is known. We will study
trigonometric identities in the next four sections. We start with the basic
identities and use them as our building blocks.

Basic Trigonometric Identities

The set of identities considered to be basic varies among mathematicians. This should not alarm you since it is simply a matter of mathematicians deciding which core of basic identities they feel should be memorized and which identities should be derived from the core set. We start with eight basic identities and add to that number in subsequent sections. The good news is that we already know six of the eight quite well. It is also quite likely that we know the other two also. You should firmly establish the following basic identities in your resource bank.

$$\sin^2(x) + \cos^2(x) = 1$$

$$\tan(x) = \frac{\sin(x)}{\cos(x)}$$

$$\sec(x) = \frac{1}{\cos(x)}$$

$$\csc(x) = \frac{1}{\sin(x)}$$

$$\cot(x) = \frac{\cos(x)}{\sin(x)}$$

$$\cot(x) = \frac{1}{\tan(x)}$$

These are identities you have worked with often. The next two identities are also basic. They are:

$$\tan^2(x) + 1 = \sec^2(x)$$

$$1 + \cot^2(x) = \csc^2(x)$$

These are easily derived from $\sin^2(x) + \cos^2(x) = 1$ by dividing both sides of this identity by $\cos^2(x)$ and $\sin^2(x)$, respectively.

There is but one more feature of these eight basic identities that we need to consider and that is the variety of forms in which they appear. For example, we need to recognize that $\sin^2(x) = 1 - \cos^2(x)$ is the same basic identity as $\sin^2(x) + \cos^2(x) = 1$ except in a slightly different format. It is obtained by subtracting $\cos^2(x)$ from both sides of $\sin^2(x) + \cos^2(x) = 1$.

EXAMPLE SET A

Write the basic identity $\tan^2(x) + 1 = \sec^2(x)$ in at least two other forms.

Solution:

Two other forms can be obtained by subtracting 1 from both sides or $\tan^2(x)$ from both sides. The first subtraction leads to:

$$\tan^2(x) = \sec^2(x) - 1$$

and the second to:

$$\sec^2(x) - \tan^2(x) = 1$$

Of course, we need to be equally adept at recognizing $1 = \sec^2(x) - \tan^2(x)$ and $\tan^2(x) - \sec^2(x) = -1$ as we are at recognizing $\sec^2(x) - \tan^2(x) = 1$. This is not a difficult task provided we are able to remember at least one form of the basic identity, say $\tan^2(x) + 1 = \sec^2(x)$. Then if we suspect that an expression containing $\tan^2(x)$ and $\sec^2(x)$ might be the same identity in another form, we can try to algebraically manipulate one to the other to check our suspicion.

EXAMPLE SET B

Write the basic identity $\sec(x) = \dfrac{1}{\cos(x)}$ in at least two other forms.

Solution:

Two other forms can be obtained either by solving for $\cos(x)$ or by multiplying both sides by $\cos(x)$. The first leads to:

$$\cos(x) = \frac{1}{\sec(x)}$$

and the second to:

$$\sec(x)\cos(x) = 1$$

Generating Identities

Our goal concerning identities is to learn how to prove them. The details of that topic are taken up in the next section. Here, the goal is to get a feel for identities in preparation for learning the strategy to prove them.

The process of generating identities can be a lot of fun and, in addition, has no set of rules. The overall idea is to start with a *known* identity and change its form by performing acceptable algebraic operations on it. The easiest way to learn how to generate identities is to consider one or two examples.

EXAMPLE SET C

Produce at least four new identities (different from the ones above) starting with the basic identity $\sec(x) = \dfrac{1}{\cos(x)}$.

Solution:

We cannot possibly go wrong if we perform acceptable algebraic operations on this basic identity. The following decisions regarding which operations will be performed are purely arbitrary.

We can divide both sides of the original identity by $\cos(x)$ to obtain:

$$\frac{\sec(x)}{\cos(x)} = \frac{1}{\cos^2(x)}$$

We can use this new identity and replace $\cos(x)$ by $\dfrac{1}{\sec(x)}$ on the left, simplify, and then replace $\cos^2(x)$ by $1 - \sin^2(x)$ on the right to obtain:

$$\sec^2(x) = \frac{1}{1 - \sin^2(x)}$$

We can use this new identity and replace $\sec^2(x)$ by $\tan^2(x) + 1$ and also replace $\sin^2(x)$ by $\dfrac{1}{\csc^2(x)}$ and simplify the right side to obtain:

$$\tan^2(x) + 1 = \frac{1}{1 - \dfrac{1}{\csc^2(x)}}$$

$$\tan^2(x) + 1 = \frac{\csc^2(x)}{\csc^2(x) - 1}$$

Notice that if we replace $\csc^2(x) - 1$ by $\cot^2(x)$ in this last identity, we produce yet another identity. You can check your results by graphing both sides of the equation in the same graphing window. You will need to be careful inputting the right side as you must use $\csc(x) = \dfrac{1}{\sin(x)}$, which will require numerous parentheses.

Also, notice that the allowable set of replacement values for these last two identities has been restricted further. The values $\{0, \pm\pi, \pm 2\pi, \ldots\}$ are no longer allowable. This is because the identity we used in the substitution, $\sin(x) = \dfrac{1}{\csc(x)}$, has a further restriction on the variable, namely, $x \neq k\pi$ where k is an integer. Such restrictions carry over to the new identity when the substitution is made. We need to carry these restrictions to the new identity when we multiply or divide by expressions like $\cot(x)$, which can sometimes be zero or undefined.

We can add 1 to both sides of the original identity and simplify the right side to obtain:

$$sec(x) + 1 = \frac{1}{\cos(x)} + 1$$

$$sec(x) + 1 = \frac{1 + \cos(x)}{\cos(x)}$$

We should understand that every time we perform an acceptable algebraic operation on an identity we have the same identity but in a different form. The new form can be slightly different than the previous one or it can be significantly different depending on the expression and the operation used.

While the previous example can almost suffice in showing us how we might generate a whole host of identities, we will consider another example. The next example will increase the complexity of the generated identities.

EXAMPLE SET D

Start with the basic identity $\sin^2(s) = 1 - \cos^2(s)$ and produce several identities with different form.

Solution:

The algebraic operations we have chosen to apply to this identity are no better than any others you might choose on your own.

We might start by multiplying both sides by $1 + \cos^2(s)$ and performing the operations shown below. Remember that each new equation is an identity equivalent to the previous one but in a different form.

$$\sin^2(s)[1 + \cos^2(s)] = [1 - \cos^2(s)][1 + \cos^2(s)]$$

$$\sin^2(s)[1 + \cos^2(s)] = 1 - \cos^4(s)$$

$$\sin^2(s) + \sin^2(s)\cos^2(s) = 1 - \cos^4(s)$$

$$\sin^2(s) + \sin^2(s)\frac{1}{\sec^2(s)} = 1 - \frac{1}{\sec^4(s)}$$

$$\frac{\sin^2(s)\sec^2(s) + \sin^2(s)}{\sec^2(s)} = \frac{\sec^4(s) - 1}{\sec^4(s)}$$

Before the list of generated identities gets too long, we pause for a moment to be sure you have followed each of the algebraic operations. We now continue on with a decision to replace $\sec^2(s)$ by $\tan^2(s) + 1$ and move on with a few additional operations.

$$\frac{\sin^2(s)[\tan^2(s) + 1] + \sin^2(s)}{\tan^2(s) + 1} = \frac{\sec^4(s) - 1}{\sec^4(s)}$$

$$\frac{\sin^2(s)\tan^2(s) + 2\sin^2(s)}{\tan^2(s) + 1} = \frac{\sec^4(s) - 1}{\sec^4(s)}$$

We can see that this process can go on indefinitely. For instance, we could replace $\tan^2(s)$ by $\dfrac{1}{\cot^2(s)}$ and simplify. We could replace one of the $\sec^2(s)$ in $\sec^4(s)$ by $\dfrac{1}{\cos^2(s)}$ and simplify. And on and on and on.

We might ask why anyone would consider generating other forms of a known identity. The primary reason is to become familar with identities and the algebraic operations that give them a different form.

EXERCISES

In exercises 1–12, determine if the given expressions are basic identities.

1. $\cot(x) = \dfrac{1}{\tan(x)}$.

2. $\cot^2(s) - \csc^2(s) = 1$.

3. $\cot^2(s) - \csc^2(s) = -1$.

4. $\csc^2(r) - \cot^2(r) = 1$.

5. $\csc^2(w) - \cot^2(w) = -1$.

6. $\csc^2(t) - 1 = \cot^2(t)$.

7. $\csc^2(t) - 1 = -\cot^2(t)$.

8. $\sin^2(2z) + \cos^2(2z) = 1$.

9. $\tan^2(5s) - \sec^2(5s) = -1$.

10. $\csc(3w)\sec(3w) = 1$.

11. $\sin^2(s) + \cos^2(s) = 1$.

12. $\tan(2t)\cot(2t) = 1$.

In exercises 13–18, write the given basic identity in at least two other forms.

13. $1 + \cot^2(x) = \csc^2(x)$.

14. $\tan(s) = \dfrac{\sin(s)}{\cos(s)}$.

15. $\cot(t) = \dfrac{\cos(t)}{\sin(t)}$.

16. $\cot(w) = \dfrac{1}{\tan(w)}$.

17. $\csc(r) = \dfrac{1}{\sin(r)}$.

18. $\sin^2(3z) + \cos^2(3z) = 1$.

In exercises 19–24, use the given basic identity to produce five identities with different forms. You will get a great deal more benefit from these problems by using a large variety of algebraic operations and also by increasing the complexity of the expressions used with each operation.

19. $\tan^2(s) = \sec^2(s) - 1$.

20. $\csc^2(w) = \cot^2(w) + 1$.

21. $\tan(3s) = \dfrac{1}{\cot(3s)}$.

22. $\tan(5z) = \dfrac{\sin(5z)}{\cos(5z)}$.

23. $\cot(2y) = \dfrac{\cos(2y)}{\sin(2y)}$.

24. $\sec(u) = \dfrac{1}{\cos(u)}$.

In exercises 25 and 26, accept the fact that the given expression is an identity and use it to produce five identities with different forms.

25. $\dfrac{\sin(u)}{\csc(u)} + \dfrac{\cos(u)}{\sec(u)} = 1$.

26. $\csc(s) - \sin(s) = \cos(s)\cot(s)$.

8.5 Proving Identities

Introduction

The Idea of Proof

Guidelines to Proving Identities

Summary of Proving Trigonometric Identities

Introduction

We created numerous identities in the previous section from known identities using acceptable algebraic operations. Now we learn how to prove that an equation we suspect is an identity is indeed an identity, that is, prove that it is true for all allowable replacement values. This process is called verifying or proving an identity.

The Idea of Proof

Given an equation we suspect is an identity, the process is to manipulate one side of the equation using acceptable algebraic operations and trigonometric substitutions so that it becomes exactly the same as the other side. This process establishes the truth of the equation for all allowable values of the variable because both sides are only different but equivalent forms of the same expression.

EXAMPLE SET A

Verify the following identity.

$$[1 + \cos(t)][1 - \cos(t)] = \sin^2(t)$$

Solution:

We can graph the functions on each side of the equal sign to see if this might be an identity. Both functions will have the same graph if the equation is an identity.

We note that the two graphs coincide, at least in the standard Trig window, as seen in Window 8.1. We can use our graphing calculators in this way to build our confidence that the equation is likely to be an identity for all allowable replacement values. Now we try to manipulate one side of the equation so that it is transformed into the other side to verify our suspicions.

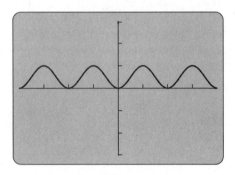

Window 8.1

One approach to proving an identity is to change the more complicated expression to the simpler one. Here, the left side seems to be more complicated. How can we simplify the left side? We can multiply it out as follows.

$$[1 + \cos(t)][1 - \cos(t)] \overset{?}{=} \sin^2(t)$$

$$1 - \cos^2(t) \overset{?}{=} \sin^2(t)$$

$$\sin^2(t) \overset{\checkmark}{=} \sin^2(t)$$

Thus,

$$[1 + \cos(t)][1 - \cos(t)] = \sin^2(t)$$

and, therefore, the suspected identity is indeed an identity. Whereby, we have reached into our resource bank to recall the basic identity $\cos^2(t) + \sin^2(t) = 1$ in the form $1 - \cos^2(t) = \sin^2(t)$. The left hand side has been transformed into the same expression as on the right and so we have proven the identity.

This example was relatively easy but, nonetheless, it gives several important ingredients that can be used in proving identities. The first is using our graphing calculators to initially decide that the equation might be an identity. While this is not a mathematical proof, it can give us confidence in our attack. Our approach to verifying an equation we suspect is an identity can vary from problem to problem. *There are no set rules we can follow that will guarantee success in proving identities.* The best we can do is to formulate guidelines that will help us in this effort and get a lot of practice. For instance, it is generally easier to transform the side that is more complicated. This is just a guideline because this may not always be the best approach although it usually is. In addition, there can be a difficulty in deciding which side is more complicated and there are cases in which both sides are equally complicated. Another ingredient is using algebraic manipulations (multiplying out the left hand side). Finally, the recognition and use of one of the forms of a basic identity allows us to make a substitution and complete the proof.

Guidelines to Proving Identities

Guidelines for Proving Identities

We need to recognize that guidelines are just that: tools that can help us accomplish a task. Guidelines are not likely to cover all situations but usually work in many of the problems we confront and frequently can give insight into situations they do not cover. With this in mind, we list several ideas to consider when proving an identity.

- Graph the equation to see if it might be an identity.
- Work with the side you think is the most complicated. If you are not sure or feel they are equally complicated, choose and simplify either side.
- Use algebraic operations and basic identities in your attempt to transform one side into the other.
- Change all trigonometric functions into sines and cosines, if appropriate.
- Whenever you feel lost or are unsure that the steps you have taken will be successful, graph your most recent expression along with the original expression on the opposite side. If they coincide, you gain confidence that you are proceeding correctly. If they differ, you will know you made a mistake, and you can proceed to look for and correct your error.

We need to expand on a couple of these ideas. The algebraic operations we use should include, but not be limited to, the following:

Expanding an expression

Factoring

Combining fractions

Separating fractions

Simplifying complex fractions

Multiplying the numerator and denominator of a fraction by the same expression

At the same time that we consider performing an algebraic operation, we need to look for opportunities to use the basic identities to help us in our proof. That is why it is essential that we have a solid knowledge of the basic identities.

There is a temptation to always change all of the trigonometric functions to sines and cosines. This can often be a useful step. However, this can also lead to a much longer and more complicated proof or even make it impossible to prove. We can help avoid this problem by always considering all the options listed in the guidelines and getting sufficient experience.

Let us get a feel for using these guidelines by considering some examples.

EXAMPLE SET B

Prove that the following equation is an identity:

$$\sin(\alpha)\cot(\alpha) = \cos(\alpha)$$

Solution:

We first plot the functions on each side and find that they are the same (at least in the window we are observing them). After studying the equation for a while, we might decide that the left side is more complicated and, in addition, since the right side involves cosine only, we probably would decide to change everything to sines and cosines initially. With these decisions, we have:

$$\sin(\alpha)\cot(\alpha) \overset{?}{=} \cos(\alpha)$$

$$\sin(\alpha)\frac{\cos(\alpha)}{\sin(\alpha)} \overset{?}{=} \cos(\alpha)$$

$$\cos(\alpha) \overset{\checkmark}{=} \cos(\alpha)$$

Since this transformation results in an expression identical to the right side, we have verified the identity.

EXAMPLE SET C

Prove that the following equation is an identity:

$$\cos^2(s) - \sin^2(s) = \cos^4(s) - \sin^4(s)$$

Solution:

A graph of the function on each side of the equation convinces us that it probably is an identity. What kinds of algebraic operations come to mind when we look at both sides of this equation? We might see that the right side is of the form $x^4 - y^4$ and is factorable. We might also see in our mind's eye that one of the factors will be $\cos^2(s) + \sin^2(s)$, which can be replaced by 1. Here, we are thinking several steps ahead as we might in chess. We might also notice that the left side is factorable as well, but its factors, $\cos(s) - \sin(s)$ and $\cos(s) + \sin(s)$, do not contain a basic identity as does one of the factors on the right side.

Let's write down the steps we've thought about.

$$\cos^2(s) - \sin^2(s) \stackrel{?}{=} \cos^4(s) - \sin^4(s)$$

$$\cos^2(s) - \sin^2(s) \stackrel{?}{=} [\cos^2(s) - \sin^2(s)][\cos^2(s) + \sin^2(s)]$$

$$\cos^2(s) - \sin^2(s) \stackrel{?}{=} [\cos^2(s) - \sin^2(s)] \cdot 1$$

$$\cos^2(s) - \sin^2(s) \stackrel{\checkmark}{=} \cos^2(s) - \sin^2(s)$$

Thus,

$$\cos^2(s) - \sin^2(s) = \cos^4(s) - \sin^4(s)$$

We have shown, using allowable algebraic operations and basic identities, how the right side is transformed into the left side.

We have used a vertical format for proving these identities. But what is the logical foundation for stating that this format shows that one side is always (for the allowable replacement values) equal to the other? Let's change to a horizontal format to illuminate the logic.

$$\cos^4(s) - \sin^4(s) = [\cos^2(s) - \sin^2(s)][\cos^2(s) + \sin^2(s)] = [\cos^2(s) - \sin^2(s)] \cdot 1$$
$$= \cos^2(s) - \sin^2(s)$$

The equal signs indicate that each expression is equivalent to the previous expression. The transitive property of equality guarantees that the first expression must be equivalent to the last expression for all allowable values of the variable. This horizontal display of the expressions is very unwieldy and difficult to follow though it clearly shows the logic underlying the proof process.

We can help to keep this logical framework in mind by slightly modifying the vertical format used above. We will place the equal signs at the far right if we are transforming only the right side and at the far left if we are transforming the left side. We introduce a vertical line between the two sides of the equation to emphasize that these equal signs apply only to one side or the other. This new format will easily allow us to restructure the proof as a single equation, as indeed it is. Our proof in the last example is displayed below in this new vertical format.

$$\cos^2(s) - \sin^2(s) = \cos^4(s) - \sin^4(s)$$

$$\left| \begin{array}{l} (\cos^2(s) - \sin^2(s))(\cos^2(s) + \sin^2(s)) = \\[2mm] (\cos^2(s) - \sin^2(s)) \cdot 1 = \\[2mm] \cos^2(s) - \sin^2(s) = \end{array} \right.$$

The next example demonstrates a number of approaches and techniques. You may wish to take a short break before reading it.

EXAMPLE SET D

Verify the following identity.

$$1 + \frac{1}{\cos(x)} = \frac{\tan^2(x)}{\sec(x) - 1}$$

Solution:

The statement of the problem indicates that we already have an equation that we suspect is an identity. If we have trouble verifying this identity, we can check that it is a suspect by graphing.

The left side has only the cosine function and involves fractions. The right side is a single fraction but involves two functions, neither of which is the cosine function. Which is the more complex? Our first attempt will be to use the right side since tangents and secants can be transformed into expressions that contain cosines.

$$1 + \frac{1}{\cos(x)} = \frac{\tan^2(x)}{\sec(x) - 1}$$

$$\left| \begin{array}{l} \dfrac{\dfrac{\sin^2(x)}{\cos^2(x)}}{\dfrac{1}{\cos(x)} - 1} = \\[8mm] \dfrac{\dfrac{\sin^2(x)}{\cos^2(x)}}{\dfrac{1 - \cos(x)}{\cos(x)}} = \end{array} \right.$$

$$\left|\ \frac{\sin^2(x)}{\cos^2(x)} \cdot \frac{\cos(x)}{1 - \cos(x)} =\right.$$

$$\left.\frac{\sin^2(x)}{\cos(x)} \cdot \frac{1}{1 - \cos(x)} =\right.$$

$$\left.\frac{1 - \cos^2(x)}{\cos(x)} \cdot \frac{1}{1 - \cos(x)} =\right.$$

$$\left.\frac{[1 + \cos(x)][1 - \cos(x)]}{\cos(x)} \cdot \frac{1}{1 - \cos(x)} =\right.$$

$$\left.\frac{1 + \cos(x)}{\cos(x)} =\right.$$

$$\left.\frac{1}{\cos(x)} + 1 =\right.$$

Thus, by the transitive property of equality, the identity is verified.

We can see that this was a rather long and complicated proof. Perhaps this is a situation where changing to sines and cosines was too hastily done. Let's work with the right side again, trying a different approach.

$$1 + \frac{1}{\cos(x)} = \frac{\tan^2(x)}{\sec(x) - 1}$$

$$\left|\ \frac{\sec^2(x) - 1}{\sec(x) - 1} =\right.$$

$$\left.\frac{[\sec(x) + 1][\sec(x) - 1]}{\sec(x) - 1} =\right.$$

$$\left.\sec(x) + 1 =\right.$$

$$\left.\frac{1}{\cos(x)} + 1 =\right.$$

This approach requires that we recognize the basic identity $\sec^2(x) - 1 = \tan^2(x)$ and imagine how using this identity in a substitution would work with the denominator. Substituting $\sec^2(x) - 1$ for $\tan^2(x)$ would at least transform the right side into an expression containing only the secant function. This step gives us a factoring opportunity if we recognize $\sec^2(x) - 1$ as the difference of two squares. Finally, we need to cancel the common factors and transform the secant function into cosines.

This was a very long example but it allowed us to study several approaches to verifying identities. The example illustrates that keeping basic identities in mind is as important as are the algebraic operations we can use. We can become more adept at working with identities if we have the flexibility to consider various approaches to verifying an identity. In most identities, like this last one, we can start with either side even though it is usually more efficient to start with the more complicated side. In addition, there are frequently several approaches that work in verifying an identity. Looking several steps ahead can also be useful in proving identities. Verifying identities can be a complicated process. We need lots of practice to become comfortable proving them.

Using the graphing calculator to check intermediate steps is very effective in giving us confidence that we have not made algebraic or substitution errors.

Summary of Proving Trigonometric Identities

We have indicated that there are no set rules for proving identities. The following are some more guidelines that can be helpful. Some initial considerations might be:

- Which side, if any, seems more complicated?
- Would it help to change everything to sines and cosines?

We have already discussed the fact that it usually pays to start with the more complicated side if it clearly has one. Changing all functions to sines and cosines is frequently helpful but should be done only after some thinking. We should hesitate doing so in situations where we see the possibility of using basic trigonometric identities that do not contain sines and cosines.

The steps we take after the initial start listed above are likely to be:

- What algebraic operations are available for the expression we are considering?
- Do any of the basic identities appear in their various forms?

There are numerous algebraic operations we can perform. It is important to think beyond what our previous training has ingrained in us. For instance, we are more inclined to combine two fractions than we are to separate a single fraction into two or more fractions. Yet, this latter operation is now as important as the former. We give almost no consideration to multiplying an expression by a fraction like $\dfrac{1 - \cos(t)}{1 - \cos(t)}$, which can be very helpful in certain situations. The reason is that we have not used this proceedure very much in the past. However, it is an operation that needs to come to mind especially when we are out of ideas. Additional help comes from noting that every time we use a reciprocal identity like $\sec(t) = \dfrac{1}{\cos(t)}$ or an identity like $\tan(t) = \dfrac{\sin(t)}{\cos(t)}$, we introduce a fraction, which gives rise to the opportunity of combining fractions or simplifying complex fractions.

Our graphing calculators continue to help us persist in completing proofs of identities. We should definitely consider their use any time we feel we are stuck or have made a mistake. We simply graph our current expression against the starting expression to either convince ourselves we are on the right track or to tell ourselves a mistake has been made. Such help is very important in our initial contact with proving identities. The graph will either give us the confidence to go on or, equally important, find the step in which we made a mistake and correct it.

One final note concerns the style of our proofs. We have promoted proofs that transform one side of a suspected identity into the other side. Another method of proving identities will be acceptable: that method consists of transforming one side of a suspected identity using the guidelines listed above. If you then reach a stage whereby you feel you have exhausted the use of your resource bank and have not yet arrived at an expression matching the other side, you should then transform the other side *independently*. The following example illustrates this approach.

EXAMPLE SET E

Verify the following identity.

$$\frac{\sin^2(t)}{1 + \cos(t)} = \frac{\tan(t) - \sin(t)}{\tan(t)}$$

Solution:

We begin by studying the equation. We might note that the left side consists of only sines and cosines. This could lead us to change the right side to sines and cosines.

$$\frac{\sin^2(t)}{1 + \cos(t)} = \frac{\tan(t) - \sin(t)}{\tan(t)}$$

$$= \frac{\dfrac{\sin(t)}{\cos(t)} - \sin(t)}{\dfrac{\sin(t)}{\cos(t)}}$$

$$\frac{\dfrac{\sin(t) - \sin(t)\cos(t)}{\cos(t)}}{\dfrac{\sin(t)}{\cos(t)}} =$$

$$\frac{\sin(t) - \sin(t)\cos(t)}{\cos(t)} \cdot \frac{\cos(t)}{\sin(t)} =$$

$$\frac{\sin(t) - \sin(t)\cos(t)}{\sin(t)} =$$

$$\frac{\sin(t)[1 - \cos(t)]}{\sin(t)} =$$

$$1 - \cos(t) =$$

We might feel stuck at this stage (a perfect time to consider multiplying by 1 in the form of a fraction). Our calculator would assure us that we have not made a mistake. So we start transforming the left side independently.

$$\frac{\sin^2(t)}{1 + \cos(t)}$$

$$= \frac{1 - \cos^2(t)}{1 + \cos(t)}$$

$$= \frac{[1 - \cos(t)][1 + \cos(t)]}{1 + \cos(t)}$$

$$= 1 - \cos(t)$$

We have finished proving the identity because, by transforming both sides independently, we have arrived at two expressions on either side that are identical. This method can give us insight into what we could have done when we first got stuck at $1 - \cos(t)$. If we were now to go back to the place where we were stuck and continued on with the one side, we would have a rigorous mathematical proof that uses the transitive property of equality at every step. However, this method of transforming both sides independently is acceptable.

Do we now see what to do to the right side of the equation at the place we got stuck? The fairly natural steps of factoring and cancelling we used as we moved to the left side are telling. In other words, if we stayed on the right side, we could have multiplied $1 - \cos(t)$ by $\dfrac{1 + \cos(t)}{1 + \cos(t)}$, the factor that divided out on the left side. Carrying through with this step by multiplying the numerators and recognizing a basic identity would finish the proof using only one side.

It should be mentioned that using the method of transforming both sides of an equation independently is quite different from treating the expression as if the equation were true. There is a problem if we start to treat the equation at any stage as though it were true. For example, if we multiply both sides by the same nonzero expression, we would no longer be proving the identity but rather asserting that two equations are equivalent. The reason may be a little subtle but it boils down to the fact that once that step has been taken, we can no longer use the transitive property that is at the heart of our identity proofs.

EXERCISES

In exercises 1–10, use your calculator to find the equations that might be identities. Verify these suspects.

1. $\cos^2(\theta)[\sec^2(\theta) - 1] = \sin^2(\theta)$.

2. $1 - 2\sin^2(\alpha) = 2\cos^2(\alpha) - 1$.

3. $[1 + \sin^2(t)][1 + \tan^2(t)] = 1$.

4. $[1 + \sin(y)][1 - \sin(y)] = \cos^2(y)$.

5. $\dfrac{\cos(w) - \sin(w)}{\cos(w)} = 1 - \tan(w)$.

6. $[1 - \cos^2(\alpha)][1 + \cot^2(\alpha)] = 1$.

7. $\sin(z) + \cos(z) = 1$.

8. $\dfrac{\cot^2(s)}{\csc^2(s)} = \dfrac{\cos^2(s)}{\sin^4(s)}$.

9. $\cos(t) + \dfrac{\sin^2(t)}{\cos(t)} = \sec(t)$.

10. $\tan(\theta) + \cot(\theta) = \dfrac{1}{\cos(\theta)\sin(\theta)}$.

In exercises 11–40, verify the identity. Remember to use your graphing calculator whenever you are concerned about having made a mistake or anytime you wish to regain confidence that you are headed in the right direction. The primary approach is the use of algebraic operations together with looking for basic identities in their various forms to help you.

11. $\cos^2(\alpha)[\sec^2(\alpha) - 1] = \sin^2(\alpha)$.

12. $\dfrac{\sec^2(w) - 1}{\sec^2(w)} = 1 - \cos^2(w)$.

13. $\tan^2(\theta) - \sin^2(\theta) = \tan^2(\theta)\sin^2(\theta)$.

14. $\dfrac{\sin(\beta)}{\csc(\beta)} + \dfrac{\cos(\beta)}{\sec(\beta)} = 1$.

15. $\dfrac{\sin^2(z)}{\tan^2(z) - \sin^2(z)} = \cot^2(z)$.

16. $\dfrac{1}{\csc(\theta) - \sin(\theta)} = \dfrac{\tan(\theta)}{\cos(\theta)}$.

17. $\dfrac{\cos(s) + \sin(s)}{\sin(s)} = 1 + \cot(s)$.

18. $\dfrac{1 + \sin(t)}{\cos(t)} = \dfrac{\cos(t)}{1 - \sin(t)}$.

    Hint: Consider $\dfrac{a}{b} = \dfrac{ak}{bk}$.

19. $\dfrac{1 - \cos(z)}{\sin(z)} = \dfrac{\sin(z)}{1 + \cos(z)}$.

20. $\dfrac{1 + \sec(w)}{\sec(w)} = \dfrac{\sin^2(w)}{1 - \cos(w)}$.

21. $\dfrac{1}{\sec(\theta) - \tan(\theta)} = \sec(\theta) + \tan(\theta)$.

22. $\cos^2(\theta) - \sin^2(\theta) = 2\cos^2(\theta) - 1$.

23. $\cos^2(\beta) - \sin^2(\beta) = 1 - 2\sin^2(\beta)$.

24. $\sec(y) - \dfrac{\cos(y)}{1 + \sin(y)} = \tan(y).$

25. $\dfrac{1 + \cos^2(t)}{\sin^2(t)} = 2\csc^2(t) - 1.$

26. $\dfrac{\sin(w) + \cos(w)}{\sec(w) + \csc(w)} = \dfrac{\sin(w)}{\sec(w)}.$

27. $\dfrac{\cos(z)}{1 - \sin(z)} = \sec(z) + \tan(z).$

28. $\dfrac{\tan^2(\alpha)}{\sec(\alpha) + 1} = \dfrac{1 - \cos(\alpha)}{\cos(\alpha)}.$

29. $\cot(\theta)\csc(\theta) = \dfrac{1}{\sec(\theta) - \cos(\theta)}.$

30. $\tan^4(s) + \tan^2(s) = \sec^4(s) - \sec^2(s).$

31. $\tan(w)\sec(w) - \sin(w) = \tan^2(w)\sin(w).$

32. $[\tan(v) + \cot(v)]^2 = \sec^2(v) + \csc^2(v).$

33. $\dfrac{\csc(\theta) + 1}{\csc(\theta) - 1} = \dfrac{1 + \sin(\theta)}{1 - \sin(\theta)}.$

34. $\sec(z) + \tan(z) = \dfrac{1}{\sec(z) - \tan(z)}.$

35. $\dfrac{2\tan(t)}{1 - \tan^2(t)} + \dfrac{1}{2\cos^2(t) - 1} = \dfrac{\cos(t) + \sin(t)}{\cos(t) - \sin(t)}.$

36. $\dfrac{\sin(\theta)\cot(\theta) + \cos^2(\theta)}{1 + \cos(\theta)} = \cos(\theta).$

37. $\dfrac{\sec(w)\tan(w) - \sin(w)}{\sin^3(w)} = \sec^2(w).$

38. $\dfrac{3}{1 - \cos(\alpha)} - \dfrac{3}{1 + \cos(\alpha)} = 6\cot(\alpha)\csc(\alpha).$

39. $[\cos^4(s) + 1 - \sin^4(s)]\sec^2(s) = 2.$

40. $[\tan(t) + \cot(t)]^2\cos^2(t) - 1 = \cot^2(t).$

8.6 Sum and Difference Identities

Introduction

The Cosine Sum Identity

The Cosine Difference Identity

The Sine Sum Identity

The Sine Difference Identity

Introduction

We are familiar with many identities that involve sums from other branches of mathematics. For example,

$$2^{a+b} = 2^a \cdot 2^b$$

$$(a + b)^2 = a^2 + 2ab + b^2$$

$$m(a + b) = ma + mb$$

In this section, we examine trigonometric identities that involve sums.

The Cosine Sum Identity

**The Cosine Sum
Identity**

Are there sum formulas in trigonometry? In particular, is there an identity involving $\cos(a+b)$? As it turns out, there is.

$$\cos(a+b) = \cos(a)\cos(b) - \sin(a)\sin(b)$$

We can use our graphing calculator to determine that this equation is true for several pairs of values. We can evaluate both sides of the equation with $a = 2$ and $b = 3$. We find that both sides are equal to 0.2836621855. We can also check the equation further by graphing both sides with $a = x$ and $b = 3$. Once again, the check works because the graphs of the two sides are the same. These calculator activities are convincing, but they do not prove that it is an identity.

Our derivation of this identity is based on the unit circle concept. We begin by placing arc lengths with measure a and b on the unit circle in standard position as in Figure 8.18.

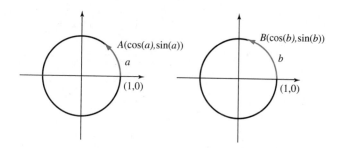

Figure 8.18

We have labeled the endpoints of the arc lengths based on our knowledge of the unit circle. Similarly, we can now draw an arc length of measure $a+b$ as in Figure 8.19.

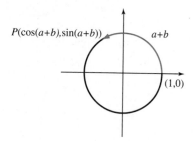

Figure 8.19

One approach to the derivation of this sum identity is to compare the expression for the lengths of two equal chords subtended by two arcs whose lengths are the same as in Figure 8.20. The arc length $-a$ is introduced to obtain two equal arcs.

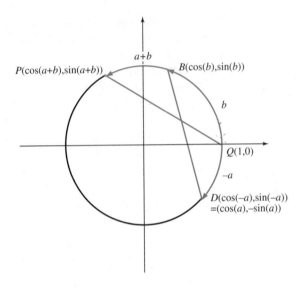

Figure 8.20

The two arc lengths PBQ and BQD are equal because they have the same length, namely, $a + b$. From geometry, equal arcs on the same circle subtend equal chords. We now use the distance formula to determine the lengths of the chords PQ and BD as follows.

$$PQ = \sqrt{[\cos(a + b) - 1]^2 + [\sin(a + b) - 0]^2}$$

$$= \sqrt{\cos^2(a + b) - 2\cos(a + b) + 1 + \sin^2(a + b)}$$

and using the identity $\sin^2(a + b) + \cos^2(a + b) = 1$, we obtain

$$= \sqrt{2 - 2\cos(a + b)}$$

and

$$BD = \sqrt{[\cos b - \cos a]^2 + [\sin b - (-\sin a)]^2}$$

$$= \sqrt{[\cos b - \cos a]^2 + [\sin b + \sin a]^2}$$

$$= \sqrt{\cos^2 b - 2\cos a \cos b + \cos^2 a + \sin^2 b + 2\sin a \sin b + \sin^2 a}$$

and using the identity $\sin^2(s) + \cos^2(s) = 1$, we obtain

$$= \sqrt{1 - 2\cos(a)\cos(b) + 1 + 2\sin(a)\sin(b)}$$

$$= \sqrt{2 - 2\cos(a)\cos(b) + 2\sin(a)\sin(b)}$$

Since PQ and BD are equal, their squares will be equal. Thus,

$$2 - 2\cos(a + b) = 2 - 2\cos(a)\cos(b) + 2\sin(a)\sin(b)$$

$$-2\cos(a + b) = -2\cos(a)\cos(b) + 2\sin(a)\sin(b)$$

and dividing both sides by -2 yields

$$\cos(a + b) = \cos(a)\cos(b) - \sin(a)\sin(b)$$

This completes the derivation. How can this identity be used?

EXAMPLE SET A

Expand $\cos(\frac{\pi}{2} - s)$ and simplify.

Solution:
First, we rewrite the expression using algebra so that it fits the cosine sum identity.

$$\cos\left(\frac{\pi}{2} - s\right) = \cos\left(\frac{\pi}{2} + (-s)\right)$$

Now expand $\cos(\frac{\pi}{2} + (-s))$ using the sum identity.

$$\cos\left(\frac{\pi}{2} + (-s)\right) = \cos\left(\frac{\pi}{2}\right)\cos(-s) - \sin\left(\frac{\pi}{2}\right)\sin(-s)$$

$$= 0 \cdot \cos(-s) - 1 \cdot \sin(-s)$$

$$= -(-\sin(s))$$

$$= \sin(s)$$

where we used the identity: $\sin(-t) = -\sin(t)$.

EXAMPLE SET B

Find the exact value of $\cos(\frac{5\pi}{12})$.

Solution:

This is not one of the familiar arc lengths that we studied in Chapter 6. Is it possible to rewrite $\dfrac{5\pi}{12}$ as a sum of two familiar arc lengths? Some of the familiar arc lengths, in terms of 12ths, are:

$$\frac{\pi}{6} = \frac{2\pi}{12}$$

$$\frac{\pi}{4} = \frac{3\pi}{12}$$

$$\frac{\pi}{3} = \frac{4\pi}{12}$$

The first two of these add to $\dfrac{5\pi}{12}$. Thus, we can use the sum identity as follows:

$$\cos\left(\frac{5\pi}{12}\right) = \cos\left(\frac{\pi}{6} + \frac{\pi}{4}\right)$$

$$= \cos\left(\frac{\pi}{6}\right)\cos\left(\frac{\pi}{4}\right) - \sin\left(\frac{\pi}{6}\right)\sin\left(\frac{\pi}{4}\right)$$

$$= \frac{\sqrt{3}}{2} \cdot \frac{\sqrt{2}}{2} - \frac{1}{2} \cdot \frac{\sqrt{2}}{2}$$

$$= \frac{\sqrt{6}}{4} - \frac{\sqrt{2}}{4}$$

$$= \frac{\sqrt{6} - \sqrt{2}}{4}$$

Therefore, $\cos\left(\dfrac{5\pi}{12}\right) = \dfrac{\sqrt{6} - \sqrt{2}}{4}$. This exact value could not be obtained without the sum identity. We can rewrite this exact value as $\cos\left(\frac{5\pi}{12}\right) \approx$ 0.2588190451 using our graphing calculators. We would probably use the approximation value in most applied situations. However, working such problems gives us the opportunity to learn the sum identities more thoroughly and build our skills in using them.

The sum identity for cosine needs to be added to our resource bank of basic identities.

The Cosine Difference Identity

Other sum and difference identities can be derived from this basic identity. For example,

$$\cos(a - b) = \cos(a + (-b))$$

$$= \cos(a) \cos(-b) - \sin(a) \sin(-b)$$

$$= \cos(a) \cos(b) - \sin(a)(-\sin(b))$$

$$= \cos(a) \cos(b) + \sin(a) \sin(b)$$

The Cosine Difference Identity

Here we have used the identities $\cos(-s) = \cos(s)$ and $\sin(-s) = -\sin(s)$. Thus, we have the difference identity for cosine:

$$\cos(a - b) = \cos(a) \cos(b) + \sin(a) \sin(b)$$

Notice that the only difference between this identity and the cosine sum identity is the sign change between the two products. For this reason, we will not add this identity to our resource bank of basic identities.

The Sine Sum Identity

Is there a sum identity for the sine function? Let's try to derive such an identity. We could attempt a similar approach to the one used for the cosine function. However, perhaps we can take some time to think about other approaches as we should in every problem-solving situation. We might recall from the first example in this section how we were able to write the sine function in terms of the cosine as in $\sin(t) = \cos(\frac{\pi}{2} - t)$. Notice that it involves an expression that can use the cosine difference identity just derived. Using this identity, we obtain:

$$\sin(a + b) = \cos\left[\frac{\pi}{2} - (a + b)\right]$$

$$= \cos\left[(\frac{\pi}{2} - a) - b\right]$$

$$= \cos\left(\frac{\pi}{2} - a\right) \cos(b) + \sin\left(\frac{\pi}{2} - a\right) \sin(b)$$

We stop here and note that we know what we can replace the expression $\cos\left(\frac{\pi}{2} - a\right)$ with from the first example in this section. But what do we do with the expression $\sin\left(\frac{\pi}{2} - a\right)$? We can use the same logic we used at the start of this derivation, namely

$$\sin\left(\frac{\pi}{2} - a\right) = \cos\left[\frac{\pi}{2} - \left(\frac{\pi}{2} - a\right)\right]$$

$$= \cos\left[\frac{\pi}{2} - \frac{\pi}{2} + a\right]$$

$$= \cos(a)$$

We can now finish the derivation as follows

$$\cos\left(\frac{\pi}{2} - a\right)\cos(b) + \sin\left(\frac{\pi}{2} - a\right)\sin(b) = \sin(a)\cos(b) + \cos(a)\sin(b)$$

The Sine Sum Identity

Thus, $\sin(a + b) = \sin(a)\cos(b) + \cos(a)\sin(b)$.

It is important to note that if we expand $\cos(\frac{\pi}{2} - (a+b))$ using the cosine difference identity, we will simply arrive at the same expression we started with, namely $\sin(a+b)$. Rewriting $\frac{\pi}{2} - (a+b)$ as $(\frac{\pi}{2} - a) - b$ was an essential step in the derivation.

The sum identity for the sine function should be added to our resource bank of basic identities.

The Sine Difference Identity

The difference identity for the sine function is derived in the same way as for the cosine function.

$$\sin(a - b) = \sin(a + (-b))$$

$$= \sin(a)\cos(-b) + \cos(a)\sin(-b)$$

$$= \sin(a)\cos(b) + \cos(a)(-\sin(b))$$

$$= \sin(a)\cos(b) - \cos(a)\sin(b)$$

The Sine Difference Identity

Thus, $\sin(a - b) = \sin(a)\cos(b) - \cos(a)\sin(b)$. Again, we will *not* add this to our set of basic identities since it only differs from the sine sum identity by the sign between the two products.

The sum and difference identities for the sine and cosine functions are summarized next:

$$\cos(a + b) = \cos(a)\cos(b) - \sin(a)\sin(b)$$

$$\cos(a - b) = \cos(a)\cos(b) + \sin(a)\sin(b)$$

$$\sin(a+b) = \sin(a)\cos(b) + \cos(a)\sin(b)$$

$$\sin(a-b) = \sin(a)\cos(b) - \cos(a)\sin(b)$$

A careful examination of these identities shows that the difference identities are easily written using the corresponding sum identities by simply changing the signs between the products.

EXAMPLE SET C

Find the exact value of $\sin(\frac{\pi}{12})$.

Solution:

We note, once again, that $\frac{\pi}{12}$ is not one of the special arcs we have studied. However, recalling the special arcs we do know written in 12ths, we can see that $\frac{\pi}{12} = \frac{\pi}{4} - \frac{\pi}{6}$. We can now use the sine difference identity to find the answer.

$$\sin\left(\frac{\pi}{12}\right) = \sin\left(\frac{\pi}{4} - \frac{\pi}{6}\right)$$

$$= \sin\left(\frac{\pi}{4}\right)\cos\left(\frac{\pi}{6}\right) - \cos\left(\frac{\pi}{4}\right)\sin\left(\frac{\pi}{6}\right)$$

$$= \frac{\sqrt{2}}{2} \cdot \frac{\sqrt{3}}{2} - \frac{\sqrt{2}}{2} \cdot \frac{1}{2}$$

$$= \frac{\sqrt{6}}{4} - \frac{\sqrt{2}}{4}$$

$$= \frac{\sqrt{6} - \sqrt{2}}{4}$$

EXAMPLE SET D

Verify that $\sin(t + 2\pi) = \sin(t)$.

Solution:

We should already know that this expression is equal to $\sin(t)$ because of the periodicity of the sine function. However, we can now prove this using the sine sum identity.

$$\sin(t + 2\pi) = \sin(t)\cos(2\pi) + \cos(t)\sin(2\pi)$$

$$= \sin(t) \cdot 1 + \cos(t) \cdot 0$$

$$= \sin(t)$$

We need to use our own special memorizing techniques to add the sum identities for the sine and cosine functions to the core set of basic identities. We must have these identities in our resource bank because they are so necessary in proving identities.

EXAMPLE SET E

Verify that $\cos(t + t) = \cos^2(t) - \sin^2(t)$.

Solution:

We can prove this identity by using the cosine sum identity.

$$\cos(t + t) = \cos(t)\cos(t) - \sin(t)\sin(t)$$

$$= \cos^2(t) - \sin^2(t)$$

Note: $\cos(t + t) = \cos(2t)$. We will study this function in the next section.

EXERCISES

In exercises 1–10, find both approximate and exact values of the given expression.

1. $\sin(\pi/12)$ 2. $\cos(7\pi/12)$

3. $\tan(5\pi/12)$ 4. $\sin(105°)$

5. $\sin(75°)$ 6. $\cos(15°)$

7. $\cos(11\pi/12)$ 8. $\sec(5\pi/12)$

9. $\sin(195°)$ 10. $\cos(165°)$

In exercises 11–22, verify each identity.

11. $\cos\left(\dfrac{\pi}{2} - s\right) = \sin(s)$

12. $\sin\left(\dfrac{\pi}{2} - s\right) = \cos(s)$

13. $\sin\left(\dfrac{\pi}{2} + s\right) = \cos(s)$

14. $\cos\left(\dfrac{\pi}{2} + s\right) = -\sin(s)$

15. $\cos(\pi - s) = -\cos(s)$

16. $\sin(\pi - s) = \sin(s)$

17. $\cos(\pi + s) = -\cos(s)$

18. $\sin(\pi + s) = -\sin(s)$

19. $\sin(2\pi - s) = -\sin(s)$

20. $\cos(2\pi - s) = \cos(s)$

21. $\tan(s + \pi) = \tan(s)$

22. $\tan(\pi - s) = -\tan(s)$

23. Given $\sin(s) = 4/5$ and $\cos(t) = 5/13$ and that s and t are in the first quadrant, find exact and approximate values of each of the following.

    (a) $\sin(s + t)$
    (b) $\cos(s - t)$
    (c) $\tan(s + t)$

24. Given $\cos(s) = 2/3$ and $\sin(t) = 3/4$ with s in quadrant IV and t in quadrant II, find exact and approximate values of each of the following.

    (a) $\cos(s + t)$
    (b) $\sin(s - t)$
    (c) $\tan(s - t)$

25. Given $\sin(s) = -8/17$ and $\cos(t) = -1/3$ with s in quadrant IV and t in quadrant II, find exact and approximate values of each of the following.

    (a) $\cos(s - t)$
    (b) $\sin(s + t)$
    (c) $\tan(s + t)$

26. Given $\tan(s) = 3/4$ and $\sin(t) = 2/5$ with s in quadrant III and t in quadrant II, find exact and approximate values of each of the following.

    (a) $\cos(s - t)$
    (b) $\sin(s + t)$
    (c) $\tan(s - t)$

In exercises 27–30, write the given expression in terms of one trigonometric function of one angle.

27. $\cos(15°)\cos(41°) - \sin(15°)\sin(41°)$

28. $\sin(22°)\cos(33°) - \cos(22°)\sin(33°)$

29. $\sin(18°)\cos(26°) + \cos(18°)\sin(26°)$

30. $\cos(52°)\cos(34°) + \sin(52°)\sin(34°)$

8.7 The Double- and Half-Angle Identities

Introduction

The Double-Angle Identities

Finding All the Trigonometric Values Given Any One of Them

The Half-Angle Identities

Introduction

We have one last set of basic identities that we need to add to our core set of identities. These are called the double-angle identities and the half-angle identities. The double-angle identities are relatively easy to derive while the half-angle identities are a bit more difficult. We can benefit from learning how to derive them on our own but, whether we learn to derive them or only memorize them, they will be a part of our core set of basic identities.

The Double-Angle Identities

The goal is to write the sine or cosine of a double angle (say 2α) in terms of a single angle (α). When this identity is needed, first recall that $2\alpha = \alpha + \alpha$. We then apply the sum identities to complete the derivation. The derivation for the cosine double-angle identity is as follows:

$$\cos(2\alpha) = \cos(\alpha + \alpha)$$

$$= \cos(\alpha)\cos(\alpha) - \sin(\alpha)\sin(\alpha)$$

$$= \cos^2(\alpha) - \sin^2(\alpha)$$

so that $\cos(2\alpha) = \cos^2(\alpha) - \sin^2(\alpha)$. The double-angle sine identity is derived in the same manner.

$$\sin(2\alpha) = \sin(\alpha + \alpha)$$

$$= \sin(\alpha)\cos(\alpha) + \cos(\alpha)\sin(\alpha)$$

$$= 2\sin(\alpha)\cos(\alpha)$$

These two double-angle identities:

$$\cos(2\alpha) = \cos^2(\alpha) - \sin^2(\alpha)$$

and

The Sine Double-Angle Identity

$$\sin(2\alpha) = 2\sin(\alpha)\cos(\alpha)$$

need to be part of our resource bank, either memorized or as derivable formulas.

There are two other forms of the cosine double-angle identity that tend to be used more frequently than the original one we just derived. The other two forms are useful because they allow you to write the cosine double-angle identity in terms of only cosines or only sines. With this idea in mind, we can employ the basic identity $\cos^2(\alpha) + \sin^2(\alpha) = 1$ in some of its other forms. The form containing only cosines is derived as follows:

$$\cos(2\alpha) = \cos^2(\alpha) - \sin^2(\alpha)$$

$$= \cos^2(\alpha) - [1 - \cos^2(\alpha)]$$

$$= \cos^2(\alpha) - 1 + \cos^2(\alpha)$$

$$= 2\cos^2(\alpha) - 1$$

The form containing only sines is similarly derived:

$$\cos(2\alpha) = \cos^2(\alpha) - \sin^2(\alpha)$$

$$= [1 - \sin^2(\alpha)] - \sin^2(\alpha)$$

$$= 1 - 2\sin^2(\alpha)$$

The Cosine Double-Angle Identity

Thus, the three equivalent forms of the cosine double-angle identity are:

$$\cos(2\alpha) = \begin{cases} 2\cos^2(\alpha) - 1 \\ \cos^2(\alpha) - \sin^2(\alpha) \\ 1 - 2\sin^2(\alpha) \end{cases}$$

We need to add these other two forms to our core of basic identities if we cannot derive them from the original identity on our own. This again shows how the number of basic identities in our core set can vary.

We can get some practice with these new identities by considering some examples.

EXAMPLE SET A

Verify the identity $\cos^4(\alpha) - \sin^4(\alpha) = \cos(2\alpha)$.

Solution:
We should not hesitate to use our calculators anytime during a proof if we are unsure or stumped. There is also an important observation we should consider when working with trigonometric expressions in which the arguments of the trigonometric functions are different. That is certainly the case in this example where the left side contains a single angle while the right side contains a double angle. Such situations will normally require us to get the arguments of the trigonometric functions to be the same. For the problem at hand, we might decide to work on the left side as it looks more complicated. We might also see that the left side is a difference of two squares, which gives us the option of factoring it.

$$\cos^4(\alpha) - \sin^4(\alpha) = \cos(2\alpha)$$

$$= [\cos^2(\alpha) + \sin^2(\alpha)][\cos^2(\alpha) - \sin^2(\alpha)]$$

$$= 1 \cdot [\cos^2(\alpha) - \sin^2(\alpha)]$$

$$= \cos(2\alpha)$$

We examine another important type of problem in the next example.

EXAMPLE SET B

Given that $\sin(\theta) = \frac{3}{5}$ and that θ is in the first quadrant, find exact values for $\cos(2\theta)$ and $\sin(2\theta)$.

Solution:
We know the value for $\sin(\theta)$ and can, therefore, get the value of $\cos(\theta)$ using the basic identity $\cos^2(\theta) = 1 - \sin^2(\theta)$. Using this identity and the fact that θ is in the first quadrant, we can obtain $\cos(\theta) = \frac{4}{5}$ (since $\cos(\theta)$ is positive). We can now use the double-angle identities to complete the problem.

$$\cos(2\theta) = 1 - 2\sin^2(\theta)$$

$$= 1 - 2\left(\frac{3}{5}\right)^2$$

$$= 1 - \frac{18}{25}$$

$$= \frac{7}{25}$$

Now for the second part.

$$\sin(2\theta) = 2\sin(\theta)\cos(\theta)$$

$$= 2\left(\frac{3}{5}\right)\left(\frac{4}{5}\right)$$

$$= \frac{24}{25}$$

EXAMPLE SET C

Given that $\sin(\theta) = \frac{3}{5}$ and that θ (in degrees) is in the first quadrant, find approximate values for $\cos(2\theta)$ and $\sin(2\theta)$.

Solution:
This is also an important example because it shows how our calculators can be used to rapidly find the solutions to such problems when only approximations are required. Window 8.2 shows a possible computation.

Sin $^{-1}$(3/5)
                              36.86989765
Ans*2
                              73.73979529
cos (Ans)
                                        .28
sin (73.73979529)
                                        .96

Window 8.2

Thus, $\cos(2\theta) \approx 0.28$ and $\sin(2\theta) \approx 0.96$. Check to see that these values agree with the exact values we obtained earlier by performing the divisions of the exact solutions.

It is important to be able to solve problems that require either exact answers or approximate answers. It is tempting to use our calculators because we can obtain relatively accurate answers quite rapidly and often that will be all that is required. However, problems like that of the previous example that request exact answers are very useful because they build both our problem-solving skills and our resource bank. Developing good problem-solving skills and a large resource bank will richly reward us in our careers.

Finding All the Trigonometric Values Given Any One of Them

There is yet another important concept to store in our resource bank. We need to realize that if we are given the value of any one trigonometric function, as in $\csc(\beta) = 5$, we can find the values for the remaining trigonometric functions of the angle β as well as the values for all six trigonometric functions for the angle 2β and, as we will see shortly, the values for all six trigonometric functions for the angle $\frac{\beta}{2}$. The next example illustrates this idea.

EXAMPLE SET D

Given that $\tan(\beta) = 2$ and that β is in the first quadrant, find exact values for $\sin(\beta)$, $\cos(\beta)$, $\csc(\beta)$, $\sec(\beta)$, $\cot(\beta)$, $\sin(2\beta)$, $\cos(2\beta)$, $\tan(2\beta)$, $\cot(2\beta)$, $\csc(2\beta)$, and $\sec(2\beta)$.

Solution:

Once we know the value of $\sin(\beta)$, we know the value of its reciprocal function $\csc(\beta)$, and similarly for the two other reciprocal function pairs $\cos(\beta)$, $\sec(\beta)$ and $\tan(\beta)$, $\cot(\beta)$. The question is, how do we find a value for at least one of each pair? Clearly, if $\tan(\beta) = 2$, then $\cot(\beta) = \frac{1}{2}$. Can we recall a basic identity that involves the tangent function and some other function? Perhaps our resource bank will help us recall that $\tan^2(\beta) + 1 = \sec^2(\beta)$, which we can put in the form $\sec(\beta) = \pm\sqrt{\tan^2(\beta) + 1}$. We choose the positive value since β is in the first quadrant. Once this identity is recalled, we can go forward with the solution.

$$\sec(\beta) = \sqrt{\tan^2(\beta) + 1}$$

$$= \sqrt{2^2 + 1}$$

$$= \sqrt{5}$$

then

$$\cos(\beta) = \frac{1}{\sqrt{5}}$$

At this point, we can get the value of $\sin(\beta)$ using $\sin^2(\beta) + \cos^2(\beta) = 1$ which, when calculated, is $\sin(\beta) = \frac{2}{\sqrt{5}}$. We can also get the value of $\csc(\beta)$ using the reciprocal identity.

We still have the task of finding the values of the six trigonometric functions for 2β. This is a reasonable task now that we have the values of $\sin(\beta)$ and $\cos(\beta)$.

$$\cos(2\beta) = 2\cos^2(\beta) - 1$$

$$= 2\left(\frac{1}{\sqrt{5}}\right)^2 - 1$$

$$= \frac{2}{5} - 1 = -\frac{3}{5}$$

and

$$\sin(2\beta) = 2\sin(\beta)\cos(\beta)$$

$$= 2\left(\frac{2}{\sqrt{5}}\right)\left(\frac{1}{\sqrt{5}}\right) = \frac{4}{5}$$

We probably know by now that once we know the values of the sine and cosine functions for an angle, we can find the value of the remaining trigonometric functions for that angle. The problem is, therefore, essentially finished. We will see how we can also obtain the values of the six trigonometric functions for $\dfrac{\beta}{2}$ once we study the half-angle identities.

The Half-Angle Identities

Look at the cosine double-angle identity, $\cos(2\alpha) = 2\cos^2(\alpha) - 1$, for a few moments. We know that the angle on the left is twice the angle on the right. But that means that the angle on the right is half the angle on the left. Using this reversal, we can derive the half-angle identities from the cosine double angle identity using any of its various forms. We will find that the cosine double-angle identity in the forms containing (1) only the cosine function or (2) only the sine function will serve our purpose best. In the first case, we solve for $\cos(\alpha)$ and in the second, we solve for $\sin(\alpha)$.

$$\cos(2\alpha) = 2\cos^2(\alpha) - 1$$

$$1 + \cos(2\alpha) = 2\cos^2(\alpha)$$

$$\frac{1 + \cos(2\alpha)}{2} = \cos^2(\alpha)$$

$$\pm\sqrt{\frac{1 + \cos(2\alpha)}{2}} = \cos(\alpha)$$

This series of algebraic operations transforms our original identity into

$$\cos(\alpha) = \pm\sqrt{\frac{1 + \cos(2\alpha)}{2}}$$

Since this expression is an identity, we can replace the variable with any number or legitimate expression. Replace α by $\dfrac{\theta}{2}$ to obtain

$$\cos\left(\frac{\theta}{2}\right) = \pm\sqrt{\frac{1 + \cos(\theta)}{2}}$$

This last identity is called the cosine half-angle identity. We obtain a similar result working with the other cosine double-angle identity.

$$\cos(2\alpha) = 1 - 2\sin^2(\alpha)$$

$$2\sin^2(\alpha) = 1 - \cos(2\alpha)$$

$$\sin^2(\alpha) = \frac{1 - \cos(2\alpha)}{2}$$

$$\sin(\alpha) = \pm\sqrt{\frac{1 - \cos(2\alpha)}{2}}$$

The Half-Angle Identities

We again replace α by $\dfrac{\theta}{2}$ to obtain $\sin\left(\dfrac{\theta}{2}\right) = \pm\sqrt{\dfrac{1 - \cos(\theta)}{2}}$. This last identity is called the sine half-angle identity. The sine and cosine half-angle identities are:

$$\cos\left(\frac{\theta}{2}\right) = \pm\sqrt{\frac{1 + \cos(\theta)}{2}}$$

$$\sin\left(\frac{\theta}{2}\right) = \pm\sqrt{\frac{1 - \cos(\theta)}{2}}$$

We need to add these two identities to our core set of basic identities. Note that these identities have the possibility of being positive or negative. This may appear to be a problem at first but the potential difficulty soon disappears as we realize that the correct sign for the answer is determined by what quadrant the angle $\dfrac{\theta}{2}$ is in.

It should also be mentioned at this point that although we have been using the word angle to describe all the identities in this section, they are equally true for arcs and numbers. That is, $\cos(2t) = \cos^2(t) - \sin^2(t)$ is an identity whether t is thought of as an angle, an arc, or a number. It is usually called the cosine double-angle identity only because it seems to be more comfortable to say.

We examine a use of the half-angle identities in the next example.

EXAMPLE SET E

Find the exact value of $\sin(15°)$.

Solution:
We notice that 15° is half of 30°, which is an angle whose trigonometric values we know. We proceed to use the sine half-angle identity with this information.

$$\sin(15°) = \sqrt{\frac{1 - \cos(30°)}{2}}$$

$$= \sqrt{\frac{1 - \frac{\sqrt{3}}{2}}{2}}$$

$$= \sqrt{\frac{\frac{2 - \sqrt{3}}{2}}{2}}$$

$$= \sqrt{\frac{2 - \sqrt{3}}{4}}$$

Note that we could also have used the sine difference identity since $\sin(15°) = \sin(45° - 30°)$.

If we are going to change or verify an identity in which the numerical multipliers in the arguments of the trigonometric functions are not the same, then, at some point, we need to change them to be the same. Changing them at the start is usually more beneficial but, as always, it depends on the problem. The following example illustrates this idea.

EXAMPLE SET F

Verify the following identity.

$$2\sin^2(2\theta) + \cos(4\theta) = 1$$

Solution:

We note that the cosine function has an argument (4θ) that is twice the argument of the sine function (2θ). While we could change both arguments to single angles, it will be far easier to simply write $\cos(4\theta)$ in terms of 2θ using the cosine double-angle identity as follows:

$$2\sin^2(2\theta) + \cos(4\theta) = 1$$

$$= 2\sin^2(2\theta) + 1 - 2\sin^2(2\theta)$$

$$= 1$$

Thus, $2\sin^2(2\theta) + \cos(4\theta) = 1$ and the identity is verified.

EXERCISES

In exercises 1–10, find both approximate and exact values of the given expressions.

1. $\sin(\pi/12)$ 2. $\cos(\pi/8)$

3. $\cos(7\pi/12)$ 4. $\tan(3\pi/8)$

5. $\sin(15°)$ 6. $\cos(15°)$

7. $\tan(75°)$ 8. $\csc(-3\pi/8)$

9. $\cos(7.5°)$. (Hint: Use the results of exercise 6.)   10. $\sin(\pi/24)$. (Hint: Use the results of exercise 1.)

In exercises 11–16, write the expression as a single trigonometric function of kt, where k is a natural number.

11. $2\sin(3t)\cos(3t)$ 12. $2\cos^2(5t) - 1$

13. $4 - 8\sin^2(2t)$ 14. $2\cos^2\left(\dfrac{t}{2}\right) - 1$

15. $\sin(t)\cos(t)$

16. $4\cos\left(\dfrac{t}{2}\right)\sin\left(\dfrac{t}{2}\right)$

17. Given $\sin(\theta) = 4/5$ and $\cos(\theta) < 0$, find $\sin(2\theta)$, $\cos(2\theta)$, $\tan(2\theta)$, $\sin\left(\dfrac{\theta}{2}\right)$, $\cos\left(\dfrac{\theta}{2}\right)$, and $\tan\left(\dfrac{\theta}{2}\right)$.

18. Given $\cos(\theta) = -12/13$ and $\sin(\theta) > 0$, find $\sin(2\theta)$, $\cos(2\theta)$, $\tan(2\theta)$, $\sin\left(\dfrac{\theta}{2}\right)$, $\cos\left(\dfrac{\theta}{2}\right)$, and $\tan\left(\dfrac{\theta}{2}\right)$.

19. Given $\tan(t) = 3/4$ and $\sin(t) < 0$, find $\sin(2t)$, $\cos(2t)$, $\tan(2t)$, $\sin\left(\dfrac{t}{2}\right)$, $\cos\left(\dfrac{t}{2}\right)$, and $\tan\left(\dfrac{t}{2}\right)$.

20. Given $\cos(2\theta) = -2/3$ and $180° \le 2\theta \le 270°$, find $\sin(2\theta)$, $\sin(\theta)$, $\cos(\theta)$, $\tan(\theta)$, $\sin\left(\dfrac{\theta}{2}\right)$, $\cos\left(\dfrac{\theta}{2}\right)$, and $\tan\left(\dfrac{\theta}{2}\right)$.

21. Given $\sin(2\alpha) = 3/5$ and $\pi/2 \le 2\alpha \le \pi$, find $\cos(2\alpha)$, $\sin(\alpha)$, $\cos(\alpha)$, $\sin\left(\dfrac{\alpha}{2}\right)$, $\cos\left(\dfrac{\alpha}{2}\right)$, and $\tan\left(\dfrac{\alpha}{2}\right)$.

22. Given $\sin\left(\dfrac{\theta}{2}\right) = 1/3$ and $\cos\left(\dfrac{\theta}{2}\right) < 0$, find $\cos\left(\dfrac{\theta}{2}\right)$, $\sin(\theta)$, $\cos(\theta)$, $\tan(\theta)$, $\sin(2\theta)$, and $\cos(2\theta)$.

Verify the identities in exercises 23–30.

23. $2\sin^2(\theta) = 1 - \cos(2\theta)$.

24. $\dfrac{\sin(2t)}{1 + \cos(2t)} = \tan(t)$.

25. $\dfrac{1 - \cos(2s)}{2\sin(s)\cos(s)} = \tan(s)$.

26. $\dfrac{2}{\sin(2t)} = \tan(t) + \cot(t)$.

27. $\sin(3t) = 3\sin(t) - 4\sin^3(t)$.

28. $\cos(3t) = 4\cos^3(t) - 3\cos(t)$.

29. $\sin\left(\dfrac{x}{2}\right)\cos\left(\dfrac{x}{2}\right) = \dfrac{\sin(x)}{2}$.

30. $[\cos\left(\dfrac{t}{2}\right) - \sin\left(\dfrac{t}{2}\right)]^2 = 1 - \sin(t)$.

8.8 Conditional Trigonometric Equations

Introduction

Approximating Solutions to Conditional
Trigonometric Equations

Finding Exact Solutions to Conditional
Trigonometric Equations

Introduction

We have considered trigonometric equations that were identities for most of this chapter. Now we will consider conditional trigonometric equations— equations that are true for only some of the replacement values. In contrast to many algebraic equations we have solved, most conditional trigonometric equations have infinite solution sets because trigonometric functions are periodic. For example, we know there are infinitely many solutions to the simple trigonometric equation $\sin(t) = \frac{1}{2}$. In solving (not proving) such equations, we will use the techniques we have developed earlier in this book beginning with a method that uses our graphing calculators.

Approximating Solutions to Conditional Trigonometric Equations

Let's solve the following equation:

$$2\cos^2(x) + \cos(x) = 1$$

First, we transform the equation by subtracting 1 from both sides of the equation to produce an expression that is set equal to 0.

$$2\cos^2(x) + \cos(x) - 1 = 0 \tag{8.1}$$

Next, we graph, in radian mode, the left side as a function to obtain the graph illustrated in Window 8.3. Use the TRACE feature to approximate the first zero to the right of the origin. An approximate answer using the ZOOM Trig feature is $x = 1$. We notice that the next two positive zeros are near $x = \pi$ and $x = 5.2$, respectively. We also notice that there are three matching zeros to the left of the origin. Does the function appear to repeat in some fixed period? The expression in equation 8.1, suggests that the function is periodic with a period that is at least 2π because the $\cos(x)$ appears. Use the TRACE feature to scroll to the right to observe that the function does appear to be periodic with a period of about 6. That means that there are an infinite number of zeros. We

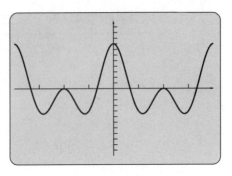

Window 8.3

need only determine the location of the zeros in one period to generate all the zeros by adding multiples of the function's period. The approximate solution set, therefore, is

$$\{1.05 + k2\pi\} \quad \cup \quad \{\pi + k2\pi\} \quad \cup \quad \{5.24 + k2\pi\}$$

Technically, this should be written as

$$\{x|x = 1.05 + k2\pi\} \quad \cup \quad \{x|x = \pi + k2\pi\} \quad \cup \quad \{x|x = 5.24 + k2\pi\}$$

where k is an integer. We will, however, use the first abbreviated form of the solution set. As we know, we can obtain much greater accuracy using the ZOOM feature, even though this approach will not provide exact answers.

Finding Exact Solutions to Conditional Trigonometric Equations

We can also approach this problem from an algebraic point of view. Recall the original equation that we wished to solve.

$$2\cos^2(x) + \cos(x) = 1$$

This is an equation involving the variable x. This variable appears only as the argument of a cosine function, namely, it appears in both $\cos^2(x)$ and $\cos(x)$. We can, therefore, view this as an equation in the more complicated variable, $\cos(x)$. In this variable, the equation is quadratic because the highest power of $\cos(x)$ is 2. We have several methods of solving quadratic equations. The factoring method involves less manipulation if it can be used. Most methods require that we place the quadratic equation in standard form. Rewriting the equation in standard form gives

$$2\cos^2(x) + \cos(x) - 1 = 0$$

Can we factor the left side of this equation? Recognizing $\cos(x)$ as the variable, we have a trinomial expression that might be factorable as the product of two binomial expressions. With this in mind, we proceed as follows:

$$2\cos^2(x) + \cos(x) - 1 = 0$$

$$[2\cos(x) \quad 1][\cos(x) \quad 1] = 0$$

Finally, we need to check whether the possible sign choices will provide factors whose product is the original expression.

$$[2\cos(x) - 1][\cos(x) + 1] = 0$$

We can check to see that this is correct by multiplying out the factors:

$$[2\cos(x) - 1][\cos(x) + 1]$$

$$= 2\cos(x)\cos(x) + 1 \cdot 2\cos(x) - 1 \cdot \cos(x) - 1 \cdot (1)$$

$$= 2\cos^2(x) + \cos(x) - 1$$

We need to become accustomed to viewing $\cos(x)$ in the same way that we view familiar variables like x, y, and z.

Now we set each factor to 0 and solve for $\cos(x)$.

$$[2\cos(x) - 1][\cos(x) + 1] \quad = \quad 0$$

$$2\cos(x) - 1 = 0 \quad \text{or} \quad \cos(x) + 1 = 0$$

$$2\cos(x) = 1 \quad \text{or} \quad \cos(x) = -1$$

$$\cos(x) = \frac{1}{2} \quad \text{or} \quad \cos(x) = -1$$

Now that we have solved the equation for $\cos(x)$, we return to our original goal of solving for x. We have solved similar equations where a single trigonometric function is equal to a constant. We know $\frac{\pi}{3}$ is one of the infinite number of solutions of the equation $\cos(x) = \frac{1}{2}$. The solution set for this equation is based on the periodicity of the cosine function and is

$$\left\{\frac{\pi}{3} + k2\pi\right\} \quad \text{where } k \text{ is an integer}$$

We also know that $\dfrac{5\pi}{3}$ is a solution of $\cos(x) = 1/2$ so that all of the solutions involving this angle are: $\left\{\frac{5\pi}{3} + k2\pi\right\}$ where k is an integer. Likewise, we know π is one of the infinite number of solutions of the equation $\cos(x) = -1$. The solution set for this equation is

$$\{\pi + k2\pi\} \quad \text{where } k \text{ is an integer}$$

Sometimes, this is written

$$\{(2k + 1)\pi\} \quad k \text{ is an integer}$$

Although we could use either of these equivalent descriptions, we will usually use the first one.

The solution to the original equation is the combination of these solutions and is written as

$$\left\{\frac{\pi}{3} + k2\pi\right\} \cup \{\pi + k2\pi\} \cup \left\{\frac{5\pi}{3} + k2\pi\right\} \quad \text{where } k \text{ is an integer}$$

We should check these answers against the results we obtained using our graphing calculators.

We are now ready to consider solution techniques for a variety of trigonometric equations.

EXAMPLE SET A

Find exact solutions to the following trigonometric equation:

$$2\sin^2(x) - \cos(x) - 1 = 0$$

Solution:

We notice immediately that there are two different trigonometric functions in this equation. We solve most conditional trigonometric equations by rewriting them in terms of a single trigonometric function. Can we change the equation so that there is only one such function? The basic identities we learned previously will be a handy resource in such problems. We know that $\sin^2(x) = 1 - \cos^2(x)$ and so we can rewrite the original equation as:

$$2[1 - \cos^2(x)] - \cos(x) - 1 = 0$$

We see that this new equation is quadratic in $\cos(x)$, therefore, we proceed as we did in the previous problem.

$$2[1 - \cos^2(x)] - \cos(x) - 1 = 0$$

$$2 - 2\cos^2(x) - \cos(x) - 1 = 0$$

$$-2\cos^2(x) - \cos(x) + 1 = 0$$

$$2\cos^2(x) + \cos(x) - 1 = 0$$

This turns out to be the same equation we solved in the previous example and has solutions

$$\left\{\frac{\pi}{3} + k2\pi\right\} \cup \{\pi + k2\pi\} \cup \left\{\frac{5\pi}{3} + k2\pi\right\} \quad \text{where } k \text{ is an integer}$$

We can now see that this example and the previous one have the same equations to solve although the equations appear in different forms.

EXAMPLE SET B

Find exact solutions to the following trigonometric equation.

$$\cos(2x) = \sin(x) - \sin^2(x) - 1$$

Solution:

This problem does not have a single trigonometric function and, in addition, another problem surfaces. The arguments of each function are not the same. As with identities of this type, the arguments need to be the same. The cosine double angle identity will allow us to make them the same. Although any of the three forms we learned will do, a little thought tells us that the form containing only sines is probably best since its use yields an equation in a single trigonometric function.

$$\cos(2x) = \sin(x) - \sin^2(x) - 1$$

$$1 - 2\sin^2(x) = \sin(x) - \sin^2(x) - 1$$

$$2 - \sin^2(x) - \sin(x) = 0$$

$$\sin^2(x) + \sin(x) - 2 = 0$$

We again arrive at an expression that looks factorable. Factoring gives us:

$$[\sin(x) - 1][\sin(x) + 2] = 0$$

Setting each factor equal to 0 and solving yields:

$$\sin(x) - 1 = 0 \quad \text{or} \quad \sin(x) + 2 = 0$$

$$\sin(x) = 1 \quad \text{or} \quad \sin(x) = -2$$

Now that we have solved for the variable $\sin(x)$, we can find the solution set for x. We are familiar with the solutions of $\sin(x) = 1$, but what about

$\sin(x) = -2$? We need to recall that the range of the sine function is $[-1, 1]$. Therefore, $\sin(x) = -2$ has no solution. We can now write the complete solution set as

$$\left\{ \frac{\pi}{2} + k2\pi \right\} \quad \text{where } k \text{ is an integer.}$$

We can, if we wish, check this result with our graphing calculators.

The next example gives us an opportunity to look at an approach that does not use the idea of obtaining an expression with a single trigonometric function.

EXAMPLE SET C

Find exact solutions to the following trigonometric equation.

$$\sin(2x) + \sin(x) = 0$$

Solution:
We notice here that the functions are the same but the arguments are different. We reach into our resource bank to pull out the sine double-angle identity. This results in:

$$\sin(2x) + \sin(x) = 0$$

$$2\sin(x)\cos(x) + \sin(x) = 0$$

At this point, we might notice that $\sin(x)$ is a factor of the left side and that the right side is 0. We factor the left side to see if this helps.

$$\sin(x)[2\cos(x) + 1] = 0$$

We see that this does indeed help since setting each factor equal to 0 gives two equations each of which contains a single trigonometric function. The details of these steps are:

$$\sin(x) = 0 \quad \text{or} \quad 2\cos(x) + 1 = 0$$

$$\sin(x) = 0 \quad \text{or} \quad \cos(x) = -1/2$$

The solutions to both of these equations should be quite familiar.

$$\{k\pi\} \quad \cup \quad \left\{ \frac{2\pi}{3} + k2\pi \right\} \quad \cup \quad \left\{ \frac{4\pi}{3} + k2\pi \right\} \quad \text{where } k \text{ is an integer}$$

EXAMPLE SET D

Find exact solutions to the following trigonometric equation.

$$\cos(2x) = -1$$

Solution:

We may be tempted to use the cosine double angle formula from our resource bank because we have done so in a previous example. We have, however, solved equations like this for the argument of the function. In this case, the argument is $2x$ not x itself, but that should not deter us. We should note that there is no mixture of angle multipliers. Under these conditions, we need to ask ourselves, "The cosine of what is -1?" The answer yields:

$$2x = \pi \quad \text{or in general} \quad 2x = \pi + k2\pi, \text{ where } k \text{ is an integer.}$$

We have the solution set for $2x$, now we need the solution set for x. To find it, we divide both sides of the equation by 2 with care.

$$x = \frac{\pi + k2\pi}{2} = \frac{\pi}{2} + k\pi, \text{ where } k \text{ is an integer.}$$

Thus, the solution is

$$\left\{ \frac{\pi}{2} + k\pi \right\}, \text{ where } k \text{ is an integer.}$$

EXERCISES

In exercises 1–18, find all solutions that lie in the interval $[0, 2\pi)$. For each problem, find both *approximate* solutions and *exact* solutions.

1. $4\sin(x) - 2 = 0$

2. $2\cos(x) = \sqrt{3}$

3. $2\sin(x) = \sqrt{2}$

4. $\tan(x) + 1 = 0$

5. $\tan^2(x) = 1$

6. $3\tan^2(x) - 1 = 0$

7. $2 - 8\cos^2(x) = 0$

8. $\sin^2(x) + \sin(x) - 1 = 0$

9. $2\cos^2(x) + 3\cos(x) + 1 = 0$

10. $\sin^2(x) + \sin(x) - 2 = 0$

11. $2\sin^2(x) - 3\cos(x) + 1 = 0$

12. $2\cos^2(x) - \sin(x) - 1 = 0$

13. $\sin(2x) - \sin(x) = 0$

14. $2\sin(3x) = -1$

15. $\tan(4x) = -1$

16. $\cos(2x) = \sin(x)$

17. $\tan^4(x) - 2\sec^2(x) + 3 = 0$

18. $\cot(3x) = \sqrt{3}$

In exercises 19–25, find all solutions that lie in the interval $[0°, 360°)$. For each problem, find both *approximate* solutions and *exact* solutions.

19. $2\sin(x)\cos(x) + \sin(x) = 0$

20. $2\cos^2(x) + 3\cos(x) - 2 = 0$

21. $\sec(x) - 2\cos(x) = 1$

22. $2\sin(x) + \csc(x) - 3 = 0$

23. $\cos(2x) + \cos(x) = 0$

24. $\sin^2(2x) = 1$

25. $4\cos^2(2x) = 1$

In exercises 26–33, approximate all solutions.

26. $9\cos^2(x) + 6\cos(x) - 8 = 0$

27. $5\sin^2(x) - 11\sin(x) + 2 = 0$

28. $2\cos(2x) + 3\cos(x) + 1 = 0$

29. $\tan(x) - 3\cot(x) = 2$

30. $6\sin(2x)\cos(2x) = -1$

31. $\sin^2(x) + 5\cos(x) + 2 = 0$ Hint: Use the quadratic formula.

32. $3\tan^2(x) - \tan(x) - 3 = 0$ Hint: Use the quadratic formula.

33. $10\cos^2(x) - 4\cos(x) - 5 = 0$ Hint: Use the quadratic formula.

34. Graph the function $y = \sin(x) + \sin(2x)$ from 0 to 4. Is this graph periodic? If so, what is the period? From the graph find the first four zeros of the function. Compare the results with the results you obtain by solving analytically.

Summary and Projects

Summary

Solving Right Triangles   The key idea in solving right triangles is to involve two given pieces of information (aside from the 90° angle) with one unknown piece of information. We then choose the appropriate trigonometric function that involves these three parts and proceed to solve for the unknown. The resources we need to solve right triangles are trigonometric ratios like:

$$\sin(\textit{an acute angle}) = \frac{\text{side opposite}}{\text{hypotenuse}}$$

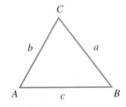

Solving Non-Right Triangles   We use either the law of sines or the law of cosines to solve non-right triangles. We use the law of sines whenever we know any two angles and any side or two sides and an angle opposite one of them. The latter can give rise to the ambiguous case, which sometimes has two possible solutions. Referring to the figure on the left, the law of sines is:

$$\frac{a}{\sin(A)} = \frac{b}{\sin(B)} = \frac{c}{\sin(C)}$$

We use the law of cosines when we know three sides of the triangle or two sides and the included angle. Referring to the figure on the left, the law of cosines is:

$$c^2 = a^2 + b^2 - 2ab\cos(C)$$

Proving Trigonometric Identities   There are many things to remember when proving or verifying trigonometric identities. It is important to remember that our graphing calculator can be of great help at any stage to convince us that the left and right side of the identity give the same graph. The approach to verifying the identity can be summarized by stating that we transform one side of the identity into the other side by performing algebraic operations in conjunction with the use of basic identities whenever they appear in their various forms. In addition, if there is a mix of angles (arguments), we need to use basic identities to get the angles to be of the same multiplicity. The basic identities are listed below:

$$\sec(s) = \frac{1}{\cos(s)} \qquad \csc(s) = \frac{1}{\sin(s)} \qquad \csc(s) = \frac{1}{\sin(s)}$$

$$\tan(s) = \frac{\sin(s)}{\cos(s)} \qquad \cot(s) = \frac{\cos(s)}{\sin(s)}$$

$$\cos^2(s) + \sin^2(s) = 1$$

$$\tan^2(s) + 1 = \sec^2(s)$$

$$\cot^2(s) + 1 = \csc^2(s)$$

$$\cos(a + b) = \cos(a)\cos(b) - \sin(a)\sin(b)$$

$$\sin(a + b) = \sin(a)\cos(b) + \cos(a)\sin(b)$$

$$\sin(2s) = 2\sin(s)\cos(s)$$

$$\cos(2s) = \cos^2(s) - \sin^2(s) = 2\cos^2(s) - 1 = 1 - 2\sin^2(s)$$

$$\sin\left(\frac{s}{2}\right) = \pm\sqrt{\frac{1 - \cos(s)}{2}}$$

$$\cos\left(\frac{s}{2}\right) = \pm\sqrt{\frac{1 + \cos(s)}{2}}$$

Solving Conditional Trigonometric Equations   Again, use the graphing calculator to obtain a quick approximate solution(s). The usual approach is to write the expression in terms of only one trigonometric function, if necessary. We use the basic identities to accomplish this. The exception to this approach occurs when we are able to factor the expression and obtain a product of two or more trigonometric expressions each of which contains just one trigonometric function. This also requires that the other side of the equation be 0. The factoring approach works very well in such cases and we follow through by setting each trigonometric factor equal to 0 and solving. There is one more important aspect to solving conditional trigonometric equations. We need to be able to view the trigonometric functions as a single variable. If we are not able to do this, we should make a substitution for the trigonometric function. For example, consider the following equation:

$$\cos^2(x) - 2\cos(x) - 3 = 0$$

Once we are able to view $\cos(x)$ as a single variable, it is relatively easy to see that we can factor the trigonometric expression and obtain $[\cos(x) - 3][\cos(x) + 1] = 0$ and proceed by setting each factor to 0 and solving. If we do not see that it can factor, we are still in a position to analyze the problem and consider using the quadratic formula, for instance. All is not lost if we cannot view the trigonometric functions as a variable at ÆrstWe can use a substitution until we gain this capability. In the above equation, we can substitute $z = \cos(x)$ to obtain:

$$z^2 - 2z - 3 = 0$$

We can solve this equation by the methods we learned in intermediate algebra. All we need to remember, if we use this approach, is that after we solve for z, we replace z by $\cos(x)$ and Ænisbolving the resulting trigonometric equations.

Projects

1. Using your resource bank from geometry, explain why three parts of the triangle must be known to uniquely solve the triangle. Be sure to discuss why one of the pieces of information must be the length of a side. Include drawings to illustrate cases in which two sides of two right triangles are equal but the triangles are not congruent.

2. The area of a triangle can be obtained by taking one-half the product of any two sides and the sine of the included angle. Derive this formula using the fact that the area of a triangle is one-half the base times the height.

3. Derive the tangent difference identities $\tan(a + b)$ and $\tan(a - b)$ using the fact that $\tan(s) = \dfrac{\sin(s)}{\cos(s)}$.

4. Evaluate $\dfrac{\tan(35°) + \tan(55°)}{1 - \tan(35°)\tan(55°)}$ on your graphing calculator. Explain why you get an error message on most calculators and a wrong answer on the TI-81.

5. Derive the tangent double- and half-angle identities. (Hint: One approach is to use the fact that $\tan(s) = \dfrac{\sin(s)}{\cos(s)}$.)

CHAPTER 9

Systems of Equations and Matrices

9.1   Systems of Two Linear Equations in Two Unknowns

Introduction

The Graphing Method

The Elimination by Addition Method

The Logic of the Elimination by Addition Method

Parallel and Coincident Lines

The Elimination by Substitution Method

Introduction

Engineers, financial analysts, and scientists use systems of equations in solving a variety of problems. Such systems consist of two or more equations considered together (or simultaneously). We will study the solutions of such systems including those cases where there are many solutions or no solution. Systems of linear equations are covered first and systems of nonlinear equations and inequalities last. Between these two topics we introduce matrices and several methods of solving systems with matrices.

The Graphing Method

Recall that $f(x) = mx + b$ is the general representation of the linear function. We have an infinite number of ordered pairs of real numbers that are in the solution set of such functions since the domain of this function contains all real numbers. Replacing $f(x)$ by y gives the following representation of the linear function as an equation.

$$y = mx + b$$

For example, the equation $2x - 3y = 6$ can be put into the form $y = mx + b$ as $y = \frac{2}{3}x - 2$. Recall also that such equations always graph as straight lines with slope equal to m and y-intercept $(0, b)$. Thus, the slope of the line represented by $y = \frac{2}{3}x - 2$ is $m = \frac{2}{3}$ and it intersects the y-axis at $(0, -2)$. Once again, the solution set consists of an infinite number of ordered pairs of real numbers that satisfy the equation.

Consider now a system of two linear equations in two unknowns:

$$\begin{cases} 2x + 3y = 7 \\ 3x - 5y = -2 \end{cases}$$

While each of these equations considered separately has an infinite number of solutions, taken together they have at most one solution. We can visualize this one solution by recalling that each of these equations represents a linear function and graphs into a straight line in a two-dimensional rectangular coordinate system. Window 9.1 is the graph of this system. The two lines intersect

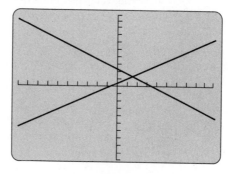

Window 9.1

in a point since they are not parallel or the same line. The point of intersection is the only ordered pair of numbers that belongs to the solution sets of both equations; that is, substituting the ordered pair of numbers appropriately in each equation results in true statements. We need to understand that an ordered pair of numbers must satisfy both equations to be deemed a solution to the system of equations.

EXAMPLE SET A

Determine the solution for the following system of equations:

$$\begin{cases} 2x + 3y = 7 \\ 3x - 5y = -2 \end{cases}$$

Solution:

One approach is to graph each function on our graphing calculator and then use the ZOOM feature to estimate the solution. Remember you will need to solve for y in each equation in order to enter the function rule. The solution, with an error of at most 0.01, is $(1.53, 1.32)$. This example illustrates that the graphical method gives approximate results only.

The Elimination by Addition Method

Another method for solving systems of linear equations entails eliminating variables using algebraic manipulation. To simplify the discussion, let's consider two linear equations in two unknowns x and y. The idea is to generate equations that are equivalent to the original equations with the coefficients of, say x, having the same numerical values but opposite signs. Once we accomplish this, we can add the two equations together to obtain one equation in one variable, in this case, y. We can then solve this equation for y which, in turn, is the y-coordinate of the ordered pair representing the point of intersection of the two lines. Let's apply this method to the previous example.

EXAMPLE SET B

Determine the solution for the system of equations for the previous example using the elimination by addition method.

Solution:

$$\begin{cases} 2x + 3y = 7 \\ 3x - 5y = -2 \end{cases}$$

Recall that multiplying both sides of an equation by the same nonzero number results in an equivalent equation (both have identical solution sets). Taking note of the coefficients of x, we can multiply the first equation by 3 and the second equation by -2 to obtain

$$\begin{cases} 6x + 9y = 21 \\ -6x + 10y = 4 \end{cases}$$

Adding these equations will eliminate x and yield the following equation in y:

$$19y = 25$$

Solving for y gives $y = 25/19$. We could substitute this value in either of the original equations and solve for x or we could use the elimination method again to eliminate y and solve for x. Since we are studying the elimination method, let's continue with it. Again, observing the equations, we see that multiplying both sides of the first equation by 5 and the second equation by 3 will yield equivalent equations with the coefficients of y being equal and opposite in sign.

$$\begin{cases} 10x + 15y = 35 \\ 9x - 15y = -6 \end{cases}$$

Adding these equations eliminates y and yields one equation in x, namely, $19x = 29$. Solving for x gives $x = 29/19$. Our solution to this system of equations is, therefore, $(29/19, 25/19)$. This ordered pair is the exact solution. We can check this result using paper and pencil or using our graphing calculators. We can also check that the solution we obtained using the graphing method was quite reasonable.

The Logic of the Elimination by Addition Method

The elimination by addition method is based on the logic of solving equations. To use this method, we first place both equations in standard form with the x and y terms on one side and the constant term on the other. We then create two equations equivalent to the original equations by multiplying by appropriate numbers to cause the coefficients of either x or y to be the same numerically but opposite in sign. To do this, we first must decide which variable we are going to eliminate, say x. We then multiply the first equation by the second equation's x-coefficient and multiply the second equation by the first equation's x-coefficient. If both these coefficients have the same sign, we simply use the negative of one or the other. When we add these two new equations, the x term is eliminated.

We now have one equation in one unknown, namely y. We can easily find the value of y that solves this linear equation, but how do we interpret this value of y? Focusing on the single intersection point of the original system (Window 9.1), we see that there is only one ordered pair of numbers that satisfy both equations simultaneously. If we assume that these are the numbers we are using for x and y in the two equations we are adding, then equals added to equals are equal. Thus, the value of y that solves the one-variable equation must be the y coordinate of the intersection point. We have seen how to obtain the x-coordinate of the intersection point in the previous example.

Parallel and Coincident Lines

Problems arise when the two lines are either parallel or the same line since in such cases either the system has no solution or many solutions. These cases can sometimes be discovered using the graphical method, but we cannot always be absolutely sure that two lines are parallel (or coincident) just from the graphical representations. So, we need to know how to determine when these situations arise using algebraic manipulations.

This can be done in at least two ways. One way is to solve each of the two equations for y and observe the slope and y-intercept of each. If the slopes are the same, the lines are either parallel or coincident (the same line). If the y-intercepts are different, then the lines are parallel, otherwise they are coincident.

The second way uses information arising from the process of the elimination method. This process allows us to determine when the lines are parallel (no solution) or coincident (infinitely many solutions). In such cases, both variables are eliminated in the system when we attempt to eliminate one of the variables. We end up with either $0 = 0$ or $0 = k$ (k a nonzero real number). The equation $0 = 0$ represents the coincident lines case and $0 = k$, the parallel lines case. We can verify these statements using the notion of parallelism as in the previous paragraph.

EXAMPLE SET C

1. Solve the following system of equations.

$$\begin{cases} 3x + 5y = 7 \\ y = -\dfrac{3}{5}x - 2 \quad (*) \end{cases}$$

Solution:
Clearing fractions in the second equation and placing the system into standard form, we obtain:

$$\begin{cases} 3x + 5y = 7 \\ 3x + 5y = -10 \end{cases}$$

Multiplying both sides of the second equation by -1 results in the following system:

$$\begin{cases} 3x + 5y = 7 \\ -3x - 5y = 10 \end{cases}$$

and adding yields:

$$0 = 17$$

When we arrive at a result such as this, we state that the lines are parallel and do not intersect. We can verify this result by solving for y in the first equation as follows.

$$3x + 5y = 7$$

$$y = -\frac{3}{5}x + \frac{7}{5} \quad (\dagger)$$

Since the slopes of lines $(*)$ and $(\dagger)$ are the same, the lines are either parallel or coincident. The y-intercepts are different and so we conclude that the lines are parallel, but not coincident.

2. Solve the following system of equations.

$$\begin{cases} 2x + 5y = 3 \\ 4x = 6 - 10y \end{cases}$$

Solution:

We begin by placing the equations in standard form.

$$\begin{cases} 2x + 5y = 3 \\ 4x + 10y = 6 \end{cases}$$

Next we can multiply the first equation by -2 and add the result to the second equation. We start by multiplying the first equation by -2.

$$\begin{cases} -4x - 10y = -6 \\ 4x + 10y = 6 \end{cases}$$

Adding the equations results in:

$$0 = 0$$

A result such as this describes coincident lines. Again, to convince ourselves that this is the case, we can place both equations in slope-intercept form.

$$\begin{cases} y = -\dfrac{2}{5}x + \dfrac{3}{5} \\ y = -\dfrac{4}{10}x + \dfrac{6}{10} \end{cases}$$

We can see that these two equations have the same slope and y-intercept and are, therefore, the same line.

The Elimination by Substitution Method

Another approach to solving systems of linear equations is the substitution method. We continue to consider two linear equations in two unknowns. The idea is to solve for, say y, in one of the equations and substitute the resulting expression for y in the other equation. This results in one equation in one unknown, namely x. We obtain the x-coordinate of the point of intersection when we solve for x if the two lines intersect. We then substitute this value back into one of the original equations to obtain the value of y. An example will serve to show how this method works.

EXAMPLE SET D

Determine the solution to the following system of equations using the substitution method.

$$\begin{cases} 2x + 5y = 3 \\ 3x - y = -4 \end{cases}$$

Solution:

Solving for y in the second equation yields

$$3x + 4 = y$$

Now substitute the expression $3x + 4$ for y in the first equation and solve for x.

$$2x + 5(3x + 4) = 3$$

$$2x + 15x + 20 = 3$$

$$17x = -17$$

$$x = -1$$

We now have the x-coordinate of the point of intersection of the two lines and we can find the y-coordinate of this point by substituting this value of x into either of the original equations, or into the equation that gives y in terms of x, $y = 3x + 4$. Let's choose $3x - y = -4$ for this purpose.

$$3(-1) - y = -4$$

$$-3 - y = -4$$

$$-y = -1$$

$$y = 1$$

Checking the ordered pair $(-1, 1)$ in each equation, we convince ourselves that it is indeed the solution.

Note that we could have solved for y in the first equation and substituted in the second equation. However, this choice would have resulted in fractions, which are a bit harder to work with. There are cases in which we cannot avoid fractions in the substitution method but when we can, it is usually helpful to do so. If fractions cannot be avoided, the elimination by addition method is usually more convenient.

The process of substitution will give rise to a contradiction ($5 = 7$) or an identity ($5 = 5$) if the system of equations represents parallel lines or coincident lines. We can solve both equations for y if we suspect that the lines are parallel or coincident.

One use we can make of systems of equations is in describing mathematically and solving word problems that have exactly two unknowns.

EXAMPLE SET E

The sum of two numbers is 38 and their difference is 20. Find the numbers.

Solution:
We can begin as follows:

$$\text{Let} \quad x = \text{the larger number}$$

$$\text{and} \quad y = \text{the smaller number}$$

Now setting up the mathematical model that describes the problem leads to the following system of equations:

$$\begin{cases} x + y = 38 \\ x - y = 20 \end{cases}$$

We observe that y has the same coefficient in both equations but opposite in sign. We can eliminate y by adding and obtain:

$$2x = 58$$

Solving for x gives us the larger number, which is 29. Substituting 29 for x in either of the original equations and solving for y gives us the smaller number, which is 9. It is easy to check the solution.

EXERCISES

In exercises 1–8, solve the given system of equations.

1. $\begin{cases} x + 2y = 8 \\ 3x - 2y = -4 \end{cases}$

2. $\begin{cases} x + 2y = 5 \\ 5x - 2y = 1 \end{cases}$

3. $\begin{cases} 2x + y = -3 \\ 2x - 3y = 1 \end{cases}$

4. $\begin{cases} 3x + 2y = 3 \\ 3x + 4y = 9 \end{cases}$

5. $\begin{cases} -x + y = -5 \\ 2x + 3y = 5 \end{cases}$

6. $\begin{cases} 2x + y = -7 \\ -3x + y = 3 \end{cases}$

7. $\begin{cases} 4x - 15y = -1 \\ 2x + 3y = -4 \end{cases}$

8. $\begin{cases} 4x - 3y = -7 \\ 8x - 9y = -28 \end{cases}$

In exercises 9–14, solve the given system of equations by the substitution method.

9. $\begin{cases} x + 2y = 4 \\ 3x - 5y = -7 \end{cases}$

10. $\begin{cases} 2x - y = 6 \\ 7x - 3y = 4 \end{cases}$

11. $\begin{cases} 2x - 3y = 6 \\ 4x - 5y = 11 \end{cases}$

12. $\begin{cases} 13x + 5y = 7 \\ -9x + 2y = 5 \end{cases}$

13. $\begin{cases} -5x + 2y = 7 \\ 2x + 3y = 4 \end{cases}$

14. $\begin{cases} 7x - 3y = 10 \\ 4x + y = 7 \end{cases}$

15. Find the number of girls in a class of 32 if there are three times as many boys as girls.

16. Find the number of girls in a class of 30 if there are two boys for every three girls.

17. The sum of two numbers is 11 and twice one number added to five times the other is 49. Find the numbers.

18. Two numbers differ by six and three times one number is equal to the other. Find the numbers.

In exercises 19-26, solve the given system of equations.

19. $\begin{cases} 2x + 3y = 41 \\ 5x - 2y = 5 \end{cases}$

20. $\begin{cases} 3x + 2y = 8 \\ 5x - 5y = 3 \end{cases}$

21. $\begin{cases} 3u - 2v = 7 \\ 6u - 4v = -9 \end{cases}$

22. $\begin{cases} 6s - 8t = 6 \\ 3s - 4t = 3 \end{cases}$

23. $\begin{cases} 8m - 11n = 5 \\ 5m - n = 4 \end{cases}$

24. $\begin{cases} 7a + 3b = 10 \\ 2b - a = 7 \end{cases}$

25. $\begin{cases} \dfrac{1}{3}x + \dfrac{2}{5}y = 1 \\ \dfrac{3}{5}x - \dfrac{5}{7}y = 1 \end{cases}$

26. $\begin{cases} \dfrac{4}{3}x - \dfrac{2}{3}y = 1 \\ \dfrac{3}{2}x - 2y = 2 \end{cases}$

27. A linear function of the form $f(x) = mx + b$ passes through the points $(2, 3)$ and $(-2, 5)$. Find the function.

28. A linear function of the form $g(x) = mx + b$ passes through the points $(-2, 5)$ and $(-3, 2)$. Find the function.

29. Betty and Bob spent $6.00 for $3\frac{3}{4}$ pounds of ice cream and cake for Johnny's birthday party. The ice cream costs $1.10 a pound and the cake costs $1.80 a pound. How many pounds of ice cream did they buy if they purchased a total of $3\frac{3}{4}$ pounds of ice cream and cake?

30. Jamaican and Hawaiian Kona coffees are to be mixed together to create ten pounds of Kava Cup Blend coffee. If Jamaican coffee sells for $4.70 a pound and Hawaiian Kona sells for $3.20 a pound, how much of each coffee must be mixed to obtain a blend selling for $4.00 a pound?

31. Describe a real-life situation that can generate the following systems of equations:

(a) $$4x + 10y = 98$$
$$15x + 2y = 190$$

(b) $$x - 3y = 0$$
$$4x + 7y = 114$$

Solve each system.

32. It's the end of her trip to India, and Noriko has 781 rupees in her pocket. Not wishing to bring foreign currency home, she buys 9 key-chains and 13 pens as souvenirs using all her rupees. The key-chains are 2 1/2 times as expensive as the pens. How much does each pen and key-chain cost?

33. (Refer to the previous exercise) While shopping for the souvenirs Noriko spies six silk scarves and four jewelry boxes that she wants. Each box costs 16 more than twice the cost of each scarf. She exchanges $100 for rupees and buys the scarves and boxes, leaving her with no rupees. The rate

of exchange at the time is Rs 31.90 per dollar. Find the dollar price of each scarf and each jewelry box.

34. Jim is delivering his mom's homemade pies. At Kelper's he delivers 16 apple pies and five pecan pies, which come to a total of $100.75. At the second place, Ruby's Kitchen, he gives a bill for delivering 12 apple and four pecan pies that total $77. When asked how much an individual pie costs, Jim doesn't know. He sits down and writes some equations to help answer the question. What are the correct equations and their solutions?

35. (Refer to the previous exercise) Jim delivers three kinds of pies to three different restaurants each week. Papa's Cafe orders 5 cheesecakes and 9 pumpkin pies that total $67.20; Joe's Deli is handed a weekly bill of $192.85 for their order of 7 pumpkin pies and 31 lemon meringue pies; and Nina's Cafe receives 8 cheesecakes and 11 lemon meringue pies, which totals $103.35. What is the cost of each kind of pie?

36. The Acme Paint Company sent its ace painter, Chun, to do a job. When he had finished half of the job, Chun was called away to another job. A novice painter finished the job. It took 13 1/2 hours to finish the job, which cost $115.50. If Chun is paid two and a half times the novice's price ($5.50/hour), how many hours did each painter work?

9.2   Systems of Three Linear Equations in Three Unknowns

Introduction

Connecting the Methods

Generalizing the Approach

Introduction

The extension from a system of two linear equations in two unknowns to three equations in three unknowns (or more) may not be obvious at first but should become a recognizable pattern fairly soon. To that end, let us consider extending the elimination methods to a system of three linear equations in three unknowns, say x, y, and z.

The overall idea here is to choose two of the equations and eliminate one of the variables, say z. Then choose a different pair of equations and eliminate the *same* variable. These efforts result in two linear equations in two unknowns. At this point, we can incorporate any of the methods we learned for two linear equations in two unknowns to solve for the values of x

and y. Then, we can substitute these values into any one of the three original equations to find the value of z.

We can get a visual picture of the solution to three linear equations in three unknowns by first accepting the fact that a linear equation in three unknowns represents a plane in space. We then need to recall that if two non-coplanar planes intersect, their intersection is a line. The third plane will intersect this line in a point provided the plane and line of intersection are not parallel or provided the plane does not contain the line. A point in space is represented by an ordered triple of numbers. This ordered triple would then represent the intersection of the three planes and hence be the solution of the system of three equations.

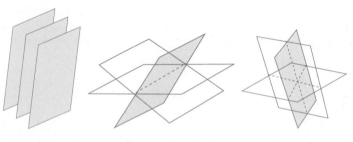

Figure 9.1

EXAMPLE SET A

Determine the solution for the following system of three linear equations in three unknowns.

$$\begin{cases} -2x + y - z = 1 \\ 5x - 2y + 3z = 3 \\ x - y - 2z = -4 \end{cases}$$

Solution:

Our first task is to decide which variable we are going to eliminate. Let's eliminate z. Choose the first two equations to start the elimination process. We note that z can be eliminated by multiplying both sides of the first equation by 3 and adding the result to the second equation. This new equation represents the line of intersection of the two planes. We now carry out these steps as follows:

$$-6x + 3y - 3z = 3$$

$$5x - 2y + 3z = 3$$

We note that this causes the numerical coefficients of z to be the same yet opposite in sign. Adding these two equations gives:

$$-x + y = 6 \qquad (9.1)$$

This equation represents the line of intersection of these two planes. We now choose two different equations from which to eliminate z. We have a choice of equations 1 and 3 or equations 2 and 3. However, we might observe that equations 1 and 3 are a more convenient choice because of the -1 coefficient for z in one of them. (A variable can be easily eliminated by addition when one of the two equations has a variable with a coefficient of 1 or -1. Only that equation needs to be multiplied by a constant to match up the variable's coefficients—one positive, one negative.) With this choice, we can multiply the first equation by -2 and then add the result to the third equation as follows.

$$4x - 2y + 2z = -2$$
$$x - y - 2z = -4$$

Adding these two equations will give the line of intersection of the two planes represented by these two equations.

$$5x - 3y = -6 \qquad (9.2)$$

The two resulting equations 9.1 and 9.2 form the system:

$$\begin{cases} -x + y = 6 \\ 5x - 3y = -6 \end{cases}$$

We can now solve this system of two linear equations in two unknowns using any of the methods learned earlier. We can then substitute the values obtained for x and y back into any one of the original equations to find the corresponding value of z. The final result of performing these operations is the ordered triple $(6, 12, -1)$. Again, we can check this solution by substituting these values in the three original equations to verify that each equation results in a true statement. Such checking would be an excellent use of our graphing calculators.

Connecting the Methods

There are numerous important reasons to understand the theory presented in the elimination (or substitution) method for a system of two linear equations in two unknowns. One reason is that we cannot use the graphing feature of our graphing calculators to solve systems with more than two equations and two unknowns. We then rely on our understanding of either the elimination or substitution method for two equations in two unknowns to extend our attack on systems with more equations.

We can also use the powerful matrix capabilities of the graphing calculator to find accurate numeric solutions of such systems of equations easily. We consider this approach after we cover the section on matrices.

There are a couple of points we should note before going on. First, the substitution method can also be used to solve the above system. For example, we could solve for z in the first equation and then substitute the expression we obtain for z in the second and third equations. This would give two equations in two unknowns and we would proceed as before.

Other possible outcomes for systems of three linear equations in three unknowns are that two of the three planes are parallel, all three planes are parallel, or the three planes intersect in the same line. Although the number of such possibilities increases with increasing numbers of equations and unknowns being considered in the system, the results are similar in nature. Unfortunately, visual intepretations disappear as we move to systems with more equations and unknowns. The good news, however, is that the extension of the theory we studied in these initial cases allows us to attack problems containing many equations. The approach also allows us to interpret those problems that yield no solution or infinitely many solutions.

We say that a system of linear equations is consistent but *dependent* when it has many solutions, consistent and *independent* when there is a unique solution, and *inconsistent* when there is no solution.

EXAMPLE SET B

1. Determine the solution for the following system of three linear equations in three unknowns.

$$\begin{cases} x + 3y - 2z = -9 \\ 2x - y - z = -6 \\ 5x + 15y - 10z = -18 \end{cases}$$

Solution:

We can eliminate x fairly easily by multiplying the first equation by -2 and adding it to the second equation. Then we would multiply the first equation by -5 and add it to the third equation. Carrying out these steps we obtain:

$$-2x - 6y + 4z = 18$$
$$2x - y - z = -6$$

and adding gives:

Working with equations one and three, we obtain:

$$-5x - 15y + 10z = 45$$

$$5x + 15y - 10z = -18$$

and adding gives:

$$0 = 27$$

This last equation is a contradiction, which leads us to conclude that the system has no solution and we classify it as inconsistent.

2. Solve the following system of equations:

$$\begin{cases} x + 3y - 2z = -9 \\ 4x + 3y - 5z = 1 \\ 2x - 3y - z = 19 \end{cases}$$

Solution:

Since the coefficient of x in the first equation is 1, eliminating x seems to be a reasonable approach. (If we study the system, we might even notice that eliminating y would be easier.) We can eliminate x by multiplying the first equation by -4 and adding the result to the second equation. Then we would multiply the first equation by -2 and add the result to the third equation. The first step leads to:

$$-4x - 12y + 8z = 36$$

$$4x + 3y - 5z = 1$$

and adding yields:

$$-9y + 3z = 37$$

We next work with equations one and three to obtain:

$$-2x - 6y + 4z = 18$$

$$2x - 3y - z = 19$$

and adding yields:

$$-9y + 3z = 37$$

We now have the following two equations in two unknowns to solve:

$$\begin{cases} -9y + 3z = 37 \\ -9y + 3z = 37 \end{cases}$$

They obviously represent the same line. This means that the three planes intersect in a line rather than in a point. This results in an infinite number of solutions, which means the system is consistent but dependent. We can also see that if we attempt to eliminate one of the variables in this system of two equations, we would obtain $0 = 0$, which leads to the same conclusion. We arrive at such a result when the two equations in two unknowns represent the same line but are not as visually obvious as in this case.

Generalizing the Approach

The approach to solving n linear equations in n unknowns follows the same pattern as for three equations in three unknowns. For example, with four linear equations in four unknowns, say x, y, z, and w, we would decide on which variable to eliminate and proceed to eliminate it using three different pairs of equations of the system. This would result in three linear equations in three unknowns, which we would solve using the methods just discussed. The cases of dependent and inconsistent systems would also follow the same pattern. The solution process becomes more and more time-consuming with much more potential for error as the number of equations and unknowns increase. Fortunately, shortcuts have been developed and these shortcuts, along with our graphing calculators or computers with appropriate software, allow us to obtain solutions more quickly and with fewer arithmetic errors. Such approaches are the topic of the next few sections.

EXERCISES

In exercises 1-20, solve the given system of equations. Describe each system as consistent and independent, inconsistent, or consistent but dependent.

1. $\begin{cases} 3x + 2y - z = 7 \\ x - 2y + z = 1 \\ 2x - y + z = 2 \end{cases}$

2. $\begin{cases} x + 2y - z = 6 \\ 2x + 3y - z = 11 \\ 2x - 4y + z = -1 \end{cases}$

3. $\begin{cases} 2x + y + 3z = -1 \\ x - y - z = 0 \\ 3x + 2y - 2z = 17 \end{cases}$

4. $\begin{cases} 2x + y - 6z = -12 \\ -x + y + z = -7 \\ 5x - 3y + 7z = 11 \end{cases}$

5. $\begin{cases} x + 2y + 3z = -5 \\ 2x + y + 2z = -1 \\ 3x - 2y - z = 7 \end{cases}$

6. $\begin{cases} x + y + 3z = 2 \\ 2x - y - 2z = -3 \\ 3x + 2y - 2z = -13 \end{cases}$

7. $\begin{cases} 2x + y + 3z = 8 \\ x + 3y + 2z = 7 \\ 3x + 2y + z = 9 \end{cases}$

8. $\begin{cases} -4x + 3y + z = -9 \\ 2x + y - 5z = 7 \\ x + y + 3z = 4 \end{cases}$

9. $\begin{cases} 4x - 3y + 2z = -10 \\ -2x + 3y + 4z = 18 \\ 2x - 4y + 3z = -6 \end{cases}$

10. $\begin{cases} -4x + 5y - 7z = 0 \\ -8x + 3y + 5z = 3 \\ 2x + z = 0 \end{cases}$

11. $\begin{cases} x + y + z = 2 \\ 7x + 3y - 2z = -9 \\ 5x - 3y + 2z = 9 \end{cases}$

12. $\begin{cases} 3x + 2y - 7z = 1 \\ -x - y + z = -5 \\ -4x - 2y + 11z = 2 \end{cases}$

13. $\begin{cases} 3x + y + 2z = 6 \\ 3x + 10y - 4z = -18 \\ -6x + 5y + 4z = 9 \end{cases}$

14. $\begin{cases} 2x + 3y - 4z = 13 \\ 4x + 3y + 8z = 7 \\ 4x - 6y + 4z = -11 \end{cases}$

15. $\begin{cases} x + y + z = 2 \\ 2x + y - z = 5 \\ 3x + 2y + z = 4 \end{cases}$

16. $\begin{cases} x + y + z = 1 \\ 2x - y + z = 1 \\ 3x + y + 2z = 3 \end{cases}$

17. $\begin{cases} x + z = 0 \\ x + y = 3 \\ 3x + y + 2z = 3 \end{cases}$

18. $\begin{cases} x + z = 0 \\ 3x + y = 2 \\ 2x + y - z = 1 \end{cases}$

19. $\begin{cases} 3x + y + 2z = -5 \\ x + y = -3 \\ x + z = -1 \end{cases}$

20. $\begin{cases} x + y + z = 3 \\ 2x + 2y + 2z = 6 \\ 5x + 5y + 5z = 15 \end{cases}$

21. A linear equation in three variables represents the equation of a plane in space. Any three planes in three-dimensional space may or may not meet at a point (there are several possibilities). When three planes do meet at a point, the 3-D coordinates (x, y, z) of the point are found by solving the equations for the planes. If the three planes do not meet at a point, unique solutions will not exist. Find the coordinates of the unique point of intersection (if any) of the following sets of three planes.

(a) $\begin{cases} 4x + 2y - z = 5 \\ 3x - y + 2z = -5 \\ 7x - 3y + z = 8 \end{cases}$

(b) $\begin{cases} x + y + z = 0 \\ 2x + 2y + 2z = 5 \\ x + z = 2 \end{cases}$

22. Susan is making three kinds of granola using oats, nut mix, and dried fruit. The first kind of granola contains 80% oats, 10% nuts and 10% dried fruit. The second kind of cereal has 70% oats, 10% nuts and 20% dried fruit. The third kind has 50% oats, 30% nuts and 20% dried fruit. The cost is $0.64 per pound for the first kind, $0.85 per pound for the second kind, and $1.11 per pound for the third kind. What are the prices per pound for each kind of raw material?

23. The *ABC* Shoe Manufacturing Company makes three kinds of shoes and sells them through their stores in Fresno, Brentwood, and San Jose. Each store receives the same kind of shipment each month (i.e., *A* number of kind *A* shoes, *B* number of kind *B* shoes, and *C* number of kind *C* shoes). The stores sell the shoes at different prices as shown in the first table. The total sales are noted in the second table.

Price Table			
City	Shoe A	Shoe B	Shoe C
Fresno	$21.00	$23.00	$24.00
Brentwood	$24.00	$30.00	$36.00
San Jose	$31.00	$38.00	$50.00

Total Sales	
Fresno	$8,310
Brentwood	$10,800
San Jose	$14,210

How many shoes of each kind were delivered?

9.3   Matrices

Introduction

Matrices and Matrix Notation

Matrix Operations

Row Operations in Solving Systems of Linear Equations

Introduction

One of the major uses of matrices is to streamline the computations involved in solving systems of equations. Another is the beauty of its shorthand representation of such systems, which allows us to describe the problem and solution of such systems in a very compact way. You will soon appreciate this compact description. Also, matrices are used to organize numerical and non-numerical data and information compactly.

Matrices and Matrix Notation

Matrix

Matrix
A **matrix** is a rectangular array of elements.

For our purposes, we only consider matrices in which the elements are real numbers. For example, the following rectangular array is a matrix.

$$B = \begin{bmatrix} -2 & 1 & -1 & 1 \\ 5 & -2 & 3 & 3 \\ 1 & -1 & -2 & -4 \end{bmatrix}$$

This matrix is given the name B and has three rows and four columns. We use this fact to say that B is a 3-by-4 (written 3×4) matrix whereby the three rows and four columns are the dimensions of the matrix. The numbers in the array are called elements of the array. For example, -2, 1, -1, and 1 are elements in the matrix and also represent the first row of B. An equally important notion is the location of an element in an array. For instance, 5 is an element in array B and is located in the second row, first column. This location will be described as b_{21}. Thus, b_{21} is the element in array B that is located in the second row and first column. Here, $b_{21} = 5$. We note that the element in the third row and third column is -2 and would be identified in general as b_{33}. A general matrix A with m rows and n columns would be written

$$A = \begin{bmatrix} a_{11} & a_{12} & \cdots & a_{1n} \\ a_{21} & a_{22} & \cdots & a_{2n} \\ \vdots & \vdots & \ddots & \vdots \\ a_{m1} & a_{m2} & \cdots & a_{mn} \end{bmatrix}$$

Matrix Operations

In general, a_{ij} represents the element in matrix A that is located in the ith row and the jth column.

Row Operations

Now that we have defined the set of operations we need to perform on equations, we define these same operations on matrices. The following operations are called elementary matrix **row operations**.

1. Interchange any two rows of a matrix.

2. Multiply any row by a nonzero real number.

3. Replace any row by the sum of itself and the multiple of any other row.

EXAMPLE SET A

Use elementary row operations on matrix B to produce a matrix with a zero in the second row, first column and zeros in the third row, first and second columns.

Solution:

First let's interchange the first and third rows. The purpose here is the convenience obtained in having a numeric coefficient of 1 in the first row. This gives:

$$\begin{bmatrix} 1 & -1 & -2 & -4 \\ 5 & -2 & 3 & 3 \\ -2 & 1 & -1 & 1 \end{bmatrix}$$

Next, replace the second row by the sum of itself and -5 times the first row to obtain a zero in the second row, first column. This yields:

$$\begin{bmatrix} 1 & -1 & -2 & -4 \\ 0 & 3 & 13 & 23 \\ -2 & 1 & -1 & 1 \end{bmatrix}$$

Note that we can get a zero in the third row, first column by replacing the third row by the sum of itself and 2 times the first row. The result of this operation is:

$$\begin{bmatrix} 1 & -1 & -2 & -4 \\ 0 & 3 & 13 & 23 \\ 0 & -1 & -5 & -7 \end{bmatrix}$$

Let us now interchange the second and third rows to obtain:

$$\begin{bmatrix} 1 & -1 & -2 & -4 \\ 0 & -1 & -5 & -7 \\ 0 & 3 & 13 & 23 \end{bmatrix}$$

We can better appreciate this operation by noting that we can now conveniently get a zero in the third row, second column by replacing the third row by the sum of itself and 3 times the second row. This results in:

$$\begin{bmatrix} 1 & -1 & -2 & -4 \\ 0 & -1 & -5 & -7 \\ 0 & 0 & -2 & 2 \end{bmatrix}$$

Row Operations in Solving Systems of Linear Equations

A matrix such as the final matrix in the previous example is said to be in *upper triangular form*. A matrix is an upper triangular matrix if all entries below the entries having equal row and column numbers ($a_{n,n}$) are all zero. We can now connect the matrix elementary row operations we have studied so far to the solution of systems of linear equations. Let us look at the system of three equations in three unknowns we solved earlier using the elimination method.

$$\begin{cases} -2x + y - z = 1 \\ 5x - 2y + 3z = 3 \\ x - y - 2z = -4 \end{cases}$$

Coefficient Matrix

Look at the coefficients of the unknowns. They form a square matrix called the **coefficient matrix**. Let's name this matrix A, which is represented by:

$$A = \begin{bmatrix} -2 & 1 & -1 \\ 5 & -2 & 3 \\ 1 & -1 & -2 \end{bmatrix}$$

Constant Matrix

The constants on the right side of the system of equations form a rectangular matrix with 3 rows and 1 column. This matrix is called the **constant matrix**. Let's name this matrix C, which is represented by:

$$C = \begin{bmatrix} 1 \\ 3 \\ -4 \end{bmatrix}$$

Augmented Matrix

The matrix formed by placing the elements from the coefficient matrix followed by the elements from the constant matrix is called the **augmented matrix**. It is given by:

$$\begin{bmatrix} -2 & 1 & -1 & 1 \\ 5 & -2 & 3 & 3 \\ 1 & -1 & -2 & -4 \end{bmatrix} \quad \text{which represents} \quad \begin{cases} -2x + y - z = 1 \\ 5x - 2y + 3z = 3 \\ x - y - 2z = -4 \end{cases}$$

You may notice that this matrix is the same as matrix B defined at the start of this section and is one kind of matrix representation of the original system of equations. We placed this matrix in upper triangular form obtaining:

$$\begin{bmatrix} 1 & -1 & -2 & -4 \\ 0 & -1 & -5 & -7 \\ 0 & 0 & -2 & 2 \end{bmatrix} \quad \text{which represents} \quad \begin{cases} 1x - 1y - 2z = -4 \\ 0x - 1y - 5z = -7 \\ 0x + 0y - 2z = 2 \end{cases}$$

Now if we interpret the last row of the matrix as $-2z = 2$ and solve for z, we obtain $z = -1$. Interpreting the second row of the matrix as $-1y - 5z = -7$, substituting -1 for z and solving for y gives $y = 12$. Finally, intepreting the first row of the matrix as $1x - 1y - 2z = -4$, substituting -1 for z and 12 for y, and solving for x gives $x = 6$. This is the same solution we obtained using the elimination method. This last method is called the *back-substitution method*. In general, it consists of placing the augmented matrix of a system of linear equations into upper triangular form and then working your way back starting with the last equation.

This matrix method essentially strips off the variables from the system of equations and works with the coefficients only. This abbreviates the work and makes it easier and more accurate. We concentrate the information in a matrix, manipulate the matrix, and then reconstitute the original system of equations in upper triangular form to find the solutions by back substititution.

The use of matrices organizes the attack on the solution of systems of linear equations. It also allows for a symbolic representation for the system of linear equations. This symbolic representation greatly reduces the amount of writing needed and, more importantly, greatly enhances the understanding of other simpler approaches to the solution of such systems, one of which will be covered later in this chapter.

EXERCISES

In exercises 1–8, determine the upper triangular matrix form.

1. $\begin{bmatrix} 3 & 2 & -1 & 7 \\ 1 & -2 & 1 & 1 \\ 2 & -1 & 1 & 2 \end{bmatrix}$

2. $\begin{bmatrix} 1 & 2 & -1 & 6 \\ 2 & 3 & -1 & 11 \\ 2 & -4 & 1 & -1 \end{bmatrix}$

3. $\begin{bmatrix} 2 & 1 & 3 & -1 \\ 1 & -1 & -1 & 0 \\ 3 & 2 & -2 & 17 \end{bmatrix}$

4. $\begin{bmatrix} 2 & 1 & -6 & -12 \\ -1 & 1 & 1 & -7 \\ 5 & -3 & 7 & 11 \end{bmatrix}$

5. $\begin{bmatrix} 1 & 2 & 3 & -5 \\ 2 & 1 & 2 & -1 \\ 3 & -2 & -1 & 7 \end{bmatrix}$

6. $\begin{bmatrix} 3 & 5 & 2 & 6 \\ 3 & 10 & -4 & -18 \\ -6 & 5 & 4 & 9 \end{bmatrix}$

7. $\begin{bmatrix} 2 & 3 & -4 & 13 \\ 4 & 3 & 8 & 7 \\ 4 & -6 & 4 & -11 \end{bmatrix}$

8. $\begin{bmatrix} 3 & 1 & 2 & -5 \\ 1 & 1 & 0 & -3 \\ 1 & 0 & 1 & -1 \end{bmatrix}$

In exercises 9–16, solve the system of equations using a matrix representation of the system and the upper triangular form. (Hint: Refer to exercise 1–8.)

9. $\begin{cases} 3x + 2y - z = 7 \\ x - 2y + z = 1 \\ 2x - y + z = 2 \end{cases}$

10. $\begin{cases} x + 2y - z = 6 \\ 2x + 3y - z = 11 \\ 2x - 4y + z = -1 \end{cases}$

related

11. $\begin{cases} 2x + y + 3z = -1 \\ x - y - z = 0 \\ 3x + 2y - 2z = 17 \end{cases}$

12. $\begin{cases} 2x + y - 6z = -12 \\ -x + y + z = -7 \\ 5x - 3y + 7z = 11 \end{cases}$

13. $\begin{cases} x + 2y + 3z = -5 \\ 2x + y + 2z = -1 \\ 3x - 2y - z = 7 \end{cases}$

14. $\begin{cases} 3x + 5y + 2z = 6 \\ 3x + 10y - 4z = -18 \\ -6x + 5y + 4z = 9 \end{cases}$

15. $\begin{cases} 2x + 3y - 4z = 13 \\ 4x + 3y + 8z = 7 \\ 4x - 6y + 4z = -11 \end{cases}$

16. $\begin{cases} 3x + y + 2z = -5 \\ x + y = -3 \\ x + z = -1 \end{cases}$

In exercises 17 and 18, use elementary row operations to obtain a matrix that has zeros in positions below entries a_{11} and a_{22} and above entries a_{22} and a_{33}. This form is sometimes referred to as diagonal or Gauss-Jordan form.

17. $\begin{bmatrix} 1 & 1 & 1 & 2 \\ 2 & 1 & -1 & 5 \\ 3 & 2 & 1 & 4 \end{bmatrix}$

18. $\begin{bmatrix} 2 & 1 & 3 & 8 \\ 1 & 3 & 2 & 7 \\ 3 & 2 & 1 & 9 \end{bmatrix}$

In exercises 19 and 20, solve the system of equations using a matrix representation of the system and the diagonal form. (Hint: Refer to exercises 17 and 18.) Check your values for x, y, and z by storing them in your graphing calculator and then evaluating the left sides of the equations.

19. $\begin{cases} x + y + z = 2 \\ 2x + y - z = 5 \\ 3x + 2y + z = 4 \end{cases}$

20. $\begin{cases} 2x + y + 3z = 8 \\ x + 3y + 2z = 7 \\ 3x + 2y + z = 9 \end{cases}$

9.4 Fundamentals of Matrix Algebra

Introduction

Preliminary Definitions

The Identity Matrix

Inner Products

Introduction

An algebraic approach to solving systems of linear equations using matrices requires that we provide definitions and algebraic rules to use in manipulating matrices. We start with the definitions.

Preliminary Definitions

Equal Matrices

Equal Matrices

Two matrices are equal if they have the same dimensions and the corresponding entries are equal.

Matrix Addition

Matrix Addition

Two matrices can be added only if they have the same dimensions. The sum of the two matrices is a matrix of the same dimension and whose entries are obtained by adding the corresponding entries of the individual matrices.

EXAMPLE SET A

Add the matrices A and B given below. Then subtract B from A.

$$A = \begin{bmatrix} 10 & 20 \\ 30 & 40 \end{bmatrix}$$

$$B = \begin{bmatrix} 1 & 2 \\ 3 & 4 \end{bmatrix}$$

Solution:

The sum of these two matrices is obtained by adding the corresponding entries and results in:

$$\begin{bmatrix} 11 & 22 \\ 33 & 44 \end{bmatrix}$$

Notice that the result is a 2×2 matrix, which is the same dimension as both A and B. Although the definition for subtraction of two matrices is not given, it is analogous to addition and is accomplished by subtracting corresponding entries. In this case, we obtain for $A - B$:

$$\begin{bmatrix} 9 & 18 \\ 27 & 36 \end{bmatrix}$$

Scalar Multiplication

Scalar Multiplication

To multiply a matrix by a scalar (real number), multiply every entry by the scalar.

EXAMPLE SET B

Multiply the matrix

$$A = \begin{bmatrix} 10 & 20 \\ 30 & 40 \end{bmatrix}$$

by the scalar (real number) 6.

Solution:
Multiplying matrix A by 6 yields:

$$\begin{bmatrix} 60 & 120 \\ 180 & 240 \end{bmatrix}$$

Square Matrix
Main Diagonal

A **square matrix** is one that has the same number of columns and rows. The **main diagonal** of a square matrix is the set of entries composed of those entries that lie on the line that passes through the upper left and lower right entries of the matrix. For example, the main diagonal of matrix A used in the example above contains the entries 10 and 40.

The Identity Matrix

The matrix multiplication identity exhibits the same property as the real number multiplicative identity 1, namely, $a \cdot 1 = a = 1 \cdot a$.

**Multiplicative
Identity Matrix**

Multiplicative Identity Matrix

The multiplicative identity matrix is a square matrix whose main diagonal entries are 1s with 0s for all other entries. We will use the notation I_n to denote the multiplicative identity matrix with dimension $n \times n$.

For example, the 3×3 identity matrix, I_3 is

$$\begin{bmatrix} 1 & 0 & 0 \\ 0 & 1 & 0 \\ 0 & 0 & 1 \end{bmatrix}$$

**Row Matrix
Column Matrix**

A **row matrix** is a matrix with only one row and one or more columns. A **column matrix** is a matrix with only one column and one or more rows. These row and column matrices are also called *row vectors* and *column vectors*, respectively.

Inner Products

Inner Product of a Row and Column Matrix

**Inner Product of a
Row and Column
Matrix**

Assume that A is a row matrix with n columns and B is a column matrix with n rows. To perform the inner product of a row and column matrix requires that the number of columns of the row matrix be the same as the number of rows of the column matrix. The inner product is then defined as that number obtained by adding up the following products: the first entry of A times the first entry of B plus the second entry of A times the second entry of B plus the third entry of A times the third entry of B, and so forth.

EXAMPLE SET C

Find the inner product of the row matrix A

$$A = \begin{bmatrix} 1 & 2 & 3 \end{bmatrix}$$

and the column matrix B

$$B = \begin{bmatrix} 4 \\ 5 \\ 6 \end{bmatrix}$$

Solution:

The inner product of A and B is

$$AB = (1 \times 4) + (2 \times 5) + (3 \times 6) = 32$$

Inner Product of Two Matrices

Inner Product of Two Matrices

Assume that A is an $m \times p$ matrix and B is a $p \times n$ matrix. To perform the inner product of A with B, namely AB, requires that the number of columns of the left matrix be equal to the number of rows of the right matrix. The dimensions of the product matrix will be $m \times n$. The inner product, C, is then an $m \times n$ matrix obtained as follows: the first entry of C is the inner product of the first row of A and the first column of B. The entry in the first row, second column of C is the inner product of the first row of A and the second column of B, and so forth for the first row of C. In general, the entry in the ith row, jth column of C is the inner product of the ith row of A and the jth column of B.

Finding the inner product of two matrices is also called matrix multiplication .

While it is difficult to describe the inner product in words and/or symbols, it may ease the burden if we generalize the idea. Think of it this way: the element in the ith row and jth column of the product matrix is obtained by calculating the inner product of the ith row of the left matrix and the jth column of the right matrix. In other words, if A is a 3×5 matrix and B is a 5×4 matrix, then the element of the inner product AB in the second row, third column is the inner product of the second row of A and the third column of B.

The above definition tells us that if A is a 3×4 matrix and B is a 4×2 matrix, we can determine the inner product AB that results in a 3×2 matrix. Notice that we cannot find the inner product BA because the number of columns of B (2) is not equal to the number of rows of A (3). In other words, the inner product of matrices is not commutative. This may be the first time you have come across a multiplication process that is not commutative. Matrices that

Compatible Matrices

have the appropriate match up of columns and rows required to perform an inner product are called **compatible matrices**. Square matrices of the same size are always compatible. Nonsquare matrices with the number of column entries of one equal to the number of row entries of the other are compatible provided they are multiplied in the correct order.

EXAMPLE SET D

Given the matrices A and B described below, find the inner product AB.

$$A = \begin{bmatrix} -2 & 1 & -1 & 3 \\ 5 & -2 & 3 & 4 \\ 1 & -1 & -2 & -5 \end{bmatrix}$$

$$B = \begin{bmatrix} 2 & -1 \\ 3 & -3 \\ 5 & 4 \\ 1 & -2 \end{bmatrix}$$

Solution:
To illustrate the method of finding the inner product, we first show the detail of the inner product of the third row of A and the second column of B to obtain the third row, second column entry of the inner product.

$$[1 \times (-1)] + [(-1) \times (-3)] + [(-2) \times 4] + [(-5) \times (-2)] = 4$$

Finding the inner products of appropriate rows and columns to obtain the corresponding entries of the result yields:

$$AB = \begin{bmatrix} -3 & -11 \\ 23 & 5 \\ -16 & 4 \end{bmatrix}$$

You should practice finding the individual inner products to be sure you understand this result.

EXERCISES

In exercises 1-6, determine the inner product of the given matrices.

1. $\begin{bmatrix} 1 & 5 & 7 \\ -2 & 1 & 3 \\ 2 & 0 & 4 \end{bmatrix} \begin{bmatrix} 1 \\ 2 \\ 3 \end{bmatrix}$

2. $\begin{bmatrix} 1 & 4 & 3 \\ 2 & 6 & -1 \\ -2 & 3 & 4 \end{bmatrix} \begin{bmatrix} -1 \\ -1 \\ 2 \end{bmatrix}$

3. $\begin{bmatrix} 3 & 2 & -2 \\ 1 & 1 & 4 \\ -1 & 2 & 3 \end{bmatrix} \begin{bmatrix} 2 & 3 \\ 1 & -1 \\ -2 & 2 \end{bmatrix}$

4. $\begin{bmatrix} 4 & 1 & -1 \\ 2 & -1 & 2 \\ 3 & 1 & 4 \end{bmatrix} \begin{bmatrix} 2 & 2 \\ 1 & -1 \\ -2 & 1 \end{bmatrix}$

5. $\begin{bmatrix} -3 & 1 & 2 \\ 2 & 3 & -5 \\ 3 & -2 & 1 \end{bmatrix} \begin{bmatrix} 1 & 0 & 0 \\ 0 & 1 & 0 \\ 0 & 0 & 1 \end{bmatrix}$

6. $\begin{bmatrix} 1 & 0 & 0 \\ 0 & 1 & 0 \\ 0 & 0 & 1 \end{bmatrix} \begin{bmatrix} 6 & -1 & -3 \\ 2 & 4 & 7 \\ 11 & -3 & 2 \end{bmatrix}$

In exercises 7 and 8, perform the inner product and use the notion of matrix equality to write a system of equations.

7. $\begin{bmatrix} 3 & 4 & -1 \\ -2 & 5 & 3 \\ 1 & -3 & 6 \end{bmatrix} \begin{bmatrix} x \\ y \\ z \end{bmatrix} = \begin{bmatrix} 1 \\ -2 \\ 5 \end{bmatrix}$

8. $\begin{bmatrix} 1 & -4 & -3 \\ 3 & -6 & 2 \\ -4 & -3 & 2 \end{bmatrix} \begin{bmatrix} x \\ y \\ z \end{bmatrix} = \begin{bmatrix} 11 \\ 7 \\ -8 \end{bmatrix}$

In exercises 9–12, use the given matrices to demonstrate that matrix multiplication is associative.

9. $\begin{bmatrix} 3 & 1 \\ -2 & 0 \end{bmatrix} \begin{bmatrix} -1 & 2 \\ 2 & 4 \end{bmatrix} \begin{bmatrix} 1 & 0 & -1 \\ -1 & 1 & 2 \end{bmatrix}$

10. $\begin{bmatrix} 1 & -1 & 2 \\ 0 & 0 & 4 \\ -2 & 1 & 1 \end{bmatrix} \begin{bmatrix} 2 & -2 \\ 1 & -1 \\ 0 & 1 \end{bmatrix} \begin{bmatrix} 1 & -2 & -3 \\ -2 & 3 & 1 \end{bmatrix}$

11. $\begin{bmatrix} 2 & 0 & -1 & 2 \\ -3 & 1 & 1 & 2 \end{bmatrix} \begin{bmatrix} 1 & -1 & 2 \\ 3 & 0 & 0 \\ 1 & 0 & 6 \\ -3 & -1 & 1 \end{bmatrix} \begin{bmatrix} 0 & -1 \\ -3 & 1 \\ 1 & 2 \end{bmatrix}$

12. $\begin{bmatrix} 2 & -6 & 0 \\ -1 & 4 & 2 \end{bmatrix} \begin{bmatrix} 4 \\ -2 \\ -2 \end{bmatrix} \begin{bmatrix} 0 & -1 & -1 & 8 \end{bmatrix}$

9.5 Solving Systems of Equations Using Matrix Algebra

Introduction

The Inverse of a Matrix

Using the Graphing Calculator to Solve Systems of Linear Equations

Introduction

In this section we look at another method of solving systems of equations. The method involves defining the inverse of a matrix and using it to solve a matrix equation.

The Inverse of a Matrix

Consider the product I_3A, where

$$I_3 = \begin{bmatrix} 1 & 0 & 0 \\ 0 & 1 & 0 \\ 0 & 0 & 1 \end{bmatrix} \quad \text{and} \quad A = \begin{bmatrix} a & b & c \\ d & e & f \\ g & h & i \end{bmatrix}$$

You can verify that $I_3A = A$ and that $AI_3 = A$. In general, wherever the matrices I and A are compatible, $IA = A$ and $AI = A$.

Now let's start by looking again at the system we considered earlier.

$$\begin{cases} -2x + y - z = 1 \\ 5x - 2y + 3z = 3 \\ x - y - 2z = -4 \end{cases}$$

Let A be the coefficient matrix given by

$$A = \begin{bmatrix} -2 & 1 & -1 \\ 5 & -2 & 3 \\ 1 & -1 & -2 \end{bmatrix}$$

and C the constant matrix

$$C = \begin{bmatrix} 1 \\ 3 \\ -4 \end{bmatrix}$$

Matrix of Unknowns

Now let X be the **matrix of unknowns** given by:

$$X = \begin{bmatrix} x \\ y \\ z \end{bmatrix}$$

With the definition of matrix multiplication given in Section 9.4, we can represent the system of equations by the matrix equation:

$$AX = C \tag{9.1}$$

You should use paper and pencil and verify that the elements resulting from the inner product on the left when set equal to the corresponding elements on the right will generate the system of equations. If the concept of A^{-1} (the multiplicative inverse of A) existed, we could now multiply both sides of the matrix equation on the left by A^{-1} (in the same way we have done in solving linear equations like $ax = c$) to obtain:

$$A^{-1}(AX) = A^{-1}C \tag{9.2}$$

If we accept the fact that matrix multiplication is associative and the multiplicative inverse exists, then $A^{-1}A = I_3$, where

$$I_3 = \begin{bmatrix} 1 & 0 & 0 \\ 0 & 1 & 0 \\ 0 & 0 & 1 \end{bmatrix}$$

Then we obtain:

$$(A^{-1}A)X = A^{-1}C$$

$$I_3X = A^{-1}C$$

$$X = A^{-1}C$$

Thus, by multiplying the constant matrix C on the left by the inverse of the coefficient matrix A, (A^{-1}), we obtain the solution set of the system of linear equations. The beauty of this symbolic representation can be appreciated when we realize that the same general matrix equation manipulation is used whether there are three equations in three unknowns or thirty equations in thirty unknowns. Imagine the amount of writing that would be required in representing the solution process for thirty equations in thirty unknowns if we didn't have the matrix symbolism to work with!

One thing that has not been mentioned so far is the fact that approaching the solution to a system of linear equations using the inverse of the coefficient matrix requires that the inverse exists. If it doesn't, this method will not work.

To generalize the procedure, consider the following system of linear equations:

$$\begin{cases} a_{11}x_1 + a_{12}x_2 + \ldots + a_{1n}x_n = c_1 \\ a_{21}x_1 + a_{22}x_2 + \ldots + a_{2n}x_n = c_2 \\ \quad\vdots \qquad \vdots \qquad \ldots \qquad \vdots \quad \ddots \ \vdots \\ a_{m1}x_1 + a_{m2}x_2 + \ldots + a_{mn}x_n = c_m \end{cases}$$

where

$$A = \begin{bmatrix} a_{11} & a_{12} & \ldots & a_{1n} \\ a_{21} & a_{22} & \ldots & a_{2n} \\ \vdots & \vdots & \ddots & \vdots \\ a_{m1} & a_{m2} & \ldots & a_{mn} \end{bmatrix}$$

is the coefficient matrix,

$$C = \begin{bmatrix} c_1 \\ c_2 \\ \vdots \\ c_m \end{bmatrix}$$

is the constant matrix, and

$$X = \begin{bmatrix} x_1 \\ x_2 \\ \vdots \\ x_n \end{bmatrix}$$

is the matrix of the unknowns (or variables).

The system can then be represented by $AX = C$ and if the inverse of A exists, then the solution is represented by $X = A^{-1}C$. We can solve this system by finding the inverse of A using the methods we will learn in the next subsection and then taking the inner product of A^{-1} and the constant matrix C. This same solution can be obtained numerically with great accuracy using the graphing calculator.

The Inverse of a Square Matrix

The Inverse of a Square Matrix

Assume that A is an $n \times n$ matrix. The inverse of A, if it exists, is the $n \times n$ matrix, written A^{-1}, such that the inner product of A and A^{-1}, in either order, is the $n \times n$ identity matrix, I_n. That is,

$$A^{-1}A = AA^{-1} = I_n$$

Singular Matrix

The inverse of a matrix, if it exists, is very useful in this approach to solving systems of linear equations. Only square matrices can have inverses. A square matrix that does not have an inverse is called a **singular matrix**.

One method of obtaining the inverse of a matrix A is to place the appropriate size identity matrix alongside and to the right of the original matrix A to form an augmented matrix. We then use elementary row operations to transform that portion of the augmented matrix representing the entries of the matrix A into the identity matrix. When this is accomplished, the portion of the augmented matrix that originally contained the entries of the identity matrix will now contain the entries of the inverse of A. This amounts to solving n equations in n unknowns n times. The details of why this process works are part of the project at the end of the chapter.

EXAMPLE SET A

Use elementary row operations to find the inverse of the matrix

$$A = \begin{bmatrix} -2 & 1 & -1 \\ 5 & -2 & 3 \\ 1 & -1 & -2 \end{bmatrix}$$

Solution:

First write the 3×3 identity matrix alongside and to the right of A as an augmented matrix.

$$\begin{bmatrix} -2 & 1 & -1 & 1 & 0 & 0 \\ 5 & -2 & 3 & 0 & 1 & 0 \\ 1 & -1 & -2 & 0 & 0 & 1 \end{bmatrix}$$

Now interchange the first and third rows for convenience.

$$\begin{bmatrix} 1 & -1 & -2 & 0 & 0 & 1 \\ 5 & -2 & 3 & 0 & 1 & 0 \\ -2 & 1 & -1 & 1 & 0 & 0 \end{bmatrix}$$

Next replace the second row by the sum of itself and -5 times the first row to get a 0 in the first entry of the second row.

$$\begin{bmatrix} 1 & -1 & -2 & 0 & 0 & 1 \\ 0 & 3 & 13 & 0 & 1 & -5 \\ -2 & 1 & -1 & 1 & 0 & 0 \end{bmatrix}$$

Now replace the third row by the sum of itself and 2 times the first row to get a 0 in the first entry of the third row.

$$\begin{bmatrix} 1 & -1 & -2 & 0 & 0 & 1 \\ 0 & 3 & 13 & 0 & 1 & -5 \\ 0 & -1 & -5 & 1 & 0 & 2 \end{bmatrix}$$

We can next interchange rows two and three for convenience.

$$\begin{bmatrix} 1 & -1 & -2 & 0 & 0 & 1 \\ 0 & -1 & -5 & 1 & 0 & 2 \\ 0 & 3 & 13 & 0 & 1 & -5 \end{bmatrix}$$

Keeping our focus on transforming the first half of the augmented matrix into the 3×3 identity matrix, I_3, we can replace the first row by the sum of itself and -1 times the second row. Also replace the third row by the sum of itself and 3 times the second row. These operations lead to:

$$\begin{bmatrix} 1 & 0 & 3 & -1 & 0 & -1 \\ 0 & -1 & -5 & 1 & 0 & 2 \\ 0 & 0 & -2 & 3 & 1 & 1 \end{bmatrix}$$

Another convenient step would be to multiply the third row by $-1/2$ to obtain:

$$\begin{bmatrix} 1 & 0 & 3 & -1 & 0 & -1 \\ 0 & -1 & -5 & 1 & 0 & 2 \\ 0 & 0 & 1 & -3/2 & -1/2 & -1/2 \end{bmatrix}$$

Now replace the first row by the sum of itself and -3 times the third row. Also replace the second row by the sum of itself and 5 times the third row to obtain:

$$\begin{bmatrix} 1 & 0 & 0 & 7/2 & 3/2 & 1/2 \\ 0 & -1 & 0 & -13/2 & -5/2 & -1/2 \\ 0 & 0 & 1 & -3/2 & -1/2 & -1/2 \end{bmatrix}$$

Finally, multiply the second row by -1 to get:

$$\begin{bmatrix} 1 & 0 & 0 & 7/2 & 3/2 & 1/2 \\ 0 & 1 & 0 & 13/2 & 5/2 & 1/2 \\ 0 & 0 & 1 & 3/2 & 1/2 & 1/2 \end{bmatrix}$$

The entries in the second half of the augmented matrix are the entries of the inverse of A and can be checked by multiplying AA^{-1} to see that the result is the identity matrix I_3. This feat will be a great deal easier after we cover the next section. Perhaps you should wait until then to check this result.

This same method can work on any size square matrix. If the inverse of the given matrix does not exist, we will obtain all 0s in one of the rows of the original matrix as we attempt to transform the original matrix into the identity matrix. Another method for determining whether or not an inverse of a matrix exists is covered in the next section.

Using the Graphing Calculator to Solve Systems of Linear Equations

EXAMPLE SET B

Use the MATRX feature of the TI graphing calculator to solve the system of linear equations given below.

$$\begin{cases} -2x + y - z = 1 \\ 5x - 2y + 3z = 3 \\ x - y - 2z = -4 \end{cases}$$

Solution:

First press the MATRX key on your calculator and then highlight EDIT using the right-arrow key. Press 1 to begin the process of entering the coefficient matrix A. If the matrix does not already indicate that it is a 3×3 matrix, press 3 and ENTER and 3 and ENTER again to accomplish this and also place the cursor at the first entry of the matrix. (Simply press ENTER twice if it does indicate 3×3.)

Note that the entries are expected by row, so type -2, 1, and -1 pressing ENTER after each to enter the first row of the matrix A. Enter the second and third rows in the same way. When you are finished entering the coefficient matrix A, press the MATRX key and highlight EDIT once again. This time press 3 to begin the process of entering the constant matrix C.

Be sure to set C as a 3×1 matrix and then enter the numbers on the right side of the system of equations (1, 3, -4) as before. Return to the computation window when you are finished.

We can always check what is contained in a matrix by pressing the 2nd key and then either 1, 2, or 3 to display [A], [B], or [C], respectively on the TI-81. On the TI-82, press MATRX and then 1, 2, and 3 under NAMES. The TI calcuator differentiates a matrix from a variable by using square brackets. We then press ENTER to display the matrix. Try this now to check to see if you entered matrix A correctly. We can correct any errors using the MATRX key and EDIT along with the arrow keys to go to the incorrect value and type in the correct value. One note for future reference—whenever all of the entries of a matrix are not visible on the screen, you can use the arrow keys to scroll across the matrix.

Now, to complete the solution process, we use the combination of the 2nd key and the X^{-1} key (just to the left of the SIN key) to display $[\mathsf{A}]^{-1}[\mathsf{C}]$ on the screen and press ENTER. The solution is displayed as a 3×1 matrix giving $x = 6, y = 12, z = -1$.

We see how incredibly fast we can find solutions to systems of linear equtions using our graphing calculator. The solution in this case is exact because the inverse of the coefficient matrix A has finite decimal values. The inverse of A can be displayed by displaying $[\mathsf{A}]^{-1}$ on the screen and pressing ENTER. Even when the inverse of the coefficient matrix contains nonending decimal numbers, we can obtain quite accurate numeric results to systems of linear equations.

The graphing calculator can also be used to check whether or not a square matrix has an inverse. Although we will not cover the meaning of the determinant of a matrix here, we will use its functionality to check on the existence of inverses of matrices. You should still have matrix A from the example above stored in your calculator. Press the MATRX key and then press 5 to display det on the screen. Now press the 2nd key and 1 to display [A] next to det. Pressing ENTER will display the determinant of matrix A, which is 2 in this case. Whenever the determinant of a matrix is 0, that indicates that the matrix does not have an inverse. This is a fast way to determine the existence of an inverse. However, the determinant feature is not really necessary if you can recognize when the calculator is telling you that an inverse does not exist. You will obtain an error message when you attempt to get the inverse of a matrix on your calculator when the inverse does not exist.

In addition to solving systems of linear equations, the TI graphing calculator can also help us put a matrix into upper triangular form. This process can be tedious even using the calculator but will generally be less so than by hand and is, more importantly, error-free provided you do not make any mistakes entering values. This process involves the options found under the MATRX menu. Refer to the TI manual if you wish to use these options.

EXERCISES

In exercises 1–4, find the inverse of the matrix if it exists or state that the matrix is singular (that is, a matrix whose inverse does not exist). You should find the inverse of at least one matrix using paper and pencil to ensure that you understand the process covered in the text.

1. $A = \begin{bmatrix} 2 & 1 & 1 \\ 1 & 3 & 2 \\ 3 & -1 & -5 \end{bmatrix}$

2. $A = \begin{bmatrix} 1 & 2 & 3 \\ 4 & 5 & 6 \\ 7 & 8 & 9 \end{bmatrix}$

3. $A = \begin{bmatrix} 1 & 1 & -1 \\ 2 & 1 & 1 \\ 1 & -2 & 3 \end{bmatrix}$

4. $A = \begin{bmatrix} 2 & -3 & 5 \\ -3 & 2 & -4 \\ -1 & -6 & -3 \end{bmatrix}$

In exercises 5–14, solve the given system of equations using the matrix algebra functions on your graphing calculator.

5. $\begin{cases} x + y + z = 2 \\ x + y - 2z = -1 \\ 2x - y - z = 1 \end{cases}$

6. $\begin{cases} x + y - z = 3 \\ x + y + z = 0 \\ x - y - z = -1 \end{cases}$

7. $\begin{cases} -x + 2y + 4z = 0 \\ 3x - y = 5 \\ 2x + 3y - 3z = -1 \end{cases}$

8. $\begin{cases} 3y - z = 4 \\ 2x - y + z = 3 \\ 3x + y - 2z = 5 \end{cases}$

9. $\begin{cases} 12x - 11z = 13 \\ 6x + 6y - 4z = 26 \\ 6x + 2y - 5z = 13 \end{cases}$

10. $\begin{cases} 5x - 3y - z = 16 \\ 2x + y - 3z = 5 \\ 3x - 2y + 2z = 5 \end{cases}$

11.
$$\begin{cases} x + y + z + s = -7 \\ x - y + z - s = -11 \\ 2x - 2y - 3z - 3s = 26 \\ 3x + 2y + z - s = -9 \end{cases}$$

12.
$$\begin{cases} x + 2y - z + s = -5 \\ 2x - 3y + 5z - 7s = -11 \\ 5x - y - 3z - 3s = 6 \\ 3x + 3y + z - s = -9 \end{cases}$$

13.
$$\begin{cases} 2x + y + z + 3s - 4t = -2 \\ x - y + z - s + 3t = 3 \\ 2x - 2y - 3z - 5s + t = 0 \\ 3x + 5y - 2z - s - 2t = -5 \\ x + 3y + 3z - 2s + 3t = 6 \end{cases}$$

14.
$$\begin{cases} x + y - z + 2s - 3t = 1 \\ 4x - 2y + z + 3s + 3t = -3 \\ 2x - y - z - 2s + t = 3 \\ x - 7y + 5z - s - 4t = -1 \\ 2x - 8y + 2z - 4s + t = 4 \end{cases}$$

15. A computer company is working on three projects. The work is being done by three people (Angela, Rob, and Sarah), all of whom are working on all three projects. To complete the first job, 131 hours are required. The second job can be finished in 112 hours, and the third job takes 151 hours. The percent of time each person spends on the jobs is arranged in the accompanying table.

Job	Angela	Rob	Sarah
Job 1	15%	40%	35%
Job 2	58%	20%	15%
Job 3	27%	40%	50%

Find approximately how many total hours each person can work to finish the jobs. (Hint: Use a calculator. Prepare a 3×3 matrix A representing the coefficients of the relevant equation. Make a 3×1 matrix C representing the right side of the equations. Then find $[A]^{-1}[C]$.)

9.6 Systems of Nonlinear Equations

Introduction

Solving a System with at Least One Linear Equation

Solving Systems with No Linear Equations

Introduction

There are many possible systems of nonlinear equations that can be considered with regard to their solutions. We limit our discussion to those in which at least one of the equations is linear plus a few select others.

Solving a System with at Least One Linear Equation

Let us first consider a system in which at least one of the equations is linear. For this purpose, consider this example:

EXAMPLE SET A

Solve the following system of equations:

$$\begin{cases} 3x + y = 2 \\ x^2 + y = 6 \end{cases}$$

Solution:
We use the substitution method and begin by solving for y in the linear equation.

$$y = 2 - 3x$$

Next we substitute this expression for y in the nonlinear equation and solve for x.

$$x^2 + (2 - 3x) = 6$$

$$x^2 - 3x - 4 = 0$$

$$(x - 4)(x + 1) = 0$$

$$x = 4 \quad \text{or} \quad x = -1$$

We now substitute these values of x in the linear equation, for convenience, and solve for the corresponding values of y. This yields

$$3(4) + y = 2$$

$$y = -10$$

and

$$3(-1) + y = 2$$

$$y = 5$$

The solutions to this system of nonlinear equations are $(4, -10)$ and $(-1, 5)$. We can understand the reasonableness of this solution if we recognize that the given equations in the system represent a straight line and a parabola, respectively. The possibilities of the intersection of a line and a parabola are:

(1) no point of intersection, (2) one point of intersection, or (3) two points of intersection. In this case, the line intersects the parabola at two points. Again, we can check to see if the ordered pairs we obtained as solutions do indeed satisfy both equations by substituting them in each of the original equations and noting if they produce true statements. It is always a good idea to check your solutions.

It is generally easier to use the substitution method when solving a nonlinear system of equations in which one of the equations is linear in at least one of the variables.

We should always consider using our graphing calculators whenever it might be appropriate. That is certainly the case in this example since we can easily graph both the line and parabola very quickly and then zoom in on the points of intersection to obtain very accurate numeric answers (see Window 9.2). Again, however, the approach taken in the above solution gives us the analytic connection to the graphical one. It is necessary to get a strong understanding of the theory in order to be able to get the overall picture of how the mathematics works and to give us the basis on which to extend our understanding of solutions to problems that may not quite fit into the patterns we happen to study.

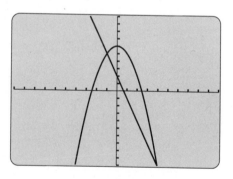

Window 9.2

Solving Systems with No Linear Equations

The approach we use on solving systems of equations in which none of the equations are linear depends on the equations given. We will consider the solution to a select set of systems. This approach can give us the insight needed to solve the same type of problems as well as prepare us to extend these approaches to problems that may be in a similar category yet are sufficiently different from the ones we studied.

EXAMPLE SET B

Solve the following system of equations. Remember, the graphing calculator can be very helpful at the beginning of any problem in which it can be incorporated even if only to give us a visual picture of the solution.

$$\begin{cases} x^2 + y^2 + 2x = 9 \\ x^2 + 4y^2 + 3x = 14 \end{cases}$$

Solution:

We first study the system and use our resource bank to look for possible approaches including those we have studied recently. We might observe that we can eliminate y^2 by multiplying the first equation by -4 and adding it to the second equation. This step eliminates the y-variable since there are no linear terms in y in either equation. We can also solve for y^2 in either equation and substitute the expression in the other equation. Let's try the elimination by addition method. First, multiply the first equation by -4 and rewrite the second equation.

$$-4x^2 - 4y^2 - 8x = -36$$

$$x^2 + 4y^2 + 3x = 14$$

Adding the two equations eliminates y^2 and gives us:

$$- 3x^2 - 5x = -22$$

Putting this quadratic equation into standard form and solving for x yields:

$$3x^2 + 5x - 22 = 0$$

$$(x - 2)(3x + 11) = 0$$

$$x = 2$$

$$x = -11/3$$

Substituting these values into either of the two original equations allows us to solve for the corresponding values of y. Those calculations result in $(2, 1)$, $(2, -1)$, $(-11/3, \sqrt{26}/3)$, and $(-11/3, -\sqrt{26}/3)$ as the solutions to the system of equations.

We take note of a few points here. Had the quadratic equation in x that resulted from eliminating y^2 been nonfactorable, we would have had to reach into our resource bank to recall the quadratic formula or the method of completing the square to solve for x. Also, the option of eliminating x is significantly more difficult than back-substituting values of y in this case. In addition, we might note that the results we obtained seem reasonable if we recognize what each equation in the system represents. We may recall that second-degree equations in two variables usually describe conic sections. In this case, the first equation represents a circle while the second represents an ellipse. We know that a circle and an ellipse can intersect in no points, one point, two points, three points, or four points as in our example. In addition, graphing these equations on our graphing calculator initially would reveal that information right from the start. For instance, Window 9.3 shows the graph of the upper halves of the equations in the previous example. You can immediately see that the graphs intersect. Recall, however, that you need to solve for y in terms of x in order to enter them in the calculator. That task can become difficult, if not impossible, in some cases if we consider higher-degree equations in our system.

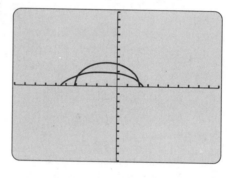

Window 9.3

Let us consider another type of system of nonlinear equations.

EXAMPLE SET C

Solve the following system of equations:

$$\begin{cases} \dfrac{3}{x^2} - \dfrac{2}{y^2} = 2 \\[2mm] \dfrac{1}{x^2} - \dfrac{1}{y^2} = -1 \end{cases}$$

Solution:

We study the system to see what approach we feel might work. Substitution does not look attractive because of the difficulty of solving for one of the variables. Elimination by addition does appear workable since we might observe that we can eliminate either variable with appropriate multiplication factors and then with addition. We might decide to eliminate y by multiplying the second equation by -2 and adding the result to the first equation. This attack leads to:

$$\begin{cases} \dfrac{3}{x^2} - \dfrac{2}{y^2} = 2 \\[2mm] \dfrac{-2}{x^2} + \dfrac{2}{y^2} = 2 \end{cases}$$

and adding gives:

$$\frac{1}{x^2} = 4$$

It is fairly easy to observe that the solutions to this equation are $x = \pm\frac{1}{2}$. Now we can substitute these values in either of the original equations or we can eliminate y. Let's use the elimination by addition process. We can multiply the second equation by -3 and add it to the first equation to obtain:

$$\begin{cases} \dfrac{3}{x^2} - \dfrac{2}{y^2} = 2 \\[2mm] \dfrac{-3}{x^2} + \dfrac{3}{y^2} = 3 \end{cases}$$

and adding yields:

$$\frac{1}{y^2} = 5$$

We can see that the solutions to this equation are $y = \pm\frac{1}{\sqrt{5}}$. The only question that remains is that of how these answers pair up. Since each variable is squared, we can observe that all pairings will work. The solution to the system is, therefore, the ordered pairs $\left(\frac{1}{2}, \frac{1}{\sqrt{5}}\right)$, $\left(-\frac{1}{2}, \frac{1}{\sqrt{5}}\right)$, $\left(\frac{1}{2}, -\frac{1}{\sqrt{5}}\right)$, and $\left(-\frac{1}{2}, -\frac{1}{\sqrt{5}}\right)$.

Studying the approach to solving these examples should give us sufficient resources to attack the same type of problems as well as problems that are similar. We need to study each problem and determine whether the substitution or addition method looks more promising. We should also look to the graphing capabilities of our calculators when the equations in the system can be readily graphed. If none of these approaches work, we may at least have gained enough experience to look to our resource bank for other possible approaches.

EXERCISES

In exercises 1-10, solve the system of equations.

1. $\begin{cases} 2x - y = 1 \\ x^2 + y^2 + 2x + 3y = 7 \end{cases}$

2. $\begin{cases} 3x - y = 1 \\ x^2 - 3x + 2y = -2 \end{cases}$

3. $\begin{cases} -x + 3y = 7 \\ y^2 - 3x - 4y = -1 \end{cases}$

4. $\begin{cases} -2x + y = -5 \\ 2x^2 + 3y^2 - 3x + 5y = 11 \end{cases}$

5. $\begin{cases} x^2 + y^2 + 2x = 9 \\ x^2 + 4y^2 + 3x = 14 \end{cases}$

6. $\begin{cases} x^2 + y^2 + 5x = 5 \\ x^2 + 4y^2 + 14x = 21 \end{cases}$

7. $\begin{cases} x^2 - 4y^2 - 2y = 8 \\ 2x^2 + y^2 + 5y = 0 \end{cases}$

8. $\begin{cases} 3x^2 - 2y^2 - 4y - 27 = 0 \\ 2x^2 - 5y^2 - 3y = 4 \end{cases}$

9. $\begin{cases} \dfrac{2}{x^2} + \dfrac{5}{y^2} = 3 \\ \dfrac{3}{x^2} - \dfrac{2}{y^2} = 1 \end{cases}$

10. $\begin{cases} \dfrac{1}{x^2} - \dfrac{3}{y^2} = 14 \\ \dfrac{2}{x^2} + \dfrac{1}{y^2} = 35 \end{cases}$

11. The equation of the orbit of earth is given by $x^2 + y^2 = 1$ where (x, y) are the coordinates of a point referred to as a set of axes that pass through the sun. A comet is spotted and the equation of its path is calculated to be $y = 0.3x^2 - 4$. Will the comet cross the earth's orbit? Solve analytically first, then check your solution by graphing the two functions together.

12. (Refer to the previous exercise) Another comet's orbit is calculated to be $y^2 = 0.2x + 0.9$. What happens this time? Does it encounter the earth's orbit? If so, what are the x and y coordinates of the points?

9.7 Systems of Inequalities

Introduction

More Complicated Systems of Inequalities

Introduction

The resource to recall here is the fact that an inequality in two variables, when viewed as an equality, will graph as a curve in the xy-coordinate plane. Further, the solution to the inequality will contain all of the points that are on one side of the curve or the other. One last thing to recall is that the curve is excluded from the solution set if it is a strict inequality ($<$ or $>$) but included otherwise.

Exclusion of the curve can be indicated by using a dashed rather than a solid curve. Then, with a system of inequalities, we shade the appropriate regions for each inequality and determine the solution (if one exists) as that shaded region common to all the inequalities. Some examples will illustrate how this method works.

EXAMPLE SET A

Solve the following system of inequalities.

$$\begin{cases} x - y > -3 \\ x + y < 7 \end{cases}$$

Solution:

First we plot the line $x - y = -3$ or $x + 3 = y$. This is a straight line that has a y-intercept of 3 and a slope of 1. The graph looks like this:

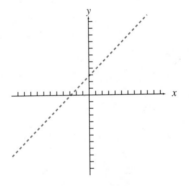

Figure 9.2

We can now use a test point to determine which side of the line to shade. The test point can be any point that does not lie on the curve. We choose $(0, 0)$ and substitute these values into the inequality to get:

$$0 - 0 > -3$$

$$0 > -3$$

Since this produces a true statement, that means the ordered pair $(0, 0)$ is in the solution set. More importantly, however, it tells us that every point on the same side of the curve as $(0, 0)$ is also in the solution set. We therefore shade the region below the line as in the graph on the left of Figure 9.3. Had the test point $(0, 0)$ produced a false statement, we would have shaded the region on the side of the curve opposite to the one containing the test point.

Now plot $x + y = 7$, which is $y = 7 - x$, and shade the appropriate region as shown in the right graph of Figure 9.3.

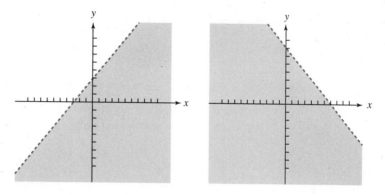

Figure 9.3

We again shade the region below the curve since a test point like $(0, 0)$ causes the inequality to be true as seen below:

$$0 + 0 < 7$$

$$0 < 7$$

All points on the same side of the curve as a test point that is in the solution set are also a part of the solution set. In both inequalities, the curve itself is dashed because the inequalities are strict inequalities and, therefore, are not included in the solution. Putting the two graphs together and noting where the shading is common, we have the following shaded region for the solution set of the given system of inequalities:

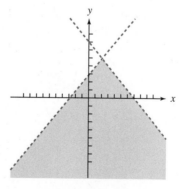

Figure 9.4

Let's look at one more example.

EXAMPLE SET B

Solve the following system of inequalities.

$$\begin{cases} -2x + y \le 5 \\ x^2 - y \le -2 \end{cases}$$

Solution:

Plotting the graph of the equation $-2x + y = 5$ and shading the appropriate side of the curve using the test point $(0,0)$ results in:

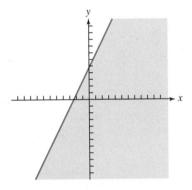

Figure 9.5

In the same way, we plot $x^2 - y = -2$ and shade the appropriate side of the curve using the test point $(0,0)$, which yields the graph on the left of Figure 9.6.

The solution is the overlapping shaded region of both these curves as in the graph on the right of Figure 9.6.

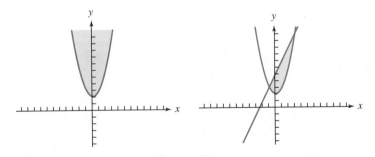

Figure 9.6

More Complicated Systems of Inequalities

These two examples are essentially all that are needed to pave the way to solving systems of inequalities. An extension of what we learned here will usually suffice in attacking other systems of inequalities. Admittedly, things can get a good deal more complicated when the number of inequalities and unknowns increases above two. However, methods similar to those given above can give us insight into successful approaches. For instance, a system of three linear inequalities in three unknowns would require that we understand that a linear equation in three variables represents a plane in three-space. The same equation made into an inequality would represent the set of points in space (ordered triples) that lie on one side of the plane or the other. We can see that just a small amount of understanding allows us to extend the ideas we learned in two-space to this case. We would complete the problem by finding the common space (if any) that results from the three linear inequalities.

Although you may not yet have studied surfaces in space, it is a good bet that you could understand the solution of a system of three inequalities in three unknowns when one or more are nonlinear. Whatever the surface representation might be, the inequality would involve the space on one side or the other of that surface. We would require that the surface be a function that is a reasonable extension of the functions of one variable you have already studied. This is the kind of extension of previously learned methods you should be trying to attain. Such an extension again requires a reasonable understanding of the theory involved along with any numeric or graphical approaches that are appropriate and helpful.

EXERCISES

In exercises 1–10, sketch the solution set for the given system of equations.

1. $\begin{cases} 3x + 2y \le 5 \\ 2x - y \ge -2 \end{cases}$

2. $\begin{cases} 5x - 4y < 5 \\ x + 2y \ge 3 \end{cases}$

3. $\begin{cases} 12x + 5y \ge -10 \\ 5x - 16y > 7 \end{cases}$

4. $\begin{cases} x + y < 1 \\ x - 2y > -2 \\ 2x + y < 2 \end{cases}$

5. $\begin{cases} x + y \ge 3 \\ x^2 + y^2 \le 16 \end{cases}$

6. $\begin{cases} x + y \ge 4 \\ x^2 + y^2 \le 25 \end{cases}$

7. $\begin{cases} y \le 2x + 3 \\ y \ge 4x^2 \end{cases}$

8. $\begin{cases} y \ge 2x \\ y \le -x^2 + 3 \end{cases}$

9. $\begin{cases} x - 2y > 3 \\ x \le -2y^2 + 4 \end{cases}$

10. $\begin{cases} 2x - y < 3 \\ \quad\ x \le 3y^2 - 2 \end{cases}$

11. Acme Tool Company has a job that is to be completed in less than six hours and should cost the client no more than $140. Two workers (Kevin and Krista) are sent. If Kevin works alone, he can complete the job in 8 hours, and if Krista works alone she can finish the job in 10 hours. Kevin is paid $20/hour and Krista is paid $12 per hour. How many hours should each worker work to get all of the $140 dollars? (Hint: Kevin does 1/8th the job in 1 hour. Assume x = number of hours worked. The two together complete the entire job. This will give you an equation. Graph it. Kevin and Krista worked less than six hours. Set the Range accordingly. Set up another equation to give the labor cost. Graph that equation. Find the point of intersection.)

Exercises 12–14 refer to the situation in exercise 11. The Acme Tool Company just received another job. The information on the three workers who could work on this job is given in the accompanying table. The time refers to how

Price Table		
Worker	**Time**	**Pay**
Abe	12 hours	$20/hour
Nancy	18 hours	$16/hour
Cy	20 hours	$12/hour

long each worker would need to work alone to complete the job (which should be completed within 10 hours and should not total more than $260 in labor). Only two people are allowed on the job.

12. Acme wants to earn the most possible amount of money on the job.

(a) If Abe and Nancy are sent, how long would each work? (Graph the two equations and find the point of intersection.)

(b) If Nancy and Cy are sent, determine how long it would take each of them.

(c) Determine how long it would take each person if Cy and Abe are sent.

13. Using the information from your calculations in exercise 12, decide which pair of workers will best meet the requirements.

14. Draw the graphs for the solutions in (c) for exercise 12. What do you conclude from them? Also, draw the lines $y = 10$ and $x = 10$. If Abe and Cy could have done the job without regard to cost, give one possible combination of their hours. (Use the TRACE feature to find the coordinates of any point on the bounding line.)

Exercises 15–18 refer to the same situation.

15. Carl is using some synthetic wood to make tables and chairs. He has 300 units of material in stock. 17 units are needed to construct a table and 5 units are required to make a chair. What are the possible combinations of the number of chairs and tables he can make from the available material? (Write down the inequality. Graph the table variable on an x-axis. Then read off the y values for the integral x values using TRACE.)

16. Suppose that Carl's profit from selling a chair is $25 and from selling a table is $250. He wants to earn a profit of at least $3,000. Draw the graph of a suitable inequality that describes the situation.

17. Combine the two graphs on the same coordinate system. What can you tell about the area that is shaded in both diagrams?

Read the graph to find the most profitable pair of combinations of chairs and tables.

18. Suppose that Carl has 210 hours available to make tables and chairs, and that he needs 15 hours to construct a table and 3 hours to make a chair. Graph the inequality for this situation. Next graph all three inequalities together. The area that is shaded in all three inequalities gives the optimum combination of tables and chairs: Carl makes the amount of money he wants with the material at hand in the time that was allotted. Read off one such combination pair.

Summary and Projects

Summary

Solving Systems of Linear Equations We can quickly and accurately solve up to six linear equations in six unknowns (or more, depending on what model of the TI graphing calculator is being used). The graphing calculator is an extremely useful tool in solving such problems, but depth of understanding comes from understanding the theory involved. Consider the following system of linear equations:

$$\begin{cases} a_{11}x_1 + a_{12}x_2 + \ldots + a_{1n}x_n = c_1 \\ a_{21}x_1 + a_{22}x_2 + \ldots + a_{2n}x_n = c_2 \\ \quad\vdots \qquad\quad \vdots \quad\ \ldots \quad\ \ \vdots\ \ddots\ \vdots \\ a_{m1}x_1 + a_{m2}x_2 + \ldots + a_{mn}x_n = c_m \end{cases} \qquad (9.1)$$

A solution to this system is a set of replacement numbers for the variables $x_1, x_2, \ldots, x_n$ that makes each and every equation true.

The Elimination by Addition Method We can solve a system of n linear equations in n unknowns by first choosing a variable to eliminate. We eliminate the chosen variable from one pair of equations by multiplying both sides of each equation by numbers that cause the coefficients of this variable to be numerically the same but opposite in sign. Then we add the two equations to eliminate the variable. By eliminating the same variable in $n - 1$ different pairs of equations, we reduce the number of equations by one. We continue in this manner until we obtain one linear equation in one unknown that we know how to solve. Then we work our way back by substitution.

The Elimination by Substitution Method We can also solve the system of equations by solving for one of the variables in one of the equations. We then substitute the expression for the variable we solve for into each of the remaining equations. This process also reduces the number of equations by one each time we apply it. We complete the rest of this approach using the same technique as in the previous method.

Many Solutions or No Solution   We need to be able to recognize those cases in which there is not a unique solution (a consistent system). We have a dependent system or an inconsistent system whenever we arrive at one of the following equations in our attempt to solve the system: $0 = 0$, $k = k$, $0 = k$, or $k = p$, where k and p are distinct integers.

Solving Systems of Linear Equations Using Matrices   We can represent a system of equations like 9.1 above using matrices. If we let A be the coefficient matrix:

$$A = \begin{bmatrix} a_{11} & a_{12} & \cdots & a_{1n} \\ a_{21} & a_{22} & \cdots & a_{2n} \\ \vdots & \vdots & \ddots & \vdots \\ a_{m1} & a_{m2} & \cdots & a_{mn} \end{bmatrix}$$

C be the constant matrix:

$$C = \begin{bmatrix} c_1 \\ c_2 \\ \vdots \\ c_m \end{bmatrix}$$

and X the matrix of unknowns:

$$X = \begin{bmatrix} x_1 \\ x_2 \\ \vdots \\ x_n \end{bmatrix}$$

then the system 9.1 above could be written:

$$AX = C$$

The solution would be obtained by multiplying both sides on the left by A^{-1} to get the solution:

$$X = A^{-1}C$$

We can accomplish this using our graphing calculators. If we were to solve the system by hand, we would need to be able to find the inverse of a matrix and then perform the inner product $A^{-1}C$.

The Inverse of a Square Matrix   Given a square matrix A, its inverse is the square matrix A^{-1}, which when multiplied by A gives the identity matrix.

That is, $AA^{-1} = A^{-1}A = I$. One method of finding the inverse of a square matrix, if it exists, is to write down the augmented matrix composed of the entries of A on the left and the entries of I on the right. We then proceed to use elementary row operations to transform A into the identity matrix I. Then the entries that were in the position of the identity matrix will be the entries of the inverse of A.

Solving Systems of Nonlinear Equations The variety of possible systems of nonlinear equations makes it difficult to have one or two approaches to their solutions. There are a couple of guidelines, however, that are useful in this regard. We will restrict the nonlinear systems to two equations in two unknowns. If one equation has a variable that has a largest exponent of 1, then solving for that variable and substituting the resulting expression for that variable in the other equation works quite well. If this is not the case, then we should look to the possibility of eliminating one of the variables using appropriate multipliers.

Projects

1. Solve the following three systems of equations using the upper triangular matrix method. Notice that the constant matrices are the columns of the 3×3 identity matrix. Use these three-column matrix solutions (in order) to create a 3×3 matrix. Show that this matrix is the inverse of the coefficient matrix (which is the same for each one of the systems of equations). Discuss the method of determining the inverse of a matrix presented in this section in light of these computations.

$$\begin{cases} -1x + 7y + 3z = 1 \\ -2x + 9y + 4z = 0 \\ -1x + 4y + 2z = 0 \end{cases}$$

$$\begin{cases} -1x + 7y + 3z = 0 \\ -2x + 9y + 4z = 1 \\ -1x + 4y + 2z = 0 \end{cases}$$

$$\begin{cases} -1x + 7y + 3z = 0 \\ -2x + 9y + 4z = 0 \\ -1x + 4y + 2z = 1 \end{cases}$$

2. Matrix algebra is the basis for the computerized animation of graphics in television and movies. Matrices are used both to store the objects that move and as the mechanism to move them on the screen. The matrix

$$A = \begin{bmatrix} 0 & 4 \\ -2 & -3 \\ 3 & 1 \\ -3 & 1 \\ 2 & -3 \\ 0 & 4 \end{bmatrix}$$

gives the points of a five-pointed star when connected. The following program connects these pairs of points with lines.

```
:PROGRAM:MATPLOT
:ClrDraw
:1→K
:Lbl 1
:Line([A](K,1),[
A](K,2),[A](K+1,
1),[A](K+1,2))
:IS>(K,5)
:Goto 1
```

You can use any 2-by-2 matrix, B, to manipulate the 6-by-2 star object. You do this by multiplying the star matrix on the right by B. Use this program to discuss the effect of the following 2-by-2 matrices on the star object.

a. $B = \begin{bmatrix} 2 & 0 \\ 0 & 2 \end{bmatrix}$

b. $B = \begin{bmatrix} 0 & 2 \\ 2 & 0 \end{bmatrix}$

c. $B = \begin{bmatrix} 1 & 0 \\ 0 & -1 \end{bmatrix}$

d. $B = \begin{bmatrix} \cos(1) & \sin(1) \\ -\sin(1) & \cos(1) \end{bmatrix}$

You should store the star object data as the matrix C. The matrix A that the MATPLOT will draw should be the product CB. You should store CB to A for each product using the statement

$$[C][B] \to [A]$$

Determine matrices B that will scale the object by a factor of 3, flip the object about the y-axis, and rotate the object $120°$. Discuss the general principles you are using in creating these matrices.

CHAPTER 10

Conics

10.1 Circles and Ellipses

Introduction

Circles

The General Representation of a Circle

Ellipses

The General Representation of an Ellipse

Introduction

Conics are geometric objects that frequently occur in nature. These objects are the circles and parabolas we have studied in this book along with ellipses and hyperbolas. Planets, moons of planets, and comets move in planar paths that are conics. Satellites are placed in elliptical orbits around the earth by booster rockets. Conics are also used in designing and building such products as radar antennas, automobile headlamps, and telescopes. They play an important role in the radar tracking of planes, ships, and satellites.

Circles

A circle is a set of points in a plane equidistant from a fixed point (also in the plane). We can graph circles using the parametric mode with the following expressions for $x(t)$ and $y(t)$.

$$x(t) = \cos(t) \quad \text{and} \quad y(t) = \sin(t) \tag{10.1}$$

The graph of this circle looks something like Window 10.1 using Zoom-Trig followed by Zoom-Square.

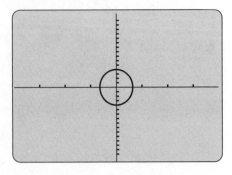

Window 10.1

This circle is centered at the origin and has a radius of 1 unit.

Notice that the circle is symmetric with respect to the origin and both the x- and y-axes. For example, the point $(\frac{1}{2}, \frac{\sqrt{3}}{2})$ has three symmetric points associated with it:

$$\left(-\frac{1}{2}, -\frac{\sqrt{3}}{2}\right) \qquad \text{symmetry about the origin}$$

$$\left(-\frac{1}{2}, \frac{\sqrt{3}}{2}\right) \qquad \text{symmetry about the } y\text{-axis, and}$$

$$\left(\frac{1}{2}, -\frac{\sqrt{3}}{2}\right) \qquad \text{symmetry about the } x\text{-axis}$$

Every non-axis point on a circle centered at the origin has three symmetric points associated with it.

We can translate the circle so that its center is at (2,3). We can accomplish this task by simply adding 2 and 3 to $x(t)$ and $y(t)$, respectively as follows:

$$x(t) = \cos(t) + 2$$

$$y(t) = \sin(t) + 3$$

The graph of the translated circle is shown in Window 10.2 using the Trig and Square options of the ZOOM menu.

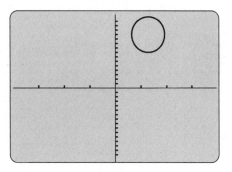

Window 10.2

We can also change the radius of the original circle in 10.1 to 4 by multiplying each of its coordinates by 4 as follows:

$$x(t) = 4\cos(t)$$

$$y(t) = 4\sin(t)$$

The resulting circle is displayed in Window 10.3.

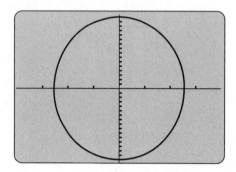

Window 10.3

We can quickly graph a circle of radius 1, centered at the origin using our graphing calculators. We can also graph a circle of any fixed radius and center by starting with the parametric representation of this unit circle and making the necessary adjustments.

EXAMPLE SET A

Sketch a graph of the circle with center at the point $(1, -2)$ and radius 5.

Solution:

The parametric representation of this circle is given by:

$$x(t) = 5\cos(t) + 1$$

$$y(t) = 5\sin(t) - 2$$

The graph of this circle is given in Window 10.4 with $-12 \le x \le 12$ and $-8 \le y \le 8$.

To display the circle so that it looks like a circle, you should use the ZOOM-Square feature if you have not already chosen the RANGE or WINDOW values in the ratio 3 to 2.

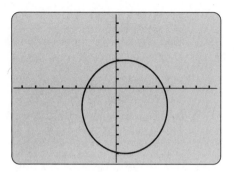

Window 10.4

The General Representation of a Circle

The general parametric representation of a circle with center at (h, k) and radius r is given by:

$$x(t) = r\cos(t) + h$$

$$y(t) = r\sin(t) + k \qquad (10.2)$$

We can connect this parametric representation of the circle to the general rectangular coordinate representation of a circle. From 10.2, we can isolate $\cos(t)$ and $\sin(t)$ as follows:

$$\frac{x - h}{r} = \cos(t)$$

$$\frac{y - k}{r} = \sin(t) \qquad (10.3)$$

Now, using the fact that $\cos^2(t) + \sin^2(t) = 1$, we can square both sides of 10.3 and add to obtain

$$\frac{(x - h)^2}{r^2} + \frac{(y - k)^2}{r^2} = 1$$

or

$$(x - h)^2 + (y - k)^2 = r^2$$

EXAMPLE SET B

Find a parametric representation and a rectangular representation of the circle with center at the point $(-2,4)$ and radius 3.

Solution:

The parametric representation will involve $\cos(t)$ and $\sin(t)$. The center will be offset from the standard origin-centered circle by the values -2 and 4. The parametric representation is:

$$x(t) = 3\cos(t) - 2$$

$$y(t) = 3\sin(t) + 4$$

The rectangular representation can be developed from 10.2 by isolating $\cos(t)$ and $\sin(t)$ and using the appropriate substitutions for h, k, and r. The rectangular form of the circle is found as follows:

$$\cos^2(t) = \left(\frac{x - (-2)}{3}\right)^2$$

$$\sin^2(t) = \left(\frac{y - 4}{3}\right)^2$$

And since

$$\cos^2(t) + \sin^2(t) = 1$$

then

$$\left(\frac{x - (-2)}{3}\right)^2 + \left(\frac{y - 4}{3}\right)^2 = 1$$

Simplifying yields

$$(x + 2)^2 + (y - 4)^2 = 9$$

Ellipses

You may know that satellites launched by NASA about the earth are placed in elliptical orbits. An ellipse has an oval shape. Use the following expressions for $x(t)$ and $y(t)$ to see the graph of an ellipse.

$$x(t) = 3\cos(t)$$

$$y(t) = 2\sin(t)$$

The graph is displayed in Window 10.5 using ZOOM-Standard, ZOOM-In with x factor and y factor set to 4 at the origin, and ZOOM-Square.

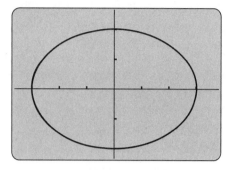

Window 10.5

Notice that the coefficients of $\cos(t)$ and $\sin(t)$ are not the same in the parametric representation of the ellipse but were equal for circles. We can see from the graph that this ellipse is symmetric with respect to the x-axis, the y-axis, and the origin. Look at your graph and determine where the ellipse crosses the x-axis and where it crosses the y-axis. Notice that these points are $(-3, 0)$ and $(3, 0)$ for the x-intercepts and $(0, -2)$ and $(0, 2)$ for the y-intercepts. The line segments joining the x-intercepts and the y-intercepts are called the axes of the ellipse. The longer of the two is called the *major* axis and the shorter is called the *minor* axis. You should interchange 3 and 2 in the parametric representation of this ellipse to see the effect this has on the graph.

As with circles centered at the origin, ellipses centered at the origin are symmetric to the origin and to both the x- and y-axes.

We can obtain the rectangular form of this ellipse using the same concept that we used for the circle. We begin by isolating $\cos(t)$ and $\sin(t)$ and using $\cos^2(t) + \sin^2(t) = 1$. The following equations show this process.

$$\frac{x}{3} = \cos(t)$$

$$\frac{y}{2} = \sin(t)$$

Since

$$\cos^2(t) + \sin^2(t) = 1$$

we obtain

$$\left(\frac{x}{3}\right)^2 + \left(\frac{y}{2}\right)^2 = 1$$

Finally,

$$\frac{x^2}{9} + \frac{y^2}{4} = 1$$

This is the standard rectangular form of an ellipse whose center is at the origin.

We can translate this ellipse in parametric form in the same way that we translated circles. If we wish to place the center of the ellipse at $(2, -1)$, we use the following parametric expressions.

$$x(t) = 3\cos(t) + 2$$

$$y(t) = 2\sin(t) - 1$$

You should graph this on your graphing calculator to see the effect of this translation.

The General Representation of an Ellipse

The standard parametric representation of an ellipse centered at the point (h, k) is given by:

$$x(t) = a\cos(t) + h$$

$$y(t) = b\sin(t) + k$$

The larger of the real numbers a and b is the semi-major axis. Semi-major means it is half the length of the major axis. The value of b is the semi-minor axis. If the larger value is associated with $\cos(t)$, the major axis is parallel to the x-axis. If the larger value is associated with $\sin(t)$, the major axis is parallel to the y-axis. We can obtain the general rectangular form of this translated ellipse. The following equations show this process.

$$\frac{x - h}{a} = \cos(t)$$

$$\frac{y - k}{b} = \sin(t)$$

Since

$$\cos^2(t) + \sin^2(t) = 1$$

we obtain

$$\left(\frac{x-h}{a}\right)^2 + \left(\frac{y-k}{b}\right)^2 = 1$$

Finally,

$$\frac{(x-h)^2}{a^2} + \frac{(y-k)^2}{b^2} = 1$$

Think about the connections of this rectangular form of the ellipse with center at (h,k), major vertices at $(\pm a, 0)$, and minor vertices at $(0, \pm b)$.

EXAMPLE SET C

Find a parametric representation and a rectangular representation of the ellipse with center at the point $(3,-2)$, semi-major axis parallel to the x-axis of length 4 and semi-minor axis of length 2.

Solution:

We recall that the standard parametric form of an ellipse centered at the point (h,k) is given by:

$$x(t) = a\cos(t) + h$$

$$y(t) = b\sin(t) + k$$

Thus, we can write the parametric form of this particular ellipse as follows:

$$x(t) = 4\cos(t) + 3$$

$$y(t) = 2\sin(t) + (-2)$$

The general equation of an ellipse with center at (h,k) and major axis parallel to the x-axis in rectangular form is

$$\frac{(x-h)^2}{a^2} + \frac{(y-k)^2}{b^2} = 1$$

Therefore, we can write the rectangular form of this particular ellipse as follows:

$$\frac{(x-3)^2}{4^2} + \frac{(y-(-2))^2}{2^2} = 1$$

or

$$\frac{(x-3)^2}{16} + \frac{(y+2)^2}{4} = 1$$

We will study the ellipse from another perspective in a later section of this chapter.

EXERCISES

For exercises 1–12, sketch the graph of the circle or ellipse and indicate the center.

1. $\dfrac{(x-2)^2}{4} + \dfrac{(y-3)^2}{2} = 1$

2. $\dfrac{(x+2)^2}{9} + \dfrac{(y-4)^2}{4} = 1$

3. $\dfrac{(x+3)^2}{16} + \dfrac{(y-2)^2}{25} = 1$

4. $\dfrac{(x-3)^2}{25} + \dfrac{(y+3)^2}{4} = 1$

5. $\dfrac{(x-3)^2}{4} + \dfrac{(y+3)^2}{4} = 1$

6. $\dfrac{(x+2)^2}{9} + \dfrac{(y-3)^2}{9} = 1$

7. $\dfrac{(x+4)^2}{5} + \dfrac{(y+7)^2}{5} = 4$

8. $\dfrac{(x-6)^2}{4} + \dfrac{(y-4)^2}{4} = 7$

9. $\dfrac{(x-2)^2}{2} + \dfrac{(y-3)^2}{18} = 2$

10. $\dfrac{(x+5)^2}{12} + \dfrac{(y+4)^2}{3} = 3$

11. $\dfrac{(x+4)^2}{2} + \dfrac{(y-3)^2}{5} = 1$

12. $\dfrac{(x+3)^2}{7} + \dfrac{(y+2)^2}{4} = 1$

For exercises 13–18, sketch the graph of the ellipse or circle and indicate the center.

13. $x(t) = 3\cos(t)$
    $y(t) = 3\sin(t)$

14. $x(t) = 4\cos(t)$
    $y(t) = 9\sin(t)$

15. $x(t) = 4\cos(t) + 2$
    $y(t) = 3\sin(t) - 1$

16. $x(t) = 5\cos(t) - 4$
    $y(t) = 2\sin(t) - 3$

17. $x(t) = 2\cos(t) + 1$
    $y(t) = 2\sin(t) - 5$

18. $x(t) = 5\cos(t) + 3$
    $y(t) = 5\sin(t) + 2$

For exercises 19–26, locate the center and radius of the circle by placing the equation in standard rectangular form using the completing the square method.

19. $x^2 + 6x + y^2 + 8y = 3$

20. $x^2 + 4x + y^2 - 6y = 5$

21. $x^2 + y^2 + 6x - 6y = 7$

22. $x^2 + y^2 - 8x + 4y = 36$

23. $4x^2 + 4y^2 - 24x + 36y = 27$

24. $2x^2 + 2y^2 - 12x + 8y = 5$

25. $x^2 + y^2 - 3x + 5y = \dfrac{15}{4}$

26. $x^2 + y^2 - 6x - 3y = -\dfrac{5}{4}$

For exercises 27–34, locate the center and endpoints of both the major and minor axes of the ellipses represented by the following equations.

27. $\dfrac{(x-2)^2}{4} + \dfrac{(y-3)^2}{16} = 1$

28. $\dfrac{(x+3)^2}{25} + \dfrac{(y-4)^2}{9} = 1$

29. $\dfrac{(x+5)^2}{16} + \dfrac{(y+2)^2}{9} = 1$

30. $\dfrac{(x-6)^2}{3} + \dfrac{(y+2)^2}{2} = 1$

31. $\dfrac{(y-2)^2}{4} + \dfrac{(x-3)^2}{16} = 1$

32. $\dfrac{(y-5)^2}{36} + \dfrac{(x+4)^2}{49} = 1$

33. $\dfrac{(y+4)^2}{12} + \dfrac{(x+4)^2}{24} = 1$

34. $\dfrac{y^2}{10} + \dfrac{(x-7)^2}{8} = 1$

For exercises 35–42, write, in paragraph form, a process to locate the center and endpoints of both the major and minor axes of the ellipses represented by the following equations. Then write the center and endpoints of both the major and minor axes.

35. $4x^2 + 9y^2 = 36$

36. $9x^2 + 12y^2 = 108$

37. $4x^2 - 8x + 9y^2 + 18y = 23$

38. $3x^2 + 12x + 6y^2 + 36y = -18$

39. $9x^2 + 4y^2 + 18x - 16y = 12$

40. $8x^2 + 3y^2 - 32x - 6y = 36$

41. $3x^2 + 4y^2 - 9x + 20y = \dfrac{-79}{4}$

42. $6x^2 + 4y^2 + 24x + 28y = -24$

43.

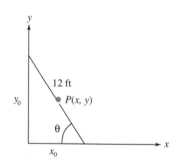

A 12-foot ladder is standing against a vertical wall, making an angle θ as shown. The variables x_0 and y_0 are (as indicated in the diagram) the horizontal distance of the foot of the ladder from the wall and the vertical height of the top of the ladder. Express x_0 and y_0 as functions of the parameter θ.

Let $P(x, y)$ be the midpoint of the ladder. Write down the functions $x = x(\theta)$ and $y = y(\theta)$. Graph this equation. What values do you set for Tmin, Tmax, etc.? What kind of curve does the equation give?

44. (Refer to the previous exercise) If instead of P you considered the point Q two feet from the base of the ladder, what would be the parametric expressions for the coordinates of Q? What graph could you get by eliminating θ? Which portion of this curve is described by the point Q as the ladder moves to different positions?

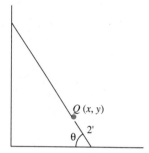

45. The parametric equation of a curve is given by

$$x(t) = 2\sin(t)$$

$$y(t) = \cos^2(t)$$

Graph the equation. What kind of a curve is it? Now, eliminate t to get the equation of the curve. Which part of the curve do you get (i.e., is it the part in the first quadrant)?

46. The parametric equation of another curve is given by

$$x(t) = 1 + \sec(t)$$

$$y(t) = 2\tan(t) \quad 0 < t < \dfrac{\pi}{2}$$

Eliminate t to get the equation of the curve. Name the curve.

47. Now that you have solved the previous two exercises, can you find a parametric representation of the ellipse

$$\dfrac{x^2}{4} + \dfrac{y^2}{9} = 1$$

48. The wooded area near Jon's home is being made into a park. Jon is planning to build a trail there, to include three especially beautiful vista points. The second of these spots is 50 yards to the east of the first one, and the third one is 50 yards east and 80 yards south of the second point. If the path is to be circular, what should be the radius of the circle? Where is the center relative to the first point? (Hint: Choose the north-south line through the first point as the y-axis and the east-west line as the x-axis.

49. As part of a landscaping design, Jan has to draw three circles (see figure). The diameter of the smallest circle is to be the same as the radius of the next-largest circle. Write the equations of the three circles after choosing appropriate axes, if the largest circle has a radius of 24 meters.

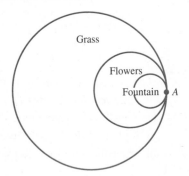

50. (Continue with the scenario in the previous exercise) In order to lay out the circles in his plan, Jan has to first determine some of the points that are going to be on the outer circle. He chooses the point A as the origin, and the line extending northward from it (shown in the figure) as the positive x-axis. He also

believes that polar coordinates will be more useful than rectangular coordinates in locating the points. Find the polar equation of the larger circle, and determine the coordinates of some of the points.

51. An artist is drawing the top half of an ellipse for a mural on the side of a building. This half-ellipse must fit inside a rectangle 18' wide by 6' high. The artist puts two nails on the base of the rectangle, 2 feet in from the sides. He attaches the two ends of a string to the two nails, puts a piece of chalk on the string and moves it around, keeping both sections of the string taut, and draws part of an ellipse. He makes sure that the top of the ellipse reaches the center point of the top of the rectangle. Will this method give him what he wants, that is, the upper half of an ellipse inside the rectangle? Deduce the equation of the ellipse and graph it together with the rectangle. (Hint: $2a$ = length of the string).

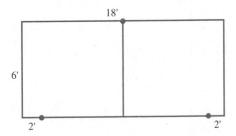

52. Modify the equation in the previous exercise so that the ends of the major axis coincide with the two bottom corners of the rectangle. How should the artist change the position of the nails and the length of the string?

10.2 Hyperbolas

Introduction

The Equation of a Hyperbola

Changing the Shape of Hyperbolas

Translating Hyperbolas

The Asymptotic Behavior of a Hyperbola

The Central Rectangle of a Hyperbola

Introduction

Hyperbolas are also in the family of conics. While there are many similarities between hyperbolas and ellipses, there is an interesting difference. Hyperbolas have asymptotes associated with them while no other conics do. In this section, we study the characteristics and behavior of hyperbolas.

The Equation of a Hyperbola

Hyperbolas can be developed from the equation

$$x^2 - y^2 = 1 \tag{10.4}$$

The functions $\cosh(x)$ and $\sinh(x)$ are called the hyperbolic cosine and hyperbolic sine functions, respectively. They satisfy 10.4 in the same way that the circular functions $\cos(x)$ and $\sin(x)$ satisfy the unit circle equation $x^2 + y^2 = 1$. For example,

$$\cosh^2(3) - \sinh^2(3) = 1$$

Although we will not develop these hyperbolic functions here, we will give their definition for your information.

The Hyperbolic Sine and Cosine Functions

The Hyperbolic Sine and Cosine Functions

The hyperbolic sine and cosine functions are written and defined as follows:

$$\sinh(x) = \frac{e^x - e^{-x}}{2} \quad \text{and} \quad \cosh(x) = \frac{e^x + e^{-x}}{2}$$

By recalling the behavior of the exponential function, we could determine the behavior of the hyperbolic sine and cosine functions. In fact, you can prove the useful identity

$$\cosh^2(x) - \sinh^2(x) = 1$$

by using the definitions. You would need to carry out the algebra in

$$\left(\frac{e^x + e^{-x}}{2}\right)^2 - \left(\frac{e^x - e^{-x}}{2}\right)^2$$

to do so. These hyperbolic functions are available on your graphing calculator in the Math menu under HYP and we will use them to graph hyperbolas. You can get a quick look at one branch of a hyperbola using your graphing calculator. First place your calculator in parametric mode. Then access the Y= menu and paste cosh in X1T and then type T. Paste sinh in Y1T and then type T. The parametric representation you have entered in your calculator is:

$$x(t) = \cosh(t)$$

$$y(t) = \sinh(t)$$

To get a representative graph of a hyperbola we might use the following RANGE or WINDOW settings: $-3 \leq T \leq 3$, $-6 \leq X \leq 6$, $-4 \leq Y \leq 4$, with a Tstep of 0.1. The graph should look something like Window 10.6.

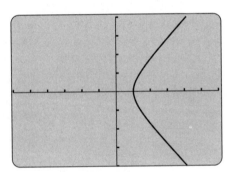

Window 10.6

This is only one branch of the hyperbola because the cosh function is always positive. We can generate the other branch by placing −X1T and Y1T in X2T and Y2T, respectively. The complete graph of the hyperbola should look something like Window 10.7.

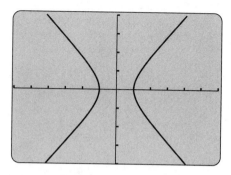

Window 10.7

We can see from the graph that this hyperbola is symmetric with respect to the x-axis, the y-axis, and the origin. The points of intersection on the x-axis are called the vertices of the hyperbola. The center of the hyperbola is the midpoint of the line segment joining the vertices (the origin in this case). The line segment joining the vertices is called the transverse axis, and the conjugate axis lies along the perpendicular bisector of the transverse axis.

Changing the Shape of Hyperbolas

We can change the shape of a hyperbola by multiplying one or the other or both of the hyperbolic functions in the parametric representation by a constant. For example, if we change Y1T to $0.5 \sinh(T)$, the graph should look something like Window 10.8.

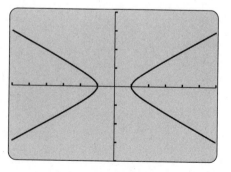

Window 10.8

Notice that the hyperbola is narrower than the previous one. Also, the vertices are at the same locations. If we now multiply the $\cosh(T)$ by 3, we will change the position of the vertices as well as the shape as shown in Window 10.9.

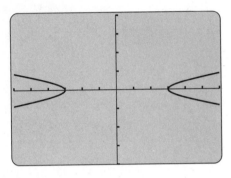

Window 10.9

Notice the new positions of the vertices. They both are now at a distance of 3 from the origin.

Translating Hyperbolas

Translating hyperbolas is similar to the translations we performed on circles and ellipses. To place a hyperbola with center at $(-2, 1)$, we change the parametric representation as follows:

$$x(t) = \cosh(t) - 2$$

$$y(t) = \sinh(t) + 1$$

To get both branches of the hyperbola this time, we need to enter $-$X1t $- 4$ for X2T since the negative sign will change the signs of both terms. Alternately, we can enter $-\cosh$ T $- 2$ for X2T. The complete graph of this hyperbola is shown in Window 10.10.

We can obtain the rectangular representation of this hyperbola using the the equation $x^2 - y^2 = 1$ in the form

$$\cosh^2(t) - \sinh^2(t) = 1 \qquad (10.5)$$

We have that

$$x(t) + 2 = \cosh(t)$$

$$y(t) - 3 = \sinh(t)$$

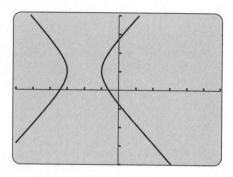

Window 10.10

by solving the parametric representation for $\cosh(t)$ and $\sinh(t)$. We now substitute these values in 10.5. We have the following results.

$$(x(t) + 2)^2 - (y(t) - 3)^2 = 1$$

or

$$(x + 2)^2 - (y - 3)^2 = 1$$

We can change the shape of the hyperbola and move the center at the same time.

EXAMPLE SET A

Find the rectangular representation of a hyperbola with center at $(-1, 4)$ and coefficient of $\cosh(t)$ equal to 3 and coefficient of $\sinh(t)$ equal to 2 in the parametric representation.

Solution:
The parametric representation is

$$x(t) = 3\cosh(t) - 1$$

$$y(t) = 2\sinh(t) + 4$$

The rectangular representation is based on these equations. First we solve each equation for its hyperbolic function.

$$\frac{x(t) + 1}{3} = \cosh(t)$$

$$\frac{y(t) - 4}{2} = \sinh(t)$$

or

$$\frac{x + 1}{3} = \cosh(t)$$

$$\frac{y - 4}{2} = \sinh(t)$$

Now we can substitute this in 10.5 as follows.

$$\left(\frac{x+1}{3}\right)^2 - \left(\frac{y-4}{2}\right)^2 = 1$$

This can be written as

$$\frac{(x+1)^2}{9} - \frac{(y-4)^2}{4} = 1$$

Try to make the connection between the multipliers of the hyperbolic functions and the denominators of the rectangular representation. Do the same for the constants that are subtracted and added to the hyperbolic function in parametric form. We can extend and generalize the results obtained in this last example to write the equations of the standard forms of a hyperbola.

$$\frac{(x-h)^2}{a^2} - \frac{(y-k)^2}{b^2} = 1$$

The center is at (h, k) and the length of the transverse axis is $2a$. The value $2b$ is called the length of the conjugate axis. You may recall from previous courses that a and b are used to construct a rectangle that helps in drawing the hyperbola. The standard form above with $\dfrac{(x-h)^2}{a^2}$ term opens horizontally.

If the hyperbola opens vertically, then the standard form is:

$$\frac{(y-k)^2}{a^2} - \frac{(x-h)^2}{b^2} = 1$$

If we consider the standard forms with center at the origin, we obtain

$$\frac{x^2}{a^2} - \frac{y^2}{b^2} = 1$$

where the x-axis contains the horizontal transverse axis and the y-axis contains the conjugate axis. The other standard form with center at the origin is:

$$\frac{y^2}{a^2} - \frac{x^2}{b^2} = 1$$

with the y-axis containing the vertical transverse axis and the x-axis containing the conjugate axis.

We can look for the vertices of these hyperbolas by setting $y = 0$ and then $x = 0$. Setting $y = 0$ in

$$\frac{x^2}{a^2} - \frac{y^2}{b^2} = 1$$

and solving for x yields $x = \pm a$. The vertices, therefore, are $(a, 0)$ and $(-a, 0)$. If we set $x = 0$, we obtain nonreal solutions for y so that there are no y intercepts.

You should carry out the same computations for

$$\frac{y^2}{a^2} - \frac{x^2}{b^2} = 1$$

to convince yourself that the vertices are $(0, a)$ and $(0, -a)$ and that there are no x intercepts.

The Asymptotic Behavior of a Hyperbola

We can use the rectangular forms to ease the task of graphing a hyperbola with pencil and paper. We begin by looking at the standard form of the hyperbola with h and k equal to zero and solving for y^2.

$$\frac{x^2}{a^2} - \frac{y^2}{b^2} = 1$$

$$\frac{y^2}{b^2} = \frac{x^2}{a^2} - 1$$

$$y^2 = b^2\left(\frac{x^2}{a^2} - 1\right)$$

$$y^2 = b^2\left(\frac{x^2}{a^2}\right) - b^2$$

$$y^2 = \frac{b^2}{a^2}x^2 - b^2$$

For large values of x, the constant b^2 term on the right will have little effect. So, we ignore it for the moment and arrive at the equation

$$y^2 = \frac{b^2}{a^2}x^2$$

This equation can be solved for y as follows.

$$y^2 = \frac{b^2}{a^2}x^2$$

$$y = \pm\frac{b}{a}x$$

or

$$y = +\frac{b}{a}x \quad \text{or} \quad y = -\frac{b}{a}x$$

These last two equations are equations of lines passing through the origin with slopes b/a and $-b/a$. The hyperbola will approach these lines for large values of x since the points on the hyperbola satisfy the equation

$$y = \pm\frac{b}{a}\sqrt{x^2 - a^2}$$

This can be seen by going back to the form

$$y^2 = \frac{b^2}{a^2}x^2 - b^2$$

and noting that

$$y^2 = \frac{b^2}{a^2}\left(x^2 - a^2\right)$$

The values of x and $\sqrt{x^2 - a^2}$ are close when x is large. These lines are called asymptotes of the hyperbola. Let's look at an example.

EXAMPLE SET B

Graph the lines $y = \frac{3}{2}x$ and $y = -\frac{3}{2}x$ and the hyperbola given below.

$$\frac{x^2}{4} - \frac{y^2}{9} = 1$$

Solution:
We can plot one branch of this hyperbola along with the two lines in parametric mode. We need to enter the following expressions in the appropriate pairs available in the Y=menu.

$$x(t) = 2\cosh(t)$$

$$y(t) = 3\sinh(t)$$

$$x(t) = t$$

$$y(t) = \frac{3}{2}t$$

$$x(t) = t$$

$$y(t) = -\frac{3}{2}t$$

The graph should look something like Window 10.11.

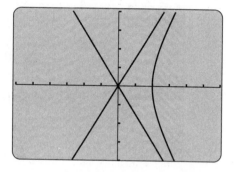

Window 10.11

You can probably visualize the other branch of the hyperbola approaching the asymptotes from the symmetry of the hyperbola.

The Central Rectangle of a Hyperbola

The central rectangle associated with the hyperbola

$$\frac{x^2}{4} - \frac{y^2}{9} = 1$$

can be constructed in the following way. First reproduce the figure in Window 10.11 on a piece of paper. Now draw a vertical line segment through the vertex $(2, 0)$ of the right branch of the hyperbola with endpoints on the two asymptotes. Then draw two horizontal line segments from these endpoints to the asymptotes in the 2nd and 3rd quadrants. Finally, complete the central rectangle by drawing a fourth side. Notice that the rectangle passes through the vertex $(-2, 0)$ of the left branch of the hyperbola. Also, the length of a horizontal side of this rectangle is the length of the transverse axis, which is

$2a$ or $2 \cdot 2 = 4$ in this case. The vertical sides are twice the length of b, which in this case is $2 \cdot 3 = 6$.

The conjugate axis is a line segment that is bisected by the transverse axis and is also the perpendicular bisector of the transverse axis. This means that we can construct the asymptotes by constructing the central rectangle based on the a and b values (and the center of the hyperbola). The asymptotes are the diagonals of this central rectangle and are very helpful in creating a rough sketch of a hyperbola using paper and pencil. Let's look at an example.

EXAMPLE SET C

Using pencil and paper, sketch the graph of

$$\frac{x^2}{16} - \frac{y^2}{25} = 1$$

Solution:
The central rectangle can be used to construct the asymptotes. We start at the origin since the hyperbola is centered there. The value of a is 4 so we locate the vertices of the hyperbola a distance of 4 to the right and left of the center (origin). With the transverse axis in place, we draw in the conjugate axis going a distance of 5 units up and 5 units down from the center. We draw in the central rectangle using the transverse and conjugate axes as guides. The central rectangle is displayed in the graph on the left of Figure 10.1. We can now construct the asymptotes by drawing in the diagonals of the central rectangle. These diagonals can now be used as guides to sketch in the hyperbola as shown in the graph on the right of Figure 10.1.

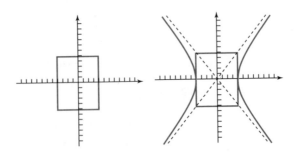

Figure 10.1

It is important to realize that the vertices of the hyperbola are associated with the value of a in the standard form of the equation. The transverse axis is on the x- or y-axis depending upon whether the standard form is an x first form or a y first form when the center is at the origin. In other words, if the form of the hyperbola is

$$\frac{x^2}{a^2} - \frac{y^2}{b^2} = 1$$

then the vertices are located at $(\pm a, 0)$. If the form is

$$\frac{y^2}{a^2} - \frac{x^2}{b^2} = 1$$

then the vertices are located at $(0, \pm a)$.

So far we have been graphing hyperbolas with center at the origin and transverse axis on the x-axis. Let's look at a hyperbola with center at $(2, 3)$ and transverse axis parallel to the y-axis.

EXAMPLE SET D

Determine the vertices and sketch the graph of the following hyperbola:

$$\frac{(y-2)^2}{36} - \frac{(x-3)^2}{4} = 1$$

Solution:

The key to describing a translated hyperbola is to first locate its center. The center is located at $(3, 2)$. The vertices will be located up and down from the center because this is a y first standard form. Since $b = 6$, the vertices are located at $(3, 2 \pm 6)$ or at the points $(3, 8)$ and $(3, -4)$. Likewise, the conjugate axis is parallel to the x-axis with endpoints $(3 \pm 2, 2)$ or at the points $(5, 2)$ and $(1, 2)$. Using these points, we can construct the central rectangle and the asymptotes as shown in the graph on the left of Figure 10.2. The sketch of this hyperbola is shown in the graph on the right of Figure 10.2.

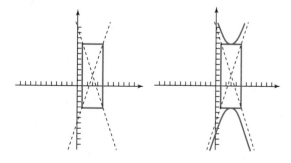

Figure 10.2

We can see that it is important to place the equation of the hyperbola in standard form. The transverse axis will be parallel to the x-axis if the standard form is the x first form. If the y term is first, the transverse axis will be parallel to the y-axis.

Often equations are presented in the general form. For example,

$$9x^2 - 4y^2 - 18x + 16y + 29 = 0$$

When we complete the square and divide both sides by the constant term, we have

$$\frac{(x-1)^2}{4} - \frac{(y-2)^2}{9} = 1$$

We can now proceed to construct the graph of this hyperbola as in the examples above.

EXERCISES

In exercises 1–6, graph the given hyperbola on your graphing calculator being sure to graph both branches. Then write the equation of the hyperbola in standard rectangular form. Finally, identify the transverse and conjugate axes and indicate their lengths.

1. $x(t) = 5\cosh(t)$
   $y(t) = 3\sinh(t)$

2. $x(t) = 2\cosh(t)$
   $y(t) = 5\sinh(t)$

3. $x(t) = 3\cosh(t)$
   $y(t) = 6\sinh(t)$

4. $x(t) = 4\cosh(t)$
   $y(t) = 2\sinh(t)$

5. $x(t) = 8\cosh(t)$
   $y(t) = 5\sinh(t)$

6. $x(t) = \cosh(t)$
   $y(t) = 3\sinh(t)$

In exercises 7–12, convert the given hyperbola to parametric form and then graph it on your graphing calculator being sure to graph both branches. Finally, identify the transverse and conjugate axes and indicate their lengths.

7. $\frac{x^2}{25} - \frac{y^2}{9} = 1$

8. $\frac{x^2}{4} - \frac{y^2}{16} = 1$

9. $\frac{y^2}{49} - \frac{x^2}{36} = 1$

10. $\frac{y^2}{4} - \frac{x^2}{9} = 1$

11. $\frac{y^2}{12} - \frac{x^2}{3} = 1$

12. $x^2 - \frac{y^2}{10} = 1$

In exercises 13–18, graph the given hyperbola on your graphing calculator being sure to graph both branches. Then write the equation of the hyperbola in standard rectangular form. Finally, identify the center of the hyperbola and the transverse and conjugate axes and indicate their lengths.

13. $x(t) = 2\cosh(t) - 3$
    $y(t) = 5\sinh(t) + 5$

14. $x(t) = 5\cosh(t) + 1$
    $y(t) = 3\sinh(t) - 2$

15. $x(t) = \cosh(t) + 2$
    $y(t) = 5\sinh(t) + 3$

16. $x(t) = 7\cosh(t) - 2$
    $y(t) = 4\sinh(t) - 1$

17. $x(t) = 6\cosh(t) - 2$
    $y(t) = 2\sinh(t) + 4$

18. $x(t) = 3\cosh(t) + 3$
    $y(t) = 4\sinh(t) - 4$

In exercises 19–24, convert the given hyperbola to parametric form and then graph it on your graphing calculator being sure to graph both branches. Finally, identify the center of the hyperbola and the transverse and conjugate axes and indicate their lengths.

19. $\dfrac{(x+3)^2}{25} - \dfrac{(y-2)^2}{9} = 1$

20. $\dfrac{(x-1)^2}{4} - \dfrac{(y-3)^2}{16} = 1$

21. $\dfrac{(y+5)^2}{36} - \dfrac{(x+3)^2}{4} = 1$

22. $\dfrac{(y-2)^2}{9} - \dfrac{(x-4)^2}{25} = 1$

23. $\dfrac{(y-4)^2}{12} - \dfrac{(x+1)^2}{8} = 1$

24. $\dfrac{(x-3)^2}{6} - \dfrac{(y-3)^2}{10} = 1$

In exercises 25–32, use pencil and paper to carefully sketch the central rectangle and asymptotes as well as the hyperbola. Then identify the center and vertices of the hyperbola and write the equations of the asymptotes. Use your graphing calculator to check your results only after you have finished the exercise using paper and pencil.

25. $\dfrac{x^2}{9} - \dfrac{y^2}{4} = 1$

26. $\dfrac{x^2}{25} - \dfrac{y^2}{16} = 1$

27. $\dfrac{y^2}{12} - \dfrac{x^2}{4} = 1$

28. $y^2 - \dfrac{x^2}{4} = 1$

29. $\dfrac{(x-3)^2}{36} - \dfrac{(y+2)^2}{49} = 1$

30. $\dfrac{(x+4)^2}{16} - \dfrac{(y+4)^2}{9} = 1$

31. $\dfrac{(y-1)^2}{25} - \dfrac{(x+2)^2}{16} = 1$

32. $\dfrac{(y+3)^2}{4} - \dfrac{(x+1)^2}{9} = 1$

In exercises 33–42, put the given hyperbola in standard form. Then identify the center and vertices of the hyperbola and write down the equations of the asymptotes.

33. $9x^2 - 4y^2 = 36$

34. $16x^2 - 12y^2 = 144$

35. $x^2 - 2y^2 = -18$

36. $16x^2 - 25y^2 = -400$

37. $16(x-1)^2 - 4(y+3)^2 - 16 = 0$

38. $9(x+2)^2 - 4(y+4)^2 + 36 = 0$

39. $4x^2 - y^2 - 32x + 4y + 56 = 0$

40. $2x^2 - y^2 + 4x - 4y = 0$

41. $9x^2 - 16y^2 + 36x + 96y = 108$

42. $4x^2 - 5y^2 - 16x + 10y + 31 = 0$

43. The arches supporting a small bridge being built will be in the form of hyperbolas, the same shape being repeated four times (see the accompanying figure which is not drawn to scale). If the axes are chosen as shown, what will be the equations of the four hyperbolas? (The equation of a similar hyperbola in the standard position is $\dfrac{y^2}{16} - \dfrac{x^2}{9} = 1$). Which branches of the hyperbolas are being used? What is the height of the arches?

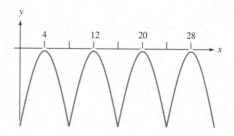

10.3 Parabolas

Introduction

The Effect of a on the Standard Parabola

The Effect of c on the Standard Parabola

The Effect of b on the Standard Parabola

The General Parabola

The Axis of a Parabola

Introduction

Parabolas are also conic sections. We have studied the transformations of the standard parabolic function $f(x) = x^2$ using translations, reflections, contractions and expansions in Chapter 2. In this chapter we extend our understanding of parabolas by considering the coefficients of the general form of the parabola, $f(x) = ax^2 + bx + c$.

We begin with the rectangular representation of a parabola:

$$y = ax^2 + bx + c \tag{10.6}$$

which is an equation that represents a parabola or parabolic function depending upon our point of view. Notice that circles, ellipses, and hyperbolas are not the graphs of functions. (They do not pass the vertical line test.)

We can use our graphing calculators to quickly display any parabolic function. For convenience we will use a slight variant of the standard graphing window, namely, we will change Xmin to -9. For example, let's graph the equation $y = \frac{1}{2}x^2 + 4x + 1$. The graph is displayed in Window 10.12.

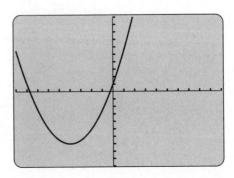

Window 10.12

This parabola intersects both the *x*- and *y*-axes. We would like to develop methods of easily graphing such parabolas with pencil and paper by determining the vertex and how wide or narrow the parabola is.

We begin by graphing the standard parabola $y = x^2$ in the same window, as shown in Window 10.13.

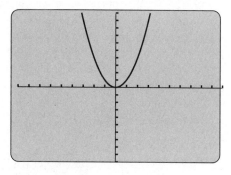

Window 10.13

We continue our study of such equations in this graphing window by considering the effects of *a*, *b*, and *c* of the parabola $y = ax^2 + bx + c$ one at a time.

The Effect of *a* on the Standard Parabola

Graph $y = x^2$ and $y = 3x^2$ in the same graphing window to observe the effect of *a* on the standard parabola. You should observe the graphs as they are constructed on your graphing calculator. The graph appears in Window 10.14.

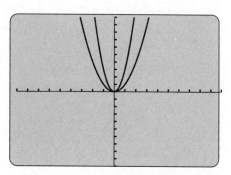

Window 10.14

Press TRACE and move along the graph of $y = x^2$ occasionally jumping to the graph of $y = 3x^2$ to see how the y-coordinates compare. Is the y coordinate of a point on the graph of $y = 3x^2$ three times the y-coordinate of $y = x^2$ for the same value of x? Notice that the graph of $y = 3x^2$ has each y coordinate moved up by a factor of 3, thereby expanding or stretching the standard graph upward, which visually has the effect of narrowing the graph.

What happens when we consider an a value of $\frac{1}{2}$? Change the equation for $y = 3x^2$ to $y = \frac{1}{2}x^2$. Graph the pair $y = x^2$ and $y = \frac{1}{2}x^2$ to see how these graphs compare point by point as before. This time each y-coordinate is half the y-coordinate of $y = x^2$, thereby contracting the standard parabola toward the x-axis, which has the visual effect of widening the parabola.

What would be the effect if a were negative in the equation $y = ax^2$? Let's compare the graphs of $y = x^2$, $y = -x^2$ and $y = -2x^2$. These graphs are displayed in Window 10.15.

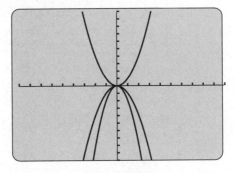

Window 10.15

You should examine these graphs using your graphing calculator and the TRACE feature to better understand the concepts involved. How would you characterize the effect of negative values of a?

The Effect of c on the Standard Parabola

Let's examine the function $y = x^2 + 2$. The graphs of $y = x^2$ and $y = x^2 + 2$ are displayed in Window 10.16.

Notice that the graph of $y = x^2 + 2$ is the same shape as the standard parabola but shifted vertically up two units. You should graph $y = x^2 - 3$ in the same window to see that this function is shifted vertically down three units. In other words, the effect of c is a vertical shift of the standard parabola as we saw in Chapter 2.

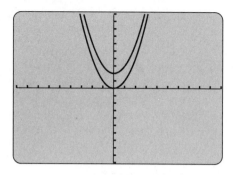

Window 10.16

The Effect of b on the Standard Parabola

Let's begin by looking at $y = x^2 + 4x$. This function along with the standard parabola is displayed in Window 10.17.

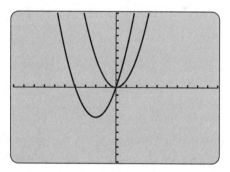

Window 10.17

Notice that the graph of $y = x^2 + 4x$ has the same shape as the standard parabola but its vertex is located at $(-2, -4)$. This is both a vertical shift and a horizontal shift. We can see why this is so by rewriting $y = x^2 + 4x$ using the notion of completing the square.

$$y = x^2 + 4x$$

$$y = x^2 + 4x + 4 - 4$$

$$y = (x + 2)^2 - 4$$

Using the function $f(x) = x^2$ we describe the change from $y = x^2$ to $y = (x+2)^2 - 4$ in the following way:

$$f(x) = x^2 \longrightarrow f(x+2) = (x+2)^2 \longrightarrow f(x+2) - 4 = (x+2)^2 - 4$$

As we recall from Chapter 2, this amounts to moving the graph $y = f(x)$ left 2 units and down 4 units. The parameter b in $y = x^2 + bx$ causes a horizontal shift of $-b/2$. But in causing this horizontal shift, the bx term creates a vertical shift as well of $b^2/4$. The situation is more complicated when the coefficient of x^2 is not 1. Let's look at an example.

EXAMPLE SET A

Find the vertex of $y = 2x^2 - 8x + 3$ and describe how the graph of this function can be derived from the standard parabola $y = x^2$.

Solution:
The method of completing the square gives a standard form we can use in analyzing the parabola. Recall that the completing the square process requires the coefficient of x^2 to be 1. This process gives:

$$y = 2x^2 - 8x + 3$$

$$y = 2(x^2 - 4x \quad) + 3$$

$$y = 2(x^2 - 4x + 4) + 3 - 8$$

$$y = 2(x - 2)^2 - 5$$

Using the function $f(x) = x^2$ we describe the change from $y = x^2$ to $y = 2(x-2)^2 - 5$ in the following way:

$$f(x) = x^2 \longrightarrow 2f(x) \longrightarrow 2f(x-2) = 2(x-2)^2 \longrightarrow 2f(x-2)^2 - 5$$

We can now see that the standard parabola will be expanded by a factor of 2, shifted to the right 2 units, and shifted down 5 units.

The General Parabola

If $g(x) = ax^2 + bx + c$ and $f(x) = x^2$, then we can determine the behavior of the graph of g using the combined effects of a, b, and c on the standard parabola $f(x) = x^2$. We complete the square and simplify below.

$$g(x) = ax^2 + bx + c$$

$$g(x) = a\left(x^2 + \frac{b}{a}x \quad \right) + c$$

$$g(x) = a\left(x^2 + \frac{b}{a}x + \frac{b^2}{4a^2}\right) + c - \frac{b^2}{4a}$$

$$g(x) = a\left(x + \frac{b}{2a}\right)^2 + \frac{4ac - b^2}{4a}$$

In this form, we can determine the vertical and horizontal shifts as well as expansions or contractions and reflections about the x-axis. This process should be placed in your resource bank. You may wish to actually memorize the formula, but the process is all you will need and, more importantly, it deepens your understanding of parabolas.

The Axis of a Parabola

It is possible to draw a careful graph of a parabola by plotting only three points provided they are the right three points. One of these three points *must* be the vertex. Once we obtain the vertex, then any other point you compute gives you a "free" point, which is a reflected point about the axis of the parabola.

The axis of any parabola of the form $y = ax^2 + bx + c$ is the vertical line passing through the vertex. It should be relatively easy to see that this line splits the parabola into two branches that are mirror images of each other about this axis. Since each computed point gives us a reflected point, then computing the vertex and two other points gives us a total of five points, which is sufficient to "lock in" the graph of the parabola. The y-intercept is easily computed by setting $x = 0$. This will be a second point (plus its reflection) provided the vertex is not on the y-axis. We then need but one more computed point, which should be chosen by its convenience and helpfulness.

EXAMPLE SET B

Use pencil and paper to determine the vertex and axis of the given parabola and then make a careful sketch of the parabola.

$$y = x^2 - 4x - 2$$

Solution:
The vertex can be obtained by completing the square.

$$y = x^2 - 4x - 2$$

$$y = x^2 - 4x + 4 - 4 - 2$$

$$y = x^2 - 4x + 4 - 6$$

$$y = (x - 2)^2 - 6$$

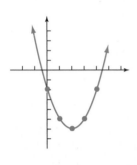

Figure 10.3

We can now read off the vertex as $(2, -6)$. The axis of the parabola is $x = 2$. The y-intercept can be easily found by setting $x = 0$, yielding $y = -2$ as the y-intercept. The reflection of this point about the axis of the parabola is $(4, -2)$. A third convenient, helpful point might be at $x = 1$. Substituting $x = 1$ into the parabola yields the point $(1, -5)$ whose reflected point is $(3, -5)$. We can now plot these five points and draw a careful sketch as shown in Figure 10.3.

EXERCISES

In exercises 1–6, describe in paragraph form the required shifting, reflecting, and expanding or contracting of the graph of $y = x^2$ needed to obtain the graphs representing the parabolas below.

1. $y = x^2 + 2x - 3$.

2. $y = x^2 - 3x - 5$.

3. $y = 3x^2 + 2x - 6$.

4. $y = \frac{1}{2}x^2 + 2x + 6$.

5. $y = -2x^2 - 5x + 2$.

6. $y = -\frac{1}{2}x^2 + 6x + 4$.

In exercises 7–16, use the completing the square technique to find the vertex and axis of the given parabola. Then find the y-intercept and one other convenient, helpful point. Use these points and the corresponding reflection points about the axis of the parabola to make a careful sketch of the parabola.

7. $y = x^2 + 4x - 3$.

8. $y = x^2 - 6x - 5$.

9. $y = -3x^2 + 9x - 3$.

10. $y = 2x^2 - 5x - 4$.

11. $y = \frac{1}{3}x^2 - x + 2$.

12. $y = -\frac{2}{3}x^2 + 4x - 6$.

13. $y = -\frac{1}{5}x^2 - 3x + 1$.

14. $y = -4x^2 + 10x - 4$.

15. $y = \frac{1}{4}x^2 + 2x + 3$.

16. $y = \frac{3}{2}x^2 - 9x - 15$.

In exercises 17–24, write an equation of the parabola from the information given. Then describe in words why your answer is not the only possible one.

17. Shift vertically up 5 units and horizontally right 2 units.

18. Shift vertically up 1 unit and horizontally left 3 units.

19. Shift vertically down 6 units and horizontally right 4 units.

20. Shift vertically down 2 units and horizontally left 1 unit.

21. Shift vertically down 3 units and expand by 4 units.

22. Shift horizontally left 4 units and contract by 3 units.

23. Shift vertically up 1 unit, horizontally right 3 units, and contract by 2 units.

24. Shift vertically down 5 units, horizontally left 4 units, and expand by 3 units.

25. Raul would like to go to Connie's house. The directions he is given tell him that the house is exactly the same distance from City Hall as it is from A Street. Looking at the map, Raul realizes he has a problem. If (x, y) represents the coordinates of a house for some suitable coordinate system (chosen by you), what kind of equation would (x, y) satisfy? If the shortest distance from City Hall to A Street is 1200 feet, write down the equation of the path along which Raul should search.

26. Raul calls Connie for further directions and finds out that the house is exactly 1200 ft from A Street and exactly 1200 ft from City Hall, due east. Where is Connie's house located? If there is a road to the house from A Street and perpendicular to it, how far along this road must Raul travel?

27. In a physics class experiment, balls are shot from a horizontal nozzle and the horizontal distance of the point where the balls strike the ground as well as the time of the fall are measured. It is found that the higher the nozzle is above the ground, the longer it takes the ball to fall. Tabulating the results, the experimenters arrive at an empirical formula relating height y to time t : $y = 16t^2$. What kind of equation does this represent? Describe it, giving the vertex (or vertices), axes of symmetry etc. Does this equation represent the equation of the path of the ball? Give reasons for your answer.

28. The same class (see the previous exercise) also finds that the horizontal distance varies with speed. In other words, the greater the velocity of projection, the farther the balls go. The formula the class arrives at is $x = ut$. Find an equation relating y (from the previous exercise) and x. This equation describes the path of the balls. What kind of equation is it? Note that the balls are projected horizontally. (Hint: Eliminate t.)

10.4   A Polar Coordinate Approach to Conics

Introduction

The Geometry of Conic Sections

Polar Coordinates and Conic Sections

Introduction

Conic sections can be described using a ratio of distances based on a fixed point and a fixed line. In this section, we will introduce polar coordinates and use them to develop equations for the conic sections based on these ratio of distances descriptions.

The Geometry of Conic Sections

We start with a geometric definition of conics.

The Conics

A conic is the set of points in a plane the ratio of whose distances from a fixed point (called the focus) and a fixed line (called the directrix) is a constant.

This constant ratio is commonly called the eccentricity and is, therefore, designated by e. We will find that:

- For an ellipse, the constant ratio is a number between 0 and 1.
- For a parabola, the constant ratio is the number 1.
- For a hyperbola, the constant ratio is a number greater than 1.

Polar Coordinates and Conic Sections

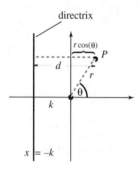

Figure 10.4

We can determine a universal form for these conics from this definition. To do this, consider the focus located at the origin and the directrix located at $x = -k$ as shown in Figure 10.4, along with an arbitrary point on the conic and pertinent labels. The constant ratio is

$$e = \frac{r}{d}$$

We can express d using k, r, and θ. The horizontal distance d from the point P to the directrix is $k + r\cos(\theta)$ as in Figure 10.4. Figure 10.5 displays the two cases resulting from θ as an acute or obtuse angle.

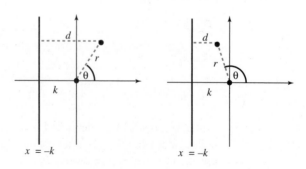

Figure 10.5

When θ is obtuse, $\cos(\theta)$ is negative and we add a negative number to k as in the right diagram in Figure 10.5. When θ is acute, $\cos(\theta)$ is positive and we add a positive number to k as in the left diagram in Figure 10.5.

Thus, we have

$$e = \frac{r}{d}$$

$$e = \frac{r}{k + r\cos(\theta)}$$

$$e(k + r\cos(\theta)) = r$$

$$ek + er\cos(\theta) = r$$

$$ek = r - er\cos(\theta)$$

$$ek = r(1 - e\cos(\theta))$$

$$\frac{ek}{(1 - e\cos(\theta))} = r$$

or,

$$r = \frac{ek}{1 - e\cos(\theta)}$$

This single equation describes ellipses, parabolas, and hyperbolas and gives r in terms of the constants e and k and the variable θ. Thus, r is a function of θ.

Let's look at an example with $e = \frac{1}{2}$.

EXAMPLE SET A

Graph the conic with

$$r = \frac{\frac{1}{2} \cdot 6}{1 - \frac{1}{2} \cdot \cos(\theta)}$$

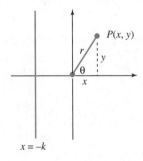

$x = -k$

Figure 10.6

Solution:

Our values of r and θ are not the Cartesian or rectangular coordinates of the point P. We can determine the rectangular coordinates of P from the r and θ values by looking at Figure 10.6. The values of $\sin(\theta)$ and $\cos(\theta)$ can be computed using right-triangle trigonometry from Figure 10.6.

$$\cos(\theta) = \frac{x}{r} \quad \text{and} \quad \sin(\theta) = \frac{y}{r}$$

Thus,

$$x = r\cos(\theta) \quad \text{and} \quad y = r\sin(\theta)$$

For any values of θ and r, the x and y values are computed using these formulas. Here, x and y can be determined by θ alone since r is a function of θ. The variable θ can be thought of as a parameter for x and y, and we can use the following formulas.

$$x(\theta) = r\cos(\theta) \tag{10.7}$$

$$y(\theta) = r\sin(\theta)$$

Where r is a function of θ.

We can enter this parametric representation of the graph on the TI-81 as follows:

```
:X1T=(.5*6)/(1−0.
5cos (T))
:Y1T=
:X2T=
:Y2T=
:X3T=X1Tcos (T)
:Y3T=X1Tsin (T)
```

Window 10.18

We use T for θ and we place the formula for r in X1T and the two functions for x and y in 10.7 as X3T and Y3T.

The two windows in Window 10.19 show the settings for the Param graphing window and the graph of the conic. The Ymax value should be

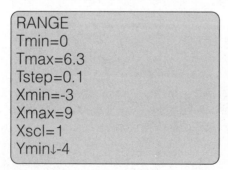

```
RANGE
Tmin=0
Tmax=6.3
Tstep=0.1
Xmin=-3
Xmax=9
Xscl=1
Ymin↓-4
```

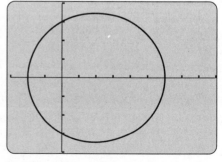

Window 10.19

set to 4 by scrolling down. The graph of this conic is an ellipse since $e = 1/2$ as indicated earlier.

The TI-82 and 85 have polar graphing capabilities that make the graphing above somewhat easier.

Parabolas and hyperbolas can be graphed in the same way.

EXAMPLE SET B

Graph the conic

$$r = \frac{1 \cdot 2}{1 - 1 \cdot \cos(\theta)}$$

where $e = 1$ and $k = 2$.

Solution:
We will use the same RANGE or WINDOW settings as in the previous example. We need to change the expression for r in X1T as follows:

:X1T = (1*2)/(1-1cos(T))

What do you get when you graph this conic? The e value is 1 and, thus, you should obtain a parabola.

The parabola obtained in the previous example opens around the focus (the origin). You may wish to try other values of k to see how this parameter effects the parabola.

We now look at an example in which the eccentricity (e) is greater than 1.

EXAMPLE SET C

Graph the conic

$$r = \frac{2 \cdot 1}{1 - 2 \cdot \cos(\theta)}$$

where $e = 2$ and $k = 1$.

Solution:
We will use the same RANGE or WINDOW settings as in the previous example. We need to change the expression for r in X1T as follows:

:X1T = (2*1)/(1-2cos(T))

You should graph this conic in Dot mode. You may wish to adjust the viewing window.

What do you get when you graph this conic? The e value is 2 and, thus, you should obtain a hyperbola.

For $e < 1$ we obtained an ellipse, for $e = 1$ we obtained a parabola, and for $e > 1$ we obtained a hyperbola. These examples do not prove that the indicated sets of e values give rise to the suggested conic; they do show that some specific values of e justify our claim.

EXAMPLE SET D

1. Graph the conic

$$r = \frac{4}{3 - 2 \cdot \cos(\theta)}$$

Solution:

We need to observe that the standard polar form of a conic requires that the left term of the denominator be 1. We can obtain a 1 in that position by dividing the numerator and denominator by whatever number is in that position, in this case 3. Dividing the numerator and denominator by 3 yields:

$$r = \frac{\frac{4}{3}}{1 - \frac{2}{3} \cdot \cos(\theta)}$$

Since the multiplier of $\cos(\theta)$ is less than 1, the conic will be an ellipse. We can graph this ellipse by changing the expression for r in X1T as follows:

:X1T = (4/3)/(1-(2/3)cos(T))

To obtain the value of k, we can set the numerator of the standard form expression of the conic to the product ek. This yields a value of $k = 2$ since $e = 2/3$.

2. Graph the conic

$$r = \frac{4}{3 + 2 \cdot \sin(\theta)}$$

and contrast it with the previous example in this set.

Solution:

We notice that the denominator contains a $+$ sign instead of a $-$ sign. Also, $\sin(\theta)$ replaces $\cos(\theta)$. Even so, the standard form requires that a

1 be located at the left term of the denominator. We again divide the numerator and denominator by 3 to obtain:

$$r = \frac{\frac{4}{3}}{1 + \frac{2}{3} \cdot \sin(\theta)}$$

We can contrast the graph of this expression with the previous one by changing the expression for r in X1T as follows:

:X1T = (4/3)/(1+(2/3)sin(T))

and graphing it. We notice that we obtain an ellipse again, but that the major axis is along the y-axis instead of the x-axis. This ellipse has the same major and minor axis lengths, which might lead us to believe that the focus is still at the origin but that the directrix must be located in a different position. Actually, the directrix is $y = 2$ in this case. You will have an opportunity to examine the remaining three standard polar coordinate forms of the conics in the exercises.

EXERCISES

In exercises 1–6, use your graphing calculator to graph the given conic in polar coordinate form. Write in paragraph form the type of conic observed, as well as the value of the eccentricity e and where you obtained its value. Also, give the location of the directrix and how you obtained it. Finally, indicate if the value of the eccentricity and the curve you observe agree with the conditions given in this section and why.

1. $r = \dfrac{6}{1 - 2 \cdot \cos(\theta)}$

2. $r = \dfrac{2}{1 - \cos(\theta)}$

3. $r = \dfrac{9}{1 - 3 \cdot \cos(\theta)}$

4. $r = \dfrac{1}{1 - \frac{1}{4} \cdot \cos(\theta)}$

5. $r = \dfrac{8}{1 - \cos(\theta)}$

6. $r = \dfrac{5}{1 - \frac{3}{5} \cdot \cos(\theta)}$

7. Derive the remaining three polar coordinate forms of the conics by placing the focus at the origin and considering the directrix to be at each of the following positions: $x = k$, $y = -k$, and $y = k$. Follow the format of the derivation for the first form in this section. You should draw the appropriate directrix and an arbitrary point on the conic. Then label r, θ, k, and d. Using the constant ratio of the distance of this point from the focus to its distance to the directrix ($e = r/d$), you then need to express d in terms of r, θ, and k, and finally solve for r. These three polar coordinate forms along with the one derived in this section are called the standard polar coordinate forms of the conics.

8. Use the previous problem as a guide to discuss and write down any connections you can make between the standard polar coordinate form of the conic and the orientation of its graph.

In exercises 9–16, place the equation in standard form, if necessary, and then explain in paragraph form what type of conic it represents and indicate the eccentricity and directrix and how you obtained them.

9. $r = \dfrac{2}{1 + \frac{1}{2} \cdot \cos(\theta)}$

10. $r = \dfrac{2}{1 - \frac{1}{2} \cdot \sin(\theta)}$

11. $r = \dfrac{2}{1 + \frac{1}{2} \cdot \sin(\theta)}$

12. $r = \dfrac{5}{1 - \sin(\theta)}$

13. $r = \dfrac{3}{1 + 4 \cdot \cos(\theta)}$

14. $r = \dfrac{4}{5 - 10\sin(\theta)}$

15. $r = \dfrac{3}{2 + \frac{1}{2} \cdot \cos(\theta)}$

16. $r = \dfrac{8}{3 + 4 \cdot \sin(\theta)}$

17. Use your graphing calculator to graph the polar coordinate ellipse

$$r = \dfrac{0.9}{1 - 0.9 \cdot \cos(\theta)}$$

and keep the display in mind. Then successively plot the ellipses obtained by using the same equation, but replace the two 0.9s with 0.8s, 0.7s, 0.6s, 0.5s, and 0.4s. Describe in your own words the effect of decreasing the eccentricity of the ellipse from a value near 1 to values moving toward 0.

There really are two foci and directrices for ellipses and hyperbolas. This should be no surprise because of the symmetric properties of these curves. For exercises 18–22, locate the center of the ellipse and then write down both foci, both major and minor vertices, and the equations of both directrices.

18. $r = \dfrac{2}{1 - \frac{1}{4} \cdot \cos(\theta)}$

19. $r = \dfrac{3}{1 - \frac{1}{2} \cdot \cos(\theta)}$

20. $r = \dfrac{3}{1 + \frac{2}{3} \cdot \sin(\theta)}$

21. $r = \dfrac{4}{1 - \frac{1}{5} \cdot \sin(\theta)}$

22. $r = \dfrac{4}{4 - 2 \cdot \cos(\theta)}$

For exercises 23–28, locate the center of the hyperbola and then write down both foci, both vertices, and the equations of both directrices.

23. $r = \dfrac{2}{1 - 4 \cdot \cos(\theta)}$

24. $r = \dfrac{2}{1 - \frac{3}{2} \cdot \sin(\theta)}$

25. $r = \dfrac{5}{1 + \frac{4}{3} \cdot \sin(\theta)}$

26. $r = \dfrac{6}{3 - 6 \cdot \cos(\theta)}$

27. $r = \dfrac{10}{5 - 10 \cdot \sin(\theta)}$

28. $r = \dfrac{1}{2 + 3 \cdot \cos(\theta)}$

Summary and Projects

Summary

Circles Circles are familiar objects. We need to keep in mind that a circle is a set of points in a plane where each of these points is the same distance from a fixed point in the plane. The general parametric representation of a circle with center (h, k) and radius r is:

$$x(t) = r\cos(t) + h$$

$$y(t) = r\sin(t) + k$$

We can derive the general rectangular representation of a circle having the same center and radius. The procedure is to isolate both $\cos(t)$ and $\sin(t)$ in these equations. Then square both sides of these equations and add. Finally, replace $\cos(t)^2 + \sin(t)^2$ with 1. This results in the following general rectangular representation of a circle:

$$(x - h)^2 + (y - k)^2 = r^2$$

Parametric Representation of an Ellipse We can quickly graph an ellipse when it is given in parametric representation. The general parametric representation of an ellipse with center (h, k) and semi-major axis a parallel to the x-axis and semi-minor axis b parallel to the y-axis is given by:

$$x(t) = a\cos(t) + h$$

$$y(t) = b\sin(t) + k$$

The semi-major axis is always the longer of the two and is named a. It is the horizontal distance from the center to the point on the ellipse either to the right or to the left. In this form, the major axis is parallel to the x-axis. The semi-minor axis is the shorter of the two and is named b. It is the vertical distance from the center to the point on the ellipse either up or down.

If the major axis is parallel to the y-axis, the parametric representation becomes:

$$x(t) = b\sin(t) + h$$

$$y(t) = a\cos(t) + k$$

Rectangular Representation of an Ellipse It is not too difficult to derive the general rectangular representation of an ellipse from the parametric. We use the same procedure as we used for the circle, namely, make use of the identity $\cos^2(t) + \sin^2(t) = 1$. We would again need to isolate $\sin(t)$ and $\cos(t)$ in each of the two parametric equations, square both sides of these equations, and add to obtain:

$$\frac{(x-h)^2}{a^2} + \frac{(y-k)^2}{b^2} = 1$$

for the ellipse with major axis parallel to the x-axis and

$$\frac{(y-k)^2}{a^2} + \frac{(x-h)^2}{b^2} = 1$$

for the ellipse with major axis parallel to the y-axis.

Hyperbolas   There are a lot of similarities between the representations of hyperbolas and ellipses although their graphs are quite different. The general parametric representation of a hyperbola requires the hyperbolic functions cosh and sinh. Using these functions, the parametric representation of a hyperbola with center (h, k) and transverse axis parallel to the x-axis is:

$$x(t) = a\cosh(t) + h$$

$$y(t) = b\sinh(t) + k$$

The vertices of the hyperbola are a distance a horizontally right and left from the center (h, k).

If the transverse axis of the hyperbola is parallel to the y-axis, then the symmetric representation is:

$$x(t) = b\sinh(t) + h$$

$$y(t) = a\cosh(t) + k$$

In this case, the vertices of the hyperbola are a distance a vertically up and down from the center (h, k).

We can obtain the general rectangular representations of the hyperbola in the same way we did for the ellipse. The only difference is that we use the identity $\cosh(t)^2 - \sinh(t)^2 = 1$ after we isolate the functions $\cosh(t)$ and $\sinh(t)$. This procedure leads to:

$$\frac{(x-h)^2}{a^2} - \frac{(y-k)^2}{b^2} = 1$$

for the hyperbola with transverse axis parallel to the x-axis and

$$\frac{(y-k)^2}{a^2} - \frac{(x-h)^2}{b^2} = 1$$

for the hyperbola with transverse axis parallel to the y-axis.

The hyperbola is the only conic that has asymptotes (lines that the graph approaches). These lines are very useful in graphing a hyperbola by hand. One of the simplest ways of obtaining and drawing the asymptotes of a hyperbola is through the central rectangle. First consider the hyperbola with center at the origin and transverse axis along the x-axis. The rectangular representation of this hyperbola is:

$$\frac{x^2}{a^2} - \frac{y^2}{b^2} = 1$$

The vertices and endpoints of the transverse axis are at $(\pm a, 0)$. The endpoints of the conjugate axis (the perpendicular bisector of the transverse axis) are at $(0, \pm b)$. We can create the central rectangle by drawing lines parallel to the x-axis through the points $(0, \pm b)$ and lines parallel to the y-axis through the points $(\pm a, 0)$. The diagonals of this central rectangle represent the asymptotes of the hyperbola. It is not difficult to determine that the equations of the asymptotes are $y = \pm \frac{b}{a} x$ for this form. If the hyperbola has center at the origin and transverse axis along the y-axis, its form is:

$$\frac{y^2}{a^2} - \frac{x^2}{b^2} = 1$$

and has asymptotes with equations $y = \pm \frac{a}{b} x$.

Perhaps the easiest way to obtain the asymptotes for hyperbolas with center at (h, k) is to work from the center. The procedure is similar to the case in which the center is at the origin. Namely, if the transverse axis is parallel to the x-axis, plot the points horizontally away from the center a distance a from the center. Then plot the points vertically away from the center a distance b. Use these points to draw the associated central rectangle and then the diagonals of this rectangle that represent the asymptotes.

Parabolas You have already studied the general representation of a parabola $y = ax^2 + bx + c$ from one perspective in Chapter 3. In this chapter, we considered the effect of the constants a, b, and c on the graph of the parabola. The effect of a is to cause the graph of the standard parabola $y = x^2$ to become more narrow ($a > 1$) or wider ($a < 1$). The effect of c is to cause the graph of the standard parabola to shift vertically (up if $c > 0$ and down if $c < 0$). The effect of b is a bit more complicated. It causes a horizontal shift (left if $b > 0$ and right if $b < 0$). However, b also causes a vertical shift in the graph of the standard parabola that is not as straight-forward as the explanation for c. Perhaps the easiest explanation comes from completing the square and focusing on the resulting form. Completing the square on $y = ax^2 + bx + c$ gives:

$$y = a\left(x + \frac{b}{2a}\right)^2 + \frac{4ac - b^2}{4a}$$

The effect of a is the same as before. However, in this form, we may be able to see that the vertex of the parabola occurs at $x = \dfrac{-b}{2a}$, which gives the corresponding y-value of $y = \dfrac{4ac - b^2}{4a}$. This form also gives us the axis of the parabola, which is the vertical line $x = \dfrac{-b}{2a}$. Once we know the vertex, we obtain a "free" point for each point computed due to the symmetry of the parabola about its axis. Thus, if we find the vertex and compute two other convenient points, we obtain a total of five points, which can nicely "lock" in the graph of the parabola.

The Polar Coordinate Representation of Conics A single equation in polar coordinates can represent ellipses, hyperbolas, and parabolas. The key to this universal representation requires acceptance of a description of conics as a set of points in the plane whose distance from a fixed point in the plane (focus) over the distance from a fixed line (directrix) is constant. This ratio is called the eccentricity and is named e. We can obtain a nice standard representation for this representation by choosing the focus at the origin. A convenient choice for the directrix is any of the following lines: $x = -k$, $x = k$, $y = -k$, $y = k$. With the first choice, $x = -k$, we can use the diagram shown in Figure 10.7 to determine the equation:

$$r = \frac{ek}{1 - e\cos(\theta)}$$

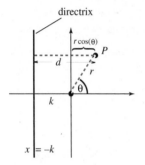

Figure 10.7

You should think about how the choice of the other three directrices produces the three other standard polar coordinate representations of the conics.

Finally, which conic is determined by which eccentricity is something you need to place in your resource bank. The association is: $0 < e < 1$ yields an ellipse, $e = 1$ a parabola, and $e > 1$ a hyperbola.

Projects

1. Find a rectangular representation of the equation of the set of points in a plane equidistant from a fixed point and fixed line in the plane. (Hint: Place the fixed point at $(c, 0)$ and the fixed line at $x = -c$. What type of curve is it?)

2. Find a rectangular representation of the equation of the set of points in a plane the difference of whose distances from two fixed points in the same plane is a constant, $2a$. (Hint: Place the two fixed points on one of the axes equidistant from the origin with one fixed point at $(c, 0)$ and the other at $(-c, 0)$. What type of curve is it?)

3. The Long Range Navigation (LORAN) system is used to determine the position of ships and planes around the globe. Two LORAN stations 400

miles apart send out a signal at the same instant (the signal travels at 980 feet/μseconds). The signals arrive at the ship *Global Magic* at different times. If we can measure the difference in times accurately, we can determine the difference in the distances the ship is from the two stations.

a. If the time differential is 300μseconds, where might the ship be?

b. If the ship is travelling parallel to the line joining the two stations and 74 miles away from it, exactly where is the ship?

c. If the ship is also able to pick up the signal from two other LORAN stations, what information would it need to know to determine its exact location with respect to these two pairs of LORAN stations?

4. An alpha particle travels on a hyperbolic path as it moves near the nucleus of a molecule. If a particle begins to move on a straight line toward a nucleus and gets to a distance of 3 angstroms of the nucleus and bounces back at an angle of about 90°, determine the equation of the path of the particle. If the velocity of the alpha particle is increased, discuss how the equation will change.

CHAPTER 11

Mathematical Induction, Sequences, and Series

11.1   Mathematical Induction

Introduction

Induction and Deduction

The Principle of Mathematical Induction

Introduction

Quite often, you might think that some mathematical conjecture is true for all numbers of a particular type. For example, a mathematics student might think that all odd numbers are prime. The first three odd numbers, 3, 5, and 7 support this conjecture; but a few examples do not prove the conjecture. Although 3, 5, and 7 are odd and prime, 9 is odd but *not* prime. In this case, the number 9 is a counterexample to the conjecture, and proves the conjecture false. How are conjectures you suspect to be true proven to be true? For example, if you suspect that the conjecture

$$1 + 2 + 3 + \cdots + n = \frac{n(n+1)}{2}$$

where n is a positive integer, is true (which it actually is), how could you prove it? Very often the method of mathematical induction is used. In this section, you explore the induction method and get practice employing it.

Induction and Deduction

In chemistry, physics, biology, and other sciences, patterns of physical or chemical behavior are often recorded in mathematical formulas. When formulas are developed using experimental data, they are called *inductive formulas*.

Induction   The process of using specific cases to form a general conclusion is **induction**.

The inductive process was used by Gregor Mendel (1822–1884) to establish genetic patterns, by Robert Boyle (1627–1691) to establish the gas laws of physics and chemistry, and by Robert Hooke (1635 – 1703) to find and verify laws for the behavior of springs.

Deduction   **Deduction** is reasoning from a general principle to a specific conclusion. In mathematics, the proof of a theorem relies on deduction as a person starts with a premise (general fact) and, through a logical sequence of steps, produces a conclusion (the specific result sought).

The Principle of Mathematical Induction

From time to time, however, the mathematician must use induction. For example, one may conjecture that a formula exists for determining the sum of the first n odd natural numbers. To develop the formula, several sums are produced and observed in hopes of recognizing a pattern.

$$1 = 1$$

$$\underbrace{1 + 3}_{2} = 4$$

$$\underbrace{1 + 3 + 5}_{3} = 9$$

$$\underbrace{1 + 3 + 5 + 7}_{4} = 16$$

$$\underbrace{1 + 3 + 5 + 7 + 9}_{5} = 25$$

$$\underbrace{1 + 3 + 5 + 7 + 9 + 11}_{6} = 36$$

Observation of the sums makes it appear that the sum of the first n odd natural numbers is n^2. We must now prove this proposition. The proof must be constructed so that it establishes the validity of the proposition for every natural number n. A proof cannot be constructed on the basis of several observations (or for that matter, a great many observations). It may be that for some larger natural number, the proposition fails to be true. (See Illuminator Set C.)

We use an inductive process in conjecturing that a formula (general principle) is true from several observations (specific cases). The method for proving formulas that have been conjectured inductively is called **mathematical induction** and it allows us to prove various propositions involving natural numbers. The method of mathematical induction is based on the following principle.

Mathematical Induction

The Principle of Mathematical Induction

The Principle of Mathematical Induction
A statement P is true for all natural numbers n if
1. The statement is true for $n = 1$.
2. Whenever P is true for $n = k$, it follows that P is true for $n = k + 1$.

We may understand the Principle of Mathematical Induction more clearly by making the following analogy. Visualize a sequence of dominoes arranged so that if one falls, it strikes the next one. We would like to know whether or not all the dominoes will fall if the first one falls. We can reach a conclusion by reasoning as follows:

1. Show that the first domino falls (show that P is true for $n = 1$).

2. a. Assume that some domino down the line, say the kth domino, falls over (assume $n = k$ is true).

    b. Show that when domino k falls over, the next domino, domino $k + 1$, is struck and falls over (if $n = k$ is true, then $n = k + 1$ is true).

Here, statement (2) of the definition is given in two parts: 2a, assume $n = k$ is true, and 2b, show true for $n = k + 1$. Figure 11.1 illustrates this idea.

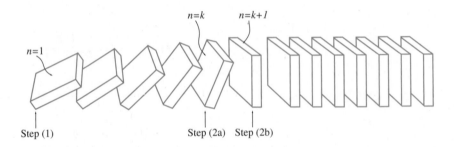

Figure 11.1

The fact that the first domino falls over and that domino $k + 1$ falls over when domino k falls over is enough to support the claim that all the dominoes fall over. Suppose $k = 1$ (the first domino). Then domino $k + 1 = 2$ (the second domino) falls over. But when the second domino falls over, the next one, the third domino falls over. Then when the third domino falls over, the next one, the fourth domino falls over. Continuing this approach, we may see more clearly the underlying logic of the induction principle.

To prove that a statement P is true for all natural numbers n using mathematical induction, we use, in order, the following three steps:

1. Show P is true for $n = 1$. Substitute 1 for n in the formula and verify its truth.

2. a. Assume P is true for $n = k$. Substitute k for n in the formula and assume it is true.

    b. Show that P is true for $n = k + 1$. Using the formula from step (2), add (or multiply) to each side the next term and by algebraic manipulation, show that a true statement results. The next term can be found by substituting $k + 1$ for n in the formula.

ILLUMINATOR SET A

Prove that the sum of the first n natural numbers is $\dfrac{n(n+1)}{2}$, that is prove that

$$1 + 2 + 3 + \cdots + n = \frac{n(n+1)}{2}$$

1. Show true for $n = 1$.

$$1 = \frac{1(1+1)}{2}$$

$$1 = \frac{1(2)}{2}$$

$$1 = 1 \qquad \text{True for } n = 1$$

2. a. Assume true for $n = k$.

   Assume that $\quad 1 + 2 + 3 + \cdots + k = \dfrac{k(k+1)}{2} \quad$ is true.

   b. Show true for $n = k + 1$. We can show the statement is true for $n = k + 1$ if we can show that

   $$1 + 2 + 3 + \cdots + k + (k+1) = \frac{(k+1)[(k+1)+1]}{2}$$

is true.

Start with the equation from step (2),

$$1 + 2 + 3 + \cdots + k = \frac{k(k+1)}{2}$$

and add the next term to each side. Then simplify the right-hand side and try to manipulate it to the desired form. The next term is $k + 1$.

$$1 + 2 + 3 + \cdots + k = \frac{k(k+1)}{2}$$

$$1 + 2 + 3 + \cdots + k + (k+1) = \frac{k(k+1)}{2} + (k+1)$$

$$= \frac{k(k+1)}{2} + \frac{2(k+1)}{2}$$

$$= \frac{k(k+1) + 2(k+1)}{2}$$

$$= \frac{k^2 + k + 2k + 2}{2}$$

$$= \frac{k^2 + 3k + 2}{2}$$

$$= \frac{(k+1)(k+2)}{2}$$

$$= \frac{(k+1)(k+1+1)}{2}$$

$$= \frac{(k+1)[(k+1)+1]}{2}$$

Thus, $1 + 2 + 3 + \cdots + n = \dfrac{n(n+1)}{2}$ is true for all natural numbers n.

ILLUMINATOR SET B

Prove that for all natural numbers n,

$$\frac{1}{1 \cdot 2} + \frac{1}{2 \cdot 3} + \frac{1}{3 \cdot 4} + \cdots + \frac{1}{n(n+1)} = \frac{n}{n+1}$$

1. Show true for $n = 1$.

$$\frac{1}{1 \cdot 2} = \frac{1}{1+1}$$

$$\frac{1}{2} = \frac{1}{2} \qquad \text{True for } n = 1$$

2. a. Assume true for $n = k$.
   Assume that

   $$\frac{1}{1 \cdot 2} + \frac{1}{2 \cdot 3} + \frac{1}{3 \cdot 4} + \cdots + \frac{1}{k(k+1)} = \frac{k}{k+1}$$

   is true.

   b. Show true for $n = k + 1$. We can show the statement is true for $n = k + 1$ if we can show that

   $$\frac{1}{1 \cdot 2} + \frac{1}{2 \cdot 3} + \frac{1}{3 \cdot 4} + \cdots + \frac{1}{(k+1)[(k+1)+1]} = \frac{k+1}{(k+1)+1}$$

   is true.

Start with the equation from step (2),

$$\frac{1}{1 \cdot 2} + \frac{1}{2 \cdot 3} + \frac{1}{3 \cdot 4} + \cdots + \frac{1}{k(k+1)} = \frac{k}{k+1}$$

and add the next term to each side. Then simplify the right-hand side and try to manipulate it to the desired form. The next term is

$$\frac{1}{(k+1)[(k+1)+1]} = \frac{1}{(k+1)(k+2)}.$$

$$\frac{1}{1 \cdot 2} + \frac{1}{2 \cdot 3} + \frac{1}{3 \cdot 4} + \cdots + \frac{1}{k(k+1)} + \frac{1}{(k+1)(k+2)}$$

$$= \frac{k}{k+1} + \frac{1}{(k+1)(k+2)}$$

$$= \frac{k(k+2)}{(k+1)(k+2)} + \frac{1}{(k+1)(k+2)}$$

$$= \frac{k(k+2)+1}{(k+1)(k+2)}$$

$$= \frac{k^2 + 2k + 1}{(k+1)(k+2)}$$

$$= \frac{(k+1)^2}{(k+1)(k+2)}$$

$$= \frac{k+1}{k+2}$$

$$= \frac{k+1}{(k+1)+1} \qquad \text{The desired form}$$

Thus,

$$\frac{1}{1 \cdot 2} + \frac{1}{2 \cdot 3} + \frac{1}{3 \cdot 4} + \cdots + \frac{1}{n(n+1)} = \frac{n}{n+1}$$

is true for all natural numbers n.

Some conjectures are false and, hence, cannot be proven true. The next example demonstrates such a conjecture.

ILLUMINATOR SET C

Prove or disprove $1 + 2 + 3 + \cdots + n = \dfrac{n^2 + n - 6}{2}$.

1. Show true for $n = 1$.

$$1 = \frac{1^2 + 1 - 6}{2}$$

$$1 = \frac{1 + 1 - 6}{2}$$

$$1 = \frac{-4}{2}$$

$$1 = -2 \qquad \text{This is a false statement.}$$

Since step (1) is not true, this conjecture is false. It is interesting to note that if step (1) is ignored and only steps (2) and (3) are looked at, this conjecture would be pronounced true. Try it!

Counterexample

When a particular case has been found that shows a conjecture to be false, that case is said to provide a **counterexample** to the conjecture. When a counterexample is found to a conjecture, the conjecture is proven false. The case $n = 1$ provides a counterexample to the conjecture of Illuminator Set C.

Not all induction proofs can be done by simply adding the next term to both sides of the statement that is assumed to be true. In fact, there are a wide variety of approaches to many problems involving induction proofs as the next three examples illustrate.

Some conjectures are *not* true for $n = 1, 2, \ldots, p$, but are true for $n \geq p$. The next example illustrates this case.

ILLUMINATOR SET D

Induction can be used to prove statements about factorials. Prove that $n! > 2^n$, for all natural numbers, $n \geq 4$.

1. Show true for $n = 4$.

$$\text{since } 4! = 4 \cdot 3 \cdot 2 \cdot 1 = 24 \text{ and } 2^4 = 16, \ 2^4 = 16$$

Thus, $n! > 2^n$, is true for $n \geq 4$.

2. a. Assume true for $n = k$. That is, assume $k! > 2^k$ is true.

   b. Show true for $n = k + 1$. We can show the statement is true for $n = k + 1$ if we can show that $(k + 1)! > 2^{k+1}$ is true.

Since $k \geq 4$, $k + 1 > 2$ and $k! > 2^k$, by assumption.

Then since $(k + 1)! = (k + 1)k!$, and we have $k! > 2^k$, multiply the left side of the inequality in step 2a by $k + 1$, the right side by 2, and simplify.

$$k! > 2^k$$

$$(k + 1)k! > 2 \cdot 2^k$$

$$(k + 1)! > 2^{k+1}$$

which is the statement we wished to show.

Thus, $n! > 2^n$ is true for all natural numbers $n \geq 4$.

ILLUMINATOR SET E

Prove that $5^n + 3$, is divisible by 4, for all natural numbers.

1. Show true for $n = 1$.

$$5^1 + 3 = 5 + 3 = 8$$

which is divisible by 4.

Thus, $5^1 + 3$, is divisible by 4

2. a. Assume true for $n = k$.
   Assume that

   $$5^k + 3$$

   is divisible by 4 is true.

   b. Show true for $n = k + 1$.
   We need to show that $5^{k+1} + 3$ is divisible by 4.

   $$5^{k+1} + 3 = 5 \cdot 5^k + 3$$

The idea is to get two terms on the right-hand side that are divisible by 4. That will mean that the sum of those two terms is divisible by 4, which in turn will mean that the left-hand side, $5^{k+1} + 3$, is divisible by 4. Since by assumption (in step 2), $5^k + 3$ is divisible by 4, we need to manipulate the right-hand side of $5^{k+1} + 3 = 5 \cdot 5^k + 3$ into an expression that involves $5^k + 3$. We'll add and subtract 12 to and from the right-hand side since adding 12 will allow the first two terms to be factored resulting in one of the factors being $5^x + 3$. The details appear below:

$$5^{k+1} + 3 = 5 \cdot 5^k + 3$$

$$= 5 \cdot 5^k + 3 + 12 - 12$$

$$= 5 \cdot 5^k + 15 - 12$$

$$= 5(5^k + 3) - 12$$

Now, the first term, $5(5^k + 3)$ is divisible by 4 by the assumption made in step 2. Since 12 is also divisible by 4, the right-hand side is divisible by 4 so that the left-hand side is divisible by 4. Thus, $5^{k+1} + 3$ is divisible by 4.

Hence, $5^n + 3$ is divisible by 4 for all natural numbers, n.

ILLUMINATOR SET F

Prove that $\ln x^n = n \ln x$ for all integers $n \geq 0$.

1. Show true for $n = 0$.

$$\ln x^0 = \ln 1 = 0 = 0 \cdot \ln x$$

Thus, $\ln x^n = n \ln x$ for $n = 1$.

2. a. Assume true for $n = k$.
   Assume that

$$\ln x^k = k \ln x$$

   b. Show true for $n = k + 1$.
   We need to show that $\ln x^{k+1} = (k + 1) \ln x$.

$$\ln x^{k+1} = \ln(x^k \cdot x)$$

$$= \ln x^k + \ln x \quad \text{by} \quad \ln(xy) = \ln(x) + \ln(y)$$

$$= k \ln x + \ln x \quad \text{by the assumption in step 2}$$

$$= (k + 1) \ln x \quad \text{factor out } \ln x$$

Thus, $\ln x^{k+1} = (k + 1) \ln x$, and we conclude that $\ln x^n = n \ln x$ for all integers $n \geq 0$.

EXERCISES

For exercises 1–41, use mathematical induction to prove each statement.

1. $2+4+6+\ldots+2n = n(n+1)$

2. $3+6+9+\ldots++3n = \dfrac{3n(n+1)}{2}$

3. $1+4+9+16+\ldots+n^2 = \dfrac{(n+1)(n)(2n+1)}{6}$

4. $2+8+18+32+\ldots+2n^2 = \dfrac{(n+1)(n)(2n+1)}{3}$

5. $1+3+5+\ldots+(2n-1) = n^2$

6. $1+2+4+\ldots+2^{n-1} = 2^n - 1$

7. $1+4+7+\ldots+(3n-2) = \dfrac{n(3n-1)}{2}$

8. $1+5+9+\ldots+(4n-3) = n(2n-1)$

9. $1+4+4^2+4^3+\ldots+4^{n-1} = \dfrac{4^n-1}{3}$

10. $1^3+2^3+3^3+\ldots n^3 = \left[\dfrac{n(n+1)}{2}\right]^2$

11. $2^3+4^3+6^3+\ldots+(2n)^3 = 2n^2(n+1)^2$

12. $1^3+3^3+5^3+\ldots+(2n-1)^3 = n^2(2n^2-1)$

13. $1^2+3^2+5^2+\ldots+(2n-1)^2 = \dfrac{n(2n-1)(2n+1)}{3}$

14. $2+2^2+2^3+\ldots+2^n = 2^{n+1}-2$

15. $2^0+2^1+2^2+2^3+\ldots+2^{n-1} = 2^n-1$

16. $3+3^2+3^3+\ldots+3^n = \dfrac{3^{n+1}-3}{2}$

17. $e^x+e^{2x}+e^{3x}+\ldots+e^{nx} = \dfrac{e^{(n+1)x}-e^x}{e^x-1}, \quad x \ne 0$

18. $\left(\dfrac{1}{2}\right)^1+\left(\dfrac{1}{2}\right)^2+\left(\dfrac{1}{2}\right)^3+\ldots+\left(\dfrac{1}{2}\right)^n = 1-\left(\dfrac{1}{2}\right)^n$

19. $\left(\dfrac{1}{3}\right)^1+\left(\dfrac{1}{3}\right)^2+\left(\dfrac{1}{3}\right)^3+\ldots+\left(\dfrac{1}{3}\right)^n = \dfrac{3^n-1}{2\cdot 3^n}$

20. $\dfrac{1}{2!}+\dfrac{2}{3!}+\dfrac{3}{4!}+\cdots+\dfrac{n}{(n+1)!} = 1-\dfrac{1}{(n+1)!}$

21. $\sin x+\sin 3x+\sin 5x+\cdots+\sin(2n-1)x = \dfrac{\sin^2 nx}{\sin x}$

22. $\cos x+\cos 3x+\cos 5x+\cdots+\cos(2n-1)x = \dfrac{\sin 2nx}{2\sin x}$

23. $\cos(x+n\pi) = (-1)^n\cos x$

24. $\sin(x+n\pi) = (-1)^n\sin x$

25. $\tan(x+n\pi) = \tan x$

26. $n \le 2^{n-1}$

27. $n < 2^n$

28. $n^2+4 < (n+1)^2, \quad n>2$

29. $n^3 > (n+1)^2, \quad n\ge 3$

30. $3n < 2^n, \quad n>4$

31. $2^n > (n+1)^2, \quad n>6$

32. $2^n > 3n, \quad n\ge 4$

33. $4^n > 2^n+3^n, \quad n\ge 2$

34. $n^2 < 2^n, \quad n\ge 5$

35. $3^n < n!, \quad n\ge 7$

36. $n! > 2^n, \quad n\ge 4$

37. $n! > 10n, \quad n\ge 5$

38. $2^{n-1} \le n!$

39. $1+n = n+1$

40. $\ln(x_1\cdot x_2\cdots x_n) = \ln x_1+\ln x_2+\cdots\ln x_n$

41. $(xy)^n = x^n y^n$

42. Prove that 3 is a factor of 4^n-1 for all natural numbers n.

43. Prove that 4 is a factor of 5^n-1 for all natural numbers n.

44. Prove that 3 is a factor of n^3+2n for all natural numbers n.

45. Prove that 5 is a factor of n^5-n for all natural numbers n.

46. Prove that 8 is a factor of $9^n + 7$ for all natural numbers n.

47. Prove that $n^2 + n$ is an even number for all natural numbers n.

48. Give a counterexample to show that $n^2 < 2^n$ is not true for all natural numbers n.

49. Give a counterexample to show that $3^n \geq n!$ is not true for all natural numbers n.

50. Prove that if the first person to possess a certain car is a legal owner, and every change of possession thereafter is a transfer of legal ownership, then every person ever to possess the car will be a legal owner. (Assume that only one person possesses the car at a time.)

51. Mr. Smith has a gene that has the property that the children of anyone who has this gene will also have the gene. Prove that every one of Mr. Smith's descendants will have the same gene.

52. Give a counterexample to the claim that any group of people big enough to field a full-sized baseball team is big enough to field a full-sized football team.

11.2 Sequences and Series

Introduction

Sequences

Graphs of Sequences

Recursion Formulas

Series

Introduction

Ordered lists of numbers occur frequently in both theoretical and applied mathematics. These lists are called sequences and are important in calculus and other branches of mathematics. In this section we first examine the concept of a sequence, establishing a definition and some notation. Then we examine the relationship between the numbers of a sequence and study what happens when the numbers of the sequence are added together.

Sequences

Examine the list of numbers $2, 5, 8, 11, \ldots$, and try to specify the next number in the list. In making the guess for this list, and for all subsequent lists, let's agree upon two conventions.

1. Our guess must be based upon some underlying pattern within the list, and

2. A likely pattern is revealed in the portion of the list displayed.

In the above list the next number is 14, as each number is 3 more than the number before it. The guess 14 would not have been so obvious if the list had appeared as $5, 2, 11, 8, \cdots$, as this list has no apparent order to it.

Examining the original list a little more closely, we see that there is a first term, a second term, a third term, a fourth term, and so on. To generalize, we shall use the notation

$$a_1, \ a_2, \ a_3, \ a_4, \ \cdots, \ a_n \cdots$$

to indicate the first, second, third, fourth, and nth terms, respectively. The quantities $a_1, a_2, a_3 \cdots$ are the *terms* of the sequence. The subscripts are called *indices* and they specify the position of a term in the sequence. The term a_n is the **general term**, and it is in the form of a rule from which any term in the sequence can be obtained. A particular term in the sequence can be obtained by substituting its index (position number) into the formula for the general term. Thus, for this list of numbers we have $a_1 = 2$, $a_2 = 5$, $a_3 = 8$, and $a_4 = 11$.

General Term

Stripping away some of the notation, we have

Term Number	Value
1	2
2	5
3	8
4	11

This table bears a striking resemblance to tables we used when working with *functions* in chapters 1 and 2. It is this resemblance that leads us to define an ordered list, or *sequence*, as a function.

Sequence

Sequence

A **sequence** is a function. The domain of a sequence is the natural numbers. The general term of a sequence is precisely the function that defines the sequence.

ILLUMINATOR SET A

Each of the following ordered lists of numbers is a sequence. Can you guess the next term? The general term?

1. $1, 2, 3, 4, \cdots$

2. $2, 4, 6, 8, 10, \cdots$

3. $-2, 3, -4, 5, \cdots$

4. $\dfrac{1}{2}, \dfrac{1}{4}, \dfrac{1}{8}, \dfrac{1}{16}, \cdots$

5. $2, 2.2, 2.22, 2.222, \cdots$

6. $7, 5, 3, 1, -1, \cdots$

7. $\dfrac{2}{3}, \dfrac{4}{9}, \dfrac{8}{27}, \dfrac{16}{81}, \cdots$

8. $1, 1, 2, 3, 5, 8, 13, \cdots$

The calculator program in Window 11.1 is convenient for computing and expressing the first 10 terms of a sequence. The sequence rule is entered as Y1 in the Y= menu. If you want more or less than 10 terms, simply change the numerical value in the line IS>(X,10) to the number you wish.

```
Prog1:SEQ
:1→X
:Lbl 1
:Disp Y1
:IS>(X,10)
:Goto 1
:End
```

Window 11.1

The important question at this point is how does one specify the function. To specify the function (that is, to determine the formula for the general term), you must determine the relationship between the index and the term value. This may or may not be possible. Once the general term a_n is specified, the sequence is expressed concisely using *brace notation* as

$$\{a_n\}_{n=1}^{n=+\infty}$$

EXAMPLE SET A

1. Specify, if possible, the general term of the sequence $2, 4, 6, 8, \cdots$.

   Solution:
   We need to look for a relation between the index (term number) of each term and the value of that term. Tables are often helpful at this point as they can help us to visualize the underlying pattern.

Index	Term	Relation
1	2	$2 \cdot \underbrace{1}_{\text{index}} = \underbrace{2}_{\text{term}}$
2	4	$2 \cdot \underbrace{2}_{\text{index}} = \underbrace{4}_{\text{term}}$
3	6	$2 \cdot \underbrace{3}_{\text{index}} = \underbrace{6}_{\text{term}}$
4	8	$2 \cdot \underbrace{4}_{\text{index}} = \underbrace{8}_{\text{term}}$

   The underlying pattern appears to be that the value of each term is exactly 2 times the value of its index. Based on the information given, we hypothesize that the general term is $a_n = 2n$.

2. Specify, if possible, the general term of the sequence $\dfrac{1}{2}, \dfrac{1}{4}, \dfrac{1}{8}, \dfrac{1}{16}, \cdots$.

   Solution:
   Each term is a fraction with numerator 1. Each denominator appears to be a power of 2. Thus, we'll look for a pattern involving $\frac{1}{2}$. The table helps make it apparent that the value of a term can be obtained by raising $\frac{1}{2}$ to the value of the index of that term.

Index	Term	Relation
1	2	$\left(\dfrac{1}{2}\right)^{\overbrace{\text{index}}^{1}} = \underbrace{\dfrac{1}{2}}_{\text{term}}$
2	4	$\left(\dfrac{1}{2}\right)^{\overbrace{\text{index}}^{2}} = \underbrace{\dfrac{1}{4}}_{\text{term}}$
3	6	$\left(\dfrac{1}{2}\right)^{\overbrace{\text{index}}^{3}} = \underbrace{\dfrac{1}{8}}_{\text{term}}$
4	8	$\left(\dfrac{1}{2}\right)^{\overbrace{\text{index}}^{4}} = \underbrace{\dfrac{1}{16}}_{\text{term}}$

Based on the information given, we hypothesize that the general term is $a_n = \left(\dfrac{1}{2}\right)^n$, or $a_n = \dfrac{1}{2^n}$.

The general term to some sequences may be difficult or impossible to determine. The next example illustrates this point.

EXAMPLE SET B

1. Specify, if possible, the general term of the sequence

$$1, \ 1, \ 2, \ 3, \ 5, \ 8, \ 13, \ \cdots$$

Solution:

This is the well-known *Fibonacci* sequence named after Leonardo Fibonacci who introduced it in the year 1202. In this sequence, each term, after the first two, is the sum of the two immediate previous terms. A formula for the general term of this sequence was so difficult to determine that one was not found until 1724 by Daniel Bernoulli. The general term of the Fibonacci sequence is

$$a_n = \frac{1}{\sqrt{5}}\left[\left(\frac{1+\sqrt{5}}{2}\right)^n - \left(\frac{1-\sqrt{5}}{2}\right)^n\right]$$

2. Specify, if possible, the general term of the sequence

$$4, \ -9, \ 6, \ -10, \ 10, \ 0, \ 2, \ \cdots$$

Solution:

This sequence was generated by having a computer produce seven random numbers. It is unlikely that a formula for the general term can be found.

Graphs of Sequences

Since a sequence is a function, we may obtain information about its behavior by examining its graph. For example, the graph of the sequence

$$\left\{\frac{n^2}{5}\right\}_{n=1}^{n=+\infty}$$

is the graph of the function

$$f(n) = \frac{n^2}{5}, \quad n = 1, \ 2, \ 3, \ \ldots$$

Since the defining expression, $\dfrac{n^2}{5}$, is defined only for natural numbers, the graph of the sequence consists of infinitely many individual points as shown in Figure 11.2.

Each of the points on the sequence graph lies on the graph of the corresponding continuous function

$$f(x) = \frac{x^2}{5}, \quad x \geq 1$$

Figure 11.2

Figure 11.3 shows the graph of $f(x)$ with the graph of $f(n)$ as a subset. The behavior of the sequence function resembles the behavior of the corresponding continuous function. In the next section, we will investigate the behavior of sequences more thoroughly.

Figure 11.3

Recursion Formulas

Quite often it is possible to relate one term of a sequence to the immediate preceding term. For example, in the sequence $1, 4, 7, 10, 13, \cdots$, we see that each term is exactly three more than the immediate preceding term. Symbolically, we express this as

$$a_n = a_{n-1} + 3$$

Recursion Formula

The formula $a_n = a_{n-1} + 3$ is an example of a **recursion formula**. Recursion formulas are used to generate a term of a sequence in terms of the preceding term. They are useful in that if the general term cannot be determined, the terms of the sequence can still be generated using the recursion formula. The advantage of the general term formula over the recursion formula is that the general term formula will generate any term from only the term number (index), whereas to generate that same term using the recursion formula, the preceding term must be generated first (and of course the preceding term to that term, and so on). Also, the general term formula is sufficient to specify a sequence whereas the recursion formula is not. For example, the recursion formula $a_n = 2a_{n-1} - 5$ does not specify a sequence until the first term of the sequence is noted. The formula $a_n = 2a_{n-1} - 5$, with $a_1 = 2$, specifies the sequence $2, -1, -7, -19, \cdots$. With $a_1 = 3$, it specifies the sequence $3, 1, -3, -11, \cdots$. Recursion formulas are used extensively in many branches of mathematics (particularly in chaos and fractal theory) and computer science.

Series

The sum of the terms of a sequence is called a series.

Series

Series
The indicated sum of the terms of a sequence is a **series**.

If the sequence consists of a finite number of terms, the series is called a *finite series*, and its sum can be determined by performing the addition. If the sequence is infinite, the series is called an *infinite series*, and the sum may or may not exist. Infinite series are considered in detail in calculus. With one exception in the next section, we will restrict our attention to finite series.

Since summations occur so frequently in mathematics, a compact notation has been developed to indicate a sum. The upper case Greek letter sigma, $\sum$, is used to indicate a sum (**s**igma for **s**um).

The summation symbol $\sum$ is used along with the general term of the sequence. The notation tells us at which term to begin the addition and at which term to stop. To illustrate,

$$\sum_{i=1}^{4} a_i{}^2$$

indicates that the first four terms of the sequence

$$\{a_n^2\}_{n=1}^{n=+\infty}$$

are to be added.

In the expression below the $\sum$ symbol, the letter i is called the *index of summation*. (The word *index* means "that which points out.") It points out the numbers in the domain that are to be used. The first number in the domain appears below the $\sum$ and is called the *initial value*, and the last number in the domain appears above the $\sum$ and is called the *terminal value*. All natural numbers between and including the initial and terminal values are used. The letter used for the index of summation can be any letter at all. It is simply a variable that represents the domain values. It should, however, match the letter used in the general term formula.

Thus, the *sigma notation* is simply an instruction to evaluate the given function (general term formula) for the specified domain values, and then to add these quantities together.

EXAMPLE SET C

Find each indicated sum.

1. $\sum_{i=1}^{5}(i+1)^2$.

Solution:

$$\sum_{i=1}^{5}(i+1)^2 = (1+1)^2 + (2+1)^2 + (3+1)^2 + (4+1)^2 + (5+1)^2$$

$$= 2^2 + 3^2 + 4^2 + 5^2 + 6^2$$

$$= 4 + 9 + 16 + 25 + 36$$

$$= 90$$

2. $\sum_{k=2}^{4}(6k-5)^{k-1}$.

Solution:

$$\sum_{k=2}^{4}(6k-5)^{k-1} = (6 \cdot 2 - 5)^{2-1} + (6 \cdot 3 - 5)^{3-1} + (6 \cdot 4 - 5)^{4-1}$$

$$= 7^1 + 13^2 + 19^3$$

$$= 7 + 169 + 6859$$

$$= 7035$$

EXAMPLE SET D

Use the sigma notation to express each series.

1. $2 + 4 + 6 + 8$.

Solution:

The general term of the sequence appears to be $2n$. Thus,

$$2 + 4 + 6 + 8 = \sum_{n=1}^{4} 2n$$

2. $1 + 8 + 27 + 64 + 125 + 216$.

Solution:

The general term of the sequence appears to be n^3. Thus

$$1 + 8 + 27 + 64 + 125 + 216 = \sum_{k=1}^{6} k^3$$

Notice that since we are using k as the variable for the index of summation, we express the general term formula using k rather than n.

The program shown (in part) in Window 11.2 can be used to find the sum of the first n terms of a series.

```
Prmg2:SERSUM
:Disp "N="
:Input N
:1→X
:0→S
:Lbl 1
:S+Y₁→S
:IS>(X,N)
```

Window 11.2

The program's last three lines are as follows:

```
:Goto 1
:Disp S
:End
```

EXERCISES

For exercises 1–8, write the next three terms of each sequence.

1. 5, 7, 9, $\cdots$

2. 6, 10, 14, $\cdots$

3. -5, 4, 13, $\cdots$

4. 4, 1, -2, $\cdots$

5. $\dfrac{1}{3}, \dfrac{1}{9}, \dfrac{1}{27}, \cdots$

6. $\dfrac{2}{5}, \dfrac{4}{5}, \dfrac{8}{5}, \cdots$

7. 1.3, 1.31, 1.313, 1.3131, $\cdots$

8. 2.7, 2.72, 2.727, 2.7272, $\cdots$

For exercises 9–21, write the first five terms of each sequence.

9. $\left\{(-1)^n\right\}_{n=1}^{n=\infty}$

10. $\left\{(-2)^n\right\}_{n=1}^{n=\infty}$

11. $\left\{\dfrac{n}{n+1}\right\}_{n=1}^{n=\infty}$

12. $\left\{\dfrac{n-1}{n+1}\right\}_{n=1}^{n=\infty}$

13. $\left\{\dfrac{n^2}{2n+1}\right\}_{n=1}^{n=\infty}$

14. $\left\{\ln\left(\dfrac{1}{n}\right)\right\}_{n=1}^{n=\infty}$

15. $\left\{1+(-1)^n\right\}_{n=1}^{n=\infty}$

16. $\left\{\dfrac{(-1)^{n+1}}{n^2}\right\}_{n=1}^{n=\infty}$

17. $\left\{\dfrac{\pi^n}{3^n}\right\}_{n=1}^{n=\infty}$

18. $\left\{\dfrac{(n+1)(n+2)}{2n^2}\right\}_{n=1}^{n=\infty}$

19. $\left\{\dfrac{\ln n}{n}\right\}_{n=1}^{n=\infty}$

20. $\{a_n\}_{n=1}^{n=\infty} = \begin{cases} (-2)^n & \text{if } n \text{ is even;} \\ (-3)^n & \text{if } n \text{ is odd} \end{cases}$

21. $\{a_n\}_{n=1}^{n=\infty} = \begin{cases} \dfrac{n+1}{n-1} & \text{if } n \text{ is even;} \\ \dfrac{n-1}{n+1} & \text{if } n \text{ is odd} \end{cases}$

22. The Fibonacci sequence is defined recursively as $a_{n+2} = a_n + a_{n+1}$, for $n \geq 1$, and $a_1 = 1$ and $a_2 = 1$. Write the first 10 terms of the sequence.

For exercises 23–34, express each sequence using the $\{a_n\}$ notation.

23. 1, 2, 3, 4, $\cdots$

24. 3, 6, 9, 12, $\cdots$

25. 4, 8, 16, $\cdots$

26. 9, 27, 81, $\cdots$

27. 0.2, 0.02, 0.002, 0.0002, $\cdots$

28. 0.31, 0.031, 0.0031, 0.00031, $\cdots$

29. $\dfrac{1}{3}, \dfrac{2}{4}, \dfrac{3}{5}, \dfrac{4}{6}, \cdots$

30. $0, \dfrac{1}{2^2}, \dfrac{2}{3^2}, \dfrac{3}{4^2}, \dfrac{4}{5^2}, \cdots$

31. $3, \dfrac{3}{2}, \dfrac{3}{4}, \dfrac{3}{8}, \dfrac{3}{16}, \cdots$

32. $-1, 2, -3, 4, -5, 6, \cdots$

33. $-2, 4, -8, 16, -32, \cdots$

34. $2, -4, 8, -16, 32, \cdots$

For exercises 35–40, find each indicated sum.

35. $\displaystyle\sum_{n=1}^{4} 2n$

36. $\displaystyle\sum_{n=1}^{5} 3n$

37. $\displaystyle\sum_{n=3}^{5} (n+1)$

38. $\displaystyle\sum_{n=1}^{6} (2n)^2$

39. $\displaystyle\sum_{n=1}^{4} (2n-1)$

40. $\displaystyle\sum_{n=1}^{5} (3n-1)$

For exercises 41–44, find and compare the sums of each series.

41. $\displaystyle\sum_{n=1}^{6} (-1)^n$

42. $\displaystyle\sum_{n=1}^{7} (-1)^n$

43. $\displaystyle\sum_{n=1}^{8} (-1)^n$

44. $\displaystyle\sum_{n=1}^{9} (-1)^n$

45. Use the information you created from exercises 37–40 to make a statement about $\displaystyle\sum_{n=1}^{\infty} (-1)^n$.

For exercises 46–51, express each series using the $\displaystyle\sum_{n=1}^{n}$ notation.

46. $1 + (-3) + 9 + (-27) + 81 + (-243)$

47. $\dfrac{1}{2} - \dfrac{1}{4} + \dfrac{1}{8} - \dfrac{1}{16} + \dfrac{1}{32} - \dfrac{1}{64}$

48. $\sin(2) + \sin(3) + \sin(4) + \sin(5) + \sin(6) + \sin(7) + \sin(8)$

49. $-\cos(3) + \cos(4) - \cos(5) + \cos(6) - \cos(7) + \cos(8) - \cos(9) + \cos(10)$

50. $0 + 1 + 2 + 3 + f(1) + f(2) + f(3) + f(4) + 2f(1) + 2f(2) + 2f(3) + 2f(4)$

51. $\dfrac{1}{2} + \dfrac{x^2}{4} + \dfrac{x^4}{8} + \dfrac{x^6}{16} + \cdots + \dfrac{x^{2n-2}}{2^n}$

52. Prove the summation property $\displaystyle\sum_{k=1}^{n} c = cn$,

    where c is a constant, and use it to evaluate $\displaystyle\sum_{k=1}^{10} 6$.

53. Prove the summation property
    $$\sum_{k=1}^{n}(a_k + b_k) = \sum_{k=1}^{n} a_k + \sum_{k=1}^{n} b_k.$$

54. A bank pays 6% annual interest compounded quarterly on its savings accounts, which means a quarter of this percentage is computed and paid on the balance every three months. On January 1, a one-time deposit is made in such an account and the account is left untouched from then on.

    (a). Write a recursion formula for the balance in the account after n quarters.

    (b). If the initial deposit is $120, what is the balance after seven quarters?

55. The world high-jump records from 1920 to 1960 show approximately this progression: 1920, 6 ft 7 in; 1930, 6 ft $8\frac{1}{2}$ in; 1940, 6 ft 10 in; 1950, 6 ft $11\frac{1}{2}$ in; 1960, 7 ft 1 in. What would you expect the record to be in 1990? In the year 2000?

56. On a number line, an electron is placed at $x = 2$. Six protons are placed along the interval $[-5, 0]$, one at every integer value of x. If the attractive force between a proton and electron is k/d^2, where d is the distance between the two and k is a constant, what is the total force from the six protons on the electron? (It is a sum of the attractions from the individual protons.) Write the answer using decimal form.

11.3 Convergence and Divergence of Sequences

Introduction

Apparent Convergence or Divergence

Intuitive Notion of Convergence

Formal Notion of Convergence

Introduction

Very often in engineering and scientific applications, it is important to know the long-term behavior of a sequence. It can be important to know if, once one is far enough out into the sequence, the terms tend to come close to some particular number. In this section, we investigate the concepts of convergence and divergence of a sequence.

Apparent Convergence or Divergence

Converge

Diverge

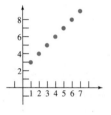

Figure 11.4

In some sequences, when one is far enough out in the sequence and as n, the term number, gets larger and larger, *all* the terms of the sequence tend to come arbitrarily close to a particular number. Such sequences are said to **converge** to that particular value. The terms of other sequences may not tend converge to any one particular number. These types of sequences are said to **diverge**.

Figure 11.4 and Figure 11.5 show the graphs of four different sequences. The sequences illustrated in Figure 11.4 and the leftmost graph in Figure 11.5 appear to be divergent. In the first sequence, the term values appear to grow without bound as n becomes larger. In the second sequence (the leftmost graph in Figure 11.5), the terms appear to oscillate about 2 and -2. The sequences illustrated in the middle and rightmost graphs of Figure 11.5 appear to be convergent. Both sequences appear to converge to 1, the first decreasing toward 1, and the second oscillating about 1, but approaching 1. We have been

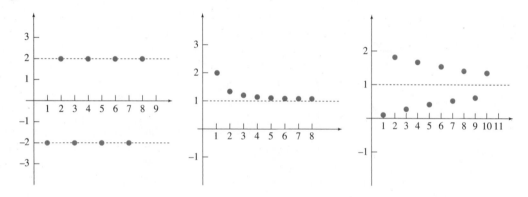

Figure 11.5

careful to say that these sequences *appear* to converge or diverge. Examining a list of terms or looking carefully at a graph *does not* prove that a sequence converges or diverges. It may be that once we are farther out into the sequence, the terms are no longer arbitrarily close to the number to which we thought the sequence converged.

Intuitive Notion of Convergence

Although the notion of convergence is expressed precisely, we can also informally define it.

Informal Definition of Convergence

Informal Definition of Convergence

Informally, we say that a sequence $\{a_n\}_{n=1}^{n=\infty}$ converges to the number L if, once we are out far enough out in the sequence, each term is arbitrarily close to the number L.

We need to discuss two concepts, the concept of *out far enough*, and the concept of *arbitrarily close*.

1. By *out far enough*, we mean that after some given term number, a particular behavior takes place. It is common practice to denote the given term number with the letter N. We can express the concept of *out far enough* by writing $n > N$, where both n and N are term numbers.

2. By *arbitrarily close* to L we mean that the difference between a_n and L can be made arbitrarily small simply by taking n large enough. It is a common practice to denote an arbitrarily small positive number with the symbol ϵ (pronounced *epsilon*). The difference between the term value a_n and L is expressed as $|a_n - L|$. We use absolute value because the term values may approach L from above or below. To express that the difference between a_n and L is less than some arbitrarily specified number ϵ, we write $|a_n - L| < \epsilon$.

It is helpful to think of determining convergence as a sort of game. Someone challenges you by arbitrarily choosing a number (ϵ). If, by going out far enough in the sequence $(n \geq N)$, you can make the difference between all the remaining term values and L less than that arbitrarily chosen number, the sequence converges. If the difference cannot be made smaller than that arbitrarily chosen number, the sequence diverges. Figure 11.6 illustrates the concept of convergence relative to this intuitive notion.

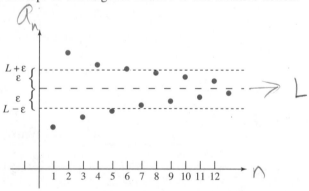

Figure 11.6

In Figure 11.6, $N = 7$, since after term number 7, every term of the sequence is within ϵ units of L. Symbolically, when $n > N$, $|a_n - L| < \epsilon$.

Formal Notion of Convergence

With this informal concept in mind, we can formally define convergence.

Convergence of a Sequence

A sequence $\{a_n\}_{n=1}^{n=+\infty}$ converges to the number L if for any positive number ϵ there is a positive number N such that when $n > N, |a_n - L| < \epsilon$.

Convergence of a Sequence

So that we may investigate the concept of convergence, we restrict our attention to certain *well-behaved* sequences. We examine *increasing sequences* that converge to a number L by increasing to it, and *decreasing sequences* that converge to a number L by decreasing to it. We also examine *alternating sequences* that converge to a number L by alternately increasing and decreasing to L. We make this restriction because it allows us to be sure that once we are out far enough into the sequence so that the terms are arbitrarily close to L, the terms will not grow large and diverge away from L.

Increasing Sequence

A sequence $\{a_n\}_{n=1}^{n=+\infty}$ is an **increasing sequence** if $a_1 < a_2 < a_3 < \cdots < a_n < \cdots$.

Increasing Sequence

The definition states that an increasing sequence is a sequence in which each term is bigger than its immediate previous term, and thus, all other previous terms. The sequence

$$1, \frac{4}{3}, \frac{6}{4}, \frac{8}{5}, \frac{10}{6}, \cdots, \frac{2n}{n+1}, \cdots$$

is an increasing sequence that converges to 2.

Decreasing Sequence

Decreasing Sequence

A sequence $\{a_n\}_{n=1}^{n=+\infty}$ is a **decreasing sequence** if $a_1 > a_2 > a_3 > \cdots > a_n > \cdots$,

---

The definition states that a decreasing sequence is a sequence in which each term is smaller than its immediate previous term and thus, all other previous terms. The sequence

$$1, \; \frac{1}{2}, \; \frac{1}{6}, \; \frac{1}{24}, \; \frac{1}{120}, \cdots, \; \frac{1}{n!}, \cdots$$

is a decreasing sequence that converges to 0. → *never will become zero*

Alternating Sequence

Alternating Sequence

An **alternating sequence** is a sequence of the form $(-1)^n\{a_n\}_{n=1}^{n=+\infty}$ and it is a sequence in which the terms alternate in sign.

The sequence

$$1, \; -0.9, \; 0.81, \; -0.729, \; 0.6561, \; -0.59049, \; \cdots, \; (-1)^{n-1}\left(\frac{9}{10}\right)^{n-1}, \; \cdots$$

$(-1)^{n-1}\left(\frac{9}{10}\right)^{n-1}$

is an alternating sequence that converges to 0.

The next example is intended to demonstrate how the calculator can be used to help us suggest whether or not a sequence converges, and if we think it converges, to which number it converges.

We can determine the apparent behavior of the sequence by constructing its graph. We will construct the graph of a sequence by placing the calculator in DOT mode, entering the defining expression as Y_1, and setting the RANGE to be $0 \le x \le n$ where $n = 95, 94, 127$ for the TI-81,TI-82, TI-85, and $-3 \le y \le 3$. (You may recall that setting $0 \le x \le 95$ directs the calculator to compute integer values of x, which, with the exception of 0, are the domain values we wish to have computed.)

EXAMPLE SET A

Construct the graph of the sequence $\left\{ \dfrac{2n}{n+1} \right\}_{n=1}^{n=\infty}$, and use the graph to (1) suggest the number L to which number it converges, and (2) determine how far out in the sequence one must be so that all the remaining terms are less than 0.04 units away from the number L found in part (1).

Solution:

1. The graph indicates that the sequence increases quickly for small values of n, but then levels off and approaches 2. This behavior can be made apparent by activating the TRACE cursor, moving it to $n = 1$, then watching the output values increase toward 2 as the cursor moves to the right.

2. We wish to know at which term all the remaining terms are within 0.04 units of 2. That is, we wish to know the value of N for which, when $n > N$,

$$\left| \frac{2n}{n+1} - 2 \right| \leq 0.04$$

We can direct the calculator to *display* this term number in the following way. Use the absolute value function abs to enter the expression $\left| \dfrac{2n}{n+1} - 2 \right|$ as, say, Y_4. Choose the TEST feature and select the $\leq$ symbol. Then type 0.04. The TEST feature computes the value of the expression and returns a 0 if the entered statement is false, and a 1 if it is true. In the graphing mode, it will construct the horizontal line $y = 0$ when the entered statement is false, and horizontal line $y = 1$ when it is true. Enter the statement and construct the graph. Activate the TRACE cursor, move it to Y_4 by pressing either the up or down arrow keys, then move it left or right until it first appears on the line $y = 1$. This value of N, 49 in this case, is the first value for which all the terms of the sequence are within 0.04 units of 2. (If you move the cursor back to the graph of the sequence, you should notice that the difference between the displayed y-value and 2 is less than 0.04.) You will find it interesting to redraw your graph so it includes the lines $y = 2 - 0.04 = 1.96$ (using, say, Y_2) and $y = 2 + 0.04 = 2.04$ (using, say, Y_3). You should reset the range to be $1.9 \leq y \leq 2.1$ (Why?). The TEST feature is no longer visible, but is still active. Using the TRACE cursor, you can then see that when $n < 49$, the sequence values are beyond 0.04 units from 2, and that when $n \geq 49$, they are within 0.04 units of 2. This graph is illustrated in Figure 11.7.

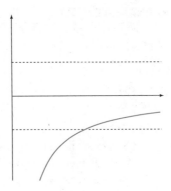

Figure 11.7

In this final example, we demonstrate how to symbolically find the term number N for which all remaining terms are within any specified positive number of units ϵ of the number L of a convergent sequence.

EXAMPLE SET B

Given that the sequence $\left\{\dfrac{2n}{n+1}\right\}_{n=1}^{n=\infty}$ converges to 2, find the term number N for which all remaining terms are within ϵ units of 2, where $\epsilon > 0$.

Solution:

We need to find n so that $\left|\dfrac{2n}{n+1} - 2\right| < \epsilon$. To do so, we'll solve this absolute value inequality for n in terms of ϵ.

$$\left|\frac{2n}{n+1} - 2\right| < \epsilon$$

$$\left|\frac{2n}{n+1} - \frac{2(n+1)}{n+1}\right| < \epsilon$$

$$\left|\frac{2n - 2n - 2}{n+1}\right| < \epsilon$$

$$\left|\frac{-2}{n+1}\right| < \epsilon$$

$$\frac{|-2|}{|n+1|} < \epsilon$$

$$\frac{2}{n+1} < \epsilon \quad \text{since } n+1 \text{ is positive}$$

$$2 < \epsilon(n+1)$$

$$\frac{2}{\epsilon} < n+1$$

$$\frac{2}{\epsilon} - 1 < n \quad \text{or}$$

$$n > \frac{2}{\epsilon} - 1$$

We let N be any positive integer such that $N > \frac{2}{\epsilon} - 1$.

Thus, if ϵ is any positive number, and if $N = \frac{2}{\epsilon} - 1$, then all the terms of the sequence for which $n \geq N$, $\left| \frac{2n}{2+1} - 2 \right| < \epsilon$. In particular, if $\epsilon = 0.04$, then

$$N = \frac{2}{0.04} - 1 = 49$$

If we let $N = 49$, then we get precisely the result we got in the last example using our calculator.

EXERCISES

In exercises 1–19, each sequence function is well-behaved in the sense that each either diverges or converges to the value to which it appears to converge graphically. Use your calculator to determine if the given sequence converges or diverges. If it converges, specify to which number it converges and the number N such that $|a_n - L| < \epsilon$ for all $n > N$ for the given ϵ. Since N is not unique, answers may vary.

1. $\dfrac{8n^3 + 5n^2 + 7}{4n^3 - 2n + 2}$, $\quad \epsilon = 0.02$

2. $\ln(n+4) - \dfrac{1}{2}\ln(n)$, $\quad \epsilon = 0.01$

3. $(-1)^n \dfrac{n+1}{n}$, $\quad \epsilon = 0.01$

4. $2 + (-1)^n \left(\dfrac{9}{10} \right)^n$, $\quad \epsilon = 0.01$

5. $-2 + (-1)^n \left(\dfrac{7}{8} \right)^n$, $\quad \epsilon = 0.01$

6. $\dfrac{10n - 1}{4n + 5}$, $\quad \epsilon = 0.05$

7. $\dfrac{2n}{2n + 1}$, $\quad \epsilon = 0.01$

8. $\dfrac{n^2 + 1}{3n(n+1)}$, $\quad \epsilon = 0.05$

9. $2 + e^{1/n}$, $\quad \epsilon = 0.02$

10. $\sin\left(\dfrac{n\pi}{2} \right)$, $\quad \epsilon = 0.05$

11. $\cos\left(\dfrac{n\pi}{2} \right)$, $\quad \epsilon = 0.04$

12. $\dfrac{3+(-1)^n \sqrt{n}}{n+2}$, $\epsilon = 0.05$. (Hint: This sequence converges, but takes a long time to do so. Keep the right arrow key depressed to move the TRACE cursor to larger and larger values of n, and until the test line $y = 1$ appears.)

13. $2.3 + \dfrac{(-1)^n}{2^n}$, $\epsilon = 0.0001$

14. $\dfrac{e^n - e^{-n}}{e^n + e^{-n}}$, $\epsilon = 0.0001$

15. $\left(1 + \dfrac{1}{n}\right)^n$, $\epsilon = 0.01$

16. $\dfrac{(n+1)!}{n!}$, $\epsilon = 0.02$

17. $\dfrac{3^n}{n^3}$, $\epsilon = 0.02$

18. $\dfrac{\ln(n)}{\ln(2n)}$, $\epsilon = 0.03$. (Hint: This sequence converges, but does so very slowly.)

19. $\dfrac{n!}{10^{6n}}$, $\epsilon = 0.02$

In exercises 20–27, each sequence converges. This time, however, you are given a value of N and you will use your calculator to specify the value of ϵ such that $|a_n - L| < \epsilon$ for all $n > N$.

20. $\dfrac{1-2n}{1+2n}$, $N = 11$

21. $\dfrac{\ln(n)}{\sqrt{n}}$, $N = 76,632$

22. $\dfrac{n!}{n^n}$, $N = 7$

23. $\dfrac{2n + \sin(n)}{n + \cos(5n)}$, $N = 279$

24. $\dfrac{(-4)^n}{n!}$, $N = 14$

25. $1.5 + \ln\left(1 + \dfrac{1}{n}\right)^n$, $N = 9$

26. $\dfrac{\ln(2n+1)}{n}$, $N = 108$

27. $\dfrac{\ln(n^2 + n)^{1/2}}{n}$, $N = 734$

28. Suppose the sequence function $A(n) = 240 + \dfrac{100,000}{n}$ represents the average cost $A(n)$ of producing n bicycles. What is the trend in the average cost per bicycle as the number of bicycles produced is increased?

29. The sequence function $\overline{C}(n) = 0.89 + \dfrac{1500}{n}$ represents the average cost $\overline{C}$ of producing n tubes of toothpaste. What is the trend in the average cost per tube as the number of tubes produced is increased?

30. The population $P(n)$ (in thousands of people) of a small valley community n years from 1990 is given by the sequence function

$$P(n) = \dfrac{32n^2 + 97n + 135}{n^2 + 5n + 60}$$

What is the expected population of the community in the long term?

31. The concentration $C(t)$ (in milligrams per liter) of a drug in the human body is related to the time t (in hours) that the drug has been in the body by the sequence function

$$C(t) = \dfrac{34t}{t^2 + 2}$$

What will be the concentration of the drug in the body in the long term?

32. The amount $A(t)$ of pollution in a lake is related to the time t in years by the sequence function

$$A(t) = \left(\dfrac{t+3}{t+1}\right)^t$$

What is the expected amount of pollution in the lake in the long term?

For each of the following convergent sequences in exercises 33–36, find a value of N in terms of ϵ such that for every $n > N$, $|a_n - L| < \epsilon$.

33. $\left\{\dfrac{5n}{5n+6}\right\}_{n=1}^{n=\infty}$ converges to 1

34. $\left\{\dfrac{14n-21}{2n+5}\right\}_{n=1}^{n=\infty}$ converges to 7

35. $\left\{\dfrac{6n^2-1}{2n^2-1}\right\}_{n=1}^{n=\infty}$ converges to 3

36. $\left\{\dfrac{10n^2-3}{2n^2-3}\right\}_{n=1}^{n=\infty}$ converges to 5

37. Do the high-jump data from exercise 55 in the previous section suggest a convergent or

divergent sequence? Does anything strike you as odd about the answer?

38. A jet pilot is trying to break the sound barrier, denoted by mach number 1. As the seconds go by, she reads her airspeed gauge: mach 0.75, 0.85, 0.90, 0.925, 0.9375, …. If this pattern holds up, will the airspeed eventually exceed mach 1? Justify your answer.

39. Does the function $A(n) = \ln n$ determine a convergent or a divergent sequence? Compare and discuss the answers you get using mathematical reasoning and using a graphing calculator.

11.4 Arithmetic and Geometric Series

Introduction

Arithmetic Sequences

Arithmetic Series

Geometric Sequences

Geometric Series

Infinite Geometric Series

Introduction

Certain sequences and series appear so often in theoretical and applied problems that they are worth studying in some detail. In this section, you are introduced to arithmetic sequences and series, and geometric sequences and series.

Arithmetic Sequences

Suppose that you start saving money at home by putting $500 into a box, and then each month you add $200 to this savings. The amount of money in the box over time can be looked upon as a sequence. For example, for a

6-month period, the accounting record would show amounts of $500, $700, $900, $1100, $1300, $1500, and $1700.

The terms that comprise the amounts after deposits can be looked upon as the terms of the sequence

$$500, \ 700, \ 900, \ 1100, \ 1300, \ 1500, \ 1700, \ldots$$

This sequence was constructed by adding the same constant value, 200 in this case, to the initial value and then to each resulting value after that.

Arithmetic Sequence

Arithmetic
Sequence

A sequence in which every pair of successive terms differ by the same number d is called an **arithmetic sequence**. The number d is called the **common difference**, and $d = a_n - a_{n-1}$.

ILLUMINATOR SET A

Each of the following sequences is arithmetic.

1. $-12, \ -7, \ -2, \ 3, \ 8, \ \ldots$ with common difference 5.
2. $3, \ 7, \ 11, \ 15, \ 19, \ \cdots,$ with common difference 4.
3. $10, \ 2, \ -6, \ -14, \ -22, \ \cdots,$ with common difference -8.

We can produce a formula for the next term of an arithmetic sequence by solving the formula for the common difference, $d = a_n - a_{n-1}$, for a_n in terms of d and a_{n-1}. (You might recognize this new formula as a *recursion formula*.)

$$d = a_n - a_{n-1}$$

$$a_n = a_{n-1} + d$$

Using this recursion formula, we can produce the formula for the general term of the sequence. We can suggest a formula after observing just a few terms, beginning with $n = 1$.

$$a_2 = a_1 + d$$

$$a_3 = a_2 + d = (a_1 + d) + d = a_1 + 2d$$

$$a_4 = a_3 + d = (a_1 + 2d) + d = a_1 + 3d$$

$$a_5 = a_4 + d = (a_1 + 3d) + d = a_1 + 4d$$

Notice that for a particular term number, the coefficient of d is always one less than that term number.

General Term of an
Arithmetic
Sequence

General Term of an Arithmetic Sequence

If the sequence $a_1,\ a_2,\ a_3,\ \ldots,\ a_n,\ \ldots,$ is an arithmetic sequence with common difference d, then the general term of the sequence is given by the formula

$$a_n = a_1 + (n-1)d$$

for every $n > 1$.

EXAMPLE SET A

Find the general term of the sequence 8, 15, 22, 29, 36, $\ldots$, and generate the 25th term of the sequence.

Solution:

We first determine if the sequence is arithmetic. Since

$$15 - 8 = 7,\ 22 - 15 = 7,\ 29 - 22 = 7,\ 36 - 29 = 7$$

the sequence is an arithmetic sequence with common difference 7. We can use the formula for producing the general term of the sequence with $a_1 = 8$ and $d = 7$

$$a_n = a_1 + (n-1)d$$

$$a_n = 8 + (n-1)7$$

$$a_n = 1 + 7n$$

Thus, the general term of the sequence is $a_n = 1 + 7n$ and the 25th term is

$$a_{25} = 1 + 7(25) = 176$$

EXAMPLE SET B

Given that the first and seventh terms of an arithmetic sequence are, respectively, 60 and 24, generate the 75th term of the sequence.

Solution:
To use the general term formula $a_n = a_1 + (n-1)d$ with $a_1 = 60$, and $n = 75$, we need to determine d. We'll make use of the values for terms a_1 and a_7.

$$a_n = a_1 + (n-1)d$$

$$a_7 = a_1 + (7-1)d$$

$$24 = 60 + 6d$$

$$-36 = 6d$$

$$d = -6$$

Thus, the general term formula is $a_n = 60 - 6(n-1) = 60 - 6n + 6 = 66 - 6n$.

$$a_{75} = 66 - 6(75) = -384$$

We conclude that the 75th term of the sequence is -384.

Arithmetic Series

Arithmetic Series

With every arithmetic sequence, a_1, a_2, a_3, $\ldots, a_n$, consisting of a finite number of terms, there is an associated series, $a_1 + a_2 + a_3 + \ldots + a_n$, called an **arithmetic series**. The sum of the first n terms of an arithmetic series is commonly denoted by S_n, and a formula for finding that sum can be produced in the following way. Notice that

$$S_n = a_1 + a_2 + a_3 + \ldots + a_{n-1} + a_n$$

$$= a_1 + [a_1 + d] + [a_1 + 2d] + \ldots + [a_1 + (n-2)d] + [a_1 + (n-1)d]$$

represents the sum of the n terms of the sequence. Now, by the commutative property of addition, the same sum should be produced by adding the terms in reverse order (right-to-left rather than left-to-right as above). That is,

$$S_n = a_n + a_{n-1} + \ldots + a_2 + a_1$$

$$= [a_1 + (n-1)d] + [a_1 + (n-2)d] + \ldots + [a_1 + 2d] + [a_1 + d] + a_1$$

Now, add corresponding quantities of these equations on each side of the "=" sign.

$$2S_n = [a_1 + a_n] + [a_2 + a_{n-1}] + \ldots + [a_{n-1} + a_2] + [a_n + a_1]$$

$$= [a_1 + a_1 + (n-1)d] + [a_1 + d + a_1 + (n-2)d] + \cdots$$

$$+ [a_1 + (n-1)d + a_1]$$

$$= [2a_1 + (n-1)d] + [2a_1 + (n-1)d] + \ldots$$

$$+ [2a_1(n-1)d] + [2a_1 + (n-1)d]$$

Since there are n like terms, we have

$$2S_n = n[2a_1 + (n-1)d]$$

$$S_n = \frac{n}{2}[2a_1 + (n-1)d]$$

Since $a_n = a_1 + (n-1)d$, we can immediately produce another formula for S_n by substituting a_n for $a_1 + (n-1)d$.

$$S_n = \frac{n}{2}[2a_1 + (n-1)d]$$

$$= \frac{n}{2}[a_1 + a_1 + (n-1)d]$$

$$= \frac{n}{2}[a_1 + 1 + a_n]$$

We have produced the following two formulas for S_n.

Sum of an Arithmetic Series

Sum of an Arithmetic Series

To find the sum of the first n terms of an arithmetic series, use

$$S_n = \frac{n}{2}[2a_1 + (n-1)d]$$

if a_1 and d are known

$$S_n = \frac{n}{2}[a_1 + a_n]$$

if both a_1 and a_n are known.

EXAMPLE SET C

1. Find the sum of the first thirty-five terms of an arithmetic series if the first term is -10, and the common difference is 21.

   Solution:
   Since we are given the first term and the common difference, we will use the formula involving a_1 and d.

   $$S_n = \frac{n}{2}[2a_1 + (n-1)d]$$

   $$S_{35} = \frac{35}{2}[2(-10) + (35-1)21]$$

   $$= 12,145$$

   Thus, the sum of the first thirty-five terms of this series is $12,145$.

2. Find the sum of every other odd integer between and including -65 and -9.

   Solution:
   This sequence of every other odd integer is arithmetic since every other odd integer differs by 4. Since we are given the first and last terms of the series, we'll use the formula involving a_1 and a_n. However, to use this formula, we'll have to determine the value of n, the number of terms to be added. We can do so by using the formula that involves the information we are given, namely a_1, a_n, and d. That formula is $a_n = a_1 + (n-1)d$.

   $$a_n = a_1 + (n-1)d$$

   $$-9 = -65 + (n-1)4$$

   $$n = 15$$

   Now, using $n = 15$, $a_1 = -65$, and $a_{15} = -9$, we get

   $$S_n = \frac{n}{2}[a_1 + a_n]$$

   $$= \frac{15}{2}[-65 + (-9)]$$

   $$= \frac{15}{2}(-74)$$

   $$= -555$$

Thus, the sum of every other odd integer between and including -65 and -9 is -555.

Geometric Sequences

Suppose you put $\$A$ into a bank that pays you 7% interest each year, and left the money there year after year. At the end of the first year, the amount A of money you would have would be the original amount $\$A$, plus the 7% interest made on the $\$A$. That is,

$$\$A + 0.07(A) = A(1 + 0.07) = A(1.07)$$

The amount A in the account at the end of two years is

$$A(1.07) + 0.07[A(1.07)] = A(1.07)[1 + 0.07] = A(1.07)^2$$

In a similar way, the amount of money in the account at the end of 3, 4, 5, and n years is

$$A(1.07)^3, \ A(1.07)^4, \ A(1.07)^5, \ A(1.07)^n$$

This sequence of amounts

$$A(1.07), \ A(1.07)^2, \ A(1.07)^3, \ A(1.07)^4, \ A(1.07)^5 \ldots, A(1.07)^n$$

is an example of a *geometric sequence*. This sequence was constructed by multiplying the initial value by the constant 1.07, multiplying that value by 1.07, and so forth.

Geometric Sequence

Geometric Sequence

A sequence in which the quotient of every pair of successive terms is the same number r is called a **geometric sequence**. The number r is called the *common ratio*, and $r = \dfrac{a_n}{a_{n-1}}$.

We can produce a formula for the general term of a geometric sequence by solving the common ratio formula for a_n in terms of a_{n-1} and r, beginning with $n = 2$.

$$r = \frac{a_n}{a_{n-1}}$$

$$r = \frac{a_2}{a_1} \longrightarrow a_2 = a_1 r$$

$$r = \frac{a_3}{a_2} \longrightarrow a_3 = a_2 r = a_1 r \cdot r = a_1 r^2$$

$$r = \frac{a_4}{a_3} \longrightarrow a_4 = a_3 r = a_1 r^2 \cdot r = a_1 r^3$$

$$r = \frac{a_5}{a_4} \longrightarrow a_5 = a_4 r = a_1 r^3 \cdot r = a_1 r^4$$

The pattern shows that for a particular term, the exponent on the common ratio, r, is one less than the term number. Observation of this pattern leads us to suggest the following formula for the general term of a geometric sequence.

General Term of a Geometric Sequence

General Term of a Geometric Sequence

If the sequence a_1, a_2, a_3, $\cdots, a_n, \cdots$, is a geometric sequence with common ratio r, then the **general term** of the sequence is given by $a_n = a_1 r^{n-1}$, for every $n > 1$.

EXAMPLE SET D

1. Find the ninth term of the geometric sequence $\frac{1}{8}, \frac{-1}{20}, \frac{1}{50}, \frac{-1}{125}, \cdots$.

   Solution:
   Since the sequence is geometric and we have the first term, we need only find the common ratio r. We'll then apply the general term formula.

$$r = \frac{a_n}{a_{n-1}} \quad \text{Using } n = 2, \text{ we have}$$

$$= \frac{a_2}{a_1}$$

$$= \frac{-1/20}{1/8}$$

$$= \frac{-2}{5}$$

Now, with $a_1 = \dfrac{1}{8}$ and $r = \dfrac{-2}{5}$, we have

$$a_9 = \frac{1}{8} \cdot \left(\frac{-2}{5}\right)^8$$

$$= \frac{1}{8} \cdot \frac{256}{390,625}$$

$$= \frac{32}{390,625}$$

2. If the first and fifth terms of a geometric sequence are 3 and 4, respectively, find the common ratio to three decimal places.

   Solution:

   $$a_n = a_1 r^{n-1} \quad \text{Using } n = 5, \text{ we have}$$

   $$a_5 = a_1 r^{5-1}$$

   $$4 = 3r^4$$

   $$r^4 = \frac{4}{3}$$

   $$r = \left(\frac{4}{3}\right)^{1/4}$$

   $$r = 1.0745699\ldots$$

   $$r = 1.075 \quad \text{to three decimal places.}$$

Geometric Series

Geometric Series

With every geometric sequence, $a_1, a_2, a_3, \ldots, a_n$, consisting of a finite number of terms, there is an associated series, $a_1 + a_2 + a_3 + \ldots + a_n$, called a **geometric series**. The sum of the first n terms of a geometric series is commonly denoted by S_n, and a formula for finding that sum can be produced in the following way. Notice that

$$S_n = a_1 + a_2 + a_3 + \ldots + a_{n-1} + a_n$$

$$= a_1 + a_1 r + a_1 r^2 + \ldots + a_1 r^{n-2} + a_1 r^{n-1}$$

represents the sum of the n terms of the series. Now, multiply each side by r.

$$rS_n = a_1r + a_1r^2 + a_1r^3 + \ldots + a_1r^{n-1} + a_1r^n$$

Subtract corresponding quantities on each side of the "=" sign.

$$S_n - rS_n = (a_1 - a_1r) + (a_1r - a_1r^2) + (a_1r^2 - a_1r^3) + \ldots$$

$$+ (a_1r^{n-2} - a_1r^{n-1}) + (a_1r^{n-1} - a_1r^n)$$

$$S_n(1 - r) = a_1 - a_1r + a_1r - a_1r^2 + a_1r^2 - a_1r^3 + \ldots +$$

$$a_1r^{n-2} - a_1r^{n-1} + a_1r^{n-1} - a_1r^n$$

$$= a_1 - a_1r^n, \quad \text{and if } r \neq 1,$$

$$S_n = \frac{a_1 - a_1r^n}{1 - r}$$

Since $a_n = a_1r^{n-1}$, or equivalently, $ra_n = a_1r^n$, we can immediately produce another formula for S_n by substituting ra_n for a_1r^n.

Sum of a Geometric Series

Sum of a Geometric Series

To find the sum of the first n terms of a **geometric series**, use

$$S_n = \frac{a_1 - a_1r^n}{1 - r}$$

if a_1 and r are known, and

$$S_n = \frac{a_1 - ra_n}{1 - r}$$

if both a_1 and a_n are known.

EXAMPLE SET E

The first four terms of a geometric sequence are $-6, 24, -96, 384, \ldots$. Find the sum of the first 14 terms of this sequence.

Solution:

We have that $n = 14$, $a_1 = -6$, and $r = \dfrac{24}{-6} = -4$. Since we do not have the 24th term, we'll use the formula involving only a_1 and r.

$$S_n = \frac{a_1 - a_1 r^n}{1 - r}$$

$$S_{14} = \frac{-6 - (-6)(-4)^{14}}{1 - (-4)}$$

$$= 322,122,546$$

Thus, the sum of the first 14 terms of the series is $322,122,546$.

Infinite Geometric Series

The formula we developed for finding S_n for a geometric series applies only to a series with a finite number of terms. Depending on the value of the common ratio, r, it may be possible to find the sum of infinitely many terms.

Consider a value for r that is between -1 and $+1$, that is $-1 < r < 1$, or equivalently, $|r| < 1$. Notice how r^n is affected as the value of n increases without bound. As n approaches infinity, r^n approaches 0. (Symbolically, as $n \to \infty$, $r \to 0$. Try raising 1/2 to larger and larger powers and verify that it gets closer and closer to 0.) With this fact in mind, consider the formula for the sum of the first n terms of a geometric series.

$$S_n = \frac{a_1 - a_1 r^n}{1 - r} = \frac{a_1}{1 - r} - \frac{a_1 r^n}{1 - r}$$

Consider the second term on the right hand side. As $n \to \infty$, $r^n \to 0$, which means that $a_1 r^n \to 0$, which, in turn, means that $\dfrac{a_1 r^n}{1 - r} \to 0$.

Thus, as $n \to \infty$, $\dfrac{a_1 - a_1 r^n}{1 - r} \to \dfrac{a_1}{1 - r}$.

Denoting the sum of an infinite geometric series by S_∞, we have produced the following formula.

Sum of an Infinite Geometric Series

If a_1 is the first term of an infinite **geometric series** and r is the common ratio, then

$$S_\infty = \frac{a_1}{1 - r}, \quad |r| < 1$$

If $|r| \geq 1$, the sequence has no sum.

EXAMPLE SET F

1. Find the value of $200 + 40 + 8 + \dfrac{8}{5} + \dfrac{8}{25} + \cdots$, if it exists.

   Solution:
   Since $a_1 = 200$ and $r = \dfrac{1}{5}$ so that $|r| < 1$, the sum exists and

   $$S_\infty = \frac{a_1}{1-r}$$

   $$= \frac{200}{1 - 1/5}$$

   $$= 250$$

2. Find the value of $-\dfrac{1}{2} + 2 - 8 + 32 - 128 + \cdots$, if it exists.

   Solution:
   Since $r = -4$ so that $|r| > 1$, the sum does not exist.

3. Rational numbers are numbers that can be expressed as the quotient of two integers, and whose decimal representations either terminate or continue indefinitely, but contain repeating blocks of digits. The number $0.\overline{64} = 0.646464\ldots$ is a rational number and therefore can be expressed as the quotient of two integers. Express $0.646464\ldots$ as the quotient of two integers.

   Solution:
   $$0.646464\ldots = 0.64 + 0.0064 + 0.000064 + \cdots$$

   $$= 0.64 + 0.64(0.01) + 0.64(0.0001) + \cdots$$

   $$= 0.64 + 0.64(0.01)^1 0.64(0.01)^2 + \cdots$$

   Now we see a geometric series with $a_1 = 0.64$, and $r = 0.01$. Thus,

   $$S_\infty = \frac{a_1}{1-r}$$

   $$= \frac{0.64}{1 - 0.01}$$

   $$= \frac{0.64}{0.99} = \frac{64}{99}$$

   Thus, $0.646464\ldots = \dfrac{64}{99}$.

EXERCISES

For exercises 1–11, (a) write the next three terms of the sequence, and (b) find the sum of the first nine terms of the sequence.

1. $5, -1, -7, -13, \ldots$

2. $-18, -15, -12, -9, \ldots$

3. $1, \dfrac{2}{3}, \dfrac{4}{9}, \dfrac{8}{27}, \ldots$

4. $\dfrac{5}{8}, -\dfrac{1}{4}, \dfrac{1}{10}, -\dfrac{1}{25}, \ldots$

5. $-\dfrac{14}{15}, \dfrac{7}{5}, -\dfrac{21}{10}, \dfrac{63}{20}, \ldots$

6. $33, 21, 9, -3, \ldots$

7. $x - 8, x - 3, x + 2, x + 7, \ldots$

8. $4\sqrt{5}, 20, 20\sqrt{5}, 100, \ldots$

9. $\dfrac{6}{y}, 6, 6y, 6y^2, \ldots$

10. $0.4, 0.004, 0.00004, 0.0000004, \ldots$

11. $92, 9.2, 0.92, 0.092, \ldots$

12. Find the sum of all odd integers between and including 1 and 99.

13. Find the sum of all even integers between and including 100 and 1000.

14. Find the sum of all multiples of 4 between and including 20 and 112.

15. Find the sum of every other even integer between and including 0 and 300.

For exercises 16-21, express each rational number as the quotient of two integers.

16. $0.\overline{44}$

17. $0.\overline{38}$

18. $0.1\overline{27}$

19. $0.8\overline{58}$

20. $23.4\overline{62}$

21. $11.73\overline{623}$

22. Find the general term of the sequence 3, 21, 39, 57, 75, ..., then produce the 20th term.

23. Find the general term of the sequence 3, 6.6, 10.2, 13.8, ..., then produce the 15th term.

24. Find the general term of the sequence $1, -\sqrt{3}, 3, -3\sqrt{3}, \ldots$, then produce the 25th term.

25. Find the general term of the sequence $1, -1, 1, -1, 1, -1 \ldots$, then produce the 250th term.

26. Given that the third and tenth terms of an arithmetic sequence are 22 and 78, respectively, find the 20th term.

27. Given that the 7th and 20th terms of an arithmetic sequence are -14 and -66, respectively, find the 15th term.

28. The first and eighth terms of a geometric sequence are 6 and 10, respectively. Find the common ratio and the 15th term.

29. A ball is dropped from a height of 4 feet. Each time it hits the floor, it bounces up 3/5 of the distance it fell. Find the distance traveled by the ball up to the instant it hits the floor for the sixth time. Find the distance traveled by the ball up to the instant it comes to rest.

30. A high school auditorium has 40 rows of seats. The first row has 25 seats, the second row has 26 seats, the third row has 27 seats, and so on. How many seats does the auditorium contain?

31. Beginning with your parents, how many ancestors do you have going back 10 generations?

32. Which is the more lucrative deal, $10,000,000 now, or 1 penny a day doubled for 30 days?

33. A new $18,000 car depreciates at the rate of 26% a year. What is the market value of the car in 5 years?

34. The population of a country is increasing at the rate of 0.2% a year. If the current population of the country is 800,000 people, what will be the population of the country in 20 years?

35. Five years ago, a woman's salary was $28,000. Since then, she has received 6% raises each year. What is her salary now?

36. Is the sequence $\{\ln 3\}_{n=1}^{n=\infty}$ an arithmetic or geometric sequence? What is the common difference or ratio?

37. Is the sequence $\left\{\ln \dfrac{3}{4^n}\right\}_{n=1}^{n=\infty}$ an arithmetic or geometric sequence? What is the common difference or ratio?

For exercises 38–44, find each indicated infinite sum, if it exists. If it does not exist, so state.

38. $4 + \dfrac{1}{2} + \dfrac{1}{16} + \dfrac{1}{128} + \cdots$

39. $1 + \dfrac{2}{3} + \dfrac{4}{9} + \dfrac{8}{27} + \cdots$

40. $\dfrac{1}{2} + \dfrac{1}{4} + \dfrac{1}{8} + \dfrac{1}{16} + \cdots$

41. $\dfrac{4}{5} + \dfrac{1}{5} + \dfrac{1}{20} + \dfrac{1}{80} + \cdots$

42. $\dfrac{1}{2} - \dfrac{1}{4} + \dfrac{1}{8} - \dfrac{1}{16} + \cdots$

43. $4 + \dfrac{16}{3} + \dfrac{64}{9} + \dfrac{256}{27} + \cdots$

44. $1 + 1.1 + 1.01 + 1.001 + 1.001 + \cdots$

45. The radioactivity of a laboratory sample is measured once a day, at noon, and the measurements are recorded as a sequence of readings. The level measured is found to equal $R_0 e^{-kt}$ where R_0 is the level measured on the first day, t is the number of days elapsed since the first reading, and k is a constant. What is the common ratio in the sequence of readings?

46. A journalist ordered by a judge in a criminal trial to disclose her sources refuses to do so. The judge finds the journalist in contempt of court and imposes fines for every week the journalist remains in contempt: $500 for the first week, $600 for the second, $700 for the third, and so on. What will the total in fines come to if the journalist finally gives in after eight weeks?

47. The St. Petersburg Paradox. Suppose you are offered the chance, for a fee, to play the following game one time: You flip a coin until it comes up heads and then receive a prize of 2^n, where n is the number of times you had to flip the coin. How much should you be willing to pay to play this game? To find out, start by writing the series {(likelihood that the first head comes on the nth flip)2^n} for n ranging from 1 to infinity. Then sum all the terms of the series to calculate the expected monetary value of the game.

11.5 The Binomial Theorem

Introduction

Binomial Patterns

Factorial and Combination Notation

The Binomial Theorem

The rth term in a Binomial Expansion

Pascal's Triangle

Introduction

By now, you know very well that $(x+4)^2 \neq x^2 + 16$, but rather $(x+4)^2 = x^2 + 8x + 16$. In this section we take a detailed look at expansions of binomial expressions. We begin by observing patterns that are common to all binomial expansions, develop a formula that generates the numerical coefficient of any term in a binomial expansion, then develop and prove a formula that generates the expansion of any binomial expression.

Binomial Patterns

We are now at a point where we can explore the underlying pattern of coefficients that exists in the sequence of expansions

$$(a+b)^1, \ (a+b)^2, \ (a+b)^3, \ (a+b)^4, \ \cdots$$

We begin by observing the first seven expansions in the sequence.

$$(a+b)^0 = 1$$

$$(a+b)^1 = a + b$$

$$(a+b)^2 = a^2 + 2ab + b^2$$

$$(a+b)^3 = a^3 + 3a^2b + 3ab^2 + b^3$$

$$(a+b)^4 = a^4 + 4a^3b + 6a^2b^2 + 4ab^3 + b^4$$

$$(a+b)^5 = a^5 + 5a^4b + 10a^3b^2 + 10a^2b^3 + 5ab^4 + b^5$$

$$(a+b)^6 = a^6 + 6a^5b + 15a^4b^2 + 20a^3b^3 + 15a^2b^4 + 6ab^5 + b^6$$

Our goal is to construct a formula that, for a specified n, generates the expansion of $(a + b)^n$. Observation of the above examples can lead us to an abundance of conclusions about the properties of such a formula. For a specified n,

1. Each expansion has $n + 1$ terms.

2. The first term is a^n and the last term is b^n.

3. The sum of the exponents of a and b in each term is n.

4. Moving left-to-right, from term to term, the exponent of a decreases by one, and the exponent of b increases by one.

5. If the coefficient of a particular term is multiplied by the exponent on a in that term, and this product is divided by the number of the term, the coefficient of the next successive term is obtained. That is, for term numbers k and $k + 1$,

$$\text{coefficient of term } k + 1 =$$

$$\frac{(\text{coefficient of term } k)(\text{exponent on } a \text{ in term } k)}{k}$$

This property allows us to compute the coefficents of terms from left-to-right.

6. The *nth* (next to last) term of the expansion is nab^{n-1}.

EXAMPLE SET A

Use the properties of the expansion of $(a + b)^n$ to write the expansion of $(a + b)^5$.

Solution:
From property 1, we know there will be $5 + 1 = 6$ terms in the expansion.
From property 2, the first term is a^5 and the last term is b^5.
From property 5, the coefficient of term 2 is $\dfrac{1 \cdot 5}{1} = 5$.
From property 4, the exponents of a and b in this term are 4 and 1, respectively.
So far we have that $(a + b)^5 = a^5 + 5a^4b + \cdots + b^5$.
We obtain the third term by again applying properties 4 and 5. The coefficient is

$$\frac{5 \cdot 4}{2} = 10$$

and the exponents on a and b are 3 and 2, respectively.
We now have $(a + b)^5 = a^5 + 5a^4b + 10a^3b^2 + \cdots + b^5$.
We construct the 4th and 5th terms in a similar way to get

$$(a + b)^5 = a^5 + 5a^4b + 10a^3b^2 + 10a^2b^3 + 5ab^4 + b^5$$

Factorial and Combination Notation

Before we produce a rule that generates all the terms of the expansion of $(a + b)^n$, we'll examine some notation that significantly simplifies the appearance of the rule.

Quite often in mathematics, it is useful to compute the product of the first n positive integers. The symbol $n!$, which is read as *n factorial*, is used to denote this product. For example

$$1! = 1$$

$$2! = 2 \cdot 1 = 2$$

$$3! = 3 \cdot 2 \cdot 1 = 6$$

$$4! = 4 \cdot 3 \cdot 2 \cdot 1 = 24$$

$$5! = 5 \cdot 4 \cdot 3 \cdot 2 \cdot 1 = 120$$

We can generalize this multiplication as follows.

The Computation of $n!$

The Computation of $n!$

If n is any positive integer, then $n! = n(n-1)(n-2)(n-3) \cdots 2 \cdot 1$.

Notice in the few factorial examples above that

$$3! = 3 \cdot \underbrace{2 \cdot 1}_{2!} = 3 \cdot 2!$$

$$4! = 4 \cdot \underbrace{3 \cdot 2 \cdot 1}_{3!} = 4 \cdot 3!$$

$$5! = 5 \cdot \underbrace{4 \cdot 3 \cdot 2 \cdot 1}_{4!} = 5 \cdot 4!$$

This is a useful factorial property as it provides a convenient way for simplifying computations involving factorials, and we generalize as follows.

n! as a Product of *n*
and $(n-1)!$

$n!$ **as a Product of** n **and** $(n-1)!$
If n is a positive integer, then $n! = n(n-1)!$.

This property is convenient as it allows us to write factorials in terms of smaller factorials. For example, since $4! = 4 \cdot \underbrace{3 \cdot 2 \cdot 1}_{3!}$, we can write 4! as $4 \cdot 3!$.

Similarly, $7! = 7 \cdot 6!$, and $12! = 12 \cdot 11 \cdot 10 \cdot 9!$.

This definition of $n!$ is good only for positive integers greater than or equal to 1. We can extend it just a bit further to include 0 if we notice that when $n = 1$,

$$n! = n(n-1)!$$

$$1! = 1(1-1)!$$

$$1 = 1 \cdot 0!$$

$$0! = 1$$

ILLUMINATOR SET A

Evaluate each factorial expression.

1. $\dfrac{7!}{4!} = \dfrac{7 \cdot 6 \cdot 5 \cdot 4!}{4!} = 7 \cdot 6 \cdot 5 = 210.$

2. $\dfrac{8!}{3!5!} = \dfrac{8 \cdot 7 \cdot 6 \cdot 5!}{3!5!} = \dfrac{8 \cdot 7 \cdot 6}{3!} = 8 \cdot 7 = 56$

The factorial expression

$$\frac{n!}{r!(n-r)!}$$

occurs so frequently in mathematics in the context of counting that it is represented by the special **binomial coefficient** symbol $\dbinom{n}{r}$. That is

$$\binom{n}{r} = \frac{n!}{r!(n-r)!}$$

The binomial coefficient is read *n choose r*, and it counts the number of ways *r* distinct objects can be chosen from *n* distinct objects without regard to order.

Binomial Coefficient

ILLUMINATOR SET B

Evaluate each binomial coefficient.

1. $\binom{6}{4} = \dfrac{6!}{4!(6-4)!} = \dfrac{6!}{4!2!} = \dfrac{6 \cdot 5 \cdot 4!}{4!2!} = \dfrac{6 \cdot 5}{2} = 15$

The expression $\binom{6}{4}$ counts the number of ways 4 distinct objects can be chosen from 6 distinct objects. There are 15 ways to choose 4 objects from 6.

2. $\binom{6}{0} = \dfrac{6!}{0!(6-0)!} = \dfrac{6!}{0!6!} = \dfrac{1}{0!} = \dfrac{1}{1} = 1$

The expression $\binom{6}{0}$ counts the number of ways 0 distinct objects can be chosen from 6 distinct objects. There is 1 way to choose 0 objects from 6.

3. $\binom{6}{6} = \dfrac{6!}{6!(6-6)!} = \dfrac{6!}{6!0!} = \dfrac{1}{0!} = \dfrac{1}{1} = 1$

The expression $\binom{6}{6}$ counts the number of ways 6 distinct objects can be chosen from 6 distinct objects. There is 1 way to choose 6 objects from 6.

Notice that $\binom{6}{6} = \binom{6}{0}$.

The Binomial Theorem

Now that we have a convenient notation, we return to the problem of constructing a formula that produces all the terms of the expansion of $(a+b)^n$.

If the six observations and conclusions we made just prior to Illuminator Set A are correct, then the first five terms in the expansion of $(a+b)^n$ are listed below. Terms 2 – 5 are obtained using Properties 5 and 4.

$$\text{1st term} := a^n$$

$$\text{2nd term} := \frac{n}{1}x^{n-1}y^1 = nx^{n-1}y$$

$$\text{3rd term} := \frac{n \cdot (n-1)}{2}x^{(n-1)-1}y^{1+1} = \frac{n(n-1)}{2}x^{n-2}y^2$$

$$\text{4th term} := \frac{n(n-1) \cdot (n-2)}{2 \cdot 3}x^{(n-2)-1}y^{2+1}$$

$$= \frac{n(n-1)(n-2)}{3!}x^{n-3}y^3$$

$$\text{5th term} := \frac{n(n-1)(n-2) \cdot (n-3)}{3! \cdot 4}x^{(n-3)-1}y^{3+1}$$

$$= \frac{n(n-1)(n-2)(n-3)}{4!} x^{n-4} y^4$$

We can express each term in a more compact form by rewriting each coefficient in a simpler form. If each coefficient is multiplied by 1 in the form $\frac{N!}{N!}$, we obtain the following binomial coefficients:

1. The coefficient of the first term is represented as follows:

First, we create a fraction.

$$1 = \frac{1}{1} = \frac{1}{0!}$$

Then,

$$\frac{1}{0!} = \frac{1}{0!} \cdot \frac{n!}{n!}$$

$$= \frac{n!}{0! n!}$$

$$= \frac{n!}{0! (n-0)!}$$

$$= \binom{n}{0}$$

2. The coefficient of the second term is represented as follows:

Again, we create a fraction.

$$n = \frac{n}{1} = \frac{n}{1!}$$

Then,

$$\frac{n}{1} = \frac{n}{1!} \cdot \frac{(n-1)!}{(n-1)!}$$

$$= \frac{n(n-1)!}{1!(n-1)!}$$

$$= \frac{n!}{1!(n-1)!}$$

$$= \binom{n}{1}$$

3. The coefficient of the third term is

$$\frac{n(n-1)}{2!} = \frac{n(n-1)}{2!} \cdot \frac{(n-2)!}{(n-2)!}$$

$$= \frac{n(n-1)(n-2)!}{2!(n-2)!}$$

$$= \frac{n!}{2!(n-2)!}$$

$$= \binom{n}{2}$$

4. The coefficient of the fourth term is

$$\frac{n(n-1)(n-2)}{3!} = \frac{n(n-1)(n-2)}{3!} \cdot \frac{(n-3)!}{(n-3)!}$$

$$= \frac{n(n-1)(n-2)(n-3)!}{3!(n-3)!}$$

$$= \frac{n!}{3!(n-3)!}$$

$$= \binom{n}{3}$$

5. The coefficient of the fifth term is

$$\frac{n(n-1)(n-2)(n-3)}{4!} = \frac{n(n-1)(n-2)(n-3)}{4!} \cdot \frac{(n-4)!}{(n-4)!}$$

$$= \frac{n(n-1)(n-2)(n-3)(n-4)!}{4!(n-4)!}$$

$$= \frac{n!}{4!(n-4)!}$$

$$= \binom{n}{4}$$

We can now write the expansion of $(a+b)^n$ in a more compact form.

$$(a+b)^n = \binom{n}{0}a^n + \binom{n}{1}a^{n-1}b + \binom{n}{2}a^{n-2}b^2$$

$$+ \binom{n}{3}a^{n-3}b^3 + \binom{n}{4}a^{n-4}b^4 + \cdots + nab^{n-1} + b^n$$

We can express the binomial formula even more compactly using summation notation.

$$(a+b)^n = \sum_{r=0}^{n} \binom{n}{r}a^{n-r}b^r$$

We now state and prove the binomial theorem.

The Binomial Theorem

If a and b are real numbers and n is any natural number, then

$$(a+b)^n = \sum_{r=0}^{n} \binom{n}{r}a^{n-r}b^r$$

Proof of the Binomial Theorem
By definition,

$$(a+b)^n = \underbrace{(a+b)(a+b) + \cdots + (a+b)}_{n \text{ factors}}$$

Each term in the expansion has the form $a^p b^q$, where p and q are nonnegative integers for which $p + q = n$, and one of a or b is chosen from each of the n factors $(a+b)$. Now, the term $a^{n-r}b^r$ is obtained by choosing b from any r of the n factors. Thus, since there are $\binom{n}{r}$ ways of choosing r items from n items, the coefficient of the term $a^{n-r}b^r$ is $\binom{n}{r}$.

EXAMPLE SET B

1. Use the binomial theorem to expand $(x+y)^4$.

Solution:

With $a = x, b = y$, and $n = 4$,

$$(x + y)^4 = \sum_{r=0}^{4} \binom{4}{r} x^{4-r} y^r$$

$$= \binom{4}{0} x^{4-0} y^0 + \binom{4}{1} x^{4-1} y^1 + \binom{4}{2} x^{4-2} y^2$$

$$+ \binom{4}{3} x^{4-3} y^3 + \binom{4}{4} x^{4-4} y^4$$

$$= 1x^4 1 + 4x^3 y^1 + 6x^2 y^2 + 4x^1 y^3 + 1x^0 y^4$$

$$= x^4 + 4x^3 y + 6x^2 y^2 + 4xy^3 + y^4$$

2. Use the binomial theorem to expand $(3x - 2y)^5$.

Solution:

With $a = 3x, b = -2y$, and $n = 5$,

$$(3x - 2y)^5 = \sum_{r=0}^{5} \binom{5}{r} (3x)^{5-r} (-2y)^r$$

$$= \binom{5}{0} (3x)^{5-0} (-2y)^0 + \binom{5}{1} (3x)^{5-1} (-2y)^1$$

$$+ \binom{5}{2} (3x)^{5-2} (-2y)^2 + \binom{5}{3} (3x)^{5-3} (-2y)^3$$

$$+ \binom{5}{4} (3x)^{5-4} (-2y)^4 + \binom{5}{5} (3x)^{5-5} (-2y)^5$$

$$= 1(3x)^5(1) + 5(3x)^4(-2y)^1 + 10(3x)^3(-2y)^2$$

$$+ 10(3x)^2(-2y)^3 + 5(3x)^1(-2y)^4 + 1(3x)^0(-2y)^5$$

$$= 1(243x^5)(1) + 5(81x^4)(-2y) + 10(27x^3)(4y^2)$$

$$+ 10(9x^2)(-8y^3) + 5(3x)(16y^4) + 1(1)(-32y^5)$$

$$= 243x^5 - 810x^4 y + 1080x^3 y^2 - 720x^2 y^3 + 240xy^4 - 32y^5$$

The rth term in a Binomial Expansion

It is often convenient to be able to find a specific term in the expansion of $(a + b)^n$. For example, we would like to be able to specify the 5th term of $(4w - 3z)^7$ without constructing the entire expansion. To develop a formula for specifying the rth term, we'll make some observations from the expansion of $(a + b)^5$, and generalize these observations into a formula.

$$(a+b)^5 = \binom{5}{0}a^5b^0 + \binom{5}{1}a^{5-1}b^1 + \binom{5}{2}a^{5-2}b^2$$

$$+ \binom{5}{3}a^{5-3}b^3 + \binom{5}{4}a^{5-4}b^4 + \binom{5}{5}a^{5-5}b^5$$

$$= a^5 + 5a^4b + 10a^3b^2 + 10a^2b^3 + 5ab^4 + b^5$$

1. *Specific*: In each term, the lower number in the binomial coefficient is 1 less than the term number.
    General: In term number r, the lower number of the binomial coefficient is $r - 1$.

2. *Specific*: In each term, the exponent of b is 1 less than the term number in which it appears.
    General: In term number r, the exponent of b is $r - 1$.

3. *Specific*: In each term, the exponent of a is $n -$ (exponent of b.)
    General: In term number r, the exponent of a is $n - (r - 1) = n - r + 1$.

These observations lead us to a formula for generating the rth term of the expansion of $(a + b)^n$.

The rth Term Formula

The rth Term Formula
If a and b are real numbers and n is a natural number, then the rth term of the expansion of $(a + b)^n$ is given by the formula $$r\text{th term} = \binom{n}{r-1}a^{n-r+1}b^{r-1}$$

EXAMPLE SET C

1. Specify the 5th term of $(w - 3z)^{12}$.

Solution:

With $a = w, b = -3z, r = 5$, and $n = 12$,

$$r\text{th term} = \binom{n}{r-1} a^{n-r+1} b^{r-1}$$

$$= \binom{12}{5-1} w^{12-5+1}(-3z)^{5-1}$$

$$= 495w^8(-3z)^4$$

$$= 495w^8(81z^4)$$

$$= 40,095w^8z^4$$

2. Find the numerical coefficient of the term involving h^9 in the expansion of $(2h^3 + 5k)^6$.

Solution:

The factor h^9 will occur by raising h^3 to the 3rd power. Since $a = 2h^3$, and the exponent on a in the rth term is $n - r + 1$, we must have that

$$3 = n - r + 1$$

$$3 = 6 - r + 1$$

$$3 = 7 - r$$

$$r = 4$$

Thus, $r = 4$ and h^9 appears in the 4th term. Then, using the rth term formula, we have

$$r\text{th term} = \binom{n}{r-1} a^{n-r+1} b^{r-1}$$

$$= \binom{6}{4-1}(2h^3)^{6-4+1}(5k)^{4-1}$$

$$= \binom{6}{3}(2h^3)^3(5k)^3$$

$$= 20(8h^9)(125k^3)$$

$$= 20,000h^9k^3$$

Thus, the numerical coefficient of the term involving h^9 is $20,000$.

Pascal's Triangle

If we display just the coefficients of each of the terms in the first several binomial expansions, we have the triangular pattern of numbers illustrated in Figure 11.8.

$$
\begin{array}{cccccccccccc}
(a+b)^0 & : & & & & & & 1 & & & & \\
(a+b)^1 & : & & & & & 1 & & 1 & & & \\
(a+b)^2 & : & & & & 1 & & 2 & & 1 & & \\
(a+b)^3 & : & & & 1 & & 3 & & 3 & & 1 & \\
(a+b)^4 & : & & 1 & & 4 & & 6 & & 4 & & 1 \\
(a+b)^5 & : & 1 & & 5 & & 10 & & 10 & & 5 & & 1 \\
\end{array}
$$

Figure 11.8

Pascal's Triangle   This triangular form is called **Pascal's Triangle** in honor of the French mathematician, Blaise Pascal (1623 – 1662), and its pattern is valid for higher powers of $(a+b)$.

As with sequences, we ask if the row of coefficients can be constructed only on the knowledge of the coefficients of the previous row. The underlying pattern that allows us to make this construction can be observed by considering only the a^3b term of the expansion of $(a+b)^4$.

$$(a+b)^4 = (a+b)^3(a+b)$$

$$= (a^3 + 3a^2b + 3ab^2 + b^3)(a+b)$$

$$= \cdots + a^3b + 3a^3b \cdots$$

$$= \cdots + 4a^3b \cdots$$

The coefficient of a^3b is the sum of the coefficients of the two terms from the expansion of $(a+b)^3$, namely, a^3 and $3a^2b$. In Pascal's triangle, a number in the fourth row (except for the end 1's) is the sum of the two numbers immediately above to the left and right in the third row. The pattern of construction of Pascal's triangle is indicated in Figure 11.9.

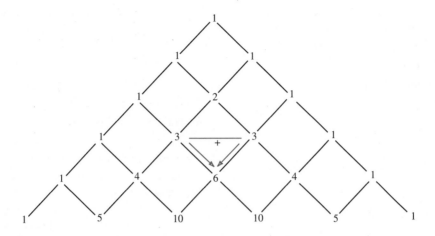

Figure 11.9

EXERCISES

For exercises 1–8, evaluate each binomial coefficient.

1. $\begin{pmatrix} 7 \\ 3 \end{pmatrix}$

2. $\begin{pmatrix} 10 \\ 2 \end{pmatrix}$

3. $\begin{pmatrix} 8 \\ 7 \end{pmatrix}$

4. $\begin{pmatrix} 4 \\ 4 \end{pmatrix}$

5. $\begin{pmatrix} n \\ n \end{pmatrix}$

6. $\begin{pmatrix} n \\ 0 \end{pmatrix}$

7. $\begin{pmatrix} n \\ n-1 \end{pmatrix}$

8. $\begin{pmatrix} n \\ n-2 \end{pmatrix}$

For exercises 9–23, use the binomial theorem to expand each binomial expression.

9. $(x+h)^3$

10. $(x+3)^4$

11. $(x-2)^5$

12. $(x+2y)^3$

13. $(x-3y)^6$

14. $(3x+2y)^5$

15. $(2x-y)^4$

16. $(\sqrt{x}+y^2)^6$

17. $(h^4 - \sqrt[4]{2k})^3$

18. $(3x+\frac{y}{3})^7$

19. $(\frac{a}{2b} - \frac{c}{3})^3$

20. $(x^2y+z^2)^4$

21. $(x^3 - 3yz^2)^3$

22. $(x^3y^4 - 2w^3z^5)^5$

23. $(x^6y - \sqrt{x})^4$

In exercises 24–27, the given numbers correspond to a specific row in Pascal's triangle. Write the next row of numbers in the triangle and indicate the exponent n of the expression $(a+b)^n$ corresponding to this new row.

24. 1 4 6 4 1

25. 1 6 15 20 15 6 1

26. $\begin{pmatrix} 3 \\ 0 \end{pmatrix} \begin{pmatrix} 3 \\ 1 \end{pmatrix} \begin{pmatrix} 3 \\ 2 \end{pmatrix} \begin{pmatrix} 3 \\ 3 \end{pmatrix}$

27. $\begin{pmatrix} 5 \\ 0 \end{pmatrix} \begin{pmatrix} 5 \\ 1 \end{pmatrix} \begin{pmatrix} 5 \\ 2 \end{pmatrix} \begin{pmatrix} 5 \\ 3 \end{pmatrix} \begin{pmatrix} 5 \\ 4 \end{pmatrix} \begin{pmatrix} 5 \\ 5 \end{pmatrix}$

For exercises 28–37, determine the numerical coefficient of the indicated term of the binomial expansion of the given expression.

28. The $x^2 y^3$ term of $(x + y)^5$.

29. The $x^3 y^2$ term of $(x + y)^5$.

30. The $a^2 b^7$ term of $(a + b)^9$.

31. The $a^3 b^4$ term of $(a + b)^7$.

32. The $a^6 b^2$ term of $(a - b)^8$.

33. The $a^5 b^3$ term of $(a - b)^8$.

34. The $a^6 c^{23}$ term of $(a + c)^{29}$.

35. The $a^{12} c^7$ term of $(a + c)^{19}$.

36. The s^2 term of $(1 + s)^{15}$.

37. The t^3 term of $(1 + t)^{13}$.

For exercises 38–45, write the binomial expansion of the given binomial expression using summation notation. For example,

$$(a + b)^2 = \sum_{r=0}^{2} \binom{2}{r} a^r b^{2-r}$$

38. $(a + b)^5$

39. $(c + d)^7$

40. $(a + 2b)^6$

41. $(a - 2b)^9$

42. $(2a - x^2)^{13}$

43. $(x^2 - 1)^{25}$

44. $(y - 3)^{52}$

45. $(2x - y^2)^{21}$

For exercises 46–49, evaluate and simplify the difference quotient $\dfrac{f(x + h) - f(x)}{h}$.

46. $f(x) = x^3 - 4$

47. $f(x) = x^4 + 2x$

48. $f(x) = 3x^5 - 2x^3 + 7x$

49. $f(x) = 2x^4 + 6x^2 - 3$

50. A club with twenty members is led by a three-member steering committee. How many different combinations of people on the steering committee are possible?

51. Pascal's triangle can be thought of as a maze in which one starts at the top and, moving downward, takes a left or right turn at every junction. Each numerical entry in the triangle gives the number of ways of getting to the corresponding junction. If for a given number of turns, every combination of left and right turns is as likely as every other, what is the likelihood that someone taking exactly three turns arrives at the leftmost junction in the fourth row?

52. Keeping in mind the information from the previous exercise, consider the following scenario. A student taking a five-question true/false quiz guesses on every question, having no idea what the correct answers are. What is the likelihood that the student will get at least four questions right?

Summary and Projects

Summary

Mathematical Induction   When you attempt to prove a mathematical statement using mathematical induction, you should understand that the task is basically to show that the statement is true for all positive integers. It will be obvious to you that the statement is true for many positive integers simply by trying as many as you like. However, is it true for *all* positive integers? The principle of mathematical induction, roughly stated, says that you must show two things: (1) the statement is true for the first positive integer, and (2) if you make the assumption that the statement is true for any arbitrary positive integer, k, you can use that assumption to show the statement is true for the next consecutive positive integer, $k + 1$.

A note of interest: you can use the same approach for mathematical statements that are not true for the first n positive integers but are true for all positive integers beyond n. The conclusion then becomes: the statement is true for all positive integers greater than n.

Sequences, Series, and Convergence

Formally, a *sequence* is a set of range values of a function whose domain values are the positive integers. Informally, it is an infinite list of numbers that is commonly written as

$$a_1, a_2, a_3, a_4, \ldots a_n, \ldots$$

or compactly as $\{a_n\}$. If we decide to add the elements of a sequence, we obtain an infinite *series*.

A sequence is said to *converge* if the terms of the sequence get closer and closer to a particular number as the domain values increase without bound. For example, the sequence

$$1, \frac{1}{2}, \frac{1}{3}, \frac{1}{4}, \ldots \frac{1}{n}, \ldots$$

converges because the terms are getting closer and closer to 0 as n increases without bound. You would need to show that it satisfies the formal definition of convergence of a sequence given in the text to prove it converges.

Arithmetic and Geometric Series

An *arithmetic series* is a series in which every term is obtained by adding a fixed value to the previous term. The sum of the first n terms of an arithmetic series,

$$a_1, a_2, a_3, a_4, \ldots a_n, \ldots$$

is given by either

$$S_n = \frac{n}{2}[2a_1 + (n-1)d] \quad \text{or} \quad S_n = \frac{n}{2}[a_1 + a_n]$$

where d is the common difference (the difference between any term and the previous term).

A *geometric series* is a series in which every term is obtained by multiplying the previous term by a fixed value. The sum of the first n terms of a geometric series, $a_1, a_2, a_3, a_4, \ldots a_n, \ldots$, is given by either

$$S_n = \frac{a_1 - a_1 r^n}{1 - r} \quad \text{or} \quad S_n = \frac{a_1 - r a_n}{1 - r}$$

where r is the common ratio (the value of any term divided by the value of the previous term).

It is also possible for an infinite geometric series to have a "sum" or to converge. This occurs when the absolute value of the common ratio is less than 1. In this case, the infinite geometric series, $a_1, a_2, a_3, a_4, \ldots a_n, \ldots$, converges to the value

$$S_\infty = \frac{a_1}{1 - r}$$

Expanding Binomial Expressions   The formula for the expansion of a binomial expression raised to a positive integer exponent n is

$$(a + b)^n = a^n + na^{n-1}b + \frac{n(n-1)}{2!}a^{n-2}b^2 + \ldots b^n$$

$$= \binom{n}{0}a^n b^0 + \binom{n}{1}a^{n-1}b^1 + \binom{n}{2}a^{n-2}b^2$$

$$+ \ldots + \binom{n}{r}a^{n-r}b^r + \ldots + \binom{n}{n}a^0 b^n$$

where $n - r + 1$ is the position of the term in the sum, $\binom{n}{r}$ denotes the number of ways n things can be chosen r at a time, and

$$\binom{n}{r} = \frac{n!}{r!\,(n - r)!}$$

Projects

1. Suppose that 16 ml of a drug is introduced into a person's bloodstream at time $t = 0$. As time passes, the kidneys and liver clean the blood by processing out the drug at the rate of 25% per hour.

   a. Construct a function that relates the hour number, $t = 1, 2, 3, \ldots$ to the amount A of drug remaining in the bloodstream. If time is considered only in hours, this function can be considered a sequence function.

   b. If time is considered to be continuous, this function is not a sequence function, but rather another type of function. Specify this other type of function.

   c. Find the first value of N such that when $t > N$, the amount of the drug in the bloodstream is less than 0.05 ml.

2. In this chapter you were introduced to the Fibonacci sequence and Pascal's triangle. The Fibonacci sequence can be found in Pascal's triangle. Find it, and write a set of directions that will show someone else how to find it.

3. When two flat sheets of glass are placed face-to-face, four reflective surfaces are formed. If a ray of light passes into the glass it can either pass straight through, or it can be reflected. If the ray has one reflection, it can be reflected in two ways. If the ray has two reflections, it can be reflected in three ways. The following figure shows these possibilities.

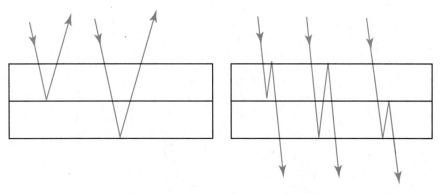

One Reflection, Two Paths   Two Reflections, Three Paths

   a. If a ray is reflected three times, there are five paths of reflection. Show these five paths on a diagram.

   b. If a ray is reflected four times, there are eight paths of reflection. Show these eight paths on a diagram.

   c. Determine how many paths of reflection there will be for a ray that is reflected 10 times. Describe how you came to your conclusion.

4. In Exercise 29, Section 11.4, you were asked to find the total distance traveled by a ball that was dropped from a height of 4 feet under the condition that each time the ball hit the floor, it bounced up 3/5 the distance it fell. The function $s(t) = s_0 - 16t^2$, where s_0 is initial height, relates the distance s an object falls to the time t it takes the object to fall that distance. Use your results of Exercise 29 and this formula to find the total amount of time it takes the ball to come to rest.

SECTION 1.1

1. $V(s) = s^3, \; s > 0$

3. $C(n) = 0.2n, \;$ integer $n \geq 0$

5. $C(n) = 0.25n, \;$ integer $n \geq 0$

7. $A(h) = 3h, \; h > 0$

9.

Input	Output
2	−6
3	−5
4	−2
5	3
a	$a^2 - 4a - 2$
b	$b^2 - 4b - 2$
x	$x^2 - 4x - 2$
$x + 3$	$x^2 + 2x - 5$
$x + h$	$x^2 + 2hx + h^2$ $- 4x - 4h - 2$
$x^2 - 4x - 2$	$x^4 - 8x^3$ $+ 8x^2 + 32x + 10$

11. $f(x) = x^2 + 1$

13. $A(b, h) = \frac{1}{2}bh, \; b > 0, \; h > 0.$

Base and height cannot be negative, since they are distances, and if either is zero, the figure is not a triangle.

15. $V(r) = \frac{4}{3}\pi r^3, \; r > 0$

The radius is a distance, and if it is zero, the object is not a sphere.

17. Answers vary.

19. $N(r, t) = rt.$ Time in hours, rate in thousands of buttons per hour (for example).

SECTION 1.2

1. 5 0 XIT

3. XIT x^2 − 3

5. $V(h) = 6h;$ 6 XIT

7.

x	$f(x)$
2	−2
3	0
4	4
5	10

9. 2 ÷ (XIT + 1)

11. (2 + XIT) ÷ 3

13. 1 + 2 ÷ XIT

15. 1 ÷ (XIT + 2) + 3

17. The graph crosses the x-axis at $\frac{3}{2}$ and the y-axis at −3.

19. The graph does not cross the horizontal axis. It crosses the vertical axis at −3.

21. The graph crosses the x-axis at 9 and the y-axis at −18.

23. Answers vary.

25. a) for $y = 1005, \; x \approx 2.42$

     b) for $y = 1250, \; x \approx 2.45$

27. $R(x) = 2x, \; P(x) = 1.25x - 500$

29. $A(x) = .075x + 50 , \; E(x) = .825x + 550$

31. \$2270

33. The claim is false

35. $U(n) = \dfrac{187500}{n} + 2.5$

37. 34.8 months

39. The sale price is $.8x$; the customer pays $.868x.$

SECTION 1.3

Answers for maximum error will vary, but must be no greater than the values given here.

1. $x = -2.78,$ max. error $= .0408$

3. $x = -0.24$ or $x = 4.24,$ max. error $= .045$

5. $x = -3.26$ or $x = 1.92,$ max. error $= .32$

7. $x = -3.26,$ max. error $= .25$

9. $x = 1.38,$ max. error $= .026$

11. $x = 23.90,$ max. error $= .0025$

13. $x = 8$, max. error $= .08$

15. $t = 1.20$ hr, max. error $= .31$ units

17. $x = 775.50$ ft, max. error must be less than .016 dollars, or around 2¢

19. $x = 9.71$, max. error $= \$200,000$

SECTION 1.4

1. 2, 12

3. 0.29

5. 25

7. a) 15

9. −2.63, 0.41

11. −4.76 or 15.76

13. $S(t, e, f) = 6t + e + 3f$

15. No

17. A straight line which appears to pass through the origin and, due to scale distortion, appears to have a positive slope less than 1.

19. $f(x) = \log_2 x$

21. $x = 7.00$

SECTION 1.5

1. $(-2.33, 0.67)$

3. $(-.41, 4), (2.41, 4).$ $f(x) > 4$ when $x < -.41$ or $x > 2.41$.

5. $x = 4$ (then $y = 8$)

7. $Q(t) = 200t$, $Q =$ no. of handsets, $t =$ hours

9. −1.08

11. It has no real zeros

13. $y = 51$, extrapolation

15. $t = 2.026$

17. $H(s) = \dfrac{s^2}{10}$

19.

Input	Output
2	1/8
3	1/18
4	1/32
5	1/50

SECTION 1.6

1.

x	$f(x)$
$-\infty < x < 1$	$f(x)$ is decreasing
$x = 1$	rel. min. at $(1, -49)$
$1 < x < \infty$	$f(x)$ is increasing
x-intercepts	$(-6, 0), (8, 0)$
y-intercept	$(1, -48)$

3.

x	$f(x)$
$-\infty < x < \infty$	$f(x)$ is increasing
x-intercept	$(0.22, 0)$
y-intercept	$(0, -1)$

5.

x	$f(x)$
$-\infty < x < -.77$	$f(x)$ is increasing
$x = -.77$	rel. max. at $(-.77, -12.40)$
$-.77 < x < -.19$	$f(x)$ is decreasing
$x = -.19$	rel. min. at $(-.19, -12.57)$
$-.19 < x < \infty$	$f(x)$ is increasing
x-intercept	$(1.47, 0)$
y-intercept	$(0, -12.5)$

7.

x	$f(x)$
$-\infty < x < .82$	decreasing
$x = .82$	rel. min. at $(0.82, 0.91)$
$.82 < x < \infty$	increasing
x-intercept	none
y-intercept	$(0, 5.3)$

9.

x	$f(x)$
$-\infty < x < -1.41$	$f(x)$ is increasing
$x = -1.41$	rel. max. at $(-1.41, 10.66)$
$-1.41 < x < 1.41$	$f(x)$ is decreasing
$x = 1.41$	rel. min. at $(1.41, -.66)$
$1.41 < x < \infty$	$f(x)$ is increasing
x-intercepts	$(-2.79, 0), (1, 0), (1.79, 0)$
y-intercept	$(0, 5)$

11.

x	$f(x)$
$-\infty < x < -2$	$f(x)$ is increasing
$x = -2$	discontinuity
$-2 < x < \infty$	$f(x)$ is increasing
x-intercept	none
y-intercept	$(0, -2.5)$

13.

x	$f(x)$
$-\infty < x < -0.53$	$f(x)$ is increasing
$x = -0.53$	discontinuity
$-0.53 < x < \infty$	$f(x)$ is increasing
x-intercept	$(32.86, 0)$
y-intercept	$(0, -2.3)$

15.

x	$f(x)$
$-\infty < x < 0$	$f(x)$ is decreasing
$x = 0$	vertical asymptote
$0 < x < \infty$	$f(x)$ is increasing
x-intercept	$(-54.60, 0), (54.60, 0)$
y-intercept	none

17.

x	$f(x)$
$-\infty < x < \infty$	$f(x)$ is decreasing
x-intercept	none
y-intercept	$(0, 160,000)$

19.

x	$f(x)$
$-\infty < x < -1.12$	$f(x)$ is increasing
$x = -1.12$	rel. max. at $(-1.12, .46)$
$-1.12 < x < 0$	$f(x)$ is decreasing
$x = 0$	rel. min. at $(0, 0)$
$0 < x < 1.12$	$f(x)$ is increasing
$x = 1.12$	rel. max. at $(1.12, .46)$
$1.12 < x < \infty$	$f(x)$ is decreasing
x--intercept	$(0, 0)$
y-intercept	$(0, 0)$

21. The ball rises, decelerating under the influence of gravity, until at $t = 1.5$ it comes to a momentary stop and then begins falling. At $t = 3$ the ball has returned to its original height. If the height of the ball is plotted as a function of time, the curve is a parabola.

23. The graph is a curve which starts at $(0, 5.9)$, rises gradually to a peak at $(3.58, 7.10)$, and then falls off steeply, hitting $y = 0$ at around $x = 10$. The percentage of pupils in private schools peaks, then, in the 1988-89 school year.

25. The concentration of the drug starts at 0, reaches a peak of .8 mg per liter after .883 hours, and then gradually tapers off, approaching zero asymptotically.

27. $A(x) = 12x - x^2$. The graph is a parabola which crosses the x-axis at $x = 0$ and $x = 12$, and peaks at $x = 6$. The maximum area of the window, represented by the peak, is 36 sq. ft.

SECTION 2.1

1. $\sqrt{13}$

3. $3\sqrt{13}$

5. $\sqrt{193}$

7. 1

9. 3

11. .0128

13. $\sqrt{(12s)^2 + (3y + 4)^2}$

15. $(4, 1)$

17. $(-2.5, -4.5)$

19. $(-171.5, -57)$

21. $(-18a, -6b)$

23. Yes

25. No

27. No

29.

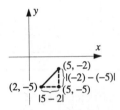

31. All points $(x, 0)$ such that
    $x^2 - 4x - 157 = 0$

33. All points $(0, y)$ such that
    $y^2 - 124y + 4420 = 0$

35. All points $(0, y)$ such that
    $y^2 - 12y + 32 = 0$

37. $x^2 - 14x + y^2 - 2y + 25 = 0$

39. $x^2 + 6x + y^2 + 10y - 110 = 0$

41. $x^2 + 4x + y^2 - 8y + 20 - \pi^2 = 0$

43. $x^2 + y^2 + 4y - 1 = 0$

45. The vertical line through $(-2, 0)$

47. The vertical line through $\left(\sqrt{10},\ 0\right)$

49. All points below and including the
    horizontal line through $(0, -4)$

51. All points on either of the two coordinate
    axes

53. The vertical lines through $(-3, 0)$ and $(4, 0)$

55. All points between the vertical lines through
    $(-1, 0)$ and $(1, 0)$

57. $7x + 4y = 65$

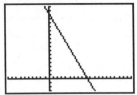

SECTION 2.2

1. No

3. Yes

5. Yes

7. Yes

9. No

11. Yes

13. Yes

15. Yes

17. No

19. Yes

21. No

23. $D = \{x | x \neq 8\}$
    $R = \{y | y \neq 0\}$

25. $D =$ All real numbers
    $R = \left\{ y | y \leq -\dfrac{15}{16} \right\}$

27. $D = \{x | x \geq -2 \text{ and } x \neq 2\}$
    $R =$ All real numbers

29. $D = \{x | x > -4\}$
    $R = \{y | y > 0\}$

31. $\{D(x) | 16.09 \leq D(x) \leq 48.26\}$

33. $\{W(t) | 6.1 \leq W(t) \leq 8.2\}$

35. $\{T(t) | 0 < T(t) \leq 68.52\}$

37. $D = \{12, 1, 2, 3\}$
    $R = \{0, 1, 2, 3, \ldots\}$

    This relation is not a function, because the
    records are kept for two consecutive days, and
    so two of the ordered pairs will have 12 as
    the first element, etc., and the second
    elements may be different.

SECTION 2.3

1.   31

3.   25

5. $-7s - 12$

7.   12

9.   16

11. $4k^3 - 6k^2 + 5k - 8$

13. $6h^2 - 8$

15. $5x + 5h - 8$

17. $7(x + h)^2 - 3(x + h) - 1$

19. $10x^2 - 11x + 4$

21. $-8x + 9$

23. $2h$

25. $2xh + 5h + h^2$

27. $14xh + 7h^2 - 8h$

29.   4

31. $2x + h - 6$

33. $10x + 5h + 9$

35. $-4x - 2h + 8$

37. $\dfrac{8}{(x + h + 5)(x + 5)}$

39.   a)   403 CD players

   b) $242,665

   c)   200 disks, or else 606 disks

41.   a)   2 sec.

   b)   214 ft.

   c)   5.7 sec.

43. $h(r) = \dfrac{.756}{\pi r^2}$

   $h(2.1) \approx .055$ cm

   There is no minimum height.

SECTION 2.4

1.   Upward displacement by 2 units

3.   Displacement to the right by 3 units

5.   Horizontal displacement to the left of 1 unit and then a vertically downward displacement of 2 units.

7.   Shift to the left of 2 units, reflection across the x-axis, then shift upwards by 3 units.

9.   Graph is contracted by a factor of $\dfrac{1}{3}$.

11.   Graph shifted to the right by 2; contracted by $\dfrac{1}{3}$; reflected across the x-axis; vertically shifted upwards by 4 units.

13. $f(x) = \dfrac{1}{4}x^2$

15. $f(x) = 4x^2$

17. $f(x) = -5x^2$

19. $f(x) = -(x + 5)^2$

21. $f(x) = -[(x - 15)^2 - 4]$

23. $f(x) = \dfrac{1}{8}[(x - 8)^2 + 8]$

25. $f(x) = 2[(x - 6)^2 - 20]$

27.

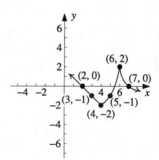

29.

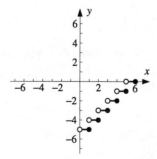

31.

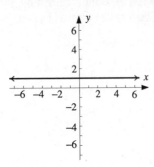

33. It has been contracted by a factor of $\frac{1}{5}$ and shifted to the right by 2 units.

35. It has been contracted by a factor of $\frac{3}{4}$, reflected across the x-axis, shifted left by 1, then shifted upward by 5 units.

37. a) $C(t) = \dfrac{2.56(t + 2) + 35}{1.8(t + 2) + 45}$

   $= \dfrac{2.56t + 40.12}{1.8t + 48.6}$

   b) $C(t) = \dfrac{2}{3}\left(\dfrac{2.56t + 35}{1.8t + 45}\right)$

   $= \dfrac{5.12t + 70}{5.4t + 135}$

39. a) $G(I) = \dfrac{5.5\left(2.5I + \sqrt{I + 30}\right)}{\sqrt{I + 30}} + 1.8$

   b) $G(I) = \dfrac{11\left(2.5I + \sqrt{I + 30}\right)}{3\sqrt{I + 30}}$

41. He did not consider fixed costs, viz., the cost of machinery, etc. Adding this cost in shifts the graph upward.

SECTION 2.5

1. $(f + g)(x) = x^2 + 5x + 1, D = \mathbf{R}$

   $(f - g)(x) = -x^2 + x + 3, D = \mathbf{R}$

   $(f \cdot g)(x) = 3x^3 + 8x^2 + x - 2, D = \mathbf{R}$

   $\left(\dfrac{f}{g}\right)(x) = \dfrac{3x + 2}{x^2 + 2x - 1}, \left\{ x | x \neq -1 \pm \sqrt{2} \right\}$

3. $(f + g)(x) = \sqrt{x(x + 3)} + \dfrac{1}{\sqrt{(x - 3)(x + 2)}}$

   $D = \{x | x > 3\} \cup \{x | x \leq -3\}$

   $(f - g)(x) = \sqrt{x(x + 3)} - \dfrac{1}{\sqrt{(x - 3)(x + 2)}}$

   $D = \{x | x > 3\} \cup \{x | x \leq -3\}$

   $(f \cdot g)(x) = \sqrt{\dfrac{x(x + 3)}{(x - 3)(x + 2)}}$

   $D = (-\infty, -3] \cup (-2, 0] \cup (3, \infty)$

   $\left(\dfrac{f}{g}\right)(x) = \sqrt{x(x + 3)(x - 3)(x + 2)}$

   $D = (-\infty, -3] \cup (-2, 0] \cup (3, \infty)$

5. a) $(f + g)(x) = \dfrac{9x}{4} - \dfrac{7}{4}$

   b) $(f - g)(-2) = \dfrac{15}{4}$

   c) $(f \cdot g)(h + 1) = \dfrac{2h^2 - 3h - 2}{4}$

   d) $\left(\dfrac{f}{g}\right)(4) = \dfrac{1}{28}$

7. $(f \circ g)(x) = 2x^2 - 2x + 5, D = (-\infty, \infty)$

   $(g \circ f)(x) = 4x^2 + 18x + 20, D = (-\infty, \infty)$

9. $(f \circ g)(x) = 4x^4 - 18x^2 + 16, D = (-\infty, \infty)$

   $(g \circ f)(x) = 2x^4 - 4x^3 - 14x^2 - 16x + 27,$

   $D = (-\infty, \infty)$

11. $(f \circ g)(x) = \dfrac{3}{5(x - 4)}, D = \{x | x \neq 4\}$

   $(g \circ f)(x) = \dfrac{3}{5x} - 4, D = \{x | x \neq 0\}$

13. $(f \circ g)(x) = \sqrt{x + 1}, D = [-1, \infty)$

   $(g \circ f)(x) = \sqrt{x} + 1, D = [0, \infty)$

15. $(f \circ g)(x) = \dfrac{5 - x}{x}, D = \{x | x \neq 0, -5\}$

   $(g \circ f)(x) = \dfrac{5}{x - 1}, D = \{x | x \neq \pm 1\}$

17. $(f \circ g)(x) = \dfrac{4x + 50}{7x + 10}, D = \left\{ x | x \neq -\dfrac{10}{7} \right\}$

   $(g \circ f)(x) = \dfrac{28x + 41}{-2x - 14}, D = \{x | x \neq -7\}$

19. $(f \circ g)(x) = \{(-5, 4)\}, \ g \circ f = \{(4, 1), (-5, 2)\}$

21. $(f \circ g \circ h)(x) = 24,$

                              $= 24$ at $x = 0$

23. $(f \circ g)(x) = x$

    $(g \circ f)(x) = x$

25. $(f \circ g)(x) = x, \ D = \{x | x \neq 1\}$

27. a) \$37,196

    b) \$54,131

    c) 3.67

29. a) 88.50 lb

    b) 24.95 lb

    c) 7.34 weeks

31. $s(x) = \dfrac{1}{2\sqrt{25 - x}}, \ D = \{x | x < 25\}$

                    $R = \{f(x) | f(x) < 0\}$

    As x approaches 25, the slope becomes infinite, i.e., the curve becomes almost vertical.

33. $M(n) = 30 + 25(n - 1) = 25n + 5$

    $C(n) = 41.25 + 6.25n$

    Total cost for 10 days = \$756.25

35. $C = \$316.25$

37. $G(x) = 7.50x + 3.50(850 - x) = 4x + 2975$

       G in dollars

    Minimum gate = \$2,975

    Maximum gate = \$6,375

39. $A = \pi r^2$

    $V = \dfrac{4}{3} \pi r^3$

    $A = \pi^{1/3} \left(\dfrac{3V}{4} \right)^{2/3}$

    $V = \dfrac{4A^{3/2}}{3\sqrt{\pi}}$

41. $C = 2\pi\sqrt{t + 1}$

    $A = \pi(t + 1)$

    $C(0) = 6.28$

    $C(60) = 49.07$

SECTION 2.6

1. $m = kn$

3. $u = kp^5$

5. $y = \dfrac{k}{x^2}$

7. $t = \dfrac{ks^2}{q^3}$

9. $y = \dfrac{kx^3 z^2}{w\sqrt{z}}$

11. $y = kx$

    $y = 5x$

    $y = 10$ when $x = 2$

13. $y = \dfrac{k}{x}$

    $y = \dfrac{400}{x}$

    $y = 10$ when $x = 40$

15. $y = kxw^2$

    $y = \dfrac{3xw^2}{50}$

    $y = \dfrac{147}{25}$ when $x = 2, \ w^2 = 7$

17. $L = 560$ pounds

19. $d = 125$ inches

21. $f = 243.8$ kilocycles

23. 180 pounds

25. The number of mutations is also tripled

27. The new force is 24 times the original force

29. The new velocity is $\dfrac{1}{\sqrt{2}}$ times the original velocity. To double the velocity, the radius has to be reduced to $\dfrac{r}{4}$.

31. y varies directly as x

33. y varies directly as the third power of x

35. y varies inversely as x

SECTION 2.7

1. Strongly linear

3. Weakly linear

5. Weakly linear

7. Nonlinear

9. Nonlinear

11. $\hat{y} = 105.6341413 - .334863856x$

$r = .989808628$

A one unit change in x means a $-\frac{1}{3}$ unit change in y.

13. Nonlinear

15. $\hat{y} = -8.173195091 + 1.004356113x$

$r = .8914348279$

A one unit change in x means a one unit change in y.

17. $\hat{y} = 24.54285714 + 50.28214286x$

$r = .9959210395$

A one unit change in x means a 50.3 unit change in y.

19. Nonlinear (although the correlation is high with a linear model, it is higher still with a logarithmic one)

SECTION 3.1

1. possible polynomial, odd degree, positive leading coefficient

3. not a polynomial

5. not a polynomial

7. not a polynomial

9. possible polynomial, even degree, negative leading coefficient

11. not a polynomial

13. possible polynomial, even degree, negative leading coefficient

15. polynomial function, at most one turning point

17. not a polynomial function

19. polynomial function, at most three turning points

21. not a polynomial function

23. odd

25. neither

27. neither

29. neither

31. neither

33. $f(x)$ has at most 3 zeros and at least 1 zero

35. $f(x)$ has at most 2 zeros and possibly none

37. $f(x)$ has at most 6 zeros and possibly none

39. $-4, 6$

41. $-.25, 3.5$

43. $-5, 8.33$

45. 4.1

47. $-4.3, -.7$

49. 1) $V(r) = \pi r^2(15 - 2r) + \frac{4}{3} \pi r^3$

2) 281.75 cu ft

3) 2.50 ft

51. 1) 3.2 hrs

2) 5.2 hrs

53. $f(x) = \begin{cases} 1 - (x - 1)^2 \text{ for } 0 \le x \le 2 \\ 1 - (x - 3)^2 \text{ for } 2 \le x \le 4 \\ 1 - (x - 5)^2 \text{ for } 4 \le x \le 6 \end{cases}$

55. No, since one might be a compression of the other

57. 2885 at ten yrs, max. pop. at 4.4 yrs

59. $\frac{13}{2}$ ft on a side

SECTION 3.2

1. $1 + 4i$

3. $8 + 6i$

5. $2 - 2i$

7. $5 - 6i$

9. $19 - 22i$

11. $-33 - 56i$

13. 29

15. $13 + 3\sqrt{3}\, i$

17. $\frac{25}{29} - \frac{10}{29} i$

19. $-\frac{22}{37} + \frac{16}{37} i$

21. $-\frac{13}{5} - \frac{4}{5} i$

23. $-\dfrac{3}{25} - \dfrac{4}{25}i$

25. $-i$

27. $\overline{z_1 + z_2}\ = \overline{a + bi + c + di}$

$= \overline{(a + c) + (b + d)i}$

$= (a + c) - (b + d)i$

$= a - bi + c - di$

$= \overline{z}_1 + \overline{z}_2$

29. $\overline{z_1 \cdot z_2}\ = \overline{(a + bi)(c + di)}$

$= \overline{ac + adi + bci - bd}$

$= ac - adi - bci - bd$

$= (a - bi)(c - di)$

$= \overline{z}_1 \cdot \overline{z}_2$

31. $z_1 + \overline{z}_1\ = a + bi + \overline{a + bi}$

$= a + bi + a - bi$

$= 2a$

33. $i = i \qquad\qquad i^5 = i$

$i^2 = -1 \qquad\quad i^6 = -1$

$i^3 = -i \qquad\quad i^7 = -i$

$i^4 = 1 \qquad\quad i^8 = 1$

The numbers $i, -1, -i, 1$ are repeated over and over.

$$i^n = \begin{cases} i & \text{for } n = 4k + 1, \\ -1 & \text{for } n = 4k + 2, \\ -i & \text{for } n = 4k + 3, \\ 1 & \text{for } n = 4k, \end{cases}$$

where k = any integer

35. no real zeros, two complex zeros

37. no real zeros, four complex zeros

39. two real zeros, four complex zeros

41. no real zeros, four complex zeros

The number of complex zeros is always even.

SECTION 3.3

1. a) $x - 2$
   b) $(x - 2)(x + 1) + 5$

3. a) $a - 2$
   b) $(a - 2)(a + 2) + 8$

5. a) $x^2 + x + 2$
   b) $(x^2 + x + 2)(x - 1) + 8$

7. a) $x^2 + 5x + 11$
   b) $(x^2 + 5x + 11)(x - 2) + 20$

9. a) $x^2 + x - 9$
   b) $(x - 2)(x^2 + x - 9) - 3$

11. a) $2x^2 - 8x + 6$
    b) $(2x^2 - 8x + 6)(x - 4)$

13. a) $x^4 + x^3 - 4x^2 - 2x + 3$
    b) $(x^4 + x^3 - 4x^2 - 2x + 3)(x + 1) - 1$

15. a) $x^4 - x^3 + x^2 - x + 1$
    b) $(x^4 - x^3 + x^2 - x + 1)(x + 1)$

17. a) $2x^2 + (3 - 4i)x - (10 + 10i)$
    b) $[2x^2 + (3 - 4i)x - (10 + 10i)]$
       $[x - (1 - 2i)] + (-20 + 10i)$

19. $(3x^2 - 5x + 2)$

21. a) a third-degree polynomial with negative leading coefficient
    b) yes; higher-degree polynomials with only three zeros, for example

SECTION 3.4

1. $4, -6$

3. $-9.8, -3.2, 5.3$

5. yes

7. no

9. $x + 3$

11. $x + \dfrac{3}{2}$ and $x + \dfrac{5}{2}$

13. $x - 4$

15. true

17. true

19. true

21. $x^2 + 3x - 10$

23. $x^3 - x^2 - 9x + 9$

25. $x^2 - 7$

27. $(x - 5)(x - 2)(x + 4)(x + 8)$

29. $x^2 - 6x + 13$

31. $x^3 - 7x^2 + 17x - 15$

33. $x^3 - 3x^2 - 10x$

35. $(x - 5)(x - 4)(x + 2)(x + 5)$;
zeros are $-5, -2, 4, 5$

37. $(x + 10)(x + 9)(x + 4)(x)(x - 2)(x - 6)$;
zeros are $-10, -9, -4, 0, 2, 6$

39. $(x + 6)(x - 4)$; zeros are $-6, 4$

41. $(4x + 3)(6x - 1)$; zeros are $-\dfrac{3}{4}, \dfrac{1}{6}$

43. $6(x - 5)(x + 5)$; zeros are $5, -5$

45. $(x - 6)(x^2 + 6x + 36)$, real zero is 6

47. $(x^2 + 4)(x^3 + 3)$, real zero is $\sqrt[3]{-3}$

49. $(x^2 + 4)(x + 2)(x - 2)(x + 3)(x - 3)$, real
zeros are $-2, 2, -3, 3$

51. $x = -1$ is a zero of even multiplicity. The
polynomial has $x + 1$ and $x - 2$ as factors,
and has odd degree; $x = 2$ is a zero.

SECTION 3.5

1. No

3. No

5. No

7. No

9. $(x - 1)(x - 2)(x - 7)$

11. $(x - 2)(x + 4)^2$

13. $(x + 4)(x + 2)(x - 1)(x - 3)$

15. $(x + 3)(x + 2)(x - 2)(x - 1)(x - 3)$

17. $(x - 2)[x - (1 - 5i)][x - (1 + 5i)]$

19. When $x = -1$, the polynomial equals zero.

21.

$$
\begin{array}{r|rrrrr}
-5 & 1 & 10 & 22 & -30 & -75 \\
   &   & -5 & -25 & 15 & 75 \\
\hline
   & 1 & 5 & -3 & -15 & 0
\end{array}
$$

23. $(4 - 3i)^2 - 8(4 - 3i) + 25$
$= 9 + 9i = 0$

25. $f(-i) = (-i)^3 + 2(-i)^2 - 3(-i) - (-i)^2 i$
$\qquad - 2(-i)i + 3i$
$\qquad = i - 2 + 3i + i - 2 + 3i$
$\qquad = -4 + 8i \neq 0$

27. $x^2 - 4$

29. $x^4 - x^3 - 10x^2 - 8x$

31. $x^2 - 8x + 17$

33. $x^2 + .9x - 4$

35. The device should be triggered at $t = 29.81$
seconds.

SECTION 3.6

1.

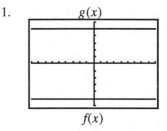

$g(x)$ is reflection of $f(x)$ across x-axis

3.

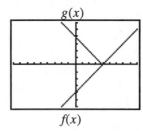

$f(x) = g(x)$ for $x \geq 4$, and for $x < 4$, graph of
$g(x)$ is reflection of $f(x)$ across x-axis.

5.

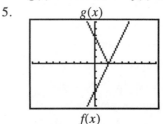

$f(x) = g(x)$ for $x \geq 2$. For $x < 2$, $g(x)$ is
reflection of $f(x)$ across x-axis.

7.

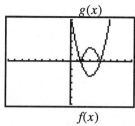

$f(x) = g(x)$ for $x \leq 3 - \sqrt{2}$ and $x \geq 3 + \sqrt{2}$.
For $3 - \sqrt{2} < x < 3 + \sqrt{2}$, $g(x)$ is reflection
of $f(x)$ across x-axis.

9.

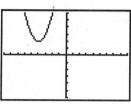

$g(x) = f(x)$

11.

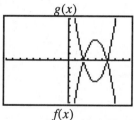

$g(x)$ is reflection of $f(x)$ across x-axis

13.

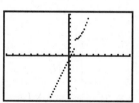

15.

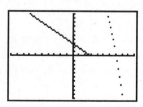

17.

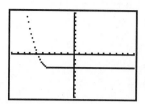

19.

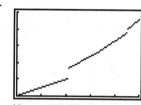

Xmax= 500, Ymax=50

21.

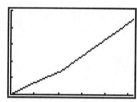

Xmax=52, Ymax=25000

23. No. Since the ball also had the velocity of the car, it falls ahead of the tree. Better strategy would be to aim before the car can cross the tree.

25. She will overshoot the ramp.

27. Solving the equation gives the time it takes for her to come down to the same height.

speed $\approx$ 112.64 ft/sec.

SECTION 3.7

1. $-3 \leq x \leq 7$

3. $x > 1$

5. $(-2 \leq x \leq 1) \cup (4 \leq x)$

7. $(-\infty, -3] \cup [-2, 2] \cup [3, \infty)$

9. $(x \leq -2) \cup (0 \leq x \leq 3)$

11. $x \neq 4$

13. $-5 < x < -2$

15. $-\dfrac{4}{3} < x < \dfrac{1}{2}$

17. $x \geq 3$

19. $[-5, -2] \cup [2, 4]$

21. $(-7 < x < -4) \cup (-1 < x < 2) \cup (5 < x)$

23. 5.51 seconds

25. $\approx$ 13.89 hrs

27. For $0 < t < 21.5$ days

29. $10.28 < x < 37.41$

SECTION 4.1

1.

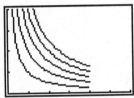

(How were graphs 1 - 7 generated on the TI-81? Hint: Use Parametric Mode.)

3.

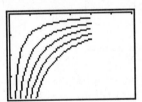

5.

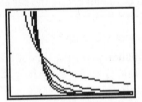

7.

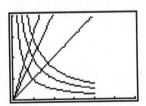

9.

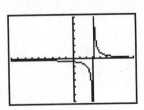

11.

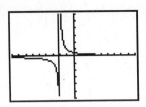

13.

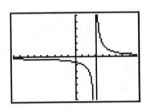

15.

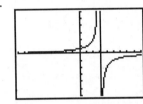

17.

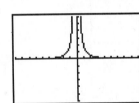

19.

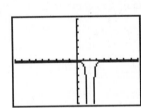

21. The graph has a vertical asymptote at $x = a$ and a horizontal asymptote at $y = 0$. It is decreasing on $(-\infty, a)$ and (a, ∞).

23. The graph has a horizontal asymptote at $y = 4$. The graph has a vertical asymptote at $x = 3$. It is decreasing on $(-\infty, 3)$ and $(3, \infty)$.

25. As x increases, the function value decreases.

x	$1/x^2$
10^1	10^{-2}
10^2	10^{-4}
10^3	10^{-6}
10^4	10^{-8}
10^5	10^{-10}
10^6	10^{-12}
10^7	10^{-14}
10^8	10^{-16}
10^9	10^{-18}

27. The numbers in the domain column are decreasing, and as x decreases toward zero, the function increases rapidly.

x	$1/x$
10^0	1
10^{-1}	10
10^{-2}	10^2
10^{-3}	10^3
10^{-4}	10^4
10^{-5}	10^5
10^{-6}	10^6
10^{-7}	10^7
10^{-8}	10^8
10^{-9}	10^9

29. $f(x) = 2 - \dfrac{160}{x}$

Amount of water gained in 120 minutes is $120\,f(x)$. The function is not defined for $x = 0$.

$x = 240$ minutes indicates the amount of time in which the full pond will be emptied because of the leak.

SECTION 4.2

1. horizontal asymptote $y = 1$, vertical asymptote $x = 6$

3. horizontal asymptote $y = 3$, vertical asymptote $x = 3$

5. horizontal asymptote $y = 0$, no vertical asymptote

7.

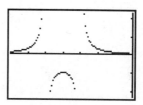

9.

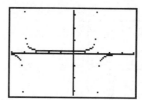

11.

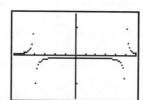

13.

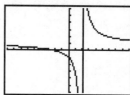

asymptotes: $x = 2, y = 1$

15.

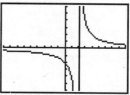

asymptotes: $x = 2, y = 0$

17.

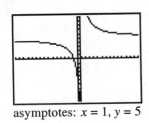

asymptotes: $x = 1, y = 5$

19.

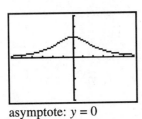

asymptote: $y = 0$

21.

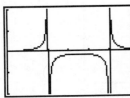

asymptotes: $x = 2, x = 5, y = 0$

23. The function is always positive, symmetric about the y-axis, has a maximum value of 2 at $x = 0$, and has a horizontal asymptote at $y = 0$.

25. The function is positive in $(3, \infty)$
The function is negative in $(-\infty, 3)$
Vertical asymptote $x = 3$
Horizontal asymptote $y = 0$
Intersects the y-axis at $y = -\dfrac{1}{3}$

27. The function is positive in $(-\infty, 2)$
The function is negative in $(2, \infty)$
Vertical asymptote $x = 2$
Horizontal asymptote $y = 0$
Intersects the y-axis at $y = \dfrac{1}{4}$

29. The function is positive in $\left(-\infty, -\dfrac{5}{2}\right)$
The function is negative in $\left(-\dfrac{5}{2}, -\infty\right)$
Vertical asymptote $x = -\dfrac{5}{2}$
Horizontal asymptote $y = 0$
Intersects the y-axis at $y = -\dfrac{3}{5}$

31. This is an upward expanded parabola, vertex at $\left(\dfrac{1}{6}, -\dfrac{61}{12}\right)$

33. a) 1.6 liters of acid, 3.4 liters of water
b) .59 liter of acid. Resulting volume is 5.59 liters.
c) 4.92 liters of 45%, app.
3.08 liters of 32%, app.

SECTION 4.3

1. As $x \to +\infty$ or $-\infty$, $f(x) \to 3x + 11$
3. As $x \to +\infty$ or $-\infty$, $h(x) \to x^2 + 4x + 11$
5. As $x \to +\infty$ or $-\infty$, $g(x) \to x - 1$
7. None. Function undefined at $x = -4$
9. None. Function undefined at $x = 2, x = -3$
11. $x = 3, x = -2$
13.

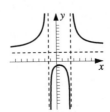

asymptotes: $x = \dfrac{1 \pm \sqrt{13}}{2}$, $y = 3$

15.

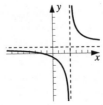

asymptotes: $x = 3, y = 1$

17.

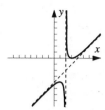

asymptotes: $x = 2$, $y = x - 4$

19.

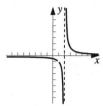

asymptotes: $x = 2$, $y = 0$; there is also a "hole" at $x = 1$

21.

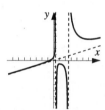

y ranges from -80 to 80

asymptotes: $x = \dfrac{3 \pm \sqrt{5}}{2}$, $y = 3x + 7$

23.

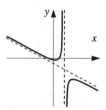

y ranges roughly from -100 to 100
asymptotes: $x = 4$, $y = -4x - 14$

25. The function has a vertical asymptote at $x = 1$ and a horizontal asymptote at $y = 0$. The function is positive for $x < 1$ and negative for $x > 1$. Intersects y-axis at $y = 1$. Does not intersect x-axis.

27. The function is linear, i.e., $h(x) = x + 1$; $x \neq 4$

29. $w(x) = \dfrac{100}{x}$, $x \neq 0$

$P(x) = 2x + \dfrac{200}{x}$

$C(x) = 6.6\left[2x + \dfrac{200}{x}\right]$

C is min. for $x = 10$

$C(10) = \$264$

A domain of $0 < x < 100$ or thereabouts would be physically reasonable. Other restrictions are also possible.

SECTION 4.4

1. $x = t$, $y = t^2 + 2$

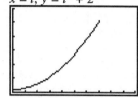

Xmax = 10, Ymax = 60

3. $x = t$, $y = 2t$

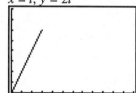

5. $x = t$, $y = 5 - t^2$

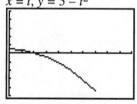

Ymin = -50, Ymax = 50

7. $x = t$, $y = t - 5$

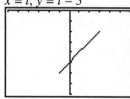

9. $x = t, y = t^2 + 5t$

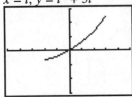

Ymin = -30, Ymax= 30

11.   Let x = horizontal displacement
        y = vertical displacement
        $x(t) = 50t, t \geq 0$
        $y(t) = 3t$

13.

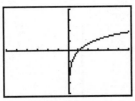

15.   Counter clockwise. Elliptical originally. A
        circle after zoom-square.

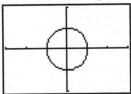

17.   Relations. Not functions by vertical line
        test.

19. $y = \sqrt{x + 1}$

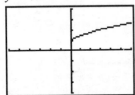

21.   a) $y = 8x - \dfrac{x^2}{25}$, a function

        b) $x = 100 \pm 5\sqrt{400 - y}$, not a function

23. $t \approx 23.5$ sec.
        $x \approx 8,272$ ft.

SECTION 4.5

1. f^{-1}

x	y
4	1
5	2
7	3
4	4

Not a function.

3. h^{-1}

x	y
2	−1
10	−2
10	−3
2	−4

Not a function.

5.

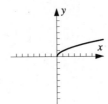

Function.

7.

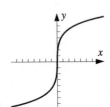

Function.

9.

Function.

11.

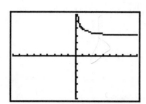

13.

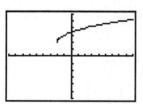

15.

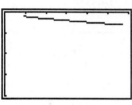

Xmax=60, Ymin=-40

17.

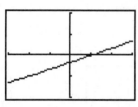

19.

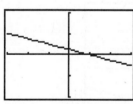

Xmin=-15, Xmax=15;
Ymin=-10, Ymax=10

21. $f^{-1}(x) = -2 + \sqrt{x + 15}, \ x \geq -15$

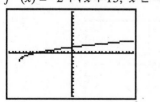

23. $h^{-1}(x) = \dfrac{3}{2} + \sqrt{x + \dfrac{33}{4}}, \ x \geq -\dfrac{33}{4}$

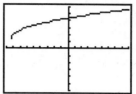

25. $g^{-1}(x) = \dfrac{4 - x}{3}$

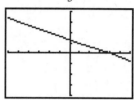

27. $f^{-1}(x) = \dfrac{1}{2} + \sqrt{\dfrac{25x + 4}{4x}}, \qquad x \neq 0$

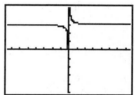

29. $(f \circ g)(x) = 2\left[\dfrac{1}{2}(x - 5)\right] + 5$

$= x - 5 + 5 = x$

$(g \circ f)(x) = \dfrac{1}{2}[(2x + 5) - 5] = \dfrac{1}{2} \cdot 2x = x$

31. $(f \circ g)(x) = \dfrac{1}{4\left(\dfrac{1 - 5x}{4x}\right) + 5}$

$= \dfrac{1}{\dfrac{1 - 5x}{x} + 5} = x$

$(g \circ f)(x) = \dfrac{1 - 5 \cdot \dfrac{1}{4x + 5}}{4 \cdot \dfrac{1}{4x + 5}}$

$= \dfrac{\dfrac{4x + 5 - 5}{4x + 5}}{\dfrac{4}{4x + 5}}$

$= \dfrac{4x}{4x + 5} \cdot \dfrac{4x + 5}{4} = x$

33. $(f \circ g)(x) = \sqrt[3]{\dfrac{1 - \dfrac{1}{3x^3+1}}{3 \cdot \dfrac{1}{3x^3+1}}} = \sqrt[3]{\dfrac{\dfrac{3x^3+1-1}{3x^3+1}}{\dfrac{3}{3x^3+1}}}$

$= \sqrt[3]{\dfrac{3x^3}{3x^3+1} \cdot \dfrac{3x^3+1}{3}} = \sqrt[3]{x^3} = x$

$(g \circ f)(x) = \dfrac{1}{\left[3 \cdot \left(\sqrt[3]{\dfrac{1-x}{3x}} \right)^3 + 1 \right]}$

$= \dfrac{1}{\left[3 \cdot \dfrac{1-x}{3x} + 1 \right]} = \dfrac{1}{\left[\dfrac{1}{x} \right]} = x$

SECTION 4.6

1. $f^{-1}(x) = 2x - 4,\ 1.5 \le x \le 4.5$

3. $h^{-1}(x) = \dfrac{1 + 3x}{x},\ x \le -\dfrac{1}{8}$ or $x \ge \dfrac{1}{2}$

5. $g^{-1}(x) = \dfrac{x + \sqrt{25x^2 + 4x}}{2x},\ \dfrac{1}{104} \le x < \infty$

7. $f^{-1}(x) = \dfrac{x-3}{2},\ R = $ All reals

9. $h^{-1}(x) = \sqrt{x + 3},\ R$ is $y \ge 0$

11. $g^{-1}(x) = -\sqrt{x + 3},\ R$ is $y \le 0$

13. $f^{-1}(x) = \dfrac{2}{x + 3},\ R$ is $y \ne 0$

15. $h^{-1}(x) = -\dfrac{3}{2} - \sqrt{x + \dfrac{29}{4}},\ R$ is $y \le -\dfrac{3}{2}$

17. $g^{-1}(x) = 4 - \sqrt{x + 14},\ R$ is $y \le 4$

19. $f^{-1}(x) = \dfrac{2 + \sqrt{5x + 9}}{5},\ R$ is $y > 1$

21. Answers may vary

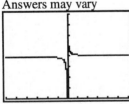

23. Answers may vary

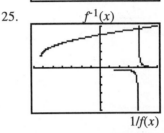

25.

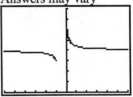

$f^{-1}(x)$

$1/f(x)$

$f^{-1}(x) = \dfrac{3 + \sqrt{25 + 4x}}{2}$

SECTION 4.7

1. $(x - 7)^{2/3}$

3. $(x^2 + 3)^{5/6} + 3$

5. $(2x + a)^{-2/3}$

7. $-4(3 - x)^{2/3}$

9. $x^{2/7}\, y^{3/7}\, z^{4/7}$

11.

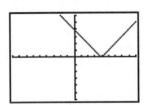

13.

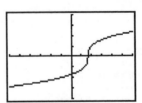

15.

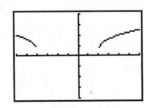

17.

19.

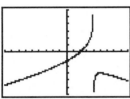

21. $x = 24$

23. No solution

25. 1

27. No solution

29. 16

31. $-3,123$

SECTION 5.1

1.

x	x^2	2^x	$x^2 > 2^x$ or $2^x > x^2$?
0	0	1	$x^2 < 2^x$
1	1	2	$x^2 < 2^x$
2	4	4	$x^2 = 2^x$
3	9	8	$x^2 > 2^x$
4	16	16	$x^2 = 2^x$
5	25	32	$x^2 < 2^x$
6	36	64	$x^2 < 2^x$
10	100	1024	$x^2 < 2^x$

3. $f(x) = \dfrac{1}{10^x}$, decay function

5. 6^x, growth function

7. e^x, growth function

9. Each is reflection of other across y-axis

11. Each is reflection of other across y-axis

13. The two graphs are identical

15. Each is reflection of other across y-axis

17.

x	$f(x) = (1 + x)^{1/x}$
1	2.0
10	2.593742460
100	2.704813829
1,000	2.716923932
10,000	2.718145927
100,000	2.718268237
1,000,000	2.718280469
1,000,000,000	2.718281827

19.

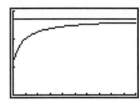

21.

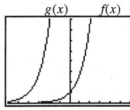

$g(x)$ $f(x)$

23.

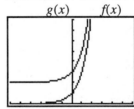

$g(x)$ $f(x)$

25.

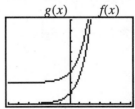

$g(x)$ $f(x)$

27.

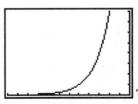

They are identical

29.

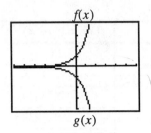

$f(x)$

$g(x)$

31.

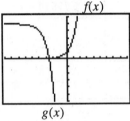

$f(x)$

$g(x)$

33. 1.39

35. 7.47

37. 6.37

39. $(f \circ g)(x) = e^{x-5}$

 $(g \circ f)(x) = e^x - 5$

 Not same

 $(f \circ g)(x) = e^x$ shifted by 5 units to the right

 $(g \circ f)(x) = e^x$ shifted by 5 units vertically down

41. $Q(t) = 9.959(.8)^t$

 $Q(8) \approx 1.669$

43. $P(t) = 5.998(1.020)^t$

 $P(15) \approx 8.07$

45. $C(t) = 2.965(.836)^x$

47. $2,377.59

49. It becomes 15.59 times the original amount

51. 16 hours

53. a) 10.64 inches

 b) 27.05 inches

 c) 28.49 inches

 d) .9 miles

55. a) 71.1° F

 b) 16.4 min.

57. 29.3 ft

59. half full or June 20th

 one quarter full on June 19th

 $N(t) = 4.2^t$

61. Once Misa starts collecting interest, she is always ahead

63. $P(n) = 2000(1 + .015)^n$, n = number of years since 1980

 $P(20) \approx 2,694$

65. Yes

67. $20,137.53. Yes.

SECTION 5.2

1. $f^{-1}(x) = \dfrac{1}{2} \log_7 x$

3. $f^{-1}(x) = \log_2 x - 4$

5. $f^{-1}(x) = \dfrac{1}{6} \log_5 (x - 2) + \dfrac{1}{2}$

7. $\log_6 36 = 2$

9. $\log_7 343 = 3$

11. $\log_3 \left(\dfrac{1}{81} \right) = -4$

13. $\log_{0.02} 0.000008 = 3$

15. $\log_{27} 9 = \dfrac{2}{3}$

17. $\log_b a = c$

19. $\log_{10} 1 = 0$

21. $\ln y = x$

23. $2^5 = 32$

25. $\left(\dfrac{1}{2} \right)^3 = \dfrac{1}{8}$

27. $4^{-2} = \dfrac{1}{16}$

29. $e^1 = e$

31. $y = 4$

33. $b = \sqrt{6}$

35. $y = -2$

37. $x = 8$

39. 7

41. e

43. 0

45. 1

47. $\frac{1}{2}$

49. 1

51. .9031

53. −.3757

55. .9000

57. ≈ 1.5

59. ≈ 2.5

61.
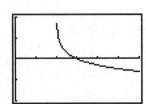

63.
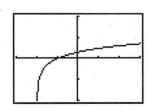

65.
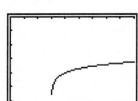

67.
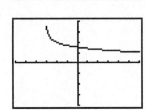

69. 10.07 yrs

71. $3,073.43

73.
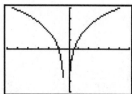

For $x > 0$, the two graphs are identical, because logarithms are exponents and $x(2x + 1) = 2x^2 + x$ (see also next section). For $x < 0$, $f(x)$ is undefined because $\ln(x)$ is undefined, whereas $g(x)$ is defined for $x < -\frac{1}{2}$, where $2x^2 + x > 0$.

75. Multiply by a suitable number $C > 1$, i.e.,
$$y = \begin{cases} C\ln x, & 1 \le x \le 5 \\ C\ln(10 - x), & 5 \le x \le 9 \end{cases}$$

77. $$y = \begin{cases} 3\ln x, & 1 \le x \le 6 \\ 3\ln(12 - x), & 6 \le x \le 11 \\ 3\ln(x - 10), & 11 \le x \le 16 \\ 3\ln(22 - x), & 16 \le x \le 21 \end{cases}$$

SECTION 5.3

1. $\log_5 14$

3. $\log_3 \left(\dfrac{x^4}{y^2 z^3} \right)$

5. $\log_e \left(\dfrac{x^3 y^{1/4} z^{2/9}}{w^{5/6}} \right)$

7. $\ln(x^{-3} y^{-2})$

9. $\log_4 \left(\dfrac{(x - 3)^2}{\sqrt{x - 4}} \sqrt{x + 2} \right)$

11. $\ln \left(\dfrac{\ln x^2}{x^2} \right)^2$

13. $\log \left(x^4 \cdot 10^{14 - x^4} \right)$

15. $2 \log_6 a + 3 \log_6 b - 2 \log_6 c - 4 \log_6 d$

17. $\ln 15 + 2 \ln x - 6 \ln(3x - 5) - 9 \ln(x - 5)$

19. $\frac{1}{8} \log 3 + \frac{1}{4} \log a + \frac{1}{2} \log b + \frac{3}{4} \log c$
$\quad - \frac{1}{4} \log 2 - \frac{1}{4} \log x - \frac{1}{32} \log y$

21. $\log(2 \log x + 3 \log y) - \log(3 \log a + 4 \log b)$

23. $\log 6 = .7782$

$\log_3 6 = 1.6309 = \dfrac{\log 6}{\log 3}$

25. $\log 6 = .7782$

$\log_5 6 = 1.1133 = \dfrac{\log 6}{\log 5}$

27. as n approaches 10 from below, $\log_n x$
approaches $\log_{10} x$ from above

29. $\log_{10} 6 = .7782$

$\log_{\frac{1}{10}} 6 = -.7782$

31. Let $x = \log_a m$ and $y = \log_a n$

Then $a^x = m$ and $a^y = n$

So $\dfrac{a^x}{a^y} = \dfrac{m}{n} = a^{x-y}$

Hence, $\log_a \left(\dfrac{m}{n}\right) = x - y$

$\qquad = \log_a m - \log_a n$

33. $\log_{10} x = \dfrac{\ln x}{\ln 10}$

35. 4.6439

37. $.3542$

39. Let $x = 1, y = 1, b = 2$

$\log_2 (1 + 1) = \log_2 2 = 1 \neq \log_2 1 + \log_2 1$
$\qquad = 0$ since $\log_2 1 = 0$

41. $\log_2 (2 \cdot 1) = \log_2 2 = 1$ and $\log_2 2 \cdot \log_2 1$
$\qquad = 1 \cdot 0 = 0$

43. $\log .5$ is negative

∴ $\log (\log .5)$ can not be calculated since $\log$
is not defined for negative numbers

45.

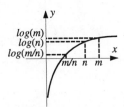

SECTION 5.4

1. $x = \dfrac{\ln 15}{\ln 6} \approx 1.51$

3. $x \approx -1.27$

5. $x \approx -.03$

7. $x \approx -5.59$

9. $x \approx -6.13$

11. $x \approx -.57$

13. no solution

15. $x \approx -2.12$ or 1.42

17. $x = 20$

19. $x = 1$

21. $x \approx 10^{7.5}$

23. $x = 100$

25. $x = \ln\left(y \pm \sqrt{y^2 - 1}\right), y > 1$

27. $x = \dfrac{1}{6} \ln \left(\dfrac{1 + y}{y - 1}\right), |y| > 1$

29. $x \approx e^{3.396}$

31. $Q(t) = 9.959\, e^{-.223t}$

33. $P(t) = 5.998\, e^{.0198t}$

35. $t = \dfrac{1}{k} \ln \dfrac{PC_0}{C - P},\ C > P$

37. $\ln Q = \ln 15 + \dfrac{1}{2} \ln P + \dfrac{3}{T}$

$\qquad = \ln 15 + \ln\sqrt{P} + \ln e^{3/T}$

$\qquad = \ln \left(15 e^{3/T}\sqrt{P}\right)$

$Q = 15 e^{3/T}\sqrt{P}$

39. 13.02 hrs

41. 16.41 min.

43. $pH = 4.74$

45. $H^+ = 4.37 \times 10^{-8}$

47. 61.7609

49. 10^{10} times

51. $100,000$ times

53. $1.58 \times 10^5\, i_0$

55. 100 times

57. 7.33 days

59. 7.925 hrs

61. $9,056.93$ years

SECTION 6.1

1. $\dfrac{3^R}{4}$

3. $\dfrac{2^R}{3}$

5. 4^R

7. 3^R

9. 6^R

11. $\dfrac{\pi^R}{3}$

13. $\dfrac{5}{18}\pi^R$

15. $-\dfrac{5}{12}\pi^R$

17. $60°$

19. $286.5°$

21. $\left(-\dfrac{1080}{\pi}\right)°$

23. Answers vary

25. Answers vary

27. Answers vary

29. Answers vary

31. I

33. I

35. II

37. II

39. III

41. III

43. IV

45. IV

31. - 45.

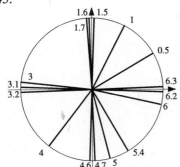

47. - 53.

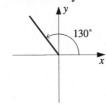

55. - 57.

59. No, when the meters read 5 the two cars will have traveled different distances. The angles are $60°$ and $30°$, respectively. The x-car is faster.

61. ≈ 377 cm

SECTION 6.2

1. $-.3556$

3. -1.8858

5. 5.7965

7. $.3556$

9. $\dfrac{5\pi^R}{6}$

11. $\dfrac{\pi^R}{4}$

13. 1^R

15. $.013\pi$

17. $60°$

19. $300°$

21. $-22.5°$

23. $27°$

25. $90°$

27. $-286.5°$

29. $271.3°$

31. Answers vary

33. Answers vary

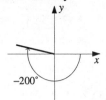

35. Answers vary

37. Answers vary

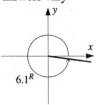

SECTION 6.3

Answers 1 - 7 are to the nearest integer

1. $\alpha \approx 1^R$

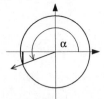

3. $\alpha \approx 5^R$

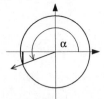

5. $\alpha \approx -1^R$

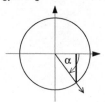

7. $\alpha \approx -1^R$

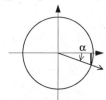

9. $2^R \approx 11.5°$
11. $149°$
13. $3.9^R \approx 223.5°$
15. $298°$
17. $40°, \text{I}$
19. $130°, \text{II}$
21. $200°, \text{III}$
23. $300°, \text{IV}$
25. Approx. $7\frac{1}{2}$ times

SECTION 6.4

1.

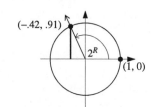

3.

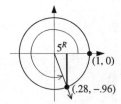

5.

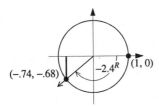

7.   see above

9.

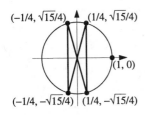

11.

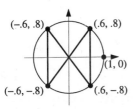

13.

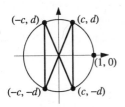

15.   (.23, .97)

17.   (−.59, .81)

19. $\left(a,\ \sqrt{1-a^2} \right)$

21.   Value of T increases. As T increases from 0 to 3.1, x decreases from 1 to −.991352. As T increases from 3.1 to 6.3, the value of x increases to .99985864. As T increases from 0 to 1.5 the value of y increases from 0 to .99749499. Then as T increases from 1.5 to 4.7 the value of y decreases to −.9999233. Then as T increases from 4.7 to 6.3 the value of y increases to .0168139.

23.

T	X	Y
0.1	.9950	.0998
−0.1	.9950	−.0998
.07	.7648	.6442
−.07	.7648	−.6442
1.3	.2675	.9636
−1.3	.2675	−.9636
1.9	−.3233	.9463
−1.9	−.3233	−.9463

25. $x = -.5000$, $y = .8660$; $(T + 2\pi)$ value is the same

27.   Mike has traveled 40 miles and has approximately 3.54 hours of riding to go.

29.   Joe has the greater angular speed, and will complete his circle about 25 minutes before Moe does.

SECTION 7.1

1. $\sin(\theta) = \dfrac{\sqrt{3}}{2}$

     $\cos(\theta) = \dfrac{1}{2}$

3. $\sin(\theta) = -\dfrac{1}{\sqrt{2}}$

     $\cos(\theta) = \dfrac{1}{\sqrt{2}}$

5. $\sin(\theta) = 1$
     $\cos(\theta) = 0$

7. $\sin(\theta) = 0.892$
     $\cos(\theta) = 0.452$

9. $\sin(\theta) = -0.993$
     $\cos(\theta) = 0.115$

11. $\sin(.98) = 0.830$
     $\cos(.98) = 0.557$

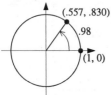

13. $\sin(5.9) = -0.372$
     $\cos(5.9) = 0.928$

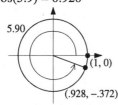

15. $\sin(-5.62) = 0.618$
     $\cos(-5.62) = 0.786$

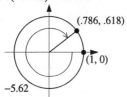

17. $\sin(15.7) = -.037$
     $\cos(15.7) = -.999$

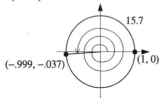

19. adj. = 5
     opp. = 8
     hyp. $= \sqrt{5^2 + 8^2} = \sqrt{89}$
     $\sin \theta = \dfrac{8}{\sqrt{89}}$, $\cos \theta = \dfrac{5}{89}$

21. adj. = 7.2
     opp. = 4.5
     hyp. ≈ 8.5
     $\sin \theta \approx .529$
     $\cos \theta \approx .847$

23. D = All real numbers
     $R = [-1, 1]$

25. $\cos(s) = -\dfrac{\sqrt{21}}{5} = -.9165$

27. $\sin(\alpha) = -\dfrac{2\sqrt{2}}{3} = -.9428$

29. $\theta = \dfrac{\pi}{4}, \dfrac{5\pi}{4}, \dfrac{9\pi}{4}, \ldots$

31. a) 1
     b) 1
     c) 4

33. Since $\dfrac{(2n + 1)\pi}{2} = \dfrac{\pi}{2} + n\pi$, the terminal
     point of the arc length $\dfrac{\pi}{2} + n\pi$ will either
     fall on the positive y-axis (if n is even) or
     the negative y-axis (if n is odd). That is, the
     terminal point on the unit circle will have
     coordinates $(0, 1)$ or $(0, -1)$. Hence
     $\sin(2n + 1)\dfrac{\pi}{2} = \pm 1$.

35.

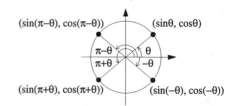

37. $\cos\left(\dfrac{\pi}{3}\right) \approx .5017962$
     max. error $\approx .0018316$

39. max. height ≈ 18.5 ft
     max. dist. = 105.7 ft

41. He is not able to see all of the window. He
     should be approx. 4.15 ft from his window
     to observe the whole window on the other
     side.

SECTION 7.2

1.

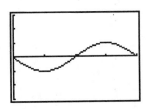

3.

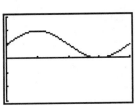

5.

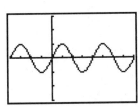

7.

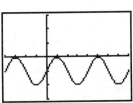

9. The value of sine function decreases from +1 to −1.

11. 1

13. $-\dfrac{\sqrt{3}}{2}$

15. $-\dfrac{\sqrt{3}}{2}$

17. $\dfrac{\sqrt{3}}{2}$

19. II, .927

21. II, .259

23. IV, .139

25. III, −.766

27. II, .527

29. II, .700

31. IV, .595

33. II, .610

SECTION 7.3

In exercises 1-3, the tick marks on the horizontal axis represent intervals of $\pi/2$ units, and the vertical axis tick marks represent 1 unit intervals.

1.

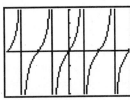

XMIN=−π, XMAX=π

3.

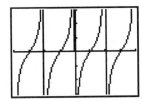

5. $-\sqrt{3}$

7. −1

9. $-\dfrac{1}{\sqrt{3}}$

11. 0

13. 19.670

15. 1255.766

17. −.953

19. .980

21. −.854

23. 1.248

25. −572.957

27. .601

29. −5.145

31. −.176

33. negative

35. positive

37. negative

39. negative

41. negative

43. positive

45. a) 45°

    b) 40.6°

    c) 50.2°

    The optimal distance is between 6 and 7 ft.

47. Range = [0, 581]

    hor. dist. = 2,770 ft

49. hor. dist. = 2,770 ft

    max. vert. dist. = 825 ft

    If the angles of projection are complementary angles, then the horizontal distance attained appears to be the same.

SECTION 7.4

1. 2

3. $\dfrac{2\sqrt{3}}{3}$

5. $\sqrt{3}$

7. $\dfrac{2\sqrt{3}}{3}$

9. $\dfrac{2\sqrt{3}}{3}$

11. 0

13. −1.068

15. 5.978

17. −.962

19. 1.081

21. −1.654

23. −.213

25. Domain: $\{x | x \neq n\pi$ for any integer $n\}$

    Range: $(-\infty, -2] \cup [2, \infty)$

27. Domain: $\{x | x \neq \dfrac{n\pi}{2}$ for any integer $n\}$

    Range: $(-\infty, \infty)$

SECTION 7.5

In exercises 1-27, the tick marks on the horizontal axis represent intervals of $\pi/2$ units, and the vertical axis tick marks represent 1 unit intervals. More than two cycles may be shown.

1. Amplitude = 4

    Period = $\dfrac{2\pi}{3}$

    Phase Shift = 0

3. Amplitude = 5

    Period = π

    Phase Shift = 0

5. Amplitude = $\dfrac{3}{2}$

    Period = $\dfrac{\pi}{2}$

    Phase Shift = 0

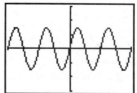

XMIN=$-\pi$, XMAX=π

7. Amplitude = $\dfrac{5}{2}$

    Period = 6π

    Phase Shift = 0

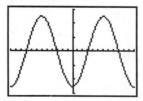

9.   Amplitude = 3

    Period = π

    Phase Shift = $-\dfrac{\pi}{6}$

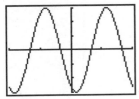

11.   Amplitude = 2

    Period = $\dfrac{\pi}{2}$

    Phase Shift = $-\dfrac{\pi}{2}$

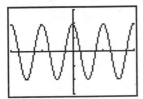

13.   Amplitude = $\dfrac{2}{5}$

    Period = 4π

    Phase Shift = $\dfrac{\pi}{8}$

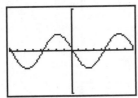

15.   Amplitude = $\dfrac{3}{5}$

    Period = $\dfrac{2\pi}{3}$

    Phase Shift = $\dfrac{\pi}{4}$

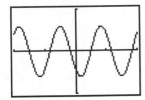

17.   Amplitude = 4

    Period = $\dfrac{4\pi}{3}$

    Phase Shift = $\dfrac{\pi}{12}$

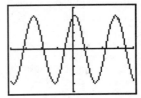

19.

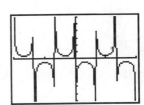

XMIN=$-\pi$, XMAX=π

21.

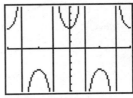

23.

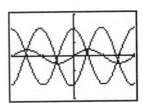

25.

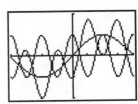

27.

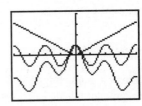

29. Yes. $y = \sin\left(\frac{2\pi}{5}x\right)$

31. $\frac{9}{2}$ up, so $y = \frac{5}{2}\sin\left(\frac{2\pi}{5}x\right) + \frac{9}{2}$

33. In the last step; then the phase shift would be 1. The equation would be
$y = \frac{5}{2}\cos(\frac{2\pi}{5}(x-1)) + \frac{9}{2}$

35. To find the "amplitude" change, measure the vertical distance between the "peaks" of the graph. Then $y = \frac{3}{2}\csc\left(\frac{\pi}{4}x\right)$.

37. 1, so $y = \frac{3}{2}\csc\left[\frac{\pi}{4}(x-1)\right] + \frac{9}{2}$

SECTION 7.6

1. $\frac{\pi}{6}$

3. $-\frac{\pi}{4}$

5. $\frac{5\pi}{6}$

7. $\frac{\pi}{6}$

9. $\frac{\pi}{4}$

11. .253

13. −.675

15. does not exist

17. 1.383

19. .423

21. $D = (-\infty, \infty)$, $R = \left(3 - \frac{\pi}{20},\ 3 + \frac{\pi}{20}\right)$

23. Let $u = \sin^{-1}(x)$. Then
$\sin(u) = x$, $\cos(u) = \sqrt{1 - x^2}$, and
$\tan(u) = \dfrac{x}{\sqrt{1 - x^2}}$. Taking
$\tan^{-1}$ of both sides of the last expression
yields $\sin^{-1}(x) = \tan^{-1}\left(\dfrac{x}{\sqrt{1 - x^2}}\right)$

25. $\sin^{-1}(\theta) + \cos^{-1}(\theta) = \frac{\pi}{2}$

In exercises 27-35, the tick marks on the horizontal axis represent intervals of 1 unit each, and the vertical axis tick marks represent π/2 unit intervals

27.

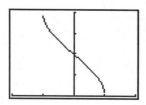

29.

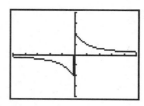

31.

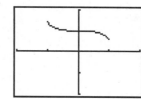

33.

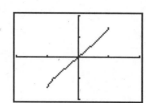

35.

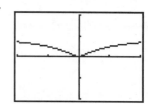

SECTION 7.7

1. Answers vary

3. Amplitude = .05
Period = 1
Phase Shift = 0
Frequency = 1

5. Amplitude = 5

   Period $= \dfrac{1}{1.397} \approx .716$

   Phase Shift = 0

   Frequency = 1.397

7. Amplitude = 103

   Period = .018

   Phase Shift = 0

   Frequency = 55.56

9. Amplitude = 32

   Period = 20,943.95

   Phase Shift = .0001

   Frequency $= 4.77 \times 10^{-5}$

11. Yes

13. No

15. No

17. The data is not quite periodic, even though there is a high every fourth quarter as well as a low every fourth quarter. The graph is not sinusoidal because the highs and lows are not at the exact same levels. Population depends on many things besides the predator/prey relationship; weather for example.

19. Answers vary

SECTION 8.1

1. $m\angle B = 26°$, $b = 5.85$, $c = 13.35$

3. $m\angle B = 75.5°$, $a = .82$, $c = 3.27$

5. $m\angle B = 49°$, $a = 2.81$, $b = 3.23$

7. $m\angle B = 18.7°$, $a = 8.39$, $b = 2.84$

9. $m\angle B = 46.5°$, $b = 26.03$, $c = 35.88$

11. $c = 3.81$, $m\angle A = 22.53°$, $m\angle B = 67.47°$

13. $a = 35.40$, $m\angle A = 42.40°$, $m\angle B = 47.60°$

15. 34 feet

17. 1.15 miles

19. 246.5 ft

21. $\cos\theta = \dfrac{6400}{6400 + h}$

To solve for h one needs angular speed. Given that it is 2π radians per 65 mins.,
$h = 1{,}251.2$ km.

23. 44.7 miles in 1 hr., 89.4 in 2 hrs., 134.2 in 3 hrs.

25. 24.3 ft

27. 179 ft

29. ≈ 13.5 ft

SECTION 8.2

1. $m\angle B = 83.3°$, $a = 2.59$, $b = 3.03$

3. $m\angle A = 82.8°$, $b = 3.03$, $c = 2.51$

5. $m\angle A = 52.6°$, $m\angle C = 74.8°$, $a = 30.3$

7. $m\angle A = 76.7°$, $m\angle B = 72.3°$, $a = 22.7$ or
   $m\angle A = 15.4°$, $m\angle B = 107.7°$, $a = 15.4$

9. $m\angle C = 42.7°$, $m\angle B = 81°$, $b = 2.26$

11. $m\angle A = 57.9°$, $m\angle A = 90°$, $a = 2.92$

13. 105.1 miles

15. 110 ft

17. 243 yds; given possible error, between 174 yds and 453 yds. Lisa could instead measure the bridge (if it is perpendicular to the river bank) and measure the angle between Lisa and Kim's houses and the line parallel to the bridge.

19. 62.7 ft

21. The other sides measure 502.3 ft and 370.8 ft. The area is 89,066 sq ft.

SECTION 8.3

1. $c = 1.9$, $m\angle A = 35.3°$, $m\angle B = 118.9°$

3. $b = 780.8$, $m\angle A = 44.1°$, $m\angle C = 38.5°$

5. $a = 5.4$, $m\angle B = 39.4°$, $m\angle C = 114.8°$

7. $b = 1740.3$, $m\angle A = 37.8°$, $m\angle C = 86.4°$

9. $m\angle A = 40.4°$, $m\angle B = 55.9°$, $m\angle C = 83.7°$

11. $m\angle A = 40.5°$, $m\angle B = 99.3°$, $m\angle C = 40.2°$

13. $m\angle A = 83.7°$, $m\angle B = 44.2°$, $m\angle C = 52.1°$

15. The helicopter should go at 25.9° away from the other station

17. 67.6 miles

19. 72.7°

SECTION 8.4

1. Yes
3. Yes
5. No
7. No
9. Yes
11. Yes
13. $\csc^2(x) - \cot^2(x) = 1$
    $\cot^2(x) = \csc^2(x) - 1$
15. $\cos(t) = \sin(t) \cdot \cot(t)$
    $\sin(t) = \dfrac{\cos(t)}{\cot(t)}$
    $\cot(t)\sec(t) = \csc(t)$
17. $\csc(r) \cdot \sin(r) = 1$
    $\sin(r) = \dfrac{1}{\csc(r)}$

19. - 25. Answers vary

SECTION 8.5

1. Yes; for $\theta \neq (2n+1)\frac{\pi}{2}$ for any integer n
3. No
5. Yes; for $\cos w \neq 0$
7. No
9. Yes; for $\cos t \neq 0$
11. $\cos^2(\alpha)[\sec^2(\alpha) - 1]$
    $= \cos^2(\alpha)\sec^2(\alpha) - \cos^2(\alpha)$
    $= 1 - \cos^2(\alpha) = \sin^2(\alpha)$
13. $\tan^2(\theta) - \sin^2(\theta) = \dfrac{\sin^2(\theta)}{\cos^2(\theta)} - \sin^2(\theta)$
    $= \sin^2(\theta)\left(\dfrac{1}{\cos^2(\theta)} - 1\right)$
    $= \sin^2(\theta)\left(\dfrac{1 - \cos^2(\theta)}{\cos^2(\theta)}\right)$
    $= \sin^2(\theta)\dfrac{\sin^2(\theta)}{\cos^2(\theta)} = \sin^2(\theta) \cdot \tan^2(\theta)$
15. $\dfrac{\sin^2(z)}{\tan^2(z) - \sin^2(z)} = \dfrac{\sin^2(z)}{\dfrac{\sin^2(z)}{\cos^2(z)} - \sin^2(z)}$
    $= \dfrac{\sin^2(z)}{\sin^2(z)\left(\dfrac{1}{\cos^2(z)} - 1\right)} = \dfrac{\cos^2(z)}{1 - \cos^2(z)}$
    $= \dfrac{\cos^2(z)}{\sin^2(z)} = \cot^2(z)$
17. $\dfrac{\cos(s) + \sin(s)}{\sin(s)} = \dfrac{\cos(s)}{\sin(s)} + \dfrac{\sin(s)}{\sin(s)}$
    $= \cot(s) + 1$
19. $\dfrac{1 - \cos z}{\sin z} = \dfrac{(1 - \cos z)(1 + \cos z)}{\sin z\,(1 + \cos z)}$
    $= \dfrac{1 - \cos^2 z}{\sin z\,(1 + \cos z)}$
    $= \dfrac{\sin^2 z}{\sin z\,(1 + \cos z)} = \dfrac{\sin z}{1 + \cos z}$
21. $\dfrac{1}{\sec\theta - \tan\theta}$
    $= \dfrac{1 \cdot (\sec\theta + \tan\theta)}{(\sec\theta - \tan\theta)(\sec\theta + \tan\theta)}$
    $= \dfrac{\sec\theta + \tan\theta}{\sec^2\theta - \tan^2\theta} = \dfrac{\sec\theta + \tan\theta}{1}$
    $= \sec\theta + \tan\theta$
23. $\cos^2\beta - \sin^2\beta = (1 - \sin^2\beta) - \sin^2\beta$
    $= 1 - 2\sin^2\beta$
25. $\dfrac{1 + \cos^2(t)}{\sin^2(t)} = \dfrac{1 + 1 - \sin^2(t)}{\sin^2(t)}$
    $= \dfrac{2 - \sin^2(t)}{\sin^2(t)} = \dfrac{2}{\sin^2(t)} - \dfrac{\sin^2(t)}{\sin^2(t)}$
    $= 2\csc^2(t) - 1$
27. From exercise 18,
    $\dfrac{\cos(z)}{1 - \sin(z)} = \dfrac{1 + \sin(z)}{\cos(z)}$
    $= \dfrac{1}{\cos(z)} + \dfrac{\sin(z)}{\cos(z)} = \sec(z) + \tan(z)$
29. $\cot(\theta) \cdot \csc(\theta) = \dfrac{\cos(\theta)}{\sin(\theta)} \cdot \dfrac{1}{\sin(\theta)}$
    $= \dfrac{\cos(\theta)}{\sin^2(\theta)} = \dfrac{\cos(\theta)}{1 - \cos^2(\theta)}$
    $= \dfrac{\cos(\theta)/\cos(\theta)}{\dfrac{1 - \cos^2(\theta)}{\cos(\theta)}} = \dfrac{1}{\sec(\theta) - \cos(\theta)}$
31. $\tan(w)\sec(w) - \sin(w)$

$$= \frac{\sin(w)}{\cos(w)} \cdot \frac{1}{\cos(w)} - \sin(w)$$

$$= \frac{\sin(w) - \sin(w) \cos^2(w)}{\cos^2(w)}$$

$$= \frac{\sin(w)[1 - \cos^2(w)]}{\cos^2(w)}$$

$$= \frac{\sin(w) \sin^2(w)}{\cos^2(w)}$$

$$= \tan^2(w) \sin(w)$$

33. $\dfrac{\csc(\theta) + 1}{\csc(\theta) - 1} = \dfrac{\dfrac{1}{\sin(\theta)} + 1}{\dfrac{1}{\sin(\theta)} - 1} \cdot \dfrac{\sin(\theta)}{\sin(\theta)}$

$$= \frac{1 + \sin(\theta)}{1 - \sin(\theta)}$$

35. $\dfrac{2 \tan(t)}{1 - \tan^2(t)} + \dfrac{1}{2 \cos^2(t) - 1}$

$$= \frac{2 \dfrac{\sin(t)}{\cos(t)} \cdot \cos^2(t)}{\left(1 - \dfrac{\sin^2(t)}{\cos^2(t)}\right) \cdot \cos^2(t)}$$

$$+ \frac{1}{\cos^2(t) + \cos^2(t) - 1}$$

$$= \frac{2 \sin(t) \cos(t)}{\cos^2(t) - \sin^2(t)}$$

$$+ \frac{1}{\cos^2(t) - \sin^2(t)}$$

$$= \frac{2 \sin(t) \cos(t) + 1}{\cos^2(t) - \sin^2(t)}$$

$$= \frac{\cos^2(t) + \sin^2(t) + 2 \sin(t) \cos(t)}{(\cos t - \sin t)(\cos t + \sin t)}$$

$$= \frac{(\cos t + \sin t)^2}{(\cos t - \sin t)(\cos t + \sin t)}$$

$$= \frac{\cos(t) + \sin(t)}{\cos(t) - \sin(t)}$$

37. $\dfrac{\sec(w) \tan(w) - \sin(w)}{\sin^3(w)}$

$$= \frac{\dfrac{1}{\cos(w)} \cdot \dfrac{\sin(w)}{\cos(w)} - \sin(w)}{\sin^3(w)}$$

$$= \frac{\dfrac{\sin(w)}{\cos^3(w)} - \sin(w)}{\sin^3(w)} \cdot \frac{\cos^2(w)}{\cos^2(w)}$$

$$= \frac{\sin(w) - \sin(w) \cdot \cos^2(w)}{\sin^3(w) \cos^2(w)}$$

$$= \frac{\sin(w) (1 - \cos^2 w)}{\sin^3(w) \cos^2(w)}$$

$$= \frac{\sin(w) \cdot \sin^2(w)}{\sin^3(w) \cos^2(w)} = \sec^2(w)$$

39. $[\cos^4(s) + 1 - \sin^4(s)]\sec^2(s)$

$$= [(\cos^2(s) + \sin^2(s))(\cos^2(s) - \sin^2(s)) + 1]\sec^2(s)$$

$$= [\cos^2(s) - \sin^2(s) + 1]\sec^2(s)$$

$$= [\cos^2(s) - \sin^2(s) + \cos^2(s) + \sin^2(s)]\sec^2(s)$$

$$= 2 \cos^2(s) \cdot \sec^2(s) = 2$$

SECTION 8.6

1. $.259, \dfrac{\sqrt{6} - \sqrt{2}}{4}$

3. $3.73, 2 + \sqrt{3}$

5. $.966, \dfrac{\sqrt{2} + \sqrt{6}}{4}$

7. $-.966, -\left(\dfrac{\sqrt{2} + \sqrt{6}}{4}\right)$

9. $-.259, \dfrac{\sqrt{2} - \sqrt{6}}{4}$

11. $\cos\left(\dfrac{\pi}{2} - s\right) = \cos\left(\dfrac{\pi}{2}\right) \cos(s)$

$$+ \sin(s) \sin\left(\dfrac{\pi}{2}\right)$$

$$= \sin(s)$$

13. $\sin\left(\dfrac{\pi}{2} + s\right) = \sin\left(\dfrac{\pi}{2}\right) \cos(s)$

$$+ \sin(s) \cos\left(\dfrac{\pi}{2}\right)$$

$$= \cos(s)$$

15. $\cos(\pi - s) = \cos(\pi) \cos(s) + \sin(\pi) \sin(s)$

$$= -\cos(s)$$

17. $\cos(\pi + s) = \cos(\pi) \cos(s) - \sin(\pi) \sin(s)$

$$= -\cos(s)$$

19. $\sin(2\pi - s) = \sin(2\pi) \cos(s) - \sin(s) \cos(2\pi)$

$$= -\sin(s)$$

21. $\tan(s + \pi) = \dfrac{\sin(s + \pi)}{\cos(s + \pi)} = \dfrac{-\sin(s)}{-\cos(s)} = \tan(s)$

23. a) $\dfrac{56}{65}$ 25. a) $\dfrac{-15 - 16\sqrt{2}}{51}$

    b) $\dfrac{63}{65}$ b) $\dfrac{8 + 30\sqrt{2}}{51}$

    c) $\dfrac{-56}{33}$ c) $\dfrac{8 + 30\sqrt{2}}{16\sqrt{2} - 15}$

27. $\cos(56°)$

29. $\sin(44°)$

SECTION 8.7

1. $.259, \dfrac{\sqrt{6} - \sqrt{2}}{4}$

3. $-.259, \dfrac{\sqrt{2} - \sqrt{6}}{4}$

5. $.259, \dfrac{\sqrt{6} - \sqrt{2}}{4}$

7. $3.73, 2 + \sqrt{3}$

9. $.991, \dfrac{\sqrt{2 + \sqrt{2 + \sqrt{3}}}}{2}$

11. $\sin(6t)$

13. $4\cos(4t)$

15. $\cos(t)$

17. $\sin(2\theta) = -\dfrac{24}{25}$ $\sin\left(\dfrac{\theta}{2}\right) = \pm\dfrac{2\sqrt{5}}{5}$

    $\cos(2\theta) = -\dfrac{7}{25}$ $\cos\left(\dfrac{\theta}{2}\right) = \pm\dfrac{\sqrt{5}}{5}$

    $\tan(2\theta) = \dfrac{24}{7}$ $\tan\left(\dfrac{\theta}{2}\right) = 2$

19. $\sin(2t) = \dfrac{24}{25}$ $\sin\left(\dfrac{t}{2}\right) = \pm\dfrac{3\sqrt{10}}{10}$

    $\cos(2t) = \dfrac{7}{25}$ $\cos\left(\dfrac{t}{2}\right) = \mp\dfrac{\sqrt{10}}{10}$

    $\tan(2t) = \dfrac{24}{7}$ $\tan\left(\dfrac{t}{2}\right) = -3$

21. $\cos(2\alpha) = -\dfrac{4}{5}$ $\sin\left(\dfrac{\alpha}{2}\right) = \sqrt{\dfrac{\sqrt{10} - 1}{2\sqrt{10}}}$

    $\sin(\alpha) = \dfrac{3}{\sqrt{10}}$ $\cos\left(\dfrac{\alpha}{2}\right) = \sqrt{\dfrac{\sqrt{10} + 1}{2\sqrt{10}}}$

    $\cos(\alpha) = \dfrac{1}{\sqrt{10}}$ $\tan\left(\dfrac{\alpha}{2}\right) = \sqrt{\dfrac{\sqrt{10} - 1}{\sqrt{10} + 1}}$

    $= \dfrac{\sqrt{10} - 1}{3}$

23. $1 - \cos(2\theta)$

    $= 1 - [\cos(\theta)\cos(\theta) - \sin(\theta)\sin(\theta)]$

    $= 1 - \cos^2(\theta) + \sin^2(\theta)$

    $= \sin^2(\theta) + \sin^2(\theta) = 2\sin^2(\theta)$

25. $\dfrac{1 - \cos(2s)}{2\sin(s)\cos(s)} = \dfrac{1 - (1 - 2\sin^2 s)}{2\sin(s)\cos(s)}$

    $= \dfrac{2\sin^2(s)}{2\sin(s)\cos(s)} = \tan(s)$

27. $\sin(3t) = \sin(t + 2t)$

    $= \sin(t)\cos(2t) + \cos(t)\sin(2t)$

    $= \sin(t)(1 - 2\sin^2 t)$

        $+ \cos(t) \cdot 2\sin(t)\cos(t)$

    $= \sin(t) - 2\sin^3(t) + 2\sin(t)\cos^2(t)$

    $= \sin(t) - 2\sin^3(t) + 2\sin(t)(1 - \sin^2 t)$

    $= 3\sin(t) - 4\sin^3(t)$

29. $\dfrac{\sin(x)}{2} = \dfrac{\sin\left(\dfrac{x}{2} + \dfrac{x}{2}\right)}{2}$

    $= \dfrac{\sin\left(\dfrac{x}{2}\right)\cos\left(\dfrac{x}{2}\right) + \cos\left(\dfrac{x}{2}\right)\sin\left(\dfrac{x}{2}\right)}{2}$

    $= \dfrac{2\sin\left(\dfrac{x}{2}\right)\sin\left(\dfrac{x}{2}\right)}{2}$

SECTION 8.8

1. $.52, 2.62; \dfrac{\pi}{6}, \dfrac{5\pi}{6}$

3. $.79, 2.36; \dfrac{\pi}{4}, \dfrac{3\pi}{4}$

5. .79, 2.36, 3.93, 5.50; $\dfrac{\pi}{4}, \dfrac{3\pi}{4}, \dfrac{5\pi}{4}, \dfrac{7\pi}{4}$

7. 1.05, 2.09, 4.19, 5.24; $\dfrac{\pi}{3}, \dfrac{2\pi}{3}, \dfrac{4\pi}{3}, \dfrac{5\pi}{3}$

9. 2.09, 3.14, 4.19; $\dfrac{2\pi}{3}, \pi, \dfrac{4\pi}{3}$

11. .815, 5.469;

$$\cos^{-1}\left(\dfrac{-3 + \sqrt{33}}{4}\right), 2\pi - \cos^{-1}\left(\dfrac{-3 + \sqrt{33}}{4}\right)$$

13. 0, 1.05, 3.14, 5.24; 0, $\dfrac{\pi}{3}, \pi, \dfrac{5\pi}{3}$

15. .58, 1.37, 2.16, 2.95, 3.73, 4.52, 5.30, 6.09; $\dfrac{3\pi}{16}, \dfrac{7\pi}{16}, \dfrac{11\pi}{16}, \dfrac{15\pi}{16}, \dfrac{19\pi}{16}, \dfrac{23\pi}{16}, \dfrac{27\pi}{16},$ $\dfrac{31\pi}{16}$

17. .79, 2.36, 3.93, 5.50; $\dfrac{\pi}{4}, \dfrac{3\pi}{4}, \dfrac{5\pi}{4}, \dfrac{7\pi}{4}$

19. 0°, 120°, 180°, 240°

21. 60°, 180°, 300°

23. 60°, 180°, 300°

25. 30°, 60°, 120°, 150°, 210°, 240°, 300°, 330°

27. .20 + 6.28n, 2.94 + 6.28n, n is any integer

29. 1.25 + 3.14n, 2.36 + 3.14n, n is any integer

31. 2.14 + 6.28n, 4.14 + 6.28n, n is any integer

33. .36 + 6.28n, 2.14 + 6.28n, n is any integer

SECTION 9.1

1. $\left(1, \dfrac{7}{2}\right)$

3. $(-1, -1)$

5. $(4, -1)$

7. $\left(-\dfrac{3}{2}, \dfrac{1}{3}\right)$

9. $\left(\dfrac{6}{11}, \dfrac{19}{11}\right)$

11. $\left(\dfrac{3}{2}, -1\right)$

13. $\left(-\dfrac{13}{19}, \dfrac{34}{19}\right)$

15. 8 girls

17. 2, 9

19. $\left(\dfrac{97}{19}, \dfrac{195}{19}\right)$

21. no solution

23. $\left(\dfrac{39}{47}, \dfrac{7}{47}\right)$

25. $\left(\dfrac{585}{251}, \dfrac{140}{251}\right)$

27. $f(x) = -\dfrac{x}{2} + 4$

29. 1.07 lbs of ice cream

31. description varies

    a) (12, 5)

    b) (18, 6)

33. Scarves are $7.00;

    Jewelry boxes are $14.50

35. Cheese cakes are $5.70, pumpkin pies are $4.30, and lemon meringue pies are $5.25.

SECTION 9.2

Systems are consistent unless otherwise indicated.

1. $x = 2, y = -1, z = -3$

3. $x = \dfrac{15}{11}, y = \dfrac{43}{11}, z = -\dfrac{28}{11}$

5. $x = 2, y = 1, z = -3$

7. $x = 2, y = 1, z = 1$

9. $x = -2, y = 2, z = 2$

11. $x = 0, y = -1, z = 3$

13. $x = \dfrac{2}{19}, y = -\dfrac{11}{19}, z = \dfrac{119}{38}$

15. $x = -3, y = 8, z = -3$

17. dependent

19. dependent

21. a) $x = 1, y = -2, z = -5$

    b) no intersection (system is inconsistent)

23. of shoe A, 150; of shoe B, 120; of shoe C, 100

SECTION 9.3

Matrix answers will not be unique.

1. $\begin{bmatrix} 1 & -2 & 1 & 1 \\ 0 & 2 & -1 & 1 \\ 0 & 0 & -1 & 3 \end{bmatrix}$

3. $\begin{bmatrix} 1 & -1 & -1 & 0 \\ 0 & 3 & 5 & -1 \\ 0 & 0 & 25 & -56 \end{bmatrix}$

5. $\begin{bmatrix} 1 & 2 & 3 & -5 \\ 0 & -3 & -4 & 9 \\ 0 & 0 & -2 & 6 \end{bmatrix}$

7. $\begin{bmatrix} 2 & 3 & -4 & 13 \\ 0 & -3 & 16 & -19 \\ 0 & 0 & -52 & 39 \end{bmatrix}$

9. $x = 2, y = -1, z = -3$

11. $x = \dfrac{15}{11}, y = \dfrac{43}{11}, z = -\dfrac{28}{11}$

13. $x = 2, y = 1, z = -3$

15. $x = \dfrac{3}{2}, y = \dfrac{7}{3}, z = -\dfrac{3}{4}$

17. $\begin{bmatrix} 1 & 0 & 0 & -3 \\ 0 & -1 & 0 & -8 \\ 0 & 0 & 1 & -3 \end{bmatrix}$

19. $x = -3, y = 8, z = -3$

SECTION 9.4

1. $\begin{bmatrix} 32 \\ 9 \\ 14 \end{bmatrix}$

3. $\begin{bmatrix} 12 & 3 \\ -5 & 10 \\ -6 & 1 \end{bmatrix}$

5. $\begin{bmatrix} -3 & 1 & 2 \\ 2 & 3 & -5 \\ 3 & -2 & 1 \end{bmatrix}$

7. $3x + 4y - z = 1$
   $-2x + 5y + 3z = -2$
   $x - 3y + 6z = 5$

9. $\begin{bmatrix} -1 & 10 \\ 2 & -4 \end{bmatrix}\begin{bmatrix} 1 & 0 & -1 \\ -1 & 1 & 2 \end{bmatrix}$

$= \begin{bmatrix} 3 & 1 \\ -2 & 0 \end{bmatrix}\begin{bmatrix} -3 & 2 & 5 \\ -2 & 4 & 6 \end{bmatrix} = \begin{bmatrix} -11 & 10 & 21 \\ 6 & -4 & -10 \end{bmatrix}$

11. $\begin{bmatrix} -5 & -4 & 0 \\ -5 & 1 & 2 \end{bmatrix}\begin{bmatrix} 0 & -1 \\ -3 & 1 \\ 1 & 2 \end{bmatrix}$

$= \begin{bmatrix} 2 & 0 & -1 & 2 \\ -3 & 1 & 1 & 2 \end{bmatrix}\begin{bmatrix} 5 & 2 \\ 0 & -3 \\ 6 & 11 \\ 4 & 4 \end{bmatrix} = \begin{bmatrix} 12 & 1 \\ -1 & 10 \end{bmatrix}$

SECTION 9.5

1. $\begin{bmatrix} \dfrac{13}{25} & -\dfrac{4}{25} & \dfrac{1}{25} \\ -\dfrac{11}{25} & \dfrac{13}{25} & \dfrac{3}{25} \\ \dfrac{2}{5} & -\dfrac{1}{5} & -\dfrac{1}{5} \end{bmatrix}$

3. $\begin{bmatrix} 1 & -\dfrac{1}{5} & \dfrac{2}{5} \\ -1 & \dfrac{4}{5} & -\dfrac{3}{5} \\ -1 & \dfrac{3}{5} & -\dfrac{1}{5} \end{bmatrix}$

5. $x = 1, y = 0, z = 1$

7. $x = \dfrac{86}{59}, y = -\dfrac{37}{59}, z = \dfrac{40}{59}$

9. no solution

11. $x = 1, y = 0\ z = -10, s = 2$

13. $x = -\dfrac{203}{1120}, y = -\dfrac{417}{1120}, z = \dfrac{1969}{1120},$
    $s = -\dfrac{109}{1120}, t = \dfrac{29}{1120}$

15. Angela works about 95 hours, Rob about 242, and Sarah about 57.

SECTION 9.6

1. $x = -\dfrac{9}{5}, y = -\dfrac{23}{5}\ ; x = 1, y = 1$

3. $x = -1, y = 2; x = 26, y = 11$

5. $x = -\dfrac{11}{3}, y = \pm\dfrac{\sqrt{26}}{3}\ ; x = 2, y = \pm 1$

7. no solution

9. $x = \pm\sqrt{\dfrac{19}{11}}, y = \pm\sqrt{\dfrac{19}{7}}$

11. No

SECTION 9.7

1.

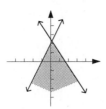

3.

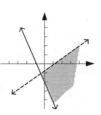

5.

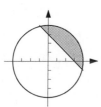

7.

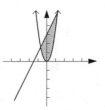

9.

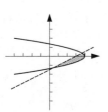

11.   Kevin 4 hrs, Krista 5 hrs

13.   Abe and Nancy earn the maximum amount in less than 10 hours.

15.

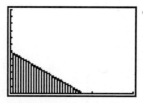

Each tick mark represents 10 tables or chairs.

Number of tables	Number of chairs
0	60
1	56
2	53
3	49
4	46
5	43
6	39
7	36
8	32
9	29
10	26
11	22
12	19
13	15
14	12
15	9
16	5
17	2

17.

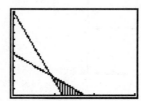

The shaded area represents combinations which allow Carl to earn more than $3,000. With 17 tables and 2 chairs he earns $4,300.

SECTION 10.1

1.

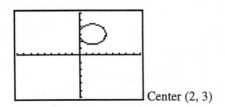

Center (2, 3)

3.

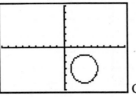

Center (–3, 2)

5.

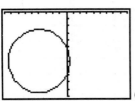

Center (3, –3)

7.

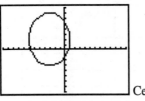

Center (–4, –7)

9.

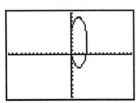

Center (2, 3)

11.

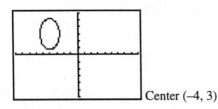

Center (–4, 3)

13.

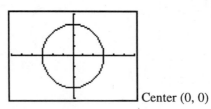

Center (0, 0)

15.

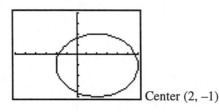

Center (2, –1)

17.

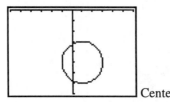

Center (1, –5)

19. $C = (-3, -4)$, $r = 2\sqrt{2}$

21. $C = (-3, 3)$, $r = 5$

23. $C = \left(3, \ -\frac{9}{2}\right)$, $r = 6$

25. $C = \left(\frac{3}{2}, \ -\frac{5}{2}\right)$, $r = \frac{7}{2}$

27. $C = (2, 3)$, major axis from $(2, -1)$ to $(2, 7)$, minor axis from $(0, 3)$ to $(4, 3)$

29. $C = (-5, -2)$, major axis from $(-9, -2)$ to $(-1, -2)$, minor axis from $(-5, -5)$ to $(-5, 1)$

31. $C = (3, 2)$, major axis from $(-1, 2)$ to $(7, 2)$, minor axis from $(3, 0)$ to $(3, 4)$

33. $C = (-4, -4)$, major axis from $\left(-4 - \sqrt{24}, \ -4\right)$ to $\left(-4 + \sqrt{24}, \ -4\right)$, minor axis from $\left(-4, \ -4 - \sqrt{12}\right)$ to $\left(-4, \ -4 + \sqrt{12}\right)$

35. $C = (0, 0)$, major axis from $(-3, 0)$ to $(3, 0)$, minor axis from $(0, -2)$ to $(0, 2)$

37. $C = (1, -1)$, major axis from $(-2, -1)$ to $(4, -1)$, minor axis from $(1, -3)$ to $(1, 1)$

39. $C = (-1, 2)$, major axis from $\left(-1, \ 2 - \frac{\sqrt{37}}{2}\right)$

 to $\left(-1, \ 2 + \frac{\sqrt{37}}{2}\right)$ minor axis from

 $\left(-1 - \frac{\sqrt{37}}{3}, \ 2\right)$ to $\left(-1 + \frac{\sqrt{37}}{3}, \ 2\right)$

41. $C = \left(\frac{3}{2}, \ -\frac{5}{2}\right)$, major axis from $\left(-\frac{1}{2}, \ -\frac{5}{2}\right)$

 to $\left(\frac{7}{2}, \ -\frac{5}{2}\right)$, minor axis from

 $\left(\frac{3}{2}, \ -\frac{5}{2} - \sqrt{3}\right)$ to $\left(\frac{3}{2}, \ -\frac{5}{2} + \sqrt{3}\right)$

43. $x_0 = 12 \cos\theta$, $y_0 = 12 \sin\theta$; $x = 6\cos\theta$,
 $y = 6\sin\theta$

45. parabola; $\frac{x^2}{4} + y = 1$; Quadrants 1 and 2

47. $x = 2\cos(t)$; $y = 3\sin(t)$

49. The origin is the center of the largest circle.
 $x^2 + y^2 - 576 = 0$
 $x^2 + y^2 - 24x = 0$
 $x^2 + y^2 - 36x + 288 = 0$

51. No, the artist will not get what he wants.

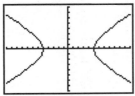

 The equation is $\frac{x^2}{85} + \frac{y^2}{36} = 1$. The right
 halves of the ellipse and rectangle are shown.
 Note the overhang.

SECTION 10.2

1.

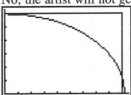

 $\frac{x^2}{25} - \frac{y^2}{9} = 1$

 transverse axis along $y = 0$, length 10
 conjugate axis along $x = 0$, length 6

3.

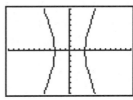

 $\frac{x^2}{9} - \frac{y^2}{36} = 1$

 transverse axis along $y = 0$, length 6
 conjugate axis along $x = 0$, length 12

5.

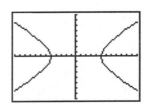

 $\frac{x^2}{64} - \frac{y^2}{25} = 1$

 transverse axis along $y = 0$, length 16
 conjugate axis along $x = 0$, length 10

7.

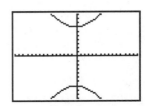

 $x(t) = 5\cosh(t)$, $y(t) = 3\sinh(t)$
 transverse axis along $y = 0$, length 10
 conjugate axis along $x = 0$, length 6

9.

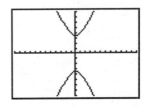

 $x(t) = 6\sinh(t)$, $y(t) = 7\cosh(t)$
 transverse axis along $x = 0$, length 14
 conjugate axis along $y = 0$, length 12

11.

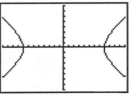

$x(t) = \sqrt{3} \sinh(t), \; y(t) = 2\sqrt{3} \coshh(t)$

transverse axis along $x = 0$, length $4\sqrt{3}$

conjugate axis along $y = 0$, length $2\sqrt{3}$

13.

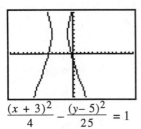

$\dfrac{(x + 3)^2}{4} - \dfrac{(y - 5)^2}{25} = 1$ center $(-3, 5)$

transverse axis along $y = 5$, length 4

conjugate axis along $x = -3$, length 10

15.

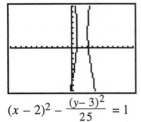

$(x - 2)^2 - \dfrac{(y - 3)^2}{25} = 1$ center $(2, 3)$

transverse axis along $y = 3$, length 2

conjugate axis along $x = 2$, length 10

17.

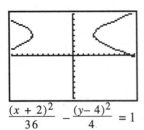

$\dfrac{(x + 2)^2}{36} - \dfrac{(y - 4)^2}{4} = 1$ center $(-2, 4)$

transverse axis along $y = 4$, length 12

conjugate axis along $x = -2$, length 4

19.

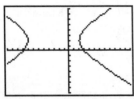

$x(t) = 5 \cosh(t) - 3, \; y(t) = 3 \sinh(t) + 2$

center $(-3, 2)$

transverse axis along $y = 2$, length 10

conjugate axis along $x = -3$, length 6

21.

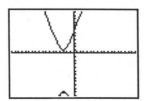

$x(t) = 2 \sinh(t) - 3, \; y(t) = 6 \cosh(t) - 5$

center $(-3, -5)$

transverse axis along $x = -3$, length 12

conjugate axis along $y = -5$, length 4

23.

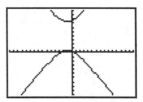

$x(t) = 2\sqrt{2} \sinh(t) - 1,$

$y(t) = 2\sqrt{3} \cosh(t) + 4$

center $(-1, 4)$

transverse axis along $x = -1$, length $4\sqrt{3}$

conjugate axis along $y = 4$, length $4\sqrt{2}$

25.

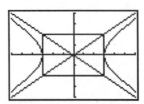

Center $(0, 0)$; vertices $(-3, 0)$, $(3, 0)$;

asymptotes $y = \dfrac{2}{3}x, \; y = -\dfrac{2}{3}x$

27.

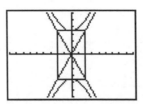

Center $(0, 0)$; vertices $\left(0, \; -2\sqrt{3}\right)$,

$\left(0, \; 2\sqrt{3}\right)$; asymptotes $y = -\sqrt{3}x, \; y = \sqrt{3}x$

29.

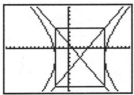

Center $(3, -2)$; vertices $(-3, -2)$, $(9, -2)$;
asymptotes $y = -\dfrac{7}{6}x + \dfrac{3}{2}$, $y = \dfrac{7}{6}x - \dfrac{11}{2}$

31.

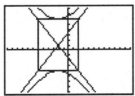

Center $(-2, 1)$; vertices $(-2, -4)$, $(-2, 6)$;
asymptotes $y = -\dfrac{5}{4}x - \dfrac{3}{2}$, $y = \dfrac{5}{4}x + \dfrac{7}{2}$

33. $\dfrac{x^2}{4} - \dfrac{y^2}{9} = 1$

Center $(0, 0)$; vertices $(-2, 0)$, $(2, 0)$;
asymptotes $y = -\dfrac{3}{2}x$, $y = \dfrac{3}{2}x$

35. $\dfrac{y^2}{9} - \dfrac{x^2}{18} = 1$

Center $(0, 0)$; vertices $(0, -3)$, $(0, 3)$;
asymptotes $y = -\dfrac{1}{\sqrt{2}}x$, $y = \dfrac{1}{\sqrt{2}}x$

37. $\dfrac{(x-1)^2}{1} - \dfrac{(y+3)^2}{4} = 1$

Center $(1, -3)$; vertices $(0, -3)$, $(2, -3)$;
asymptotes $y = -2x - 1$, $y = 2x - 5$

39. $(x-4)^2 - \dfrac{(y-2)^2}{4} = 1$

Center $(4, 2)$; vertices $(3, 2)$, $(5, 2)$;
asymptotes $y = -2x + 10$, $y = 2x - 6$

41. Degenerate hyperbola: two straight lines
with equations $y = \dfrac{3}{4}x + \dfrac{9}{2}$ and $y = -\dfrac{3}{4}x + \dfrac{3}{2}$,
intersecting at $(-2, 3)$.

43.   1) $\dfrac{(y-4)^2}{16} - \dfrac{(x-4)^2}{9} = 1$

2) $\dfrac{(y-4)^2}{16} - \dfrac{(x-12)^2}{9} = 1$

3) $\dfrac{(y-4)^2}{16} - \dfrac{(x-20)^2}{9} = 1$

4) $\dfrac{(y-4)^2}{16} - \dfrac{(x-28)^2}{9} = 1$

The bottom branches; each arch is $\dfrac{8}{3}$ ft high.

SECTION 10.3

1. Shift left one unit, down 4 units

3. Expand by 3, shift left $\dfrac{1}{3}$ unit, down $\dfrac{19}{3}$ units

5. Expand by 2, reflect across x-axis, shift left $\dfrac{5}{4}$ units, up $\dfrac{41}{8}$ units.

7.

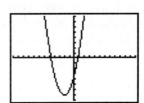

9.

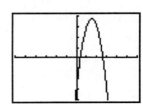

11.

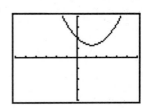

13.

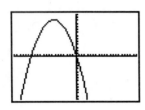

15.

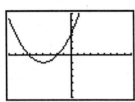

17. $y = (x - 2)^2 + 5$ or $y = x^2 - 4x + 9$

19. $y = (x - 4)^2 - 6$ or $y = x^2 - 8x + 10$

21. $y = 4x^2 - 3$

23. $y = \frac{1}{2}(x - 3)^2 + 1$

25. For an origin midway between City Hall and A Street, with City Hall in the positive y-direction, $\sqrt{x^2 + (y - 600)^2} = y + 600$

27. A parabola; $V(0,0)$, axis of symmetry $t = 0$. No, this is not the equation for the ball's path, since the independent variable is time and not horizontal displacement.

SECTION 10.4

1. hyperbola, $e = 2$, directrix $x = -3$

3. hyperbola, $e = 3$, directrix $x = -3$

5. parabola, $e = 1$, directrix $x = -8$

7. $r = \dfrac{ek}{1 + e\cos\theta}$, $r = \dfrac{ek}{1 - e\sin\theta}$,

    $r = \dfrac{ek}{1 + e\sin\theta}$

9. ellipse, $e = \frac{1}{2}$, $x = -4$

11. ellipse, $e = \frac{1}{2}$, $y = 4$

13. hyperbola, $e = 4$, $x = \frac{3}{4}$

15. ellipse, $e = \frac{1}{4}$, $x = 6$

17. The ellipse becomes smaller and more circular as $e \to 0$

19. Center $(2, 0)$; foci $(0, 0)$, $(4, 0)$;
    Major vertices $(6, 0)$, $(-2, 0)$;
    Minor vertices $\left(2, 2\sqrt{3}\right)$, $\left(2, -2\sqrt{3}\right)$;
    Directrices $x = -6$, $x = 10$

21. Center $\left(0, \frac{5}{6}\right)$; foci $(0, 0)$, $\left(0, \frac{5}{3}\right)$;
    Major vertices $\left(0, -\frac{10}{3}\right)$, $(0, 5)$;
    Minor vertices $\left(\frac{5\sqrt{2}}{\sqrt{3}}, \frac{5}{6}\right)$, $\left(-\frac{5\sqrt{2}}{\sqrt{3}}, \frac{5}{6}\right)$;
    Directrices $y = -20$, $y = \frac{65}{3}$

23. Center $\left(-\frac{8}{15}, 0\right)$; foci $(0, 0)$, $\left(-\frac{16}{15}, 0\right)$;
    Vertices $\left(-\frac{2}{3}, 0\right)$, $\left(-\frac{2}{5}, 0\right)$;
    Directrices $x = -\frac{1}{2}, -\frac{17}{30}$

25. Center $\left(0, \frac{60}{7}\right)$; foci $(0, 0)$, $\left(0, \frac{120}{7}\right)$;
    Vertices $\left(0, \frac{15}{7}\right)$, $(0, 15)$;
    Directrices $y = \frac{15}{4}$, $y = \frac{375}{28}$

27. Center $\left(0, -\frac{4}{3}\right)$; foci $(0, 0)$, $\left(0, -\frac{8}{3}\right)$;
    Vertices $\left(0, -\frac{2}{3}\right)$, $(0, -2)$;
    Directrices $y = -1$, $y = -\frac{5}{3}$

SECTION 11.1

In problems 1 through 41, we first show that the given statement is valid for $n = 1$. Then we assume that it is valid for $n = k$ and use this to show its validity for $n = k + 1$. The mathematical steps are shown. The student should provide his or her own wording.

1. $2 = 1 \cdot (1 + 1)$
    Assume $2 + 4 + \dots + 2k = k(k + 1)$
    Then $2 + 4 + \dots + 2k + 2(k + 1)$
    $= k(k + 1) + 2(k + 1)$
    $= (k + 1)(k + 2) = (k + 1)(k + 1 + 1)$

3. $\dfrac{(1 + 1)(1)(2 \cdot 1 + 1)}{6} = \dfrac{6}{6} = 1$

 Assume $1 + 4 + \ldots + k^2$

 $= \dfrac{(k + 1)(k)(2k + 1)}{6}$

 Then $1 + 4 + \ldots + k^2 + (k + 1)^2$

 $= \dfrac{(k + 1)(k)(2k + 1)}{6} + (k + 1)^2$

 $= \dfrac{(k + 1)(k)(2k + 1) + 6(k + 1)^2}{6}$

 $= \dfrac{(k + 1)(2k^2 + k + 6k + 6)}{6}$

 $= \dfrac{(k + 1)(k + 2)(2k + 3)}{6}$

 $= \dfrac{((k + 1) + 1)(k + 1)(2(k + 1) + 1)}{6}$

5. $1 = 1^2$.

 Assume

 $1 + 3 + 5 + \ldots + (2k - 1) = k^2$

 Then

 $1 + 3 + 5 + \ldots + (2k - 1) + [2(k + 1) - 1]$

 $= k^2 + 2(k + 1) - 1 = k^2 + 2k + 1$

 $= (k + 1)^2$

7. $1 = \dfrac{1(3 \cdot 1 - 1)}{2}$

 Assume $1 + 4 + 7 + \ldots + (3k - 2)$

 $= \dfrac{k(3k - 1)}{2}$

 Then

 $1 + 4 + 7 \ldots (3k - 2) + (3(k + 1) - 2)$

 $= \dfrac{k(3k - 1)}{2} + (3(k + 1) - 2)$

 $= \dfrac{k(3k - 1)}{2} + \dfrac{2(3k + 1)}{2}$

 $= \dfrac{3k^2 + 5k + 2}{2} = \dfrac{(k + 1)(3k + 2)}{2}$

 $= \dfrac{(k + 1)(3(k + 1) - 1)}{2}$

9. $1 = \dfrac{4^1 - 1}{3}$

 Assume $1 + 4 + 4^2 + \ldots + 4^{k - 1}$

 $= \dfrac{4^k - 1}{3}$

 Then $1 + 4 + 4^2 + \ldots 4^{k - 1} + 4^k$

 $= \dfrac{4^k - 1}{3} + 4^k = \dfrac{4^k - 1 + 3 \cdot 4^k}{3}$

 $= \dfrac{4 \cdot 4^k - 1}{3} = \dfrac{4^{k + 1} - 1}{3}$

11. $2(1)^2(1 + 1)^2 = 8 = 2^3$

 Assume $2^3 + 4^3 + \ldots (2k)^3 = 2k^2(k + 1)^2$

 Then $2^3 + \ldots + (2k)^3 + (2(k + 1))^3$

 $= 2k^2(k + 1)^2 + (2(k + 1))^3$

 $= 2k^2(k + 1)^2 + 2^3(k + 1)^3$

 $= 2(k + 1)^2[k^2 + 4k + 4]$

 $= 2(k + 1)^2 ((k + 1) + 1)^2$

13. $1^2 = \dfrac{1(2 - 1)(2 + 1)}{3}$

 Assume $1^2 + 3^2 + \ldots + (2k - 1)^2$

 $= \dfrac{k(2k - 1)(2k + 1)}{3}$

 Then

 $1^2 + 3^2 + \ldots + (2k - 1)^2 + [2(k + 1) - 1]^2$

 $= \dfrac{k(2k - 1)(2k + 1)}{3} + [2(k + 1) - 1]^2$

 $= \dfrac{k(2k - 1)(2k + 1)}{3} + (2k + 1)^2$

 $= (2k + 1) \left[\dfrac{k(2k - 1)}{3} + 2k + 1 \right]$

 $= (2k + 1) \dfrac{(2k^2 + 5k + 3)}{3}$

 $= \dfrac{(2k + 1)(2k + 3)(k + 1)}{3}$

 $= \dfrac{(k + 1)[2(k + 1) - 1][2(k + 1) + 1]}{3}$

15. $2^1 - 1 = 1 = 2^0$

 Assume $2^0 + \ldots + 2^{k - 1} = 2^k - 1$

 Then $2^0 + \ldots + 2^{k - 1} 2^k = 2^k - 1 + 2^k$

 $= 2(2^k) - 1 = 2^{k + 1} - 1$

17. $e^x = \dfrac{e^{(1+1)x} - e^x}{e^x - 1} = \dfrac{e^{2x} - e^x}{e^x - 1}$

    Assume $e^x + e^{2x} + \ldots + e^{kx}$

    $= \dfrac{e^{(k+1)x} - e^x}{e^x - 1}$

    Then $e^x + e^{2x} + \ldots + e^{kx} + e^{(k+1)x}$

    $= \dfrac{e^{(k+1)x} - e^x}{e^x - 1} + e^{(k+1)x}$

    $= \dfrac{e^{(k+1)x} - e^x + e^{(k+2)x} - e^{(k+1)x}}{e^x - 1}$

    $= \dfrac{e^{(k+2)x} - e^x}{e^x - 1} = \dfrac{e^{[(k+1)+1]x} - e^x}{e^x - 1}$

19. $\dfrac{3^1 - 1}{2 \cdot 3^1} = \dfrac{2}{6} = 1 - \left(\dfrac{1}{3}\right)^1$

    Assume $\left(\dfrac{1}{3}\right)^1 + \ldots + \left(\dfrac{1}{3}\right)^k = \dfrac{3^k - 1}{2 \cdot 3^k}$

    Then $\left(\dfrac{1}{3}\right)^1 + \ldots + \left(\dfrac{1}{3}\right)^k + \left(\dfrac{1}{3}\right)^{k+1}$

    $= \dfrac{3^k - 1}{2 \cdot 3^k} \left(\dfrac{1}{3}\right)^{k+1}$

    $= \dfrac{3(3^k - 1)}{3 \cdot 2 \cdot 3^k} + \dfrac{2}{2 \cdot 3^{k+1}}$

    $= \dfrac{3^{k+1} - 3 + 2}{2 \cdot 3^k} = \dfrac{3^{k+1} - 1}{2 \cdot 3^k}$

21. $\sin x = \dfrac{\sin^2 x}{\sin x}$

    Assume $\sin x + \sin 3x + \ldots + \sin(2k - 1)x$

    $= \dfrac{\sin^2 (kx)}{\sin x}$

    Then $\sin x + \sin 3x + \ldots + \sin(2k - 1)x$
    $+ \sin[2(k+1) - 1]x$

    $= \dfrac{\sin^2 kx}{\sin x} + \sin[2(k+1) - 1]x$

    $= \dfrac{\sin^2 kx + \sin x(\sin 2kx \cos x + \cos 2kx \sin x)}{\sin x}$

    $= \dfrac{1}{\sin x} [\sin^2 kx$

    $+ 2 \sin kx \cos kx \sin x \, \cos x$

    $+ (\cos^2 kx - \sin^2 kx)\sin^2 x]$

    $= \dfrac{1}{\sin x} [\sin^2 kx \cos^2 x + \cos^2 kx \sin^2 x$

    $+ 2 \sin kx \cos kx \sin x \cos x]$

    $= \dfrac{1}{\sin x} [\sin kx \cos x + \cos kx \sin x]^2$

    $= \dfrac{\sin^2(kx + x)}{\sin x} = \dfrac{\sin^2(k+1)x}{\sin x}$

23. $(-1)^1 \cos x = -\cos x = \cos(x + \pi)$,

    Assume $\cos(x + k\pi) = (-1)^k \cos x$

    Then $\cos(x + (k+1)\pi) = \cos([x + k\pi] + \pi)$

    $= \cos(x + k\pi) \cos(\pi) - \sin(x + k\pi) \sin \pi$

    $= -\cos(x + k\pi) - 0 = (-1)(-1)^k \cos(x)$

    $= (-1)^{k+1} \cos(x)$

25. $\tan(x + \pi) = \tan x$

    Assume $\tan(x + k\pi) = \tan x$

    Then $\tan[x + (k+1)\pi] = \tan[(x + k\pi) + \pi]$

    $= \tan(x + k\pi) = \tan x$

27. $1 < 2$

    Assume $k < 2^k$

    Then

    $k + 1 < 2^k + 1 \le 2^k + 2^k = 2(2^k) = 2^{k+1}$

    So $k + 1 < 2^{k+1}$

29. $3^3 = 27 > (3 + 1)^2 = 16$

    Assume $k^3 > (k + 1)^2$

    Then $(k + 1)^3 = k^3 + 3k^2 + 3k + 1 >$

    $(k + 1)^2 + 3k^2 + 3k + 1$

    $= 4k^2 + 5k + 2$

    $= k^2 + 4k + 4 + 3k^2 + k - 2$

    $> k^2 + 4k + 4$ (for $k \ge 3$)

    $= (k + 2)^2 = [(k + 1) + 1]^2$

31. For $n = 6$, $2^6 = 64 > 49 = (6 + 1)^2$

    Assume $2^k > (k + 1)^2$ for $k \ge 6$.

    Then $(k + 2)^2 = k^2 + 4k + 4$

    $= (k^2 + 2k + 1) + (2k + 3) < 2^k$

    $+ (2k + 3)$.

    And $3 < k^2 + 1$ when $k \ge 6$, so

    $2k + 3 < k^2 + 2k + 1 = (k + 1)^2 < 2^k$

    by assumption.

    $(k + 2)^2 < 2^k + (2k + 3) < 2^k + 2^k = 2^{k+1}$

33. $4^2 = 16 > 2^2 + 3^2 = 13$

Assume $4^k > 2^k + 3^k$

Then $4^{k+1} = 4 \cdot 4^k > 4(2^k + 3^k)$

$= 4 \cdot 2^k + 4 \cdot 3^k > 2 \cdot 2^k + 3 \cdot 3^k$

$= 2^{k+1} + 3^{k+1}$

35. For $n = 7$, $3 = 2187 < 5040 = 7!$

Assume $3^k < k!$ for some $k \geq 7$.

Then $3^{k+1} = 3 \; 3^k < 3(k!)$.

But $k \geq 7$ implies that $(k + 1)! = k!(k + 1)$

$> (k!)3$, so $3^{k+1} < (k+1)!$

37. $5! = 120 > 10 \cdot 5 = 50$

Assume $k! > 10k$

Then $(k + 1)! = (k + 1)k! > (k + 1) \cdot 10k$

$> 10(k + 1)$

39. $1 + 1 = 1 + 1$

Assume $1 + k = k + 1$

Then $1 + (k + 1) = (1 + k) + 1 = (k + 1) + 1$

$(1 + k) + 1 = (k + 1) + 1$

41. $(xy)^1 = x^1 y^1$

Assume $(xy)^k = x^k y^k$

Then $(xy)^{k+1} = (xy)^k \cdot xy = x^k y^k xy$

$= x^k x \cdot y^k y$

$= x^{k+1} y^{k+1}$

43. For $n = 1$, $5^1 - 1 = 4 = 4 \cdot 1$

Assume 4 is a factor of $5^k - 1$

Then there is a q such that $4q = 5^k - 1$,

so $4q + 1 = 5^k$, so $4(5q) + 5 = 5^{k+1}$. Then $4(5q + 1) + 1 = 5^{k+1}$, thus

$4(5q + 1) = 5^{k+1} - 1$

45. $1^5 - 1 = 0 = 5 \cdot 0$

Assume $k^5 - k = 5q$ for some integer q

Then $(k + 1)^5 - (k + 1)$

$= k^5 + 5k^4 + 10k^3 + 10k^2 + 5k$

$+ 1 - k - 1$

$= (k^5 - k) + 5k^4 + 10k^3 + 10k^2 + 5k$

$= 5q + 5(k^4 + 2k^3 + 2k^2 + k)$

$= 5(q + k^4 + 2k^3 + 2k^2 + k)$

47. For $n = 1$, $1^1 + 1 = 2$, which is even

Assume $k^2 + k$ is even, that is $2q = k^2 + k$.

Then $(k + 1)^2 + (k + 1)$

$= k^2 + 2k + 1 + k + 1 = (k^2 + k)$

$+ (2k + 2) = 2q + 2(k + 1)$

$= 2(q + k + 1)$.

So $(k + 1)^2 + (k + 1)$ is even.

49. For $n = 7$, $3^5 = 2{,}187 \not\geq 7! = 5{,}040$

51. Since Mr. Smith has the gene, by hypothesis all his children — the first generation — will too. Suppose everyone in the kth generation has the gene. Then every one of their children — everyone in the $(k + 1)$th generation — will too. Thus, everyone in every generation, and hence all decendants, will have the gene.

SECTION 11.2

1. $11, 13, 15$

3. $22, 31, 40$

5. $\dfrac{1}{81}, \dfrac{1}{243}, \dfrac{1}{729}$

7. $1.31313, 1.313131, 1.3131313$

9. $-1, 1, -1, 1, -1$

11. $\dfrac{1}{2}, \dfrac{2}{3}, \dfrac{3}{4}, \dfrac{4}{5}, \dfrac{5}{6}$

13. $\dfrac{1}{3}, \dfrac{4}{5}, \dfrac{9}{7}, \dfrac{16}{9}, \dfrac{25}{11}$

15. $0, 2, 0, 2, 0$

17. $\dfrac{\pi}{3}, \dfrac{\pi^2}{9}, \dfrac{\pi^3}{27}, \dfrac{\pi^4}{81}, \dfrac{\pi^5}{243}$

19. $0, \dfrac{\ln 2}{2}, \dfrac{\ln 3}{3}, \dfrac{\ln 4}{4}, \dfrac{\ln 5}{5}$

21. $0, 3, \dfrac{1}{2}, \dfrac{5}{3}, \dfrac{2}{3}$

23. $\left\{ n \right\}_{n=1}^{n=\infty}$

25. $\left\{ 2^{n+1} \right\}_{n=1}^{n=\infty}$

27. $\left\{ \dfrac{2}{10^n} \right\}_{n=1}^{n=\infty}$

29. $\left\{\dfrac{n}{n+2}\right\}_{n=1}^{\infty}$

31. $\left\{\dfrac{3}{2^{n-1}}\right\}_{n=1}^{\infty}$

33. $\left\{(-2)^n\right\}_{n=1}^{\infty}$

35. 20

37. 15

39. 16

41. 0

43. 0

45. The sum oscillates between –1 and 0

47. $\displaystyle\sum_{n=1}^{6}\dfrac{(-1)^{n+1}}{2^n}$

49. $\displaystyle\sum_{n=1}^{8}(-1)^n\cos(n+2)$

51. $\displaystyle\sum_{m=1}^{n}\dfrac{x^{2(m-1)}}{2^m}$

53. Proof by induction

$$\sum_{k=1}^{1}(a_k+b_k)=a_1+b_1=\sum_{k=1}^{1}a_1+\sum_{k=1}^{1}b_1$$

Assume $\displaystyle\sum_{k=1}^{n}(a_k+b_k)=\sum_{k=1}^{n}a_k+\sum_{k=1}^{n}b_k$

Then $\displaystyle\sum_{k=1}^{n+1}(a_k+b_k)$

$$=\sum_{k=1}^{n}(a_k+b_k)+(a_{n+1}+b_{n+1})$$

$$=\left(\sum_{k=1}^{n}a_k+a_{n+1}\right)+\left(\sum_{k=1}^{n}b_k+b_{n+1}\right)$$

$$=\sum_{k=1}^{n+1}a_k+\sum_{k=1}^{n+1}b_k$$

55. In 1990, 7 ft $5\frac{1}{2}$ in, and in the year 2000, 7 ft 7 in.

SECTION 11.3

1. converges to 2
3. diverges
5. converges to –2
7. converges to 1
9. converges to 3
11. diverges
13. converges to 2.3
15. converges to e
17. diverges
19. converges to 0
21. $\varepsilon = .09$
23. $\varepsilon = .009$
25. $\varepsilon = .052$
27. $\varepsilon = .009$
29. .89
31. 0
33. $N = \dfrac{6-6\varepsilon}{5\varepsilon}$
35. $N = \sqrt{\dfrac{2+\varepsilon}{2\varepsilon}}$
37. Divergent. What makes this odd is that one would expect the human body to be limited in how high it can jump. Odder still is the fact that since 1960 the record has risen even faster than the sequence predicts: in 1990 it stood at 8 ft even.
39. Divergent

SECTION 11.4

1. $-19, -25, -31$; $S_9 = -171$
3. $\dfrac{16}{81}, \dfrac{32}{243}, \dfrac{64}{729}$,

    $S_9 = 3\left[1-\left(\tfrac{2}{3}\right)^9\right] \approx 2.922$
5. $-\dfrac{189}{40}, \dfrac{567}{80}, -\dfrac{1701}{160}$;

    $S_9 = -14.72552$
7. $x+12, x+17, x+22$; $S_9 = 9x+108$

9. $6y^3, 6y^4, 6y^5;\ S_9 = \sum_{k=1}^{9} 6y^{k-2} = \dfrac{6(1-y^9)}{y(1-y)}$

11. $.0092,\ .00092,\ .000092;\ 102.2222221$

13. $248{,}050$

15. $11{,}400$

17. $\dfrac{38}{99}$

19. $\dfrac{286}{333}$

21. $\dfrac{1{,}161{,}887}{99{,}000}$

23. $a_n = 3 + (n-1)(3.6)$

$a_{15} = 53.4$

25. $a_n = (-1)^{n+1}$

$a_{250} = -1$

27. $a + 5 = -46$

29. $4 + 12\left(1 - \dfrac{3^5}{5^5}\right) \approx 15.07\ \text{ft}$

16 ft in total

31. $2{,}046$ ancestors

33. $\$3{,}994.21$

35. $\$35{,}349.35$

37. Arithmetic sequence, common difference is $-\ln 4$

39. 3

41. $\dfrac{16}{15}$

43. Sum does not exist

45. $r = e^{-k}$

47. You should be willing to pay an infinite amount of money.

SECTION 11.5

1. 35

3. 8

5. 1

7. n

9. $x^3 + 3x^2h + 3xh^2 + h^3$

11. $x^5 - 10x^4 + 40x^3 - 80x^2 + 80x - 32$

13. $x^6 - 18x^5y + 135x^4y^2 - 540x^3y^3$
$+\ 1215x^2y^4 - 1458xy^5 + 729y^6$

15. $16x^4 - 32x^3y + 24x^2y^2 - 8xy^3 + y^4$

17. $h^{12} - 3h^8\sqrt[4]{2k} + 3h^4\sqrt{2k} - \sqrt[4]{8k^3}$

19. $\dfrac{a^3}{8b^3} - \dfrac{a^2c}{4b^2} + \dfrac{ac^2}{6b} - \dfrac{c^3}{27}$

21. $x^9 - 9x^6yz^2 + 27x^3y^2z^4 - 27y^3z^6$

23. $x^{24}y^4 - 4x^{37/2}y^3 + 6x^{13}y^2 - 4x^{15/2}y + x^2$

25. $n = 7,\quad 1\quad 7\quad 21\quad 35\quad 35\quad 21\quad 7\quad 1$

27. $n = 6,\quad 1\quad 6\quad 15\quad 20\quad 15\quad 6\quad 1$

29. 10

31. 35

33. -56

35. 171

37. 286

39. $\sum_{r=0}^{7} \binom{7}{r} c^r b^{7-r}$

41. $\sum_{r=0}^{9} \binom{9}{r} a^r (-2b)^{9-r}$

43. $\sum_{r=0}^{25} \binom{25}{r} x^{2r} (-1)^{25-r}$

45. $\sum_{r=0}^{21} \binom{21}{r} (2x)^{21} (-y^2)^{21-r}$

47. $4x^3 + 6x^2h + 4xh^2 + h^3 + 2$

49. $8x^3 + 12x^2h + 8xh^2 + 2h^3 + 12x + 6h$

51. $\dfrac{1}{8}$

Index